Noncommutative Harmonic Analysis

Mathematical
Surveys
and
Monographs

Volume 22

Noncommutative Harmonic Analysis

Michael E. Taylor

American Mathematical Society
Providence, Rhode Island

1980 *Mathematics Subject Classification* (1985 *Revision*).
Primary 43-XX, 35-XX; Secondary 22-XX.

Library of Congress Cataloging-in-Publication Data

Taylor, Michael Eugene, 1946–
Noncommutative harmonic analysis.
(Mathematical surveys and monographs, ISSN 0076-5376; no. 22)
Bibliography: p.
Includes index.
1. Harmonic analysis. I. Title. II. Title: Noncommutative harmonic analysis. III. Series.
QA403.T29 1986 515′.2433 86-10924
ISBN 0-8218-1523-7

Printed in the United States of America
Reprinted by the American Mathematical Society, 2024.

♾ The paper used in this book is acid-free and falls within the guidelines
established to ensure permanence and durability.

This publication was typeset using $\mathcal{A}\mathcal{M}\mathcal{S}$-TEX,
the American Mathematical Society's TEX macro system.

Visit the AMS home page at URL: http://www.ams.org/

15 14 13 12 11 10 9 8 28 27 26 25 24

Contents

Introduction

This book began as lecture notes for a one semester course at Stony Brook, on noncommutative harmonic analysis. We explore some basic roles of Lie groups in linear analysis, concentrating on generalizations of the Fourier transform and the study of naturally occurring partial differential equations.

The Fourier transform is a cornerstone of analysis. It is defined by

$$\mathcal{F}f(\xi) = \hat{f}(\xi) = (2\pi)^{-n/2} \int_{R^n} f(x) e^{-ix\cdot\xi}\, dx, \tag{I.1}$$

where, given $f \in L^1$, one has $\hat{f} \in L^\infty$. If f and all its derivatives are rapidly decreasing (one says $f \in \mathcal{S}(R^n)$), so is $\hat{f}$. If you define

$$\mathcal{F}^* f(\xi) = \tilde{f}(\xi) = (2\pi)^{-n/2} \int_{R^n} f(x) e^{ix\cdot\xi}\, dx, \tag{I.2}$$

then one has the Fourier inversion formula

$$\mathcal{F}^*\mathcal{F} = \mathcal{F}\mathcal{F}^* = I \quad \text{on } \mathcal{S}, \tag{I.3}$$

and hence $\mathcal{F}$ extends uniquely to a unitary operator on $L^2(R^n)$; this is the Plancherel theorem. These facts are proved in Appendix A, at the end of this monograph.

The Fourier transform is very useful in deriving solutions to the classical partial differential equations with constant coefficients, because of the intertwining property

$$\mathcal{F}(\partial/\partial x_j) = i\xi_j \mathcal{F}, \tag{I.4}$$

which implies

$$\mathcal{F}P(D) = P(\xi)\mathcal{F}, \tag{I.5}$$

where $D_j = (1/i)\partial/\partial x_j$ and $P(\xi)$ is any polynomial. For example, the fundamental solution to the heat equation $P(t,x)$, defined by

$$\begin{aligned}(\partial/\partial t)P(t,x) &= \Delta P(t,x), \\ P(0,x) &= \delta(x),\end{aligned} \tag{I.6}$$

is transformed to an ODE with parameters for the partial Fourier transform $\hat{P}(t,\xi)$:

$$(d/dt)\hat{P}(t,\xi) = -|\xi|^2\hat{P}(t,\xi), \quad \hat{P}(0,\xi) = (2\pi)^{-n/2}, \tag{I.7}$$

which implies

$$\hat{P}(t,\xi) = (2\pi)^{-n/2}e^{-t|\xi|^2}, \tag{I.8}$$

and computing the inverse Fourier transform of this Gaussian (see Appendix A) gives

$$P(t,x) = (2\pi)^{-n}\int e^{-t|\xi|^2+ix\cdot\xi}\,d\xi = (4\pi t)^{-n/2}e^{-|x|^2/4t}. \tag{I.9}$$

It is suggestive, and quite handy, to use the operator notation

$$P(t,x) = e^{t\Delta}\delta(x), \tag{I.10}$$

where, for a decent function f, $f(\Delta)$ is defined to be $\mathcal{F}^{-1}f(-|\xi|^2)\mathcal{F}$.

From here we can apply a nice analytical device, known as the subordination identity, to derive the Poisson integral formula in the upper half space. The subordination identity is

$$e^{-yA} = \frac{1}{2}y\pi^{-1/2}\int_0^\infty e^{-y^2/4t}e^{-tA^2}t^{-3/2}\,dt, \quad \text{if } A > 0. \tag{I.11}$$

We will give a proof of (I.11) at the end of this introduction. By the spectral theorem (or the Fourier integral representation for $A = \sqrt{-\Delta}$), this identity also holds for a positive selfadjoint operator A. We can apply it to $A = \sqrt{-\Delta}$, substituting (I.9) for $e^{t\Delta}\delta(x)$, to get

$$\begin{aligned} e^{-y(-\Delta)^{1/2}}\delta(x) &= \frac{1}{2}y\pi^{-1/2}\int_0^\infty e^{-y^2/4t}e^{t\Delta}\delta(x)t^{-3/2}\,dt \\ &= (4\pi)^{-(n+1)/2}y\int_0^\infty e^{-y^2/4t}e^{-|x|^2/4t}t^{-(n+3)/2}\,dt \\ &= c_n y(y^2+|x|^2)^{-(n+1)/2}, \end{aligned} \tag{I.12}$$

where the last integral is easily evaluated using the substitution $s = 1/t$. If $n \geq 2$, then $(-\Delta)^{-1/2}e^{-y(-\Delta)^{1/2}}$ maps the space of compactly supported distributions $\mathcal{E}'(R^n)$ to the space of tempered distributions $\mathcal{S}'(R^n)$ and is obtained by integrating (I.12) with respect to y. We obtain

$$(-\Delta)^{-1/2}e^{-y(-\Delta)^{1/2}}\delta(x) = c_n'(y^2+|x|^2)^{-(n-1)/2}. \tag{I.13}$$

We can analytically continue these formulas to complex y such that $\operatorname{Re} y > 0$, and then pass to the limit of purely imaginary y, say $y = it$, to obtain a formula for the solution of the wave equation on $R \times R^n$:

$$(\partial^2/\partial t^2 - \Delta)u = 0, \qquad u(0,x) = f(x), \qquad u_t(0,x) = g(x), \tag{I.14}$$

which is

$$u(t,x) = \cos t(-\Delta)^{1/2} f + (-\Delta)^{-1/2} \sin t(-\Delta)^{1/2} g. \tag{I.15}$$

In fact we have

$$(-\Delta)^{-1/2} e^{it(-\Delta)^{1/2}} \delta(x) = c_n' \lim_{\varepsilon \downarrow 0} (|x|^2 - (t - i\varepsilon)^2)^{-(n-1)/2}. \tag{I.16}$$

Taking real and imaginary parts, we can read off the finite propagation speed, and the fact that the strict Huygens principle holds when $n \geq 3$ is odd. For example, if $n = 3$, some elementary distribution theory applied to (I.16) yields

$$(-\Delta)^{-1/2} \sin t(-\Delta)^{1/2} \delta(x) = (4\pi t)^{-1} \delta(|x| - t). \tag{I.17}$$

In the last paragraphs we have sketched one line of application of Fourier analysis, which is harmonic analysis on Euclidean space, to constant coefficient PDE. One reason such analysis works so neatly is that such operators commute with translations, i.e., with the regular representation of R^n on $L^2(R^n)$ defined by

$$R(y)u(x) = u(x+y), \qquad x, y \in R^n. \tag{I.18}$$

The Fourier transform provides a spectral representation of $R(y)$, i.e., intertwines it with

$$\mathcal{F}^{-1} R(y) \mathcal{F} u(\xi) = e^{iy\cdot\xi} u(\xi), \tag{I.19}$$

and thus any operator commuting with $R(y)$ is intertwined with a multiplication operator by $\mathcal{F}$. This restatement of (I.5) is obvious since (I.4) is just a differentiated form of (I.19). From this point of view we can look upon the functions of differential operators such as (I.12) or (I.16) as being obtained by synthesis of very simple operators (scalars) on the irreducible representation spaces into which the representation (I.18) decomposes.

Many partial differential equations considered classically, particularly boundary problems for domains with simple shapes, exhibit noncommutative groups of symmetries, and noncommutative harmonic analysis arises as a tool in the investigation of these equations. The connection between solving equations on domains bounded by spheres and harmonic analysis on orthogonal groups is one basic case. Sometimes symmetries are manifested not simply as groups of motions on the underlying spatial domain, but in more subtle fashions. For example, for the harmonic oscillator Hamiltonian $-\Delta + |x|^2$, on R^n, we have the rotation group SO(n) acting as a group of symmetries in an obvious manner, but actually SU(n) acts as a group of symmetries, by restricting the "metaplectic representation," introduced in Chapter 1 and discussed further in Chapter 11. As another such example, we mention the action of the "$ax + b$ group" on L^2 functions on a cone, which will be explored in Chapter 7.

Harmonic analysis, commutative and noncommutative, plays an important role in contemporary investigations of linear PDE. For example, pseudodifferential operators are certain variable coefficient versions of Fourier multipliers (convolution operators). They can be used to impose an approximate translation symmetry on elliptic operators, such as Laplace operators on general Riemannian manifolds, and are particularly effective in treating various analytical aspects of elliptic equations and elliptic boundary problems. Fourier integral operators allow one to impose a translational symmetry on an operator with simple characteristics (and real principal symbol). However, when a Fourier integral operator is applied to an operator with multiple characteristics, in cases where the characteristic set can be put in some sort of normal form via a symplectic transformation, one is often led to a problem in noncommutative harmonic analysis. These notes will not be concerned much with such aspects of harmonic analysis. Presentations in the commutative case can be found in [**121**, **234**, **239**], and an introduction to "noncommutative microlocal analysis" in [**235**]. Although we do not consider the general theory here, one thread to be followed through these notes is the use of noncommutative harmonic analysis to treat certain important model cases of operators with double characteristics.

The basic method of noncommutative harmonic analysis, generalizing the use of the Fourier transform, is to synthesize operators on a space on which a Lie group has a unitary representation from operators on irreducible representation spaces. Thus one is led to determine what the irreducible unitary representations of a given Lie group are, and how to decompose a given representation into irreducibles. This general study is far from complete, though a great deal of progress has been made on important classes of Lie groups.

These notes begin with an introductory Chapter 0, dealing with some basic concepts of Lie group representation theory. We lay down some of the foundations for the study of strongly continuous representations of a Lie group G, and representations induced on various convolution algebras, including the algebra of compactly supported distributions on G, and the universal enveloping algebra of G, which includes its Lie algebra. Members of the Lie algebra, or more general elements of the universal enveloping algebra, are typically represented by unbounded operators, a fact which requires some technical care. Spaces of smooth vectors for a representation, introduced by Gårding originally to complete some arguments of Bargmann, and also analytic vectors, provide useful tools to handle these technical problems. Chapter 0 also includes some descriptions of various types of Lie groups one runs into, such as nilpotent, solvable, semisimple, etc.

In Chapter 1, we begin our study of analysis on specific Lie groups with a look at the Heisenberg group. This nilpotent Lie group has a representation theory that is at once simple and rich in structure. Its irreducible unitary representations were classified by Stone and von Neumann many years ago, when the group was studied because of its occurrence in basic quantum theory. Much terminology associated with the group, in addition to its name (actually, many physicists call it the Weyl group), celebrates this. For example, certain of the group's irreducible unitary representations are called Schrödinger representations, and other unitarily equivalent forms are called Bargmann-Fok representations. We obtain an analogue of the Fourier inversion formula easily in this case, and postpone proving that all irreducible unitary representations are given until Chapter 5. A study of the Heisenberg group naturally introduces other important Lie groups, in particular the symplectic group, acting as a group of automorphisms of the Heisenberg group, and the unitary group, arising as a maximal compact subgroup of the symplectic group. It might seem unorthodox to treat the Heisenberg group before the unitary group or other compact Lie groups, since most of its irreducible unitary representations are infinite-dimensional. However, it is this group for which it is easiest to develop harmonic analysis in closest analogy to the development on Euclidean space R^n. This may be traced to the fact that the Heisenberg group really has the simplest representation theory, compared to compact groups like SU(n) or SO(n), for which combinatorial considerations arise even to list all their irreducible representations. That the classification is particularly simple for SU(2) and SO(3) still does not make harmonic analysis on these groups as easy as on the Heisenberg group, partly for the basic reason that integrals are easier to work with than infinite series, and, not unrelated, partly because of the richer set of automorphisms of the Heisenberg group, particularly groups of dilations.

Chapter 2 studies the unitary groups U(n), and Chapter 3 pursues some general results about compact Lie groups. The complete description of the irreducible representations of U(n) and SU(n), for general n, spills over into Chapter 3, where it is produced as a consequence of the Theorem of the Highest Weight. In fact, we give a complete proof of this theorem only for SU(n). Chapter 3 contains a number of general results about compact Lie groups, such as the Peter-Weyl theorem and the Borel-Weil theorem. We have also emphasized the study of compact group actions on eigenspaces of Laplace operators in Chapter 3. The interplay between the orthogonal group and harmonic analysis on spheres is discussed in Chapter 14. Here we have tried to include short, simple derivations for a number of classical identities from the theory of spherical harmonics.

In Chapter 5 we discuss a major general tool for constructing representations of Lie groups, the method of induced representations. We discuss Mackey's results on systems of imprimitivity and applications to some important classes of Lie groups, such as the groups of isometries of Euclidean space and some other groups that are semi-direct products. This chapter also gives a proof of the Stone-von Neumann theorem classifying the irreducible unitary representations of the Heisenberg group. Chapter 6 pursues the study of nilpotent Lie groups. The theory of the Heisenberg group plays a crucial role here; the representation theory in the general case is reduced by an inductive argument to that of the Heisenberg group.

Chapter 7 gives a brief study of harmonic analysis on cones. The basic analytical tool for this study is the Hankel transform

$$H_\nu f(\lambda) = \int_0^\infty f(r) J_\nu(\lambda r) r\, dr. \tag{I.20}$$

The map H_ν is unitary from $L^2(R^+, r\,dr)$ onto $L^2(R^+, \lambda\,d\lambda)$, and the Hankel inversion formula says

$$H_\nu(H_\nu f)(r) = f(r). \tag{I.21}$$

We produce the Hankel transform as an operator intertwining two irreducible unitary representations of the "$ax+b$ group," a two-dimensional solvable Lie group whose irreducible unitary representations were classified in Chapter 5. We remark that special cases of the Hankel transform arise in taking Fourier transforms of radial functions on R^n. In fact, if $f(x) = f(r)$ on R^n, a change of variable yields

$$\hat{f}(\xi) = (2\pi)^{-n/2} \int_0^\infty f(r)\psi_n(r|\xi|)r^{n-1}\,dr \tag{I.22}$$

where

$$\begin{aligned} \psi_n(r) &= \int_{S^{n-1}} e^{i(x\cdot\omega)r}\,d\omega \\ &= A_{n-1}\int_{-1}^1 e^{irs}(1-s^2)^{(n-3)/2}\,ds \\ &= (2\pi)^{n/2} r^{1-n/2} J_{n/2-1}(r) \end{aligned} \tag{I.23}$$

in view of the integral formula

$$J_\nu(z) = [\Gamma(\tfrac{1}{2})\Gamma(\nu+\tfrac{1}{2})]^{-1}(z/2)^\nu \int_{-1}^1 (1-t^2)^{\nu-1/2}e^{izt}\,dt \tag{I.24}$$

for the Bessel function. The unitarity of H_ν and the Hankel inversion formula, for $\nu = \frac{1}{2}n - 1$, follow from the Fourier inversion formula and Plancherel theorem. These results are not accidentally special cases of the result (I.21), but conceptually so, since Euclidean space R^n is the cone over S^{n-1}, and both sets of formulas arise from the spectral representation of the Laplace operator.

In Chapter 8 we study the group $\mathrm{SL}(2, R)$. This group covers the Lorentz group $\mathrm{SO}_e(2,1)$, and hence is important for relativistic physics. We derive the classification of the irreducible unitary representations of $\mathrm{SL}(2, R)$ achieved in the famous paper of V. Bargmann [**12**], and study a number of topics in harmonic analysis on this group, and also on the two dimensional hyperbolic space. Chapter 9 studies $\mathrm{SL}(2, \mathbf{C})$, which covers the Lorentz group $\mathrm{SO}_e(3,1)$, and describes its repesentations, given by Gelfand et al. [**71**]. More general Lorentz groups $\mathrm{SO}_e(n,1)$ are also introduced. Our study of their representations is somewhat less complete, though we do describe the principal series, which provides "almost all" the irreducible unitary representations for n odd. Chapter 10 considers the actions of $\mathrm{SO}_e(n,1)$ as groups of conformal transformations on balls and spheres. We show that the existence of a certain (nonunitary) representation of $\mathrm{SO}_e(n,1)$ on the space of harmonic functions on the unit ball in R^n implies Poisson's integral formula for the solution to the Dirichlet problem in the ball.

In Chapter 11 we return to the symplectic group $\mathrm{Sp}(n, R)$, introduced in Chapter 1, and make a detailed study of the metaplectic representation. Our study uses the Cartan decomposition of the symplectic group, a special case of a decomposition that plays an important role in the general study of semisimple Lie groups. The Cartan decomposition is useful in studying the metaplectic representation, because the action of the double cover of $U(n) \subset Sp(n, R)$ takes a particularly simple form in the Bargmann-Fok representation.

Chapter 12 is devoted to spinors. The spin groups are double covers of $SO(n)$, and more generally of $SO_e(p, q)$. We define Dirac operators on manifolds with spin structures, but do not really pursue harmonic analysis for these objects, though there is a great deal that has been done. Some of this is hinted at in the final chapter of these notes, and material given here on spinors, it is hoped, will make this intelligible.

Chapter 13 is a very brief introduction to the general theory of noncompact semisimple Lie groups, illustrated to some degree by the material of Chapters 8–11. We describe some general results on principal series and discrete series, with no proofs but references to some basic sources. Much more extensive introductions to this vast topic can be found in the last two chapters of Wallach [**253**], and in the two volume treatise of Warner [**256**]. This subject is under intensive development at present, some of which is described in the more recent reports [**137**, **140**]. Also see the new book [**136**].

The course from which these notes grew was given to students who had been through one semester with the book of Varadarajan [**246**]. The present monograph is addressed to people with a basic understanding of Fourier analysis (commutative harmonic analysis) and functional analysis, such as covered in one of the basic texts, e.g., Reed and Simon [**201**], Riesz-Nagy [**202**], or Yosida [**267**]. A concise statement of the ideal preparation would be "Chapters 6, 9, and 11 of Yosida," together with the first few chapters of a standard Lie group text, such as [**246**] or [**100**]. We have included four appendixes, one on the Fourier transform and tempered distributions, one on the spectral theorem, emphasizing the use of tempered distributions in the proof of the spectral theorem, one on the Radon transform on Euclidean space, proving some results needed in Chapter 6, and one on analytic vectors, discussing some technical analytical points often encountered in passing from a representation of a Lie algebra to a representation of a Lie group.

These notes focus on ways specific Lie groups arise in analysis. The emphasis is on multiplicity, rather than unity. We do treat some general approaches to certain classes of Lie groups, especially the classes of compact and of nilpotent Lie groups, and to a lesser extent the class of semisimple Lie groups, but our emphasis is on the particular.

One unifying theme is the relationship between harmonic analysis and linear PDE, the use of results from each of these fields to advance the study of the other. In addition to certain operators with double characteristics, mentioned earlier, we also study the "classical" PDE, the Laplace, heat, and wave equations, in various contexts. For example, we show how exact solutions to the wave equation store up a great deal of information about harmonic analysis on spheres and hyperbolic space. For the past twenty years or so it has been an active line of inquiry to see how the spectral analysis of the Laplacian on a general manifold (especially in the compact case) stores up geometric information, and how such analysis can be achieved by parametrices for the heat equation and for the wave equation.

To give another illustration of this point of view, we mention a PDE approach to the proof of the subordination identity (I.11). We begin with the observation that a direct proof of (I.12), for $n = 1$, will imply the identity (I.11), on the operator level, with A replaced by $(-d^2/dx^2)^{1/2}$, which in turn implies (I.11) for all positive real A, by the spectral theorem for $-d^2/dx^2$. Thus it suffices to obtain a proof of the Poisson integral formula for $n = 1$. But

$$\begin{aligned} e^{-y(-d^2/dx^2)^{1/2}}\delta(x) &= (2\pi)^{-1}\int_{-\infty}^{\infty} e^{-y|\xi|+ix\xi}\,d\xi \\ &= (2\pi)^{-1}\int_0^{\infty} (e^{-y\xi+ix\xi}+e^{-y\xi-ix\xi})\,d\xi \\ &= (2\pi)^{-1}[(y-ix)^{-1}+(y+ix)^{-1}] \\ &= \pi^{-1}y/(y^2+x^2). \end{aligned} \tag{I.25}$$

This proves the subordination identity (I.11). A more classical method of proof is to take the Mellin transform with respect to y of both sides of (I.11), deducing that (I.11) is equivalent to the duplication formula

$$\pi^{1/2}\Gamma(2z) = 2^{2z-1}\Gamma(z)\Gamma(z+\tfrac{1}{2})$$

for the gamma function. The subordination identity will make another appearance, in Chapter 1, §7, in a study of various PDEs on the Heisenberg group.

The reader will find that, once the material of the introductory Chapter 0 is understood subsequent chapters can be read in almost any order, and by and large depend on each other only loosely. For example, if one wanted to read about the theory of nilpotent Lie groups, one would need only §§1 and 2 of Chapter 1 and §§1 and 2 of Chapter 5, before tackling Chapter 6. In particular, later sections of chapters with numerous sections, dealing with special problems, can be bypassed without affecting one's understanding of subsequent chapters.

CHAPTER 0

Some Basic Concepts of Lie Group Representation Theory

The purpose of this introductory chapter is to provide some background for the analysis to be presented in subsequent chapters. We give precise notions of strongly continuous representations of Lie groups, and show how they give rise to various sorts of representations of convolution algebras, Lie algebras, and universal enveloping algebras. We point out some special features of irreducible representations, particularly irreducible unitary representations. We will be mainly concerned with unitary representations in this monograph, but we have included general discussions of Banach space representations in this introductory chapter, since on the general level considered here it does not matter usually; we do not hesitate to retreat to unitary representations when a more general situation would entail the slightest additional complication.

We also recall some of the order that has been imposed on the panoply of Lie groups. We assume the reader has had an exposure to Lie groups, so many of the subjects of this introductory chapter should be familiar. We have concentrated on some of the technical elementary aspects, which tend not to be covered in introductory Lie group texts, having to do with the fact that infinite-dimensional representations generally have unbounded generators. §1 gets things started, with a discussion of one parameter groups of operators and their generators.

1. One parameter groups of operators. In this section we look at representations of the Lie group R of real numbers. If B is a Banach space, a one parameter group of operators on B is a set of bounded operators

$$V(t)\colon B \to B, \qquad t \in R, \tag{1.1}$$

satisfying the group homomorphism property

$$V(s+t) = V(s)V(t), \quad \text{for all } s, t \in R, \tag{1.2}$$

and

$$V(0) = I. \tag{1.3}$$

We also require a continuity property. The appropriate requirement is *strong continuity*, which means

(1.4) $$t_j \to t \Rightarrow V(t_j)u \to V(t)u \quad \text{in } B, \text{ for each } u \in B.$$

A one parameter group of operators will by definition satisfy (1.1)–(1.4).

An important set of examples is the set of translation groups

(1.5) $$\tau_p(t)\colon L^p(R) \to L^p(R), \qquad 1 \le p < \infty,$$

defined by

(1.6) $$\tau_p(t)f(x) = f(x-t).$$

The properties (1.1)–(1.3) are clear in this case. Note that $\|\tau_p(t)\| = 1$ for each t. Also, we see that $\|\tau_p(t) - \tau_p(t')\| = 2$ if $t \neq t'$, by applying the difference to a function f with support in an interval of length $\frac{1}{2}|t-t'|$. It takes a little effort to verify the strong continuity (1.4), and we now show how to do this.

Note that $C_0^\infty(R)$ is dense in each Banach space $L^p(R)$, for $1 \le p < \infty$. Furthermore, if $f \in C_0^\infty(R)$, $t_j \to t$, then all the functions $\tau_p(t_j)f(x) = f(x-t_j)$ have support in a fixed compact set, and converge uniformly to $f(x-t)$, so clearly we have the convergence in (1.4) in L^p norm for each $f \in C_0^\infty(R)$. The proof that (1.5)–(1.6) is strongly continuous is completed by the following simple lemma.

LEMMA 1.1. *Let $T_j\colon B_1 \to B_2$ be a uniformly bounded set of linear operators between Banach spaces. Let $\mathcal{L}$ be a dense linear subspace of B_1, and suppose*

(1.7) $$T_j u \to T_0 u \quad \text{as } j \to \infty$$

in the B_2-norm, for each $u \in \mathcal{L}$. Then (1.7) holds for all $u \in B_1$.

PROOF. Given $u \in B_1, \varepsilon > 0$, pick $v \in \mathcal{L}$ such that $\|u - v\| < \varepsilon$. Suppose $\|T_j\| \le M$ for all j. Then

$$\begin{aligned} \|T_j u - T_0 u\| &= \|T_j u - T_j v + T_j v - T_0 v + T_0 v - T_0 u\| \\ &\le \|T_j u - T_j v\| + \|T_j v - T_0 v\| + \|T_0 v - T_0 u\| \\ &\le \|T_j v - T_0 v\| + 2M\|v-u\|. \end{aligned}$$

Letting $j \to \infty$, we have $\limsup_{j\to\infty} \|T_j u - T_0 u\| \le 2M\varepsilon$. Since $\varepsilon > 0$ is arbitrary, the proof is complete.

We note that a locally uniform bound on the norm

(1.8) $$\|V(t)\| \le M \quad \text{for } |t| \le 1$$

for some $M \in [1,\infty)$ holds for any strongly continuous group, as a consequence of the uniform boundedness theorem. As a consequence of (1.8), we have, for all $t \in R$,

(1.9) $$\|V(t)\| \le Me^{K|t|},$$

for some K (one could take $K = \log M$).

The groups we will mainly be considering are *unitary groups*, i.e., strongly continuous groups of operators $\mathrm{U}(t)$ on a Hilbert space H such that

$$\mathrm{U}(t)^* = \mathrm{U}(t)^{-1} = \mathrm{U}(-t). \tag{1.10}$$

Clearly, in this case, $\|\mathrm{U}(t)\| = 1$. The translation group τ_2 on $L^2(R)$, a special case of (1.5)–(1.6), is a unitary group.

A one parameter group $V(t)$ of operators on B has associated an *infinitesimal generator* A, which is an operator, often unbounded, on B, defined by

$$Au = \lim_{h\to 0} h^{-1}(V(h)u - u) \tag{1.11}$$

on the domain

$$\mathcal{D}(A) = \{u \in B\colon s\text{-}\lim_{h\to 0} h^{-1}(V(h)u - u) \text{ exists in } B\}. \tag{1.12}$$

PROPOSITION 1.2. *The infinitesimal generator A of $V(t)$ is a closed, densely defined operator. We have*

$$V(t)\mathcal{D}(A) \subset \mathcal{D}(A) \quad \text{for all } t \in R, \tag{1.13}$$

and

$$AV(t)u = (d/dt)V(t)u \quad \text{for } u \in \mathcal{D}(A). \tag{1.14}$$

If (1.9) *holds and* $\operatorname{Re}\lambda > K$, *then λ belongs to the resolvent set of A and*

$$(\lambda - A)^{-1}u = \int_0^\infty e^{-\lambda t}V(t)u\,dt, \qquad u \in B. \tag{1.15}$$

An operator A with domain $\mathcal{D}(A)$ is said to be closed provided

$$u_j \in \mathcal{D}(A),\ u_j \to u,\ Au_j \to v \text{ in } B \Rightarrow u \in \mathcal{D}(A), \qquad Au = v, \tag{1.16}$$

or equivalently, that the graph

$$\mathcal{G}_A = \{(u, Au) \in B \oplus B\colon u \in \mathcal{D}(A)\} \tag{1.17}$$

is closed in $B \oplus B$.

Our proof of Proposition 1.2 follows some notes of Nelson [**184**], as will our proof of Proposition 1.3. First, if $u \in \mathcal{D}(A)$, then, for $t \in R$,

$$h^{-1}(V(h)V(t)u - V(t)u) = V(t)h^{-1}(V(h)u - u), \tag{1.18}$$

which immediately gives (1.13), and also (1.14), if we replace $V(h)V(t)$ in (1.18) by $V(t+h)$. To show $\mathcal{D}(A)$ is dense in B, let $u \in B$, and consider

$$u_\varepsilon = \varepsilon^{-1}\int_0^\varepsilon V(t)u\,dt. \tag{1.19}$$

A brief calculation gives

$$\begin{aligned} h^{-1}(V(h)u_\varepsilon - u_\varepsilon) &= \varepsilon^{-1}\left[h^{-1}\int_\varepsilon^{\varepsilon+h} V(t)u\,dt - h^{-1}\int_0^h V(t)u\,dt\right] \\ &\to \varepsilon^{-1}(V(\varepsilon)u - u) \quad \text{as } h \to 0, \end{aligned} \tag{1.20}$$

so we see that $u_\varepsilon \in \mathcal{D}(A)$ for each $\varepsilon \neq 0$. But $u_\varepsilon \to u$ in B as $\varepsilon \to 0$, by (1.4), so $\mathcal{D}(A)$ is dense.

Next we prove (1.15). Denote the right side of (1.15) by R_λ, clearly a bounded operator on B. We recall that to say $\lambda \in \mathbf{C}$ belongs to the resolvent set of A is to say $\lambda - A\colon \mathcal{D}(A) \to B$ is one to one and onto, with $(\lambda - A)^{-1}$ bounded on B. First we show that

$$R_\lambda(\lambda - A)u = u \quad \text{for } u \in \mathcal{D}(A). \tag{1.21}$$

In fact, by (1.14), we have

$$\begin{aligned} R_\lambda(\lambda - A)u &= \int_0^\infty e^{-\lambda t} V(t)(\lambda u - Au)\,dt \\ &= \int_0^\infty \lambda e^{-\lambda t} V(t)u\,dt - \int_0^\infty e^{-\lambda t}(d/dt)V(t)u\,dt, \end{aligned}$$

and integrating the last term by parts produces (1.21).

The same sort of argument produces

$$R_\lambda\colon B \to \mathcal{D}(A), \tag{1.22}$$

$(\lambda - A)R_\lambda$ bounded on B, and

$$(\lambda - A)R_\lambda u = u \quad \text{for } u \in \mathcal{D}(A). \tag{1.23}$$

Since $(\lambda - A)R_\lambda$ is bounded on B, (1.23) holds for all $u \in B$, since $\mathcal{D}(A)$ is dense. This proves (1.15). Finally, since the resolvent set of A is nonempty, and $(\lambda - A)^{-1}$, being continuous and everywhere defined, is closed, so is A. The proof of the proposition is complete. We note that, similarly to (1.15), for $\operatorname{Re}\lambda < -K$, λ belongs to the resolvent set of A, and $(\lambda - A)^{-1}u = -\int_{-\infty}^0 e^{-\lambda t}V(t)u\,dt$.

We will write, symbolically,

$$V(t) = e^{tA}. \tag{1.24}$$

In view of the following proposition, the infinitesimal generator determines the one parameter group it is associated with uniquely. Hence we are justified in saying "A generates $V(t)$."

PROPOSITION 1.3. *If $V(t)$ and $W(t)$ are 1-parameter groups with the same infinitesimal generator, then $V(t) = W(t)$ for all $t \in R$.*

PROOF. Let $u \in B$, $\omega \in B^*$. Then, for $\operatorname{Re}\lambda$ large enough,

$$\begin{aligned} \int_0^\infty e^{-\lambda t}(V(t)u, \omega)\,dt &= ((\lambda - A)^{-1}u, \omega) \\ &= \int_0^\infty e^{-\lambda t}(W(t)u, \omega). \end{aligned} \tag{1.25}$$

Well-known uniqueness for the Laplace transform of a scalar valued function implies $(V(t)u, \omega) = (W(t)u, \omega)$ for all $t \in R^+$, given any $u \in B$ and any $\omega \in B^*$. By the Hahn-Banach theorem, this implies $V(t)u = W(t)u$ for $t \in R^+$. Since $V(-t) = V(t)^{-1}$ and $W(-t) = W(t)^{-1}$, the proof is complete.

We will not take the space to characterize which operators A in general are generators of one parameter groups, but we will characterize the generators of unitary groups, as *skew adjoint* operators. We define this notion.

For a densely defined operator A on a Hilbert space H, its adjoint A^* has domain

$$\mathcal{D}(A^*) = \{u \in H : |(u, Av)| \le C_u \|v\| \text{ for all } v \in \mathcal{D}(A)\}, \tag{1.26}$$

and we define A^* by

$$(A^*u, v) = (u, Av) \quad \text{for } u \in \mathcal{D}(A^*),\ v \in \mathcal{D}(A). \tag{1.27}$$

An operator is called selfadjoint if

$$\mathcal{D}(A) = \mathcal{D}(A^*) \quad \text{and} \quad A^* = A, \tag{1.28}$$

and skew adjoint if

$$\mathcal{D}(A) = \mathcal{D}(A^*) \quad \text{and} \quad A^* = -A. \tag{1.29}$$

Note that A is selfadjoint if and only if iA is skew adjoint.

We say that A is symmetric (resp. skew symmetric) if $\mathcal{D}(A) \subset \mathcal{D}(A^*)$ and $A^*u = Au$ (resp. $A^*u = -Au$) for $u \in \mathcal{D}(A)$. If A is symmetric, then

$$\operatorname{Im}((A \pm i)u, u) = \|u\|^2 \quad \text{for } u \in \mathcal{D}(A), \tag{1.30}$$

so clearly the maps

$$A \pm i : \mathcal{D}(A) \to H \tag{1.31}$$

are injective, with closed range if A is closed. The following result, due to von Neumann, can be found in many functional analysis books, e.g., [**49**, **192**, **193**, **256**].

THEOREM 1.4. *Let A be a closed symmetric operator. Then A is selfadjoint if and only if the maps* (1.31) *are both onto, equivalently, if and only if i and $-i$ belong to the resolvent set of A. In such a case, the operators*

$$U_1 = (A + i)(A - i)^{-1} \quad \textit{and} \quad U_2 = (A - i)(A + i)^{-1}$$

are unitary.

As shown in these references, a densely defined operator A has a closure $\overline{A}$, i.e., a linear operator whose graph $\mathcal{G}_{\overline{A}}$ is the closure of $\mathcal{G}_A$, if and only if the adjoint is densely defined (in which case $\overline{A} = A^{**}$). In particular, if A is symmetric (or skew symmetric) then A has a closure $\overline{A}$. We say A is *essentially selfadjoint* provided its closure is selfadjoint. A corollary of Theorem 1.4 is that a symmetric operator A is essentially selfadjoint if and only if $A + i$ and $A - i$ both have dense range.

Skew adjoint operators are related to unitary groups as follows.

THEOREM 1.5. *If* $\mathrm{U}(t)$ *is a unitary group, its infinitesimal generator is skew adjoint.*

PROOF. Suppose $u, v \in \mathcal{D}(A)$. Then

$$(Au, v) = \lim_{h \to 0} h^{-1}((\mathrm{U}(h) - I)u, v). \tag{1.32}$$

For each $h \neq 0$, the right side of (1.32) is equal to

$$h^{-1}(u, (\mathrm{U}(-h) - I)v), \tag{1.33}$$

which tends in the limit $h \to 0$ to $(u, -Av)$, i.e.,

$$(Au, v) = -(u, Av) \quad \text{for } u, v \in \mathcal{D}(A). \tag{1.34}$$

Hence A is skew symmetric. Now Proposition 1.2 implies A is closed and ± 1 are in the resolvent set of A, so Theorem 1.4 implies A is skew adjoint.

This is half of a theorem of Stone. The other half is the converse:

THEOREM 1.6. *If* A *is a skew adjoint operator on a Hilbert space, then* A *generates a unitary group*

$$\mathrm{U}(t) = e^{tA}. \tag{1.35}$$

This converse is intimately related to the spectral theorem, and is proved in Appendix B. Note that, if A is *bounded* and skew adjoint, the power series expansion

$$e^{tA} = \sum_{n=0}^{\infty} (1/n!) t^n A^n \tag{1.36}$$

is easily verified, via power series manipulation, to define a unitary group.

More generally, if A is any bounded operator on a Banach space B, the power series (1.36) defines a one parameter group $V(t)$ on B.

If the group $V(t)$ has a bounded infinitesimal generator A, not only is the power series (1.36) valid, but also, for any $u \in B$, $F(t) = V(t)u$ is a C^∞, indeed a real analytic function of $t \in R$, with values in B. Indeed, $F(t)$ extends to an entire analytic function of $t \in \mathbf{C}$, in this case. If the infinitesimal generator A of $V(t)$ is not bounded (equivalently, by the closed graph theorem, if A is not everywhere defined) then of course if $u \in B$ but $u \notin \mathcal{D}(A)$, then $F(t) = V(t)u$ is not even C^1. We shall say $u \in B$ is a "C^∞ vector" for V provided $F(t) = V(t)u$ is a C^∞ function of t with values in B. It is clear that

$$u \text{ is a } C^\infty \text{ vector } \Leftrightarrow u \in \mathcal{D}(A^k) \text{ for all } k. \tag{1.37}$$

Furthermore, we say u is an "analytic vector" for V provided $F(t) = V(t)u$ is a real analytic function of t. It is easy to see this condition is equivalent to having $F(t)$ extend to a strip $|\mathrm{Im}\, t| < K$, by exploiting the group property. Also,

$$u \text{ is an analytic vector } \Leftrightarrow \|A^k u\| \leq C(Ck)^k \tag{1.38}$$

for some $C = C(u)$.

For these concepts to be significant, we need to know these spaces are dense in B. Such facts are easily proved by refinements of the argument in Proposition 1.2, showing that $\mathcal{D}(A)$ is dense in B. In fact, for $u \in B$, consider

$$\hat{\rho}(-iA)u = \int_{-\infty}^{\infty} \rho(t)V(t)u\,dt. \tag{1.39}$$

Here we will take the notation $\hat{\rho}(-iA)$ to be merely symbolic for the right side of (1.39). Motivation for this notation can be found in the discussion of the spectral theorem in Appendix B. If $\rho \in C_0^\infty(R)$, then the identity

$$\begin{aligned} V(s)\hat{\rho}(-iA)u &= \int_{-\infty}^{\infty} \rho(t)V(s+t)u\,dt \\ &= \int_{-\infty}^{\infty} \rho(t-s)V(t)u\,dt \end{aligned} \tag{1.40}$$

clearly shows that $\hat{\rho}(-iA)u$ is a C^∞ vector. Note that

$$A^k\hat{\rho}(-iA)u = \int_{-\infty}^{\infty} \rho^{(k)}(t)V(t)u\,dt. \tag{1.41}$$

Now if $\rho \in C_0^\infty(R)$ is picked, so that $\int \rho(t)\,dt = 1$, set $\rho_j(t) = j\rho(jt)$, so $\rho_j \in C_0^\infty(R)$, and, by the strong continuity (1.4),

$$\int_{-\infty}^{\infty} \rho_j(t)V(t)u\,dt \to u \quad \text{as } j \to \infty. \tag{1.42}$$

It follows that, for any strongly continuous group $V(t)$ on a Banach space B, the space of smooth vectors is dense in B. More generally, the same arguments show that

$$u \in \mathcal{D}(A^k) \Rightarrow \hat{\rho}_j(-iA)u \to u \text{ in } \mathcal{D}(A^k) \tag{1.43}$$

for $k = 0, 1, 2, \dots$, where $\mathcal{D}(A^k)$ is given the usual graph norm.

We can derive results on analytic vectors by looking at (1.39) with

$$\rho(t) = \rho_\varepsilon(t) = (4\pi\varepsilon)^{-1/2}e^{-t^2/4\varepsilon}. \tag{1.44}$$

By (1.9), the formula (1.39) is well defined, with $\rho = \rho_\varepsilon$, for each $\varepsilon > 0$. Furthermore, we continue to have (1.40), i.e.,

$$V(s)\hat{\rho}_\varepsilon(-iA)u = (4\pi\varepsilon)^{-1/2}\int e^{-(t-s)^2/4\varepsilon}V(t)u\,dt. \tag{1.45}$$

Now it is clear that the right side of (1.45) is holomorphic in $s \in \mathbf{C}$. Since one has

$$\begin{aligned} V(s_1)V(s_2)\hat{\rho}_\varepsilon(-iA)u &= V(s_1)\hat{\rho}_{\varepsilon/2}(-iA)v(s_2)\hat{\rho}_{\varepsilon/2}(-iA)u \\ &= V(s_1+s_2)\hat{\rho}_\varepsilon(-iA)u \end{aligned} \tag{1.46}$$

for $s_1, s_2 \in R$, by analytic continuation, this continues to hold for $s_1, s_2 \in \mathbf{C}$, if (1.45) defines $V(s)\hat{\rho}_\varepsilon(-iA)u$ for $s \in \mathbf{C}$. Thus, for each $\varepsilon > 0$, each $u \in B$, $\hat{\rho}_\varepsilon(-iA)u$ is an analytic vector (in fact, an *entire* analytic vector) for $V(t)$. In analogy with (1.42), it is clear that for each $u \in B, \hat{\rho}_\varepsilon(-iA)u \to u$ as $\varepsilon \to 0$, so it follows that the space of analytic vectors for $V(t)$ is dense in B.

In the case when $U(t) = e^{itA}$ is a unitary group on a Hilbert space H, so A is selfadjoint, then the algebraic linear span of the ranges of the projections $E((-j,j))$, where E is the spectral measure of A, is dense in H, and each element in this linear span is an analytic vector. In the noncommutative case, the denseness of smooth and analytic vectors requires an argument generalizing that of the previous paragraph, rather than a simple application of the spectral theorem. We say more about this in the next section.

Another notation for (1.39) is $V(\rho)$:

$$V(\rho)u = \int \rho(t)V(t)u\,dt. \tag{1.47}$$

As we noted, this is well defined for $\rho \in C_0^\infty(R)$. If $V(t)$ is uniformly bounded, particularly if it is a unitary group, then (1.4) defines a bounded operator for any $\rho \in L^1(R)$. A simple calculation shows that

$$V(\rho_1 * \rho_2) = V(\rho_1)V(\rho_2) \tag{1.48}$$

where the *convolution* $\rho_1 * \rho_2$ is defined by

$$\rho_1 * \rho_2(t) = \int \rho_1(t-s)\rho_2(s)\,ds. \tag{1.49}$$

The identity (1.48) is equivalent to

$$(\hat{\rho}_1\hat{\rho}_2)(-iA) = \hat{\rho}_1(-iA)\hat{\rho}_2(-iA), \tag{1.50}$$

in the notation of (1.39), where $\hat{\rho}_j$ is the Fourier transform of ρ_j. We will make extensive use in the next section of the natural generalization of (1.47) to the case of a representation of a Lie group, so we will not dwell further on it here.

To end this section, we specify the infinitesimal generators for the groups τ_p on $L^p(R)$, $1 \le p < \infty$, given by (1.6). Clearly $u \in L^p(R)$ belongs to the domain $\mathcal{D}(A_p)$ of the generator if and only if

$$h^{-1}(f(x-h) - f(x)) \tag{1.51}$$

converges in L^p norm as $h \to 0$, to some limit. Note that the limit of (1.51) always exists in the distribution space $\mathcal{D}'(R)$, and is equal to

$$-(d/dx)f, \tag{1.52}$$

where d/dx is applied in the sense of distributions. Using the theory of distributions, one can show that

$$\mathcal{D}(A_p) = \{f \in L^p(R) \colon (d/dx)f \in L^p(R)\} \tag{1.53}$$

and

$$A_p f = -(d/dx)f \quad \text{for } f \in \mathcal{D}(A_p), \tag{1.54}$$

where in (1.53) and (1.54), d/dx is applied a priori in the distributional sense. Indeed, from what has just been said, it is clear that $\mathcal{D}(A_p)$ is *contained in* the right side of (1.53). To prove the reverse containment, one could apply the following general result.

LEMMA 1.7. *Let $V(t)$ be a one parameter group on B, with infinitesimal generator A. Let $\mathcal{L}$ be a weak* dense linear subspace of the dual space B^*. Suppose that $u, v \in B$, and that*

$$\lim_{h\to 0} h^{-1}\langle V(h)u - u, w\rangle = \langle v, w\rangle \quad \text{for all } w \in \mathcal{L}. \tag{1.55}$$

Then $u \in \mathcal{D}(A)$, and $Au = v$.

PROOF. The hypothesis (1.55) implies $\langle V(t)u, w\rangle$ is differentiable, and

$$(d/dt)\langle V(t)u, w\rangle = \langle V(t)v, w\rangle$$

for all $w \in \mathcal{L}$. Hence $\langle V(t)u - u, w\rangle = \int_0^t \langle V(s)v, w\rangle\, ds$ for all $w \in \mathcal{L}$. The weak* denseness of $\mathcal{L}$ implies $V(t)u - u = \int_0^t V(s)v\, ds$, so the convergence in B-norm of $h^{-1}(V(h)u - u) = h^{-1}\int_0^h V(s)v\, ds$ to v as $h \to 0$ is apparent.

2. Representations of Lie groups, convolution algebras, and Lie algebras. Let G be a Lie group, with identity element e, and let B be a Banach space. A representation π of G on B is a family of bounded operators

$$\pi(g)\colon B \to B, \qquad g \in G, \tag{2.1}$$

satisfying the properties of being a group homomorphism:

$$\pi(g_1 g_2) = \pi(g_1)\pi(g_2), \qquad g_j \in G, \tag{2.2}$$

and

$$\pi(e) = I, \tag{2.3}$$

and satisfying the condition of strong continuity:

$$g_j \to g \quad \text{in } G, \qquad u \in B \Rightarrow \pi(g_j)u \to \pi(g)u \quad \text{in } B. \tag{2.4}$$

If, in addition, $B = H$ is a Hilbert space, and each $\pi(g)$ is a unitary operator on H, so that

$$\pi(g)^* = \pi(g)^{-1} = \pi(g^{-1}), \tag{2.5}$$

then we say π is a unitary representation of G.

As the results on the special case $G = R$ in §1 suggest, to develop the theory of representations of the group G, it is important to be able to integrate over G. We endow G with a smooth left invariant measure, called a Haar measure, as follows. We define an n-form on G, $n = \dim G$. First, pick a nonvanishing element $\omega(e) \in \bigwedge^n T_e^*(G)$; this is determined up to a scalar multiple since $\dim \bigwedge^n T_e^*(G) = 1$. If

$$l_g\colon G \to G \tag{2.6}$$

is defined by

$$l_g(g_1) = g g_1, \tag{2.7}$$

then define $\omega(g) \in \bigwedge^n T_g^*(G)$ by

$$\omega(g) = \textstyle\bigwedge^n (Dl_{g^{-1}})^* \omega(e) \tag{2.8}$$

where

$$(2.9) \qquad Dl_{g^{-1}} : T_g(G) \to T_e(G)$$

is the natural derivative map and $(Dl_{g^{-1}})^*$ is its adjoint. This defines a left invariant differential form ω on G:

$$(2.10) \qquad l_g^*\omega = \omega \quad \text{for all } g \in G.$$

This form ω is nowhere vanishing, and is clearly determined uniquely, up to a constant factor. In particular, it defines an orientation on G. Integration of ω with respect to this orientation defines a left invariant measure, which we denote dg. Note that

$$(2.11) \qquad \int_G f(g)\, dg = \int_G f(g'g)\, dg$$

for all $g' \in G$. The measure dg depends on our choice of $\omega(e) \in \bigwedge^n T_e^*(G)$. Another choice $\omega'(e)$ would differ by a scalar factor c, and then the new form would differ from ω by the constant factor c, and hence $d'g$ would differ from dg by the constant factor $|c|$, so Haar measure is well defined by the prescription above, up to a constant factor.

In a similar fashion, using right translation, one can construct a smooth right invariant measure d_rg on G, well defined up to a constant multiple. Left and right invariant measures may or may not coincide, depending on the nature of G. We will say more about this later. If they do coincide, G is called unimodular.

The term "Haar measure" in this context is a slight misnomer. Haar solved the more delicate problem of showing that every locally compact topological group has a locally finite left invariant Baire measure. The construction given above for Lie groups is quite simple and was known long before Haar's work. These notes restrict attention to Lie groups. For foundational material on more general locally compact topological groups, see, [**111, 158, 259**].

In analogy with the translation groups defined by (1.5)–(1.6), we have the following representations of G on $B = L^p(G) = L^p(G, dg)$, $1 \le p < \infty$.

$$(2.12) \qquad \lambda_p(g)u(x) = u(g^{-1}x), \qquad u \in L^p(G).$$

As in §1, we easily see that λ_p satisfies (2.1)–(2.3). Also $\|\lambda_p(g)\| = 1$ for all $g \in G$, and $\|\lambda_p(g_1) - \lambda_p(g_2)\| = 2$ if $g_1 \ne g_2$. To see that λ_p satisfies the condition (2.4) of strong continuity, we can also argue as in §1. The convergence (2.4) in $L^p(G)$ norm for $u \in C_0^\infty(G)$ is elementary. But $C_0^\infty(G)$ is dense in $L^p(G)$ for $1 \le p < \infty$, so (2.4) follows in general, by Lemma 1.1. The representations (2.12) are called the left regular representations of G on $L^p(G)$. Similarly one has right regular representations on $L^p(G, d_rg)$:

$$(2.13) \qquad \rho_p(g)u(x) = u(xg), \qquad u \in L^p(G, d_rg).$$

Of course, the representations λ_2 and ρ_2 are unitary representations of G.

As indicated in §1, it is very useful to consider the action a representation π of G on B induces on $C_0^\infty(G)$. We set

$$\pi(f)u = \int_G f(g)\pi(g)u\,dg. \tag{2.14}$$

If $f \in C_0^\infty(G)$, or more generally if f is any integrable function on G with compact support, then (2.14) defines a bounded operator on B. If $\{\pi(g) : g \in G\}$ is uniformly bounded, in particular if π is a unitary representation of G, then (2.14) defines a bounded operator for all $f \in L^1(G)$. Note that, for $g_1 \in G$,

$$\begin{aligned} \pi(g_1)\pi(f)u &= \int_G f(g)\pi(g_1 g)u\,dg \\ &= \int_G f(g_1^{-1}g)\pi(g)u\,dg. \end{aligned} \tag{2.15}$$

It follows that

$$\begin{aligned} \pi(f_2)\pi(f_1) &= \int\int f_2(g_1)\pi(g_1)\,dg_1 f_1(g)\pi(g)\,dg \\ &= \int\int f_2(g_1)f_1(g_1^{-1})\pi(g)\,dg_1\,dg, \end{aligned} \tag{2.16}$$

i.e.,

$$\pi(f_2)\pi(f_1) = \pi(f_2 * f_1) \tag{2.17}$$

where the *convolution* $f_2 * f_1$ is defined by

$$f_2 * f_1(g) = \int_G f_2(g_1)f_1(g_1^{-1}g)\,dg_1. \tag{2.18}$$

Another immediate implication of (2.15) is that, for any $u \in B$, any $f \in C_0^\infty(G)$, $\pi(f)u$ is a C^∞ vector, where we say $v \in B$ is a C^∞ vector for a representation π provided that $F(g) = \pi(g)v$ is a C^∞ function on G, with values in B. Let us denote the space of C^∞ vectors for π by

$$C^\infty(\pi). \tag{2.19}$$

It is clear that, if $f_j \in C_0^\infty(G)$ is a sequence of functions, each satisfying $\int_G f_j(g)\,dg = 1$, supported in a sequence of small neighborhoods of e, shrinking to e, then

$$\pi(f_j)u \to u \quad \text{in } B \tag{2.20}$$

for all $u \in B$. It follows that the space $C^\infty(\pi)$ of smooth vectors for π is dense in B. (In particular, if B is finite-dimensional, every vector is a smooth vector.) More precisely, set

$$\mathcal{G}(\pi) = \{\pi(f)u : u \in B,\ f \in C_0^\infty(G)\}. \tag{2.21}$$

$\mathcal{G}(\pi)$ is called the "Gårding space" for π, following the important work of [**68**]. What we have shown is that

$$\mathcal{G}(\pi) \subset C^\infty(\pi), \tag{2.22}$$

and that $\mathcal{G}(\pi)$ is dense in B, for any (strongly continuous) representation π of G on B. From (2.15), it is clear that

$$\pi(g)\mathcal{G}(\pi) \subset \mathcal{G}(\pi) \quad \text{for all } g \in G. \tag{2.23}$$

We are now ready to define the action which a representation π of G induces on the Lie algebra $\mathfrak{g}$ of G. We identify $\mathfrak{g}$ with the tangent space T_eG. Each $X \in \mathfrak{g}$ is associated to a one parameter subgroup of G:

$$\gamma_X(t) = \exp(tX). \tag{2.24}$$

Thus,

$$V_X(t) = \pi(\gamma_X(t)) \tag{2.25}$$

is a (strongly continuous) one parameter group of operators on B. In §1 we discussed the infinitesimal generator of a one parameter group:

$$A_X = \lim_{h\to 0} h^{-1}(V_X(h) - I) \tag{2.26}$$

with domain

$$\mathcal{D}(A_X) = \{u \in B\colon \lim_{h\to 0} h^{-1}(V_X(h)u - u) \text{ exists}\}. \tag{2.27}$$

We will denote this closed densely defined operator by $\pi(X)$,

$$\pi(X)u = \lim_{h\to 0} h^{-1}(\pi(\exp hX)u - u), \tag{2.28}$$

on the domain (2.27).

A special case of (2.23) is that

$$V_X(t)\mathcal{G}(\pi) \subset \mathcal{G}(\pi). \tag{2.29}$$

We also note that

$$\mathcal{G}(\pi) \subset \mathcal{D}(\pi(X)) \tag{2.30}$$

and

$$\pi(X)\cdot\mathcal{G}(\pi) \subset \mathcal{G}(\pi). \tag{2.31}$$

Indeed, let $\mathfrak{X}_R^X$ be the *right invariant* vector field on G defined by

$$\mathfrak{X}_R^X f(g) = \lim_{h\to 0} h^{-1}[f((\exp hX)g) - f(g)] \tag{2.32}$$

for $f \in C^\infty(G)$. From (2.15) we have

$$\pi(X)\pi(f)u = -\int_G (\mathfrak{X}_R^X f)(g)\pi(g)u\,dg = -\pi(\mathfrak{X}_R^X f)u, \tag{2.33}$$

which gives (2.30)–(2.31). We note that the correspondence $X \mapsto \mathfrak{X}_R^X$ respects the Lie bracket structure as follows:

$$\mathfrak{X}_R^{[X,Y]} = -[\mathfrak{X}_R^X, \mathfrak{X}_R^Y], \tag{2.34}$$

for $X, Y \in \mathfrak{g}$. (Compare (2.44).) It follows that

$$\begin{aligned}(2.35)\qquad \pi([X,Y])\pi(f)u &= \pi([\mathfrak{X}_R^X, \mathfrak{X}_R^Y]f)u \\ &= \pi(\mathfrak{X}_R^X\mathfrak{X}_R^Y f)u - \pi(\mathfrak{X}_R^Y\mathfrak{X}_R^X f)u \\ &= \pi(X)\pi(Y)\pi(f)u - \pi(Y)\pi(X)\pi(f)u \\ &= [\pi(X), \pi(Y)]\pi(f)u,\end{aligned}$$

that is,

$$(2.36)\qquad \pi([X,Y])u = [\pi(X), \pi(Y)]u \quad \text{for } u \in \mathcal{G}(\pi).$$

In particular, if B is finite-dimensional, (2.36) holds for all $u \in B$.

It is important to know that the Gårding space $\mathcal{G}(\pi)$ is dense in other spaces besides the representation space B. For example, we claim $\mathcal{G}(\pi)$ is dense in the domain $\mathcal{D}(\pi(X))$ for each nonzero $X \in \mathfrak{g}$, i.e.,

PROPOSITION 2.1. *The operator $\pi(X)$ generating $\pi(\exp tX)$ is the closure of its restriction to $\mathcal{G}(\pi)$.*

PROOF. This is a formal consequence of (2.29)–(2.31), in view of the following general result.

PROPOSITION 2.2. *Let $V(t)$ be a group (even a semigroup) of operators on a Banach space B, with generator A. Let $\mathcal{L} \subset \mathcal{D}(A)$ be a dense linear subspace of B and suppose $V(t)\mathcal{L} \subset \mathcal{L}$ for all t. Then A is the closure of its restriction to $\mathcal{L}$.*

PROOF. By Proposition 1.2, it suffices to show $(\lambda - A)(\mathcal{L})$ is dense in B if $\operatorname{Re}\lambda$ is sufficiently large. So suppose $w \in B^*$ annihilates this range. Pick $u \in \mathcal{L}$ such that $\langle u, w\rangle \neq 0$. Now $(d/dt)\langle V(t)u, w\rangle = \langle AV(t)u, w\rangle = \langle \lambda V(t)u, w\rangle$ since $V(t)u \in \mathcal{L}$. This implies $\langle V(t)u, w\rangle = e^{\lambda t}\langle u, w\rangle$. But if λ is picked so that $\operatorname{Re}\lambda > K$ where $\|V(t)\| \le Me^{K|t|}$, this is impossible unless $\langle u, w\rangle = 0$. This proves the proposition.

Proposition 2.2 provides a very nice tool for obtaining results on essential selfadjointness of iA, when $V(t)$ is unitary, and also of all powers $(iA)^k$. See Appendix B for more on this. As slick as arguments using Proposition 2.2 can be, it is also instructive to see directly that, for an appropriate approximate identity $f_j \in C_0^\infty(G)$, if $u \in \mathcal{D}(\pi(X))$, then $\pi(f_j)u \to u$ in the topology of $\mathcal{D}(\pi(X))$, i.e., $\pi(X)\pi(f_j)u$ converges to $\pi(X)u$ in B. A direct attack on this involves comparing $\pi(X)\pi(f_j)$ with $\pi(f_j)\pi(X)$. This gives us an excuse to analyze analogues of (2.15) and (2.33), with the order of the operator factors reversed. For $\pi(f)\pi(g_1)$, $g_1 \in G$, we have, in place of (2.15),

$$(2.37)\qquad \pi(f)\pi(g_1)u = \int_G f(g)\pi(gg_1)u\,dg.$$

Now the Haar measure dg was constructed to be left invariant, not necessarily right invariant. But a right translate of dg is clearly still left invariant, so, as stated before, it must be a scalar multiple of dg. Hence, for any integrable function h on G,

$$\int_G h(gg_1)\,dg = \Delta(g_1)\int_G h(g)\,dg \tag{2.38}$$

where the factor $\Delta(g_1)$, called the modular function, defines a homomorphism

$$\Delta\colon G \to R^+ \tag{2.39}$$

of G into the multiplicative group of positive real numbers. Hence (2.37) yields

$$\pi(f)\pi(g_1)u = \Delta(g_1)\int_G f(gg_1^{-1})\pi(g)\,dg. \tag{2.40}$$

Now, for $X \in \mathfrak{g}$, $\gamma_X(t) = \exp tX$,

$$(d/dt)\Delta(\gamma_X(t))f(g\gamma_X(t)^{-1})|_{t=0} = -\mathfrak{X}_L^X f(g) + \Delta(X)f(g) \tag{2.41}$$

where $\mathfrak{X}_L^X$ is the *left invariant* vector field on G matching up with X at T_eG. Note that, since $\mathfrak{X}_L^X$ generates the flow of right translation by $\gamma_X(t)$,

$$\operatorname{div}\mathfrak{X}_L^X = \Delta(X). \tag{2.42}$$

Consequently, from (2.40) we derive, for $u \in \mathcal{D}(\pi(X))$,

$$\pi(f)\pi(X)u = \pi(-\mathfrak{X}_L^X f)u + \Delta(X)\pi(f)u = \pi((\mathfrak{X}_L^X)^* f)u, \tag{2.43}$$

where $(\mathfrak{X}_L^X)^* = -\mathfrak{X}_L^X + \Delta(X)$ is the formal adjoint of $\mathfrak{X}_L^X$, as a first order differential operator.

We remark that the Lie bracket on $\mathfrak{g}$ is defined so that

$$\mathfrak{X}_L^{[X,Y]} = [\mathfrak{X}_L^X, \mathfrak{X}_L^Y], \qquad X, Y \in \mathfrak{g}. \tag{2.44}$$

The difference in sign in (2.34) is explained as follows. Consider the diffeomorphism $\kappa\colon G \to G$ given by $\kappa(g) = g^{-1}$. This map produces a map κ^* on vector fields, and, for $X \in \mathfrak{g} = T_eG$, we have

$$\kappa^*\mathfrak{X}_L^X = -\mathfrak{X}_R^X. \tag{2.45}$$

Since κ^* preserves Lie brackets, this shows that (2.44) and (2.34) are equivalent. One could imagine reversing conventions, i.e., reversing the signs in (2.34) and (2.44), but this would mess up the signs in (2.36). This gives one reason for the convention (2.44).

Suppose a sequence $f_j \in C_0^\infty(G)$ is picked as follows. Fix some coordinate chart U about $e \in G$, identified with the origin; you could use exponential coordinates. Put Lebesgue measure on U, matching up with the coordinate expression for Haar measure at $e = 0$. Let $f_1(x) \in C_0^\infty(U)$, $\int_U f_1(x)\,dx = 1$, and set $f_j(x) = j^n f_1(jx)$, $n = \dim G$. Then $\int_U f_j(x)\,dx = 1$, and $\int_G f_j(g)\,dg = \alpha_j \to 1$ as $j \to \infty$. We now establish the following result, which provides a second proof of Proposition 2.1.

PROPOSITION 2.3. *If* $u \in \mathcal{D}(\pi(X)), X \in \mathfrak{g}$, *then*

$$\pi(f_j)u \to u \quad \text{in } \mathcal{D}(\pi(X)) \tag{2.46}$$

as $j \to \infty$.

PROOF. What we must show is that

$$\pi(X)\pi(f_j)u \to \pi(X)u \quad \text{in } B. \tag{2.47}$$

Since we know $\pi(f_j)\pi(X)u \to \pi(X)u$, consider the sequence of commutators

$$K_j u = \pi(X)\pi(f_j)u - \pi(f_j)\pi(X)u, \tag{2.48}$$

defined a priori for $u \in \mathcal{D}(\pi(X))$. We want to show that

$$K_j u \to 0 \quad \text{for } u \in \mathcal{D}(\pi(X)). \tag{2.49}$$

By (2.33) and (2.43), we have

$$K_j u = \pi((\mathfrak{X}_L^X - \mathfrak{X}_R^X)f_j) - \Delta(X)\pi(f_j). \tag{2.50}$$

Clearly each K_j extends uniquely to a bounded operator on B. In fact, we claim $\{K_j\}$ is a unformly bounded family of operators on B. This follows because $\mathfrak{X}_L^X - \mathfrak{X}_R^X$ is a vector field on G whose coefficients vanish at e. From the given form of f_j in local coordinates, it easily follows that

$$\|(\mathfrak{X}_L^X - \mathfrak{X}_R^X)f_j\|_{L^1(G)} \leq K < \infty. \tag{2.51}$$

Since all f_j are supported in a fixed compact set $\overline{U}$, and since, by the uniform boundedness theorem, we have a bound $\|\pi(g)\| \leq M$ for $g \in \overline{U}$, the uniform boundedness of $\{K_j\}$ follows from (2.51). Thus, by Lemma 1.1, it suffices to show that (2.49) holds on a dense linear subspace of B. In particular, it now suffices to show (2.47) holds for any $u = \pi(f_0)v$ belonging to $\mathcal{G}(\pi)$. But, by (2.33) and (2.17),

$$\begin{aligned} \pi(X)\pi(f_j)\pi(f_0)v &= -\pi(\mathfrak{X}_R^X f_j)\pi(f_0)v \\ &= -\pi(\mathfrak{X}_R^X f_j * f_0)v. \end{aligned} \tag{2.52}$$

As $j \to \infty, f_j \to \delta_e$ in $\mathcal{E}'(G)$, the space of compactly supported distributions on G, and it follows that

$$\mathfrak{X}_R^X f_j * f_0 \to \mathfrak{X}_R^X f_0 \quad \text{in } C_0^\infty(G). \tag{2.53}$$

This shows that $\pi(X)\pi(f_j)\pi(f_0)v \to \pi(X)\pi(f_0)v$ as $j \to \infty$, and completes the proof of Proposition 2.3.

A further denseness result is that $\mathcal{G}(\pi)$ is dense in $C^\infty(\pi)$, where the space of smooth vectors for π is given an appropriate Fréchet space topology. We shall prove this in the next section, where we look at representations of distributions, a topic suggested by the role of (2.53) in the proof of the last proposition. One consequence of the denseness of $\mathcal{G}(\pi)$ in $C^\infty(\pi)$ would be to generalize (2.36) to $u \in C^\infty(\pi)$. However, it is better to show this directly:

PROPOSITION 2.4. *We have, for $X, Y \in \mathfrak{g}$,*

$$\pi(X)\colon C^\infty(\pi) \to C^\infty(\pi), \tag{2.54}$$

and

$$\pi([X,Y]) = [\pi(X), \pi(Y)]u, \qquad u \in C^\infty(\pi). \tag{2.55}$$

PROOF. The statement that $\Phi(g) = \pi(g)u$ is a C^∞ function on G with values in B implies that, for any differential operator P with smooth coefficients, $P\Phi(g)$ is C^∞ also. Suppose $P = \mathfrak{X}_L^X$. Differentiability of $\pi(g)u$ at $g = e$ implies $u \in \mathcal{D}(\pi(X))$, and

$$\mathfrak{X}_L^X \pi(g)u = \pi(g)\pi(X)u. \tag{2.56}$$

Since the left side of (2.56) is clearly C^∞ with values in B, this proves (2.54). Furthermore, (2.56) implies

$$\begin{aligned} \pi(g)[\pi(X), \pi(Y)]u &= \pi(g)\pi(X)\pi(Y)u - \pi(g)\pi(Y)\pi(X)u \\ &= \mathfrak{X}_L^X \mathfrak{X}_L^Y \pi(g)u - \mathfrak{X}_L^Y \mathfrak{X}_L^X \pi(g)u \\ &= \mathfrak{X}_L^{[X,Y]} \pi(g)u = \pi(g)\pi([X,Y])u, \end{aligned}$$

and setting $g = e$ gives (2.55).

The following is a useful characterization of the space of smooth vectors for a representation π. We recall that if A_1, A_2 are (possibly unbounded) operators on B with domains $\mathcal{D}(A_1), \mathcal{D}(A_2)$, then the domain of $A_1 A_2$ is defined to be

$$\mathcal{D}(A_1 A_2) = \{u \in \mathcal{D}(A_2)\colon A_2 u \in \mathcal{D}(A_1)\}. \tag{2.57}$$

PROPOSITION 2.5. *If π is a representation of G on B, then $u \in B$ belongs to $C^\infty(\pi)$ if and only if*

$$u \in \mathcal{D}(\pi(X_{j_1}) \cdots \pi(X_{j_k})) \tag{2.58}$$

for all $X_{j_l} \in \mathfrak{g}$, any k.

PROOF. As basically noted in the proof of Proposition 2.4, $\pi(g)u$ is C^1 if and only if $u \in \mathcal{D}(\pi(X_j))$ for all $X_j \in \mathfrak{g}$, in which case $\mathfrak{X}_L^X \pi(g)u$ is given by (2.56) for all $X \in \mathfrak{g}$. Reasoning by induction on m shows that $\pi(g)u$ is C^m if and only if (2.58) holds for all $k \le m$, in which case

$$\mathfrak{X}_L^{X_{j_1}} \cdots \mathfrak{X}_L^{X_{j_k}} \pi(g)u = \pi(g)\pi(X_{j_1}) \cdots \pi(X_{j_k})u. \tag{2.59}$$

This proves the proposition.

A class of vectors more restricted than smooth vectors is the class of *analytic vectors*, discussed in Appendix D. A vector $u \in B$ is an analytic vector for a representation π of G provided $\pi(g)u$ is an analytic function on G with values in B. Alternatively, one can impose appropriate estimates on (2.59) at $g = e$, to define the concept of an analytic vector for a Lie algebra representation. Analytic vectors are useful objects to have, to show Lie algebra representations can be exponentiated to Lie group representations, as explained in Appendix D. On the other hand, it is an important fact that for any Lie group representation π of G on B, there is a dense subspace of B consisting of analytic vectors. This is also proved in Appendix D. It involves approximating $u \in B$ by a sequence

$$\pi(f_j)u = \int_G f_j(g)\pi(g)u\,dg \tag{2.60}$$

where f_j is a sequence, not in $C_0^\infty(G)$ as before, but of strongly peaked functions which are analytic on G, with appropriate estimates. It is harder to construct such f_j than in the case of $C_0^\infty(G)$, but one can let $f_j(g) = h(t_j, g), t_j \downarrow 0$, where $h(t,g)$ is the fundamental solution to the heat equation $(\partial/\partial t - \Delta)h = 0, t > 0$, where Δ is the Laplace operator on G, endowed with some left invariant Riemannian metric. For details, see Appendix D.

Once such a left invariant metric is imposed, one can show that, for any $t > 0$, $h(t,g)$ decreases faster than any exponential $e^{-K\,\mathrm{dist}(g,e)}$, as $g \to \infty$. If π is a strongly continuous representation of G on a Banach space B, it is easy to show

$$\|\pi(g)\| \le Ce^{M\,\mathrm{dist}(g,e)} \tag{2.61}$$

for some $C, M < \infty$, in analogy with (1.9). Furthermore, because G with a left invariant Riemannian metric is a homogeneous space, one can obtain an estimate

$$\mathrm{vol}\,B_R(e) \le Ce^{KR} \quad \text{for all } R, \tag{2.62}$$

for some $C, K < \infty$, where $B_R(e) = \{g \in G \colon \mathrm{dist}(g,e) \le R\}$. It follows that

$$\pi(f) = \int_G f(g)\pi(g)\,dg \tag{2.63}$$

is well defined for any continuous f on G satisfying an estimate of the form

$$|f(g)| \le Ce^{-L\,\mathrm{dist}(g,e)} \tag{2.64}$$

where $L > M + K$. One can generalize this in various ways. As already stated, if π is uniformly bounded, particularly if it is unitary, (2.63) is well defined for any $f \in L^1(G)$.

3. Representations of distributions and universal enveloping algebras. We want to extend the representation $\pi(f)$ considered in §2 for $f \in C_0^\infty(G)$ to the case where f is a compactly supported distribution on G, i.e., $f \in \mathcal{E}'(G)$. As we will see, this extension will specialize to a representation of the universal enveloping algebra of $\mathfrak{g}$. As before, π is a strongly continuous representation of G on a Banach space B. For simplicity we will assume B is reflexive. We will first define

$$\pi(k)\colon C^\infty(\pi) \to B \tag{3.1}$$

for $k \in \mathcal{E}'(G)$. Recall that to say $u \in C^\infty(\pi)$ is to say

$$\varphi_u(g) = \pi(g)u \tag{3.2}$$

is a C^∞ function of $g \in G$ with values in B. Given $w \in B^*$, set

$$\varphi_{u,w}(g) = \langle \pi(g)u, w \rangle. \tag{3.3}$$

Clearly $\varphi_{u,w} \in C^\infty(G)$. We want

$$\langle \pi(k)u, w \rangle = \langle \varphi_{u,w}, k \rangle \tag{3.4}$$

to define $\pi(k)u$. Note that the right side of (3.4) is well defined. Since k acts as a continuous linear functional on $C^\infty(G)$, for some C, m and some compact $\overline{U}$, we have

$$|\langle \varphi_{u,w}, k \rangle| \leq C\|\varphi_{u,w}\|_{C^m(\overline{U})} \leq C\|\varphi_u\|_{C^m(\overline{U})}\|w\|_{B^*}. \tag{3.5}$$

If B is reflexive, this estimate shows that (3.4) defines a unique element $\pi(k)u \in B$. It is clear that, for $u \in C^\infty(\pi)$ fixed, the map

$$k \mapsto \pi(k)u \tag{3.6}$$

is a continuous linear map of $\mathcal{E}'(G)$ into B. We can show that, for $u \in C^\infty(\pi)$, $\pi(k)u$ belongs to a subspace of B, as follows.

Note that, for $g \in G$, from (2.15) and a limiting argument, we have

$$\langle \pi(g)\pi(k)u, w \rangle = \langle \varphi_{u,w}, \lambda(g)k \rangle = \langle \pi(\lambda(g)k)u, w \rangle \tag{3.7}$$

where $\lambda(g) \colon \mathcal{E}' \to \mathcal{E}'$ is the natural extension of $\lambda(g) \colon C_0^\infty \to C_0^\infty$ defined by

$$\lambda(g)f(g_1) = f(g^{-1}g_1). \tag{3.8}$$

In fact, (3.9) defines $\lambda(g) \colon C^\infty \to C^\infty$, and on $\mathcal{E}'$ we define

$$\lambda(g) = \lambda(g^{-1})^*. \tag{3.9}$$

It is clear that (3.7) is a smooth function of $g \in G$, for any $w \in B^*$. In other words, for $k \in \mathcal{E}'(G)$,

$$\pi(k) \colon C^\infty(\pi) \to C_w^\infty(\pi) \tag{3.10}$$

where we define $C_w^\infty(\pi)$, the space of "weakly smooth" vectors, to consist of all $u \in B$ such that, for each $w \in B^*$, the function $\varphi_{u,w}$ on G defined by (3.3) belongs to $C^\infty(G)$. We have the following useful result.

LEMMA 3.1.

$$C_w^\infty(\pi) = C^\infty(\pi).$$

PROOF. Let $\Phi(x)$ be a function from a closed set $\overline{U}$ in G (which we take to be a ball in R^n, in some coordinate system) to B which is weakly C^1. That is, for $w \in B^*$, $x, x+y \in U$,

$$\langle \Phi(x+y), w \rangle = \langle \Phi(x), w \rangle + \sum y_j D_j \langle \Phi(x), w \rangle + R_1(x, y, w) \tag{3.11}$$

where $|y|^{-1}R_1(x,y,w) \to 0$ as $y \to 0$ in R^n. The uniform boundedness theorem implies

$$|y|^{-1}\|\Phi(x+y) - \Phi(x)\| \leq C_1. \tag{3.12}$$

This implies Φ is strongly Lipschitz (in particular, strongly continuous). This gives $|D_j\langle\Phi(x),w\rangle| \le C_1\|w\|$, and hence there are unique $\Psi_j(x) \in B$ such that $D_j\langle\Phi(x),w\rangle = \langle\Psi_j(x),w\rangle$; $\|\Psi_j(x)\| \le C_1$.

Now if Φ is weakly C^2, then each Ψ_j is weakly C^1 and by the argument given Ψ_j is continuous. Furthermore, if we write

$$\begin{aligned}\langle\Phi(x+y),w\rangle = \langle\Phi(x),w\rangle &+ \sum y_j\langle\Psi_j(x),w\rangle \\ &+ D^2\langle\Phi(x),w\rangle\cdot y\cdot y + R_2(x,y,w)\end{aligned} \tag{3.13}$$

where $|y|^{-2}R_2(x,y,w) \to 0$ as $y \to 0$ and apply the uniform boundedness theorem, we get

$$|y|^{-2}\|\Phi(x+y) - \Phi(x) - \sum y_j\Psi_j(x)\| \le C_2. \tag{3.14}$$

This implies Φ is strongly differentiable, with derivative $(\Psi_1(x),\dots,\Psi_n(x))$. Thus, if Φ is weakly C^2 it is strongly C^1. It follows easily that if Φ is weakly C^k then it is strongly C^{k-1}, and this proves the lemma.

Consequently we can elevate (3.10) to

$$\pi(k)\colon C^\infty(\pi) \to C^\infty(\pi) \quad \text{for } k \in \mathcal{E}'(G). \tag{3.15}$$

It is now time to topologize the linear space $C^\infty(\pi)$. For $U \subset G$ relatively compact, $X_1,\dots,X_n$ a basis for $\mathfrak{g}$, and $\alpha = (\alpha(1),\dots,\alpha(N)) \in \{1,\dots,n\}^N$, consider the seminorms

$$\begin{aligned}p_{U,\alpha}(u) &= \sup_{g\in U}\|\mathfrak{X}_R^{X_{\alpha(1)}}\cdots\mathfrak{X}_R^{X_{\alpha(N)}}\pi(g)u\|_B \\ &= \sup_{g\in U}\|\pi(X_{\alpha(N)})\cdots\pi(X_{\alpha(1)})\pi(g)u\|_B.\end{aligned} \tag{3.16}$$

The collection of these seminorms is easily seen to define a complete metric topology on $C^\infty(\pi)$, making $C^\infty(\pi)$ a Fréchet space. Note that an equivalent family of seminorms defining the topology of $C^\infty(\pi)$ is

$$\begin{aligned}p'_{U,\alpha}(u) &= \sup_{g\in U}\|\mathfrak{X}_L^{X_{\alpha(1)}}\cdots\mathfrak{X}_L^{X_{\alpha(N)}}\pi(g)u\|_B \\ &= \sup_{g\in U}\|\pi(g)\pi(X_{\alpha(1)})\cdots\pi(X_{\alpha(N)})u\|_B.\end{aligned} \tag{3.17}$$

Consequently, another equivalent family of seminorms is

$$p''_\alpha(u) = \|\pi(X_{\alpha(1)})\cdots\pi(X_{\alpha(N)})u\|_B. \tag{3.18}$$

The estimate (3.5) shows that the map (3.1) is continuous. By the closed graph theorem, we deduce the following.

PROPOSITION 3.2. *For each $k \in \mathcal{E}'(G)$, the map (3.15) is continuous.*

Furthermore, the continuity of (3.6) from $\mathcal{E}'(G)$ to B together with the closed graph theorem yields

PROPOSITION 3.3. *For each $u \in C^\infty(\pi)$, the map*

$$\mathcal{E}'(G) \to C^\infty(\pi) \tag{3.19}$$

given by

$$k \mapsto \pi(k)u \tag{3.20}$$

is continuous.

With this result, we easily deduce

PROPOSITION 3.4. *The Gårding space $\mathcal{G}(\pi)$ is dense in $C^\infty(\pi)$.*

PROOF. Let $f_j \in C_0^\infty(G)$ be such that

$$f_j \to \delta_e \quad \text{in } \mathcal{E}'(G). \tag{3.21}$$

Then, for any $u \in C^\infty(\pi)$,

$$\pi(f_j)u \to \pi(\delta_e)u = u \tag{3.22}$$

in the topology of $C^\infty(\pi)$. Since $\pi(f_j)u \in \mathcal{G}(\pi)$, the proof is complete.

Note that if we set

$$k = \mathfrak{X}_R^X \delta_e \tag{3.23}$$

so that, for $f \in C^\infty(G)$,

$$\langle f, k\rangle = -\mathfrak{X}_R^X f(e), \tag{3.24}$$

we have, for $\varphi_{u,w}$ given by (3.3), $u \in C^\infty(\pi)$,

$$\langle \varphi_{u,w}, k\rangle = -\mathfrak{X}_R^X \langle \pi(g)u, w\rangle|_{g=e} = -\langle \pi(X)u, w\rangle. \tag{3.25}$$

In other words,

$$\pi(\mathfrak{X}_R^X \delta_e)u = -\pi(X)u \quad \text{for } u \in C^\infty(\pi). \tag{3.26}$$

Note the resemblance between (3.26) and (2.33). Both results are special cases of

$$\pi(X)\pi(k)u = -\pi(\mathfrak{X}_R^X k)u, \qquad u \in C^\infty(\pi), k \in \mathcal{E}'(G). \tag{3.27}$$

Indeed, such results are special cases of the behavior of $\pi(k_1 * k_2)$, where we put an operation of convolution on $\mathcal{E}'(G)$ as follows.

Recall the convolution on $C_0^\infty(G)$ is given by

$$f_2 * f_1(g) = \int_G f_2(y) f_1(y^{-1}g)\, dy = \int_G f_1(y^{-1}) f_2(gy)\, dy. \tag{3.28}$$

We can rewrite this as

$$f_2 * f_1(g) = (\lambda(g^{-1})f_2, \check{f}_1) = (f_2, \lambda(g)\check{f}_1), \tag{3.29}$$

where

$$\lambda(g)f(y) = f(g^{-1}y) \tag{3.30}$$

and we set

$$\check{f}_1(y) = \overline{f_1(y^{-1})}. \tag{3.31}$$

It is clear that (3.29) is well defined if

$$f_1 \in C^\infty(G), \qquad f_2 \in \mathcal{E}'(G), \tag{3.32}$$

or if

$$f_1 \in C_0^\infty(G), \qquad f_2 \in \mathcal{D}'(G), \tag{3.33}$$

and we have bilinear maps $(f_1, f_2) \mapsto f_2 * f_1$:

$$C^\infty(G) \times \mathcal{E}'(G) \to C^\infty(G) \tag{3.34}$$

and

$$C_0^\infty(G) \times \mathcal{D}'(G) \to C^\infty(G). \tag{3.35}$$

These also restrict to a bilinear map

$$C_0^\infty(G) \times \mathcal{E}'(G) \to C_0^\infty(G). \tag{3.36}$$

We can extend to $f_1 \in \mathcal{E}'(G)$ by duality. To do this, we need to consider the adjoint of the convolution operator

$$K_{f_2}: C^\infty(G) \to C^\infty(G), \tag{3.37}$$

defined by

$$K_{f_2}(f_1) = f_2 * f_1, \qquad f_2 \in \mathcal{E}'(G). \tag{3.38}$$

Let $h \in C_0^\infty(g)$. If $(f_1, h) = \int_G f_1(g)\overline{h(g)}\, dg$, we have, for $f_2 \in C_0^\infty(G)$,

$$(f_2 * f_1, h) = \int\int f_1(y^{-1}) f_2(gy)\overline{h(g)}\, dg\, dy. \tag{3.39}$$

In order to compute this, let us note that, as a simple consequence of (2.38), $\Delta(g)$ is the Radon-Nikodým derivative of $d_r g$ with respect to dg, i.e.,

$$\int_G f(g^{-1})\, dg = \int_G f(g)\, d_r g = \int_G f(g)\Delta(g)\, dg. \tag{3.40}$$

Then we have

$$\begin{aligned} (f_2 * f_1, h) &= \int\int f_1(y) f_2(g^{-1}y^{-1})\overline{h(g^{-1})}\Delta(g)\Delta(y)\, dg\, dy \\ &= \int\int f_1(y)\overline{(\Delta\check{f}_2)(yg)h(g^{-1})}\, dg\, dy \\ &= \int f_1(y)\overline{(\Delta\check{f}_2) * h(y)}\, dy. \end{aligned} \tag{3.41}$$

In other words, with

$$f_2^{\#}(g) = \Delta(g)\check{\bar{f}}_2(g) = \Delta(g)\overline{f_2(g^{-1})}, \tag{3.42}$$

we have

$$K_{f_2}^* = K_{f_2^{\#}}, \quad \text{for } f_2 \in C_0^\infty(G). \tag{3.43}$$

This identity continues to hold for $f_2 \in \mathcal{E}'(G)$, in the context of (3.37), provided we define the map $f_2 \mapsto f_2^{\#}$ for $f_2 \in \mathcal{E}'(G)$ by the identity

$$(f, f_2^{\#}) = (\overline{\check{f}}, f_2), \qquad f \in C^\infty(G). \tag{3.44}$$

Thus, for any $f_2 \in \mathcal{E}'(G)$,

$$K_{f_2} : \mathcal{E}'(G) \to \mathcal{E}'(G) \tag{3.45}$$

is defined via (3.37) by

$$K_{f_2} = K_{f_2^{\#}}^*. \tag{3.46}$$

This gives our convolution product

$$\mathcal{E}'(G) \times \mathcal{E}'(G) \to \mathcal{E}'(G). \tag{3.47}$$

It is clear that, for $f_1 \in C_0^\infty(G)$ fixed, (3.36) extends the convolution $C_0^\infty(G) \times C_0^\infty(G) \to C_0^\infty(G)$ from $f_2 \in C_0^\infty(G)$ to $f_2 \in \mathcal{E}'(G)$, continuously in f_2. Also, for $f_2 \in \mathcal{E}'(G)$ fixed, the map (3.47) extends the convolution (3.36) from $f_1 \in C_0^\infty(G)$ to $f_1 \in \mathcal{E}'(G)$, continuously in f_1. Thus simple limiting arguments verify, for example, the associative law

$$(k_1 * k_2) * k_3 = k_1 * (k_2 * k_3), \qquad k_j \in \mathcal{E}'(G), \tag{3.48}$$

for convolutions, as a consequence of such an identity for $k_j \in C_0^\infty(G)$.

With the convolution of compactly supported distributions defined, we easily are able to derive their basic properties vis a vis representations. First, we see how $\pi(k)$ for $k \in \mathcal{E}'(G)$ acts on the Gårding space $\mathcal{G}(\pi)$.

LEMMA 3.5. *For $u \in B, f \in C_0^\infty(G), k \in \mathcal{E}'(G)$, we have*

$$\pi(k)\pi(f)u = \pi(k * f)u. \tag{3.49}$$

PROOF. Note that $k * f \in C_0^\infty(G)$, by (3.36). Pick $k_j \in C_0^\infty(G)$ such that $k_j \to k$ in $\mathcal{E}'(G)$. Then (2.17) implies

$$\pi(k_j)\pi(f)u = \pi(k_j * f)u \tag{3.50}$$

for each j. We know the left side of (3.50) tends to $\pi(k)\pi(f)u$ as $j \to \infty$, since $\pi(f)u \in C^\infty(\pi)$. Meanwhile, $k_j * f \to k * f$ in $C_0^\infty(G)$, so the right side tends to $\pi(k * f)u$. This proves the lemma.

Now we can generalize (2.17) to compactly supported distributions.

PROPOSITION 3.6. *For $u \in C^\infty(\pi), k_1, k_2 \in \mathcal{E}'(G)$, we have*

$$\pi(k_1)\pi(k_2)u = \pi(k_1 * k_2)u. \tag{3.51}$$

PROOF. Suppose $u = \pi(f)v, v \in B, f \in C_0^\infty(G)$. By Lemma 3.5 we have

$$\begin{aligned}\pi(k_1)\pi(k_2)u &= \pi(k_1)\pi(k_2 * f)v = \pi(k_1 * k_2 * f)v \\ &= \pi(k_1 * k_2)\pi(f)v.\end{aligned} \tag{3.52}$$

Hence (3.51) holds for $u \in \mathcal{G}(\pi)$. Since $\mathcal{G}(\pi)$ is dense in $C^\infty(\pi)$, on which the operators $\pi(k_j)$ are continuous, this completes the proof.

Let $X \in \mathfrak{g}$, with associated left and right invariant vector fields $\mathfrak{X}_L^X$ and $\mathfrak{X}_R^X$. If $k_j \in \mathcal{E}'(G)$, then

$$\mathfrak{X}_R^X(k_1 * k_2) = (\mathfrak{X}_R^X k_1) * k_2 \tag{3.53}$$

and

$$\mathfrak{X}_L^X(k_1 * k_2) = k_1 * (\mathfrak{X}_L^X k_2). \tag{3.54}$$

These identities follow for $k_j \in C_0^\infty(G)$ by a simple computation from the definition (3.28), and for $k_j \in \mathcal{E}'(G)$ they follow by a limiting argument. In particular, since

$$k * \delta_e = \delta_e * k = k \tag{3.55}$$

for any $k \in \mathcal{E}'(G)$, where δ_e is the point mass at e:

$$\langle f, \delta_e \rangle = f(e), \tag{3.56}$$

we have

$$\mathfrak{X}_R^X k = (\mathfrak{X}_R^X \delta_e) * k \tag{3.57}$$

and

$$\mathfrak{X}_L^X k = k * (\mathfrak{X}_L^X \delta_e). \tag{3.58}$$

These results can be generalized. Indeed, suppose P_R is any right invariant differential operator. Generalizing (3.53) and (3.57), we have

$$P_R k = (P_R \delta_e) * k. \tag{3.59}$$

Similarly, if P_L is any left invariant differential operator, we have

$$P_L k = k * (P_L \delta_e). \tag{3.60}$$

Note that $P_R \delta_e$ and $P_L \delta_e$ are distributions supported at $\{e\}$. Conversely, let ν be any distribution in $\mathcal{E}'(G)$ which is supported at $\{e\}$. A basic result in the theory of distributions is that ν must be obtained from δ_e and its derivatives. In a local coordinate system U about e,

$$\nu = \sum_{|\alpha| \le m} a_\alpha D^\alpha \delta_e = P(D)\delta_e. \tag{3.61}$$

Now the differential operator $P(D)$ can be written as a linear combination, with coefficients in $C^\infty(U)$ (all on the *right*), of products of vector fields, either of the form $\mathfrak{X}_L^{X_j}$ or of the form $\mathfrak{X}_R^{X_j}$, where $\{X_1, \ldots, X_n\}$ is some basis of $\mathfrak{g}$. It follows that, for some constants a'_α, a''_α,

$$\nu = \sum_{k \le m} a'_\alpha \mathfrak{X}_R^{X_{\alpha(1)}} \cdots \mathfrak{X}_R^{X_{\alpha(k)}} \delta_e = P_R \delta_e \tag{3.62}$$

and

$$\nu = \sum_{k \le m} a''_\alpha \mathfrak{X}_L^{X_{\alpha(1)}} \cdots \mathfrak{X}_L^{X_{\alpha(k)}} \delta_e = P_L \delta_e. \tag{3.63}$$

Here P_R and P_L are right and left invariant differential operators on $C^\infty(G)$, which in fact are polynomials in right (resp. left) invariant vector fields. We have proved the following result.

PROPOSITION 3.7. *The following three classes of operators coincide:*
(a) *left invariant differential operators P_L with C^∞ coefficients;*
(b) *polynomials over $\mathbf{C}$ in the left invariant vector fields;*
(c) *convolution operators $Pk = k * \nu$, for $\nu \in \mathcal{E}'(G)$ supported at $\{e\}$.*
We have an analogous statement for right invariant differential operators.

Denote the set of left invariant differential operators on G by

$$\mathcal{D}_L(G). \tag{3.64}$$

Its algebraic structure of course agrees with the algebraic structure of convolution on $\mathcal{E}'(G)$, restricted to distributions supported at $\{e\}$, with the factors reversed, since

$$P_L P'_L k = P_L(k * P'_L \delta_e) = k * P'_L \delta_e * P_L \delta_e. \tag{3.65}$$

Consequently, if for $P_L \in \mathcal{D}_L(G)$ we define

$$\pi(P_L)u = \pi(P_L^* \delta_e)u, \tag{3.66}$$

for $u \in C^\infty(\pi)$, we see that (3.66) defines a representation of $\mathcal{D}_L(G)$ on $C^\infty(\pi)$, i.e., $\pi(P_L P'_L) = \pi(P_L)\pi(P'_L)$. Note that the special case $P_L = \mathfrak{X}_L^X$ yields

$$\begin{aligned} \pi(\mathfrak{X}_L^X)u &= \pi((\mathfrak{X}_L^X)^* \delta_e)u = -\pi(\mathfrak{X}_L^X \delta_e + \Delta(X)\delta_e)u \\ &= -\pi(\mathfrak{X}_R^X \delta_e)u = \pi(X)u \end{aligned} \tag{3.67}$$

for $u \in C^\infty(\pi)$, by formulas (2.43) and (3.24).

The algebra $\mathcal{D}_L(G)$ is intimately related to the *universal enveloping algebra* $\mathfrak{U}(\mathfrak{g})$, associated to the Lie algebra $\mathfrak{g}$ as follows. Form the tensor algebra

$$\bigotimes \mathfrak{g}_{\mathbf{C}} = \mathbf{C} \oplus \mathfrak{g}_{\mathbf{C}} \oplus \mathfrak{g}_{\mathbf{C}} \otimes \mathfrak{g}_{\mathbf{C}} \oplus \cdots \tag{3.68}$$

where $\mathfrak{g}_{\mathbf{C}}$ is the complexification of $\mathfrak{g}$. Let J be the two-sided ideal in $\bigotimes \mathfrak{g}_{\mathbf{C}}$ generated by all elements of the form

$$X \otimes Y - Y \otimes X - [X, Y]. \tag{3.69}$$

Then we set

(3.70) $$\mathfrak{U}(\mathfrak{g}) = \bigotimes \mathfrak{g}_{\mathbf{C}}/J.$$

The correspondence

(3.71) $$X_{\alpha(1)} \otimes \cdots \otimes X_{\alpha(k)} \mapsto \mathfrak{X}_L^{X_{\alpha(1)}} \cdots \mathfrak{X}_L^{X_{\alpha(k)}}$$

gives rise to a homomorphism of algebras

(3.72) $$\bigotimes \mathfrak{g}_{\mathbf{C}} \to \mathcal{D}_L(G).$$

By virtue of the characterization of the Lie bracket on $\mathfrak{g}$, $\mathfrak{X}_L^{[X,Y]} = [\mathfrak{X}_L^X, \mathfrak{X}_L^Y]$, we see that J is in the kernel of this homomorphism, so we have a homomorphism of algebras

(3.73) $$\alpha\colon \mathfrak{U}(\mathfrak{g}) \to \mathcal{D}_L(G).$$

Note that $\mathfrak{U}(\mathfrak{g})$ and $\mathcal{D}_L(G)$ possess filtrations

(3.74) $$\mathfrak{U}(\mathfrak{g}) = \bigcup_{k\geq 0} \mathfrak{U}^k(\mathfrak{g}), \qquad \mathcal{D}_L(G) = \bigcup_{k\geq 0} \mathcal{D}_L^k(G),$$

where $\mathfrak{U}^k(\mathfrak{g})$ is the image of $\bigoplus_{l\leq k} \bigotimes^l \mathfrak{g}_{\mathbf{C}}$ under the natural projection $\bigotimes \mathfrak{g}_{\mathbf{C}} \to \mathfrak{U}(\mathfrak{g})$, and $\mathcal{D}_L^k(G)$ consists of left invariant differential operators of order $\leq k$. Clearly α preserves the filtration

(3.75) $$\alpha\colon \mathfrak{U}^k(\mathfrak{g}) \to \mathcal{D}_L^k(G).$$

For a closer study of the relation between $\mathfrak{U}(\mathfrak{g})$ and $\mathcal{D}_L(G)$, it is convenient to pick a basis $X_1, \ldots, X_n$ of $\mathfrak{g}$, and define linear maps

(3.76) $$\beta\colon \mathcal{P}_n \to \mathfrak{U}(\mathfrak{g}), \qquad \gamma\colon \mathcal{P}_n \to \mathcal{D}_L(G),$$

as follows. $\mathcal{P}_n$ is the set of polynomials in n variables. We set

(3.77) $$\beta(a_{i_1\cdots i_n} t_1^{i_1} \cdots t_n^{i_n}) = a_{i_1\cdots i_n} X_1^{(i_1)} \otimes \cdots \otimes X_n^{(i_n)}$$

where

(3.78) $$X_l^{(i_l)} = X_l \otimes \cdots \otimes X_l \quad (i_l \text{ factors}),$$

and we set

(3.79) $$\gamma(a_{i_1\cdots i_n} t_1^{i_1} \cdots t_n^{i_n}) = a_{i_1\cdots i_n} (\mathfrak{X}_L^{X_1})^{i_1} \cdots (\mathfrak{X}_L^{X_n})^{i_n}.$$

From (3.71) it follows that the following is a commutative diagram.

(3.80) $$\begin{array}{ccc} \mathfrak{U}(\mathfrak{g}) & \xrightarrow{\alpha} & \mathcal{D}_L(G) \\ {\scriptstyle\beta}\nwarrow & & \nearrow{\scriptstyle\gamma} \\ & \mathcal{P}_n & \end{array}$$

Note that $\mathcal{P}_n$ has a natural filtration; $\mathcal{P}_n = \bigcup_{k\geq 0} \mathcal{P}_n^k$ where $\mathcal{P}_n^k$ consists of polynomials of degree $\leq k$, and β and γ preserve filtrations

(3.81) $$\beta\colon \mathcal{P}_n^k \to \mathfrak{U}^k(\mathfrak{g}), \qquad \gamma\colon \mathcal{P}_n^k \to \mathcal{D}_L^k(G).$$

The following result, which complements Proposition 3.7, is known as the Poincaré-Birkhoff-Witt theorem.

PROPOSITION 3.8. *The maps α, β, γ are all linear isomorphisms. In particular, α is an isomorphism of algebras.*

PROOF. From Proposition 3.7 we know α is surjective. To complete the proof, it will suffice to show that γ is injective and β is surjective.

So let p be a polynomial in $\ker\gamma$. Say $p \in \mathcal{P}_n^k$, with leading term

$$p(t) = \sum_{|\alpha|=k} a_\alpha t_1^{\alpha_1} \cdots t_n^{\alpha_n} + \cdots. \tag{3.82}$$

Consider the differential operator $\gamma(p)$, in an exponential coordinate system centered at e. $\gamma(p)$ is an operator of order $\leq k$, and its leading term at e is precisely

$$\sum_{|\alpha|=k} a_\alpha D_1^{\alpha_1} \cdots D_n^{\alpha_n}. \tag{3.83}$$

If $\gamma(p) = 0$, then (3.83) must vanish. Thus actually $p \in \mathcal{P}_n^{k-1}$. An inductive argument shows $p = 0$ if $p \in \ker\gamma$.

Since γ is injective, commutativity of (3.80) implies β is also injective. To see β is surjective, let $T \in \mathfrak{U}(\mathfrak{g})$; say $T \in \mathfrak{U}^k(\mathfrak{g})$, e.g.,

$$T = \sum_{l\leq k} a_\alpha X_{\alpha(1)} \otimes \cdots \otimes X_{\alpha(l)} \mod J. \tag{3.84}$$

Using the identities

$$X_i \otimes X_j = X_j \otimes X_i + [X_i, X_j] \mod J \tag{3.85}$$

we can reorder the factors $X_{\alpha(j)}$ in (3.84) so $\alpha(j)$ is increasing, to write

$$T = \sum_{|\alpha|=k} b_\alpha X_1^{(\alpha_1)} \otimes \cdots \otimes X_n^{(\alpha_n)} + T' \mod J \tag{3.86}$$

where $X_l^{(\alpha_l)}$ is given by (3.78), and $T' \in \mathfrak{U}^{k-1}(\mathfrak{g})$. Then, clearly,

$$T = \beta\left(\sum_{|\alpha|=k} b_\alpha t_1^{\alpha_1} \cdots t_n^{\alpha_n}\right) \mod \mathfrak{U}^{k-1}(\mathfrak{g}). \tag{3.87}$$

Again by induction, we construct a polynomial p in $\mathcal{P}_n^k$, with leading term given above, so that $T = \beta(p)$. Thus β is surjective. Hence β is bijective.

Since β and α are surjective, commutativity of (3.80) implies γ is surjective, so γ is also bijective. Finally, since β and γ are isomorphisms, so is $\alpha = \gamma \circ \beta^{-1}$, and the proposition is proved.

In view of (3.67), the representation of $\mathcal{D}_L(G)$ defined by (3.66) coincides with the natural representation induced from $\mathfrak{g}$ to $\mathfrak{U}(\mathfrak{g})$. This representation of the universal enveloping algebra of $\mathfrak{g}$ is an important tool in understanding the representation of G from which it arises.

As mentioned in §2, there is use for considering $\pi(f)$ for f not necessarily compactly supported, for example, for $f \in L^1(G)$, if π is unitary. Similarly, we can define $\pi(k)$ for certain distributions, not compactly supported, by various limiting procedures. Since specific classes of distributions for which it is natural to do this vary strongly from group to group, we will not discuss any general results here.

4. Irreducible representations of Lie groups. Let π be a (strongly continuous) representation of a Lie group G on a Banach space B. We say π is *topologically irreducible*, or simply irreducible, provided B has no closed proper subspace invariant under $\pi(g)$ for all $g \in G$. We will mainly be interested in the case when $B = H$ is a Hilbert space, and π is a unitary representation. In that case, if $E \subset H$ is a closed proper invariant subspace, it is clear that its orthogonal complement is also invariant, so a unitary representation which is not irreducible can be written as a direct sum of smaller representations. In case H is also finite-dimensional, this process of reduction could be carried out a finite number of times and would stop, and any finite-dimensional unitary representation would be broken into a finite direct sum of irreducible representations.

This need not happen for nonunitary representations. Consider the representation of R on R^2 given by

$$\pi(t) = \exp t \begin{pmatrix} 0 & 1 \\ 0 & 0 \end{pmatrix} = \begin{pmatrix} 1 & t \\ 0 & 1 \end{pmatrix}. \tag{4.1}$$

It is clear that the linear span of $\binom{1}{0}$ is invariant, but has no invariant complementary subspace. For a certain class of Lie groups, the semisimple ones, it can be shown that any finite-dimensional representation is completely reducible to a direct sum of irreducible representations. This result, known as Weyl's "unitary trick," involves reducing to the case of compact groups, via an analytic continuation. See Chapter 13 for a brief discussion of this.

For a unitary representation, the following version of *Schur's lemma* is often useful.

PROPOSITION 4.1. *A group G of unitary operators on a Hilbert space H is irreducible if and only if, for any bounded linear operator A on H,*

$$UA = AU \quad \textit{for all } U \in G \Rightarrow A = \lambda I. \tag{4.2}$$

This can be proved as a simple consequence of the spectral theorem; see Appendix B for such a proof.

As one example of a situation where Proposition 4.1 applies, let π be a unitary representation of a Lie group G, and let Z denote the *center* of G. Thus, for each $g_0 \in Z, \pi(g_0)$ must commute with all $\pi(g), g \in G$. By Proposition 4.1, this implies for π irreducible:

$$\pi(g_0) = \lambda(g_0)I, \quad \text{for } g_0 \in Z, \tag{4.3}$$

where $\lambda(g_0)$ is a complex number of absolute value 1 for each $g_0 \in Z$. The Heisenberg group $\mathbf{H}^n$, studied in Chapter 1, has a one-dimensional center. Many noncommutative Lie groups, however, have no center, or a very small center. It is important that other operators arise which commute with $\pi(g)$ for all $g \in G$, and hence act as scalars if π is an irreducible unitary representation.

For example, suppose $f_0 \in C_0^\infty(G)$ belongs to the center of the convolution algebra $C_0^\infty(G)$, i.e.,

$$f_0 * f = f * f_0 \quad \text{for all } f \in C_0^\infty(G). \tag{4.4}$$

Then, if π is a unitary representation of G on H,

$$\pi(f_0)\pi(f) = \pi(f)\pi(f_0) \quad \text{for all } f \in C_0^\infty(G). \tag{4.5}$$

Given $g \in G$, pick a sequence $f_j \in C_0^\infty(G)$ converging in $\mathcal{E}'(G)$ to δ_g, and uniformly bounded in L^1-norm. It follows that

$$\pi(f_j)u \to \pi(g)u \tag{4.6}$$

for all $u \in C^\infty(\pi)$; since the operators $\pi(f_j)$ are uniformly bounded, (4.6) holds for all $u \in H$. Hence (4.5) implies

$$\pi(f_0)\pi(g) = \pi(g)\pi(f_0) \quad \text{for all } g \in G. \tag{4.7}$$

Again, if π is irreducible, Proposition 4.1 applies to yield

$$\pi(f_0) = \alpha(f_0)I \tag{4.8}$$

for all f_0 in the center of $C_0^\infty(G)$, where $\alpha(f_0) \in \mathbf{C}$.

Let us see under what conditions an element f_0 can belong to the center of $C_0^\infty(G)$. Comparing the formulas

$$\begin{aligned} f_0 * f(g) &= \int_G f(y^{-1}g)f_0(y)\,dy = \int_G f(y^{-1})f_0(gy)\,dy \\ &= \int_G f(y)f_0(gy^{-1})\Delta(y)\,dy \end{aligned} \tag{4.10}$$

and

$$f * f_0(g) = \int f(y)f_0(y^{-1}g)\,dy, \tag{4.11}$$

we see that, for f_0 to belong to the center of $C_0^\infty(G)$, we must have

$$f_0(gy^{-1})\Delta(y) = f_0(y^{-1}g) \quad \text{for all } y, g \in G. \tag{4.12}$$

In particular, $f_0(y)\Delta(y^{-1}) = f_0(y)$ for all y, so $\Delta(y^{-1}) = 1$ on $\operatorname{supp} f_0$. But if $\Delta(y) = 1$ on any open set it is identically 1. Thus, for a nonzero element f_0 in the center of $C_0^\infty(G)$ to exist, G must be unimodular, and the requirement (4.12) is equivalent to

$$f_0(y) = f_0(g^{-1}yg) \quad \text{for all } y, g \in G. \tag{4.13}$$

Such functions exist in fair profusion when G is compact. However, in many cases, such as many noncompact semisimple Lie groups, for no $y_0 \in G, y_0 \neq e$, is it the case that

$$\{g^{-1}y_0g \colon g \in G\} \tag{4.14}$$

has compact closure in G. In such cases, the center of the convolution algebra $C_0^\infty(G)$ is zero.

We can also look for elements in the center of the convolution algebra $\mathcal{E}'(G)$. In analogy with (4.12), the condition for $k_0 \in \mathcal{E}'(G)$ to belong to the center is

$$A_g^* k_0 = \Delta(g^{-1}) M_\Delta k_0 \quad \text{for all } g \in G, \tag{4.15}$$

where

$$A_g^* : \mathcal{E}'(G) \to \mathcal{E}'(G) \tag{4.16}$$

is the adjoint of

$$A_g : C^\infty(G) \to C^\infty(G), \tag{4.17}$$

defined by

$$A_g f(y) = f(g^{-1} y g). \tag{4.18}$$

If G is connected, (4.15) holds provided it is valid for all g in some neighborhood U of e. This in turn holds provided (4.15) works for all g in one parameter subgroups of G, which in turn holds if and only if

$$\mathfrak{X}_L^X k_0 - \mathfrak{X}_R^X k_0 = -\Delta(X) M_\Delta k_0 \quad \text{for all } X \in \mathfrak{g}. \tag{4.19}$$

If, for all $y_0 \neq e$, the set (4.14) does not have compact closure in G, we are forced to conclude that any k_0 in the center of $\mathcal{E}'(G)$ must be supported at $\{e\}$, i.e.,

$$k_0 = P_L \delta_e \tag{4.20}$$

for some $P_L \in \mathcal{D}_L(G)$, identified with the universal enveloping algebra $\mathfrak{U}(\mathfrak{g})$ by Proposition 3.8. In such a case, (4.19) is equivalent to $\mathfrak{X}_R^X \delta_e * P_L \delta_e = P_L \delta_e * (\mathfrak{X}_L^X \delta_e + \Delta(X)\delta_e)$. Now generally $\mathfrak{X}_R^X \delta_e = \mathfrak{X}_L^X \delta_e + \Delta(X)\delta_e$, so (4.20) defines an element of the center of $\mathcal{E}'(G)$ if and only if

$$\mathfrak{X}_L^X \delta_e * P_L \delta_e = P_L \delta_e * \mathfrak{X}_L^X \delta_e \quad \text{for all } X \in \mathfrak{g},$$

which is equivalent to having

$$\mathfrak{X}_L^X P_L = P_L \mathfrak{X}_L^X \quad \text{for all } X \in \mathfrak{g}. \tag{4.21}$$

In view of Proposition 3.7, this is equivalent to saying

$$P_L \text{ belongs to the center of } \mathcal{D}_L(G). \tag{4.22}$$

In other words, the elements of the center $\mathcal{E}'(G)$ supported at $\{e\}$ correspond precisely to the elements of the center $\mathfrak{z}(\mathfrak{g})$ of the universal enveloping algebra $\mathfrak{U}(\mathfrak{g})$. We formalize this last part.

PROPOSITION 4.2. *The center $\mathfrak{z}(\mathfrak{g})$ of $\mathfrak{U}(\mathfrak{g})$ is naturally contained in the center of $\mathcal{E}'(G)$, if G is connected.*

The good news is: Even those nasty noncompact semisimple Lie groups mentioned before have enough elements in the center of their universal enveloping algebras to be of some use. But there is a problem. Given such $P_0 \in \mathfrak{z}(\mathfrak{g})$, and a strongly continuous irreducible unitary representation π of G, there may be no a priori guarantee that $\pi(P_0)$ is bounded. Hence Proposition 4.1 is not applicable. What is required is a substantial technical improvement of Schur's lemma, which we now discuss, largely following Kirillov [**134**].

We use the fact that an irreducible unitary representation satisfies the following condition of *complete irreducibility*. A representation π of G on a Banach space B is said to be completely (topologically) irreducible provided that

$$\Pi = \text{algebraic linear span of } \{\pi(g) \colon g \in G\} \tag{4.23}$$

is dense in the algebra $\mathcal{L}(B)$ of all bounded linear transformations on B, in the *strong* operator topology. That is to say, given any $T \in \mathcal{L}(B)$, any finite collection $\{u_1, \ldots, u_N\} \subset B$, and any $\varepsilon > 0$, there exists $T_0 \in \Pi$ such that

$$\|T_0 u_j - T u_j\|_B < \varepsilon \quad \text{for } j = 1, \ldots, N. \tag{4.24}$$

It is easy to see that this condition implies topological irreducibility. For general representations, the converse need not hold, but it does hold for unitary representations, as we now show. This is a simple consequence of von Neumann's *double commutant theorem*, which states the following. Let $B = H$ be a Hilbert space, and let $\mathfrak{A} \subset \mathcal{L}(H)$ be an algebra of bounded operators. Suppose $\mathfrak{A}$ is *selfadjoint*, i.e., $T \in \mathfrak{A} \Rightarrow T^* \in \mathfrak{A}$. Let $\mathfrak{A}'$, the commutant of $\mathfrak{A}$, denote the set of bounded operators on H commuting with all operators in $\mathfrak{A}$:

$$\mathfrak{A}' = \{T \in \mathcal{L}(H) \colon TS = ST \text{ for all } S \in \mathfrak{A}\}. \tag{4.25}$$

Let $\mathfrak{A}'' = (\mathfrak{A}')'$ denote the commutant of $\mathfrak{A}'$. Von Neumann's theorem states

$$\mathfrak{A}'' = \overline{\mathfrak{A}}_s = \overline{\mathfrak{A}}_w, \tag{4.26}$$

where $\overline{\mathfrak{A}}_s$ denotes the closure of $\mathfrak{A}$ in the strong operator topology (and $\overline{\mathfrak{A}}_w$ denotes the closure in the weak operator topology). For a proof of the double commutant theorem see [**47**] or [**50**].

PROPOSITION 4.3. *If π is an irreducible unitary representation of G on H, then π is completely irreducible.*

PROOF. Let Π be given by (4.23). By Proposition 4.1, Π' consists of only multiples of the identity. Hence $\Pi'' = \mathcal{L}(H)$. But Π is clearly a selfadjoint algebra of operators (e.g., $\pi(g)^* = \pi(g^{-1})$), so the double commutant theorem implies $\overline{\Pi}_s = \Pi'' = \mathcal{L}(H)$. This completes the proof.

One can define an infinite sequence of notions of irreducibility, connecting topological irreducibility to complete topological irreducibility, as follows. We say a representation π of G on B is *k-irreducible* provided that, for any two sets of k' elements of B, $\{u_1, \ldots, u_{k'}\}$ (linearly independent) and $\{v_1, \ldots, v_{k'}\}$, and any $\varepsilon > 0$, there exists $T \in \Pi$, given by (4.23), such that

$$\|T u_j - v_j\|_B < \varepsilon \quad \text{for } 1 \le j \le k', \text{ if } k' \le k. \tag{4.27}$$

It is easy to see that π is completely irreducible if and only if it is k-irreducible for all k.

There is a useful alternative characterization of k-irreducibility. Let I_k denote the trivial representation of G on $\mathbf{C}^k$. The standard basis of $\mathbf{C}^k$ gives natural isomorphisms

$$B \otimes \mathbf{C}^k \approx \mathcal{L}(\mathbf{C}^k, B) \approx B \oplus \cdots \oplus B \quad (k \text{ terms}). \tag{4.28}$$

A representation π of G on B gives rise to representations $\pi \otimes I_k$ of G on $B \otimes \mathbf{C}^k$.

PROPOSITION 4.4. *A representation π of G on B is k-irreducible if and only if every closed subspace of $B \otimes \mathbf{C}^k$ that is invariant under $\pi \otimes I_k$ is of the form $B \otimes V$, for some linear subspace V of $\mathbf{C}^k$.*

PROOF. Look on $\pi \otimes I_k$ as acting on $B \oplus \cdots \oplus B$, as $\pi \oplus \cdots \oplus \pi$. With $u = (u_1, \dots, u_k)$, $(\pi \otimes I_k)(g)u = (\pi(g)u_1, \dots, \pi(g)u_k)$. The condition that π is k-irreducible implies the property that, for any $u = (u_1, \dots, u_k)$ such that $\{u_1, \dots, u_k\}$ is linearly independent in B, the linear span of $(\pi \otimes I_k)(g)u$ for $g \in G$ is dense in $B \otimes \mathbf{C}^k$. In other words, if $u = (u_1, \dots, u_k)$ is contained in some proper closed invariant subspace of $B \otimes \mathbf{C}^k$, then $\{u_1, \dots, u_k\}$ must be linearly dependent. In such a case, reordering the u_j, we can assume without loss of generality that $\{u_1, \dots, u_{k_0}\}$ forms a basis for the linear span of $\{u_1, \dots, u_k\}$. Then, with

$$(\pi \otimes I_k)(g)u = (\pi(g)u_1, \dots, \pi(g)u_{k_0}) \oplus (\pi(g)u_{k_0+1}, \dots, \pi(g)u_k), \tag{4.29}$$

we see that, if π is k-irreducible, the first factor spans a dense linear subspace of $B \otimes \mathbf{C}^{k_0}$ as g runs over G, and the second factor is a fixed linear function of the first factor, of the following form. With α_{jl} complex numbers such that

$$u_j = \sum_{l=1}^{k_0} \alpha_{jl} u_l \quad \text{for } k_0 + 1 \le j \le k, \tag{4.30}$$

we have

$$\pi(g)u_j = \sum_{l=1}^{k_0} \alpha_{jl} \pi(g) u_l \quad \text{for } k_0 + 1 \le j \le k. \tag{4.31}$$

This implies the linear span of $\{(\pi \otimes I_k)(g)u \colon g \in G\}$ has closure of the form $B \otimes V$ with V a linear subspace of $\mathbf{C}^k$, and the proof is easily completed.

It is clear that a representation π of G on B is topologically irreducible if and only if it is 1-irreducible. The following is the improved Schur's lemma.

PROPOSITION 4.5. *Let π be a 2-irreducible representation of G on B. Let A be a (possibly unbounded) linear operator on B, with dense domain $\mathcal{L}$. Suppose that $\pi(g)\mathcal{L} \subset \mathcal{L}$ for each $g \in G$. Also suppose A is closable, i.e., the closure of the graph $\mathcal{G}_A$ of A is the graph of a linear operator. Then, if*

$$\pi(g)Au = A\pi(g)u \quad \textit{for all } u \in \mathcal{L},\ g \in G, \tag{4.32}$$

it follows that $\overline{A}$ is a multiple of the identity

$$\overline{A} = \lambda I. \tag{4.33}$$

PROOF. The graph $\mathcal{G}_A = \{(u, Au) \in B \oplus B: u \in \mathcal{L}\}$ is a linear subspace of $B \oplus B$, and (4.32) implies $\mathcal{G}_A$ is invariant under the action $\pi \otimes I_2$ of G on $B \oplus B$. Hence $\overline{\mathcal{G}}_A$ is invariant and is a proper invariant subspace of $B \oplus B$. By Proposition 4.4, $\overline{\mathcal{G}}_A$ must be of the form $B \otimes V$ for some proper linear subspace V of $\mathbf{C}^2$; hence $\dim_{\mathbf{C}} V = 1$. Such spaces are precisely the graphs of operators of the form (4.33) (together with $0 \oplus B$, which is not a graph), so the proof is complete.

A densely defined operator A is closable if and only if its adjoint A^* is densely defined. In particular, if $B = H$ is a Hilbert space, then A is closable if it is symmetric. In general, it might not be a priori clear that such a symmetric A even has a selfadjoint extension, so Proposition 4.5 is a strong result. A consequence of principal interest is

PROPOSITION 4.6. *Let π be an irreducible unitary representation of G on H. Suppose $k \in \mathcal{E}'(G)$ belongs to the center of $\mathcal{E}'(G)$. Then $\pi(k)$ is a scalar multiple of the identity*

$$\pi(k) = \lambda I. \tag{4.34}$$

PROOF. Regard $\pi(k)$ as an operator on H with domain $\mathcal{L} = C^\infty(\pi)$. We know that $\pi(g)\mathcal{L} \subset \mathcal{L}$ for all g. Also, if k is in the center of $\mathcal{E}'(G)$,

$$\begin{aligned} \pi(g)\pi(k) &= \pi(\delta_g)\pi(k) = \pi(\delta_g * k) \\ &= \pi(k * \delta_g) = \pi(k)\pi(\delta_g) = \pi(k)\pi(g) \end{aligned} \tag{4.35}$$

for all $g \in G$, these operators acting on $C^\infty(\pi)$. Finally, note that, for $k \in C_0^\infty(G)$,

$$\pi(k)^* = \int_G \overline{k(g)}\pi(g^{-1})\,dg = \int_G \overline{k(g^{-1})}\Delta(g)\pi(g)\,dg = \pi(k^\#), \tag{4.36}$$

and more generally, or $k \in \mathcal{E}'(G)$,

$$\pi(k)^* \supseteq \pi(k^\#), \tag{4.37}$$

where $k^\# \in \mathcal{E}'(G)$ is defined by (3.44). In particular, the domain of $\pi(k)^*$ contains $C^\infty(\pi) = \mathcal{L}$, and so $\pi(k)$ is closable. Since any irreducible unitary representation is 2-irreducible, Proposition 4.5 applies, and we are done.

It is a general fact that any Lie group has lots of irreducible unitary representations. We will sketch the argument for this here, referring to [**111**, **134**], for details. It exploits the relation between unitary representations of a Lie group (or even a locally compact group) G and *positive definite* functions on G. A bounded continuous function $p(g)$ is said to be positive definite provided the operation of convolution on the right by p is a positive semidefinite operator, i.e., if for all $u \in C_0^\infty(G)$,

$$(u * p, u)_{L^2(G)} = \int_G (u * p)(g)\overline{u(g)}\,dg \geq 0. \tag{4.38}$$

One connection with unitary representations is the following. If π is a unitary representation of G on H, $\xi \in H$ a unit vector, consider

$$p(g) = (\pi(g)\xi, \xi). \tag{4.39}$$

In this case, a short calculation gives

$$(u * p, u)_{L^2(G)} = (\pi(\bar{u})^*\pi(\bar{u})\xi, \xi) = \|\pi(\bar{u})\xi\|_H^2, \tag{4.40}$$

so p is positive definite. Conversely, let $p \in C(G)$ satisfy (4.38). We put an "inner product" on $\mathcal{H}_0 = C_0^\infty(G)$ by

$$((u, v)) = (u * p, v)_{L^2(G)} = \int\int u(y)p(y^{-1}g)\overline{v(g)}\, dy\, dg. \tag{4.41}$$

Since p is continuous, we can throw in more general u and v, including at least $\mathcal{H}_1 = \{$compactly supported measures on $G\}$. Let $\mathcal{N} \subset \mathcal{H}_1$ be the set of elements u such that $((u, u)) = 0$. Form the quotient $\mathcal{H}_2 = \mathcal{H}_1/\mathcal{N}$, and complete $\mathcal{H}_2$ with respect to the induced norm, to get a Hilbert space H. A dense subspace of H is given by equivalence classes of compactly supported measures on G, or even by equivalence classes containing elements of $C_0^\infty(G)$. G acts on H via its left regular representation on $C_0^\infty(G)$. Note that, if $\lambda_g u(x) = u(g^{-1}x)$ for $u \in C_0^\infty(G)$, then

$$\begin{aligned} ((\lambda_g u, \lambda_g v)) &= (\lambda_g(u * p), \lambda_g v)_{L^2(G)} \\ &= (u * p, u)_{L^2(G)} = ((u, v)). \end{aligned} \tag{4.42}$$

This action extends naturally to measures, and we get an action of G on H by unitary operators; strong continuity can be checked, and we have associated a unitary representation π_p of G to a positive definite continuous function p on G. This sketches the famous Gelfand-Naimark-Segal construction. Note that, if $\xi \in H$ is taken to be the image of the point mass δ_e, we have

$$p(g) = (\delta_g * p, \delta_e) = ((\delta_g, \delta_e)) = ((\pi_p(g)\xi, \xi))_H. \tag{4.43}$$

In this case, ξ has the property that $\{\pi(g)\xi : g \in G\}$ is dense in H; such a vector is called a *cyclic vector*.

It can be shown that, if π is any unitary representation of G, with a cyclic vector ξ (a separable H is a sum of cyclic subspaces; note that π is irreducible if and only if all nonzero vectors are cyclic), and we consider $p(g) = (\pi(g)\xi, \xi)$, then π is unitarily equivalent to the representation above. Thus we have a surjective map

$$\text{GNS}\colon \mathcal{P} \to \mathcal{C} \tag{4.44}$$

where $\mathcal{P}$ is the set of continuous positive definite functions on G and $\mathcal{C}$ is the set of equivalence classes of cyclic unitary representations of G. We could replace $\mathcal{P}$ in (4.44) by $\mathcal{P}_1 = \{p \in \mathcal{P} : \|p\|_{L^\infty} = p(e) \le 1\}$, which is a convex subset of $L^\infty(G)$, which is also *compact* in the weak* topology. It is not empty; after all, the regular representation exists: applying (4.39) to the regular representation, with $\xi = \overline{\check{v}} \in L^2(G)$ gives

$$p(g) = (\check{v} * v)(g) \tag{4.45}$$

as a rich class of positive definite functions on G.

Now it can be shown that irreducible representations correspond precisely to *extreme* points of $\mathcal{P}_1$, under the mapping (4.44). By the Krein-Milman theorem, $\mathcal{P}_1$ has a lot of extreme points, of which it is the closed convex hull. In such a fashion the existence of a lot of irreducible unitary representations is established. One can also relate the decomposition of a given representation π into irreducibles to the expression of $p(g) = (\pi(g)\xi, \xi)$, for some cyclic vector ξ, as a barycenter of the set of extreme points of $\mathcal{P}_1$, with respect to some probability measure, via Choquet's theorem. For more on this, see [**111, 134**], and references given there.

The approach this affords to representation theory is somewhat abstract. An effective way to decompose representations does not follow easily. There may be many different ways to express an element $p \in \mathcal{P}_1$ as the center of mass of extreme points, so the uniqueness of a decomposition of a given representation into irreducibles is left open. A related matter is that the equivalence relation on the set of extreme points of $\mathcal{P}_1$, specifying when the associated representations are equivalent, is not given explicitly. The set of equivalence classes of irreducible unitary representations, the natural quotient of the set of extreme points of $\mathcal{P}_1$, could conceivably have a very messy structure.

In fact, it turns out that Lie groups (and other locally compact groups) fall into two classes, "type I" and "nontype I." The type I groups have unique decomposition properties and other nice representation theoretic behavior. Groups studied in this monograph are of type I, usually. Nontype I groups are related to exotic von Neuman algebras, and used to be regarded as hopeless from a representation theoretic point of view. Recent advances of A. Connes on the theory of von Neumann algebras have stimulated interest in representations of nontype I groups; see Sutherland [**232**]. We will say nothing further about type I and nontype I groups, referring to [**168, 256**], and particularly [**161**] for an extended discussion.

5. Varieties of Lie groups. A very important representation of a Lie group G, which plays a key role in the study of the structure of G as well as its representation theory, is the *adjoint representation* of G on its Lie algebra $\mathfrak{g}$, defined as follows. For $g \in G$, define

$$C_g : G \to G \tag{5.1}$$

by

$$C_g(y) = gyg^{-1}, \tag{5.2}$$

so G acts as a group of inner automorphisms on itself. Identifying T_eG, on which the derivative $DC_g(e)$ acts, with $\mathfrak{g}$, we define

$$\mathrm{Ad}(g) = DC_g(e). \tag{5.3}$$

This is equivalent to the identity

$$\exp(t\,\mathrm{Ad}(g)X) = g\exp(tX)g^{-1}, \qquad X \in \mathfrak{g}, \tag{5.4}$$

since both sides of (5.4) clearly define one parameter subgroups of G, with the same initial direction at $t = 0$. The adjoint representation of G induces a Lie algebra representation of $\mathfrak{g}$ into $\mathrm{End}(\mathfrak{g})$, given by the usual Lie bracket,

$$\mathrm{ad}\,Y(X) = [Y, X], \tag{5.5}$$

which follows by setting $g = \exp sY$ in (5.4) and evaluating the s derivative at $s = 0$. Consequently,

$$\mathrm{Ad}(\exp Y) = e^{\mathrm{ad}\,Y}, \qquad Y \in \mathfrak{g}. \tag{5.6}$$

If G is connected, a subspace $\mathfrak{g}_1$ of $\mathfrak{g}$ is invariant under $\mathrm{Ad}(g)$ for all $g \in G$ if and only if it is invariant under $\mathrm{ad}\,X$ for all $X \in \mathfrak{g}$. The statement

$$X \in \mathfrak{g},\ Y \in \mathfrak{g}_1 \Rightarrow [X, Y] \in \mathfrak{g}_1 \tag{5.7}$$

is the statement that $\mathfrak{g}_1$ is an *ideal* in the Lie algebra $\mathfrak{g}$. If a Lie algebra $\mathfrak{g}$ has no proper ideals (and $\mathfrak{g} \neq R$), $\mathfrak{g}$ is said to be *simple*. Simple Lie algebras are in a sense the furthest away from the Lie algebras of the commutative groups R^n, all of whose linear subspaces are not only subalgebras but ideals.

Some order can be imposed on the variety of Lie algebras by considering various properties they can share or fail to share with the commutative Lie algebras. Of course, the adjoint representation is trivial on R^n and zero on its Lie algebra. On the other hand,

$$\mathrm{ad}\colon \mathfrak{g} \to \mathrm{End}(\mathfrak{g}) \tag{5.8}$$

is clearly injective if $\mathfrak{g}$ is simple. In fact the kernel of ad is an ideal $\mathfrak{z}$ in $\mathfrak{g}$, the *center* of $\mathfrak{g}$:

$$\mathfrak{z} = \{X \in \mathfrak{g} \colon [X, Y] = 0 \text{ for all } Y \in \mathfrak{g}\}. \tag{5.9}$$

The map (5.8) is injective precisely when the center of $\mathfrak{g}$ is 0.

From some points of view, the closest a nonabelian Lie algebra can come to being abelian is to have the property that $\mathrm{ad}\,X$ is a nilpotent transformation for each $X \in \mathfrak{g}$, i.e.,

$$X \in \mathfrak{g} \Rightarrow \exists k = k(X) \quad \text{such that} \quad (\mathrm{ad}\,X)^k = 0. \tag{5.10}$$

In such a case, we say $\mathfrak{g}$ is a *nilpotent* Lie algebra. A (connected) Lie group with such a Lie algebra is said to be a nilpotent Lie group. A nilpotent Lie algebra has a lot of ideals. In fact, there exist ideals $\mathfrak{g}_j$, with $\dim \mathfrak{g}_j = j$, $j = 0, 1, \ldots, \dim \mathfrak{g}$, such that

$$[\mathfrak{g}, \mathfrak{g}_j] \subset \mathfrak{g}_{j-1}. \tag{5.11}$$

This result, a consequence of Engel's theorem, is proved in Chapter 6. Note in particular that $\mathfrak{g}_1$ is nonzero and is contained in the center of $\mathfrak{g}$. Another characterization of nilpotent Lie algebras is the following. Let

$$\mathcal{D}\mathfrak{g} = [\mathfrak{g}, \mathfrak{g}] = \text{linear span of } \{[X,Y] : X, Y \in \mathfrak{g}\}. \tag{5.12}$$

By the Jacobi identity, $[\mathfrak{g}, \mathfrak{g}]$ is an ideal in $\mathfrak{g}$ (maybe not proper). More generally, if $\mathfrak{h}$ is an ideal in $\mathfrak{g}$, then

$$[\mathfrak{g}, \mathfrak{h}] = \text{linear span of } \{[X,Y] : X \in \mathfrak{g}, Y \in \mathfrak{h}\} \tag{5.13}$$

is an ideal in $\mathfrak{g}$. Set

$$\mathfrak{g}_{(0)} = \mathfrak{g}, \qquad \mathfrak{g}_{(1)} = [\mathfrak{g}, \mathfrak{g}], \qquad \mathfrak{g}_{(j)} = [\mathfrak{g}, \mathfrak{g}_{(j-1)}]. \tag{5.14}$$

Then $\mathfrak{g}$ is nilpotent if and only if some $\mathfrak{g}_{(K)} = 0$. The simplest nonabelian situation would be when $\mathfrak{g}_{(1)} \neq 0$ but $\mathfrak{g}_{(2)} = 0$. The $\mathfrak{g}$ is said to be a step two nilpotent Lie algebra. The most basic example is the Lie algebra of the Heisenberg group $\mathbf{H}^n$, studied in Chapter 1. As noted, a nilpotent Lie algebra $\mathfrak{g}$ always has a nontrivial center $\mathfrak{z}$. It is easy to see that $\mathfrak{g}/\mathfrak{z}$ is nilpotent also. $\mathfrak{g}$ is put together from these simpler pieces. This fact permits inductive arguments to work in analyzing the representation theory of nilpotent Lie groups, as shown in Chapter 6, after certain requisite tools are developed in Chapter 5.

More generally, one considers Lie algebras $\mathfrak{g}$ with the following property: $\mathfrak{g}$ has a nontrivial center $\mathfrak{z}, \mathfrak{g}/\mathfrak{z}$ has a nontrivial center $\mathfrak{z}_1$, and so forth, until one reaches a commutative quotient. The inverse images of these respective centers in $\mathfrak{g}$ form a chain of ideals

$$\mathfrak{g} = \mathfrak{g}_1 \supseteq \mathfrak{g}_2 \supseteq \cdots \supseteq \mathfrak{g}_K \supseteq \mathfrak{g}_{K+1} = 0 \qquad (\mathfrak{g}_K = \mathfrak{z}) \tag{5.15}$$

such that $\mathfrak{g}_j/\mathfrak{g}_{j+1}$ is abelian, for $1 \leq j \leq K$. In such a case, one says $\mathfrak{g}$ is a *solvable* Lie algebra, and G is a solvable Lie group. Another characterization is the following. With $\mathcal{D}\mathfrak{g}$ given by (5.12), set

$$\mathcal{D}^p\mathfrak{g} = \mathcal{D}(\mathcal{D}^{p-1}\mathfrak{g}). \tag{5.16}$$

Then $\mathfrak{g}$ is solvable if and only if $\mathcal{D}^p\mathfrak{g} = 0$ for some p. The simplest example of a (nonnilpotent) solvable Lie algebra is the two-dimensional Lie algebra of $\mathrm{Aff}(R^1)$, the group of affine transformations of the real line, also called the "$ax+b$ group," studied in Chapters 5 and 7. The representation theory of solvable Lie groups is also amenable to inductive methods, with considerably more effort than required in the nilpotent case. Only a few specific examples are treated in these notes. One complication that arises is that some solvable groups are not "type I." Very detailed results on when such a group is type I and if so how its representation theory goes are given in [**9, 10**].

From the characterization (5.15) of solvable Lie algebras, it follows that a Lie algebra $\mathfrak{g}$ contains a (nonzero) solvable ideal if and only if it contains a (nonzero) abelian ideal. A Lie algebra with no such ideals is said to be *semisimple*. Clearly every simple Lie algebra is semisimple. It can be proved that every semisimple Lie algebra is a direct sum of simple ideals. See [**100**, **129**, **246**].

If $\mathfrak{g}$ is a Lie algebra, and $\mathfrak{a}$ and $\mathfrak{b}$ are two solvable ideals, clearly $\mathfrak{a}+\mathfrak{b}=\mathfrak{c}$ is an ideal, and since $\mathfrak{c}/\mathfrak{a} \approx \mathfrak{b}/\mathfrak{b}\cap\mathfrak{a}$, $\mathfrak{c}/\mathfrak{a}$ is solvable, so $\mathfrak{c}$ can also be shown to be solvable. Hence any Lie algebra $\mathfrak{g}$ has a unique maximal solvable ideal

$$\mathfrak{q} = \operatorname{rad}\mathfrak{g}. \tag{5.17}$$

Clearly $\mathfrak{g}/\operatorname{rad}\mathfrak{g}$ contains no solvable ideals, i.e., it is semisimple. An important theorem of *Levi* is that there exists a subalgebra $\mathfrak{m}$ of $\mathfrak{g}$ such that

$$\mathfrak{q}+\mathfrak{m}=\mathfrak{g}, \qquad \mathfrak{q}\cap\mathfrak{m}=0. \tag{5.18}$$

In such a case,

$$\mathfrak{m} \approx \mathfrak{g}/\operatorname{rad}\mathfrak{g} \tag{5.19}$$

is semisimple, and we have the *Levi decomposition*

$$\mathfrak{g} \approx \mathfrak{q}\times\mathfrak{m}, \tag{5.20}$$

of a general Lie algebra $\mathfrak{g}$ as a *semidirect product* of a solvable ideal $\mathfrak{q}$ and a semisimple Lie algebra $\mathfrak{m}$ (which acts on $\mathfrak{q}$). For a proof, see [**129**, **246**]. Important examples of Lie groups which have natural semidirect product structures include the Euclidean groups $\mathrm{E}(n)$, which are semidirect products of $\mathrm{SO}(n)$ and R^n, where $\mathrm{SO}(n)$ acts on R^n in the natural fashion, and also the Poincaré groups, semidirect products $\mathrm{SO}(n,1)\times R^{n+1}$, where the Lorentz group $\mathrm{SO}(n,1)$ acts on R^{n+1} so as to preserve the Lorentz metric. Both $\mathrm{SO}(n)$ and $\mathrm{SO}(n,1)$ are semisimple. The representation theory of the Euclidean and Poincaré groups is discussed in Chapter 5.

In addition to the adjoint representation, another representation of G of fundamental importance is the *coadjoint representation*, a representation of G on the dual space $\mathfrak{g}'$ defined by

$$\mathrm{Ad}^*(g) = \mathrm{Ad}(g^{-1})^*, \tag{5.21}$$

i.e.,

$$\langle X, \mathrm{Ad}^*(g)w\rangle = \langle \mathrm{Ad}(g^{-1})X, w\rangle, \qquad X\in\mathfrak{g},\ w\in\mathfrak{g}'. \tag{5.22}$$

The adjoint and coadjoint representations are not necessarily equivalent, although sometimes they are. An intertwining operator between the two arises via the *Killing form*

$$B(X,Y) = \operatorname{tr}(\operatorname{ad}X\operatorname{ad}Y), \tag{5.23}$$

which is a symmetric bilinear form $B\colon \mathfrak{g}\times\mathfrak{g}\to R$. Since $B(X,\cdot)$ acts as a linear form on $\mathfrak{g}$, (5.23) induces a linear map

$$\beta\colon \mathfrak{g}\to\mathfrak{g}'. \tag{5.24}$$

It is easy to see that, for $g\in G$,

$$B(\mathrm{Ad}(g)X, \mathrm{Ad}(g)Y) = B(X,Y), \tag{5.25}$$

which is equivalent to the intertwining property

$$\beta \circ \mathrm{Ad} = \mathrm{Ad}^* \circ \beta. \tag{5.26}$$

Note that rewriting (5.25) as

$$B(\mathrm{Ad}(g)X, Y) = B(X, \mathrm{Ad}(g^{-1})Y) \tag{5.27}$$

and differentiating gives

$$B(\mathrm{ad}\, ZX, Y) = -B(X, \mathrm{ad}\, ZY), \qquad X, Y, Z \in \mathfrak{g}. \tag{5.28}$$

The map (5.24) is an isomorphism if and only if B is nonsingular. Often β is far from an isomorphism. For example, if $\mathfrak{g}$ is abelian, B is identically zero. More generally, a theorem of Cartan states that a Lie algebra $\mathfrak{g}$ is solvable if and only if

$$B(X, [Y, Z]) = 0 \quad \text{for all } X, Y, Z \in \mathfrak{g}. \tag{5.29}$$

For a proof, see [**129**, **246**]. In such a case, B is clearly degenerate. Another important theorem of Cartan, proved in these references, is that B is nondegenerate, i.e., nonsingular, if and only if $\mathfrak{g}$ is semisimple. In such a case it can be shown that, if $\mathfrak{g}_1$ is an ideal, then B restricted to $\mathfrak{g}_1 \times \mathfrak{g}_1$, which coincides with the Killing form of $\mathfrak{g}_1$, is nonsingular, and the orthogonal complement $\mathfrak{g}_2$ with respect to B of $\mathfrak{g}_1$ is an ideal: $\mathfrak{g} = \mathfrak{g}_1 \oplus \mathfrak{g}_2$. Inductively, such a semisimple Lie algebra $\mathfrak{g}$ is a sum of simple ideals.

Semisimple Lie groups arise as symmetry groups of spaces with the most natural and pleasing sorts of symmetries. The paradigm examples of such spaces, whose perfection has been admired since the time of the Greeks, are the spheres S^n. The group of isometries of S^n is $\mathrm{O}(n+1)$, the orthogonal group on R^{n+1}, which has two connected components; the connected component of the identity is $\mathrm{SO}(n+1)$; its Lie algebra $\mathrm{so}(n+1)$ is semisimple (for $n+1 \geq 3$); in fact it is simple except for $n+1 = 4$; $\mathrm{so}(4) \approx \mathrm{so}(3) \oplus \mathrm{so}(3)$. These compact groups act on R^{n+1}, provided with a positive definite (Euclidean) metric. Noncompact analogues, made significant by relativity theory, are the groups $\mathrm{O}(n,1)$ of linear transformations on R^{n+1}, preserving the indefinite, nondegenerate forms $x_1^2 + \cdots + x_n^2 - x_{n+1}^2$. The group $\mathrm{O}(n,1)$ also acts as a group of isometries of hyperbolic space $\mathcal{H}^n$, which, as opposed to S^n, has constant negative curvature. S^n sits as an ideal boundary of $\mathcal{H}^{n+1}$, and the group $\mathrm{O}(n+1,1)$ induces a group of *conformal* transformations on S^n. These facts and some of their implications are investigated in Chapter 10. More generally than $\mathrm{O}(n,1)$, we can consider $\mathrm{O}(p,q)$, the group of linear transformations on R^{p+q} preserving the form $x_1^2 + \cdots + x_p^2 - x_{p+1}^2 - \cdots - x_{p+q}^2$. Linear transformations of determinant one on a complex vector space preserving a nondegenerate Hermitian inner product form the semisimple groups $\mathrm{SU}(p,q)$. These and other semisimple Lie groups, compact and noncompact, are studied in Chapters 2–3 and 8–13. One other series of semisimple groups we mention here is $\mathrm{Sp}(n,R)$, the group of linear transformations on R^{2n} preserving a certain nondegenerate skew-symmetric bilinear form on R^n, the symplectic form. This group, the symplectic group, arises in Chapter 1 as a group of automorphisms of the Heisenberg group $\mathbf{H}^n$, and is studied further in Chapter 11. Interestingly, there is a complete classification of the semisimple Lie algebras, while solvable Lie algebras seem to exist in profusion beyond classification.

Some of the semisimple Lie groups we just mentioned are compact. An important theorem of Weyl is that, if a connected Lie group G is semisimple, then G is compact if and only if the Killing form (5.23) is *negative definite.* In such a case, the negative of the Killing form induces a bi-invariant Riemannian metric on G. Generally, averaging any left invariant metric on a compact Lie group G produces a bi-invariant metric. Some compact Lie groups are not semisimple, such as the tori $\mathbf{T}^k$, and also the unitary groups $\mathrm{U}(n)$, which contain the simple subgroups $\mathrm{SU}(n)$ but have the unitary scalars cI, $|c| = 1$, as a one-dimensional center. Their Lie algebras belong to a class slightly larger than semisimple, called *reductive.* Generally, a reductive Lie algebra is one for which $[\mathfrak{g}, \mathfrak{g}] = \mathcal{D}\mathfrak{g}$ is semisimple, in which case it can be shown that $\mathfrak{g}$ is the direct sum of its center and $\mathcal{D}\mathfrak{g}$. Examples of noncompact reductive Lie algebras include $\mathrm{gl}(n, \mathbf{C})$ and $\mathrm{u}(p, q)$, whose derived algebras $\mathcal{D}\mathfrak{g}$ are, respectively, the semisimple Lie algebras $\mathrm{sl}(n, \mathbf{C})$ and $\mathrm{su}(p, q)$.

From the point of view of a student of multidimensional Fourier series, i.e., harmonic analysis on the torus, the natural generalization is the study of harmonic analysis on compact Lie groups. The discreteness of the representation theory carries over, and a fairly complete representation theory exists for compact Lie groups, some of which is discussed in Chapter 3.

We will say a little about when a Lie group G is unimodular, i.e., when its left invariant Haar measure is also right invariant. Recall the modular function $\Delta(g)$, defined by (2.38) and (3.40). Since dg and $d_r g$ are gotten by left and right translation of some nonvanishing element $\omega \in \bigwedge^n T_e^* G, n = \dim G$, it is clear that, at g, they differ by the factor which $\mathrm{Ad}^*(g)$ induces on $\bigwedge^n T_e^*$, so

$$\Delta(g) = |\det \mathrm{Ad}(g)|. \tag{5.30}$$

Classes of groups we can be sure are unimodular include compact groups. This is clear since the image of G under the homomorphism $\Delta\colon G \to R^+$ must be a compact subgroup of R^+, hence $\{1\}$. Also, any semisimple Lie group is unimodular. Indeed, since $\mathrm{Ad}(g)$ preserves the nondegenerate metric $B(X, X)$ in this case, it a fortiori preserves a volume element on $\mathfrak{g}$. Using (5.6), we can write

$$\Delta(\exp X) = e^{\mathrm{tr}\,\mathrm{ad}\,X}. \tag{5.31}$$

If $\mathfrak{g}$ is nilpotent, by (5.11) we can choose a basis of $\mathfrak{g}$ with respect to which $\mathrm{ad}\,X$ is strictly upper triangular, so clearly $\mathrm{tr}\,\mathrm{ad}\,X = 0$, and we see that any nilpotent Lie group is unimodular. Solvable Lie groups need not be unimodular; indeed, the "$ax + b$ group," $\mathrm{Aff}(R^1)$, is not.

Lie groups are mainly classified by their Lie algebras. Indeed, a connected and simply connected Lie group $\tilde{G}$ is uniquely determined by its Lie algebra

$\mathfrak{g}$. If $\alpha: \mathfrak{g} \to \mathfrak{h}$ is a Lie algebra homomorphism and $\mathfrak{h}$ is the Lie algebra of a Lie group H, then α necessarily exponentiates to a Lie group homomorphism $\alpha: \tilde{G} \to H$. This follows from an even stronger statement of the influence the algebra structure on $\mathfrak{g}$ has on the group structure of $\tilde{G}$, the Campbell-Hausdorff formula, which says

$$(5.32) \qquad (\exp X)(\exp Y) = \exp \Phi(X, Y)$$

where $\Phi(X,Y)$ is given by a certain convergent power series, for X and Y small. This power series is of a "universal" sort:

$$(5.33) \qquad \Phi(X,Y) = X + Y + \tfrac{1}{2}[X,Y] + \cdots$$

where the terms homogeneous of degree k are sums of $k-1$ fold Lie brackets of X and Y, with coefficients which do not depend on $\mathfrak{g}$. This result is also useful for exponentiating infinite-dimensional representations of $\mathfrak{g}$, when one has a dense space of analytic vectors to work with; see Appendix D for more on this. We refer to [**129**, **246**] for a proof of the Campbell-Hausdorff formula, and the precise expression for the general term in (5.33), which is fairly complicated.

We end this section with a list, by no means exhaustive, of some of the garden variety Lie groups in dimensions one through ten.

Dimension	*Groups*
1	$R^1 \to S^1 = \mathrm{U}(1) = \mathrm{SO}(2)$
2	$R^2 \to \mathbf{T}^2$, $\mathrm{Aff}(R^1)$
3	$R^3 \to \mathbf{T}^3$, $\mathrm{SU}(2) \to \mathrm{SO}(3)$, $\mathrm{Sp}(1,R) = \mathrm{SL}(2,R) \to \mathrm{SO}_e(2,1)$, $\mathrm{E}(2), \mathbf{H}^1$
4	$R^4 \to \mathbf{T}^4, \mathrm{U}(2)$, $\mathrm{GL}(2,R)$
5	$R^5 \to \mathbf{T}^5, \mathbf{H}^2$
6	$R^6 \to \mathbf{T}^6, \mathrm{SL}(2,\mathbf{C}) \to \mathrm{SO}_e(3,1)$, $\mathrm{Spin}(4) \to \mathrm{SO}(4)$, $\mathrm{SO}(2,2)$, $\mathrm{Aff}(R^2), \mathrm{E}(3)$, $\mathrm{Sp}(1,R) \times_\alpha \mathbf{H}^1, R^3 \times_\wedge \bigwedge^2 R^3$
7	$R^7 \to \mathbf{T}^7$, $\mathrm{SU}(2) \times_\rho \mathbf{C}^2$, $\mathbf{H}^3$
8	$R^8 \to \mathbf{T}^8$, $\mathrm{SU}(3)$, $\mathrm{SU}(2,1)$, $\mathrm{SL}(3,R)$, $\mathrm{GL}(2,\mathbf{C})$
9	$R^9 \to \mathbf{T}$, $\mathrm{U}(3)$, $\mathrm{GL}(3,R)$, $\mathrm{U}(2,1)$, $\mathbf{H}^4$
10	$R^{10} \to \mathbf{T}^{10}$, $\mathrm{SO}(5)$, $\mathrm{SO}_e(4,1)$, $\mathrm{SO}_e(3,2) = \mathrm{Sp}(2,R)$, $\mathrm{Sp}(2)$, $R^4 \times_\wedge \bigwedge^2 R^4$, $\mathrm{E}(4)$

Arrows indicate covering groups. Two of the groups on this list are semidirect products, with Lie algebras of the form $\mathfrak{g} = R^n \times_\wedge \bigwedge^2 R^n$, having Lie bracket

$$[(x,a),(y,b)] = (0, x \wedge y), \qquad x, y \in R^n,\ a, b \in \textstyle\bigwedge^2 R^n.$$

These are the nilpotent Lie algebras, free of Step 2, studied in Chapter 6, §2. The semidirect product $\mathrm{Sp}(1,R) \times_\alpha \mathbf{H}^1$ in dimension six is a special case of a semidirect product $\mathrm{Sp}(n, R) \times_\alpha \mathbf{H}^n$, where the symplectic group acts as a group of automorphisms of the Heisenberg group $\mathbf{H}^n$, as discussed in Chapter 1. As for the semidirect product listed in dimension seven, SU(2) acts on $\mathbf{C}^2$ in the natural fashion.

We hope that the reader of this monograph will become comfortable and familiar with such groups as listed above, and various natural generalizations, and some of their diverse roles in analysis.

CHAPTER 1

The Heisenberg Group

It is on the Heisenberg group $\mathbf{H}^n$ that harmonic analysis will be pursued first, and furthest, in analogy with Fourier analysis on R^n. A primary goal of this chapter is to develop harmonic analysis on $\mathbf{H}^n$ far enough to solve natural classes of PDEs that arise on $\mathbf{H}^n$ and $R \times \mathbf{H}^n$, analogous to the Laplace, heat, and wave equations discussed in the introduction for R^n. Some of the rich and beautiful structures that arise naturally in this pursuit will stimulate numerous investigations in subsequent chapters. The operators $\mathcal{L}_\alpha$ and P_α which we study in §§6–8, analogues of the Laplacian, are not elliptic, but, excluding exceptional values of α, possess the property of hypoellipticity, discussed in §§6 and 7. We call such operators "subelliptic."

1. Construction of the Heisenberg group $\mathbf{H}^n$. Here we want to construct the group of unitary operators on $L^2(R^n)$ generated by the n-dimensional group of translations

$$\tau_p u(x) = u(x+p), \qquad p \in R^n, \tag{1.1}$$

and the n-dimensional group of multiplications

$$m_q u(x) = e^{iq\cdot x}u(x), \qquad q \in R^n. \tag{1.2}$$

We will see that such a group is a $(2n+1)$-dimensional Lie group; $\mathbf{H}^n$ will be its universal covering group. Note that the infinitesimal generators of τ_p and m_q are easily identified. In analogy with (1.54) of Chapter 0, we have

$$\tau_p = e^{ip\cdot D} \tag{1.3}$$

where

$$(p\cdot D)u(x) = (1/i)\sum_{j=1}^{n} p_j \partial u/\partial x_j. \tag{1.4}$$

Also, clearly,

$$m_q = e^{iq\cdot X} \tag{1.5}$$

where

$$(q \cdot X)u(x) = \sum_{j=1}^{n} q_j x_j u(x). \tag{1.6}$$

Note that τ_p and m_q both preserve $C_0^\infty(R^n)$:

$$\tau_p C_0^\infty(R^n) \subset C_0^\infty(R^n); \qquad m_q C_0^\infty(R^n) \subset C_0^\infty(R^n). \tag{1.7}$$

It follows that $p \cdot D$ and $q \cdot X$, defined with domains $C_0^\infty(R^n)$, have unique selfadjoint extensions, equal to their closures. (See Proposition 2.2 of Chapter 0.) Also, if we define the Schwartz space $S(R^n)$ of rapidly decreasing functions by

$$S(R^n) = \left\{ u \in C^\infty(R^n) \colon \sup_x (1 + |x|)^N |D^\alpha u(x)| < \infty \text{ for all } N, \alpha \right\}, \tag{1.8}$$

where

$$D^\alpha u = D_1^{\alpha_1} \cdots D_n^{\alpha_n} u \tag{1.9}$$

with $D_j = (1/i)\partial/\partial x_j$, we see that

$$\tau_p S(R^n) \subset S(R^n) \quad \text{and} \quad m_q S(R^n) \subset S(R^n). \tag{1.10}$$

We begin to investigate the group structure of (1.1)–(1.2) by comparing $\tau_p m_q$ and $m_q \tau_p$. Indeed, we have

$$\tau_p m_q u(x) = (m_q u)(x + p) = e^{iq \cdot (x+p)} u(x + p), \tag{1.11}$$

and

$$m_q \tau_p u(x) = e^{iq \cdot x} (\tau_p u)(x) = e^{iq \cdot x} u(x + p). \tag{1.12}$$

Comparing (1.11) and (1.12), we get the identity

$$e^{ip \cdot D} e^{iq \cdot X} = e^{iq \cdot p} e^{iq \cdot X} e^{ip \cdot D}, \tag{1.13}$$

known as the Weyl commutator relations, the identity

$$[p \cdot D, q \cdot X] = -i(p \cdot q)I \tag{1.14}$$

being known as the Heisenberg commutator relations.

From this it is clear that the group generated by (1.1) and (1.2) consists precisely of all operators of the form

$$e^{is} e^{iq \cdot X} e^{ip \cdot D}, \qquad q, p \in R^n, \ s \in R. \tag{1.15}$$

We prefer to express such operators in the form

$$e^{i(t + q \cdot X + p \cdot D)}, \qquad q, p \in R^n, \ t \in R. \tag{1.16}$$

In fact, we have

PROPOSITION 1.1. *The operator $q \cdot X + p \cdot D$ is essentially selfadjoint on $C_0^\infty(R^n)$, and a fortiori on $\mathcal{S}(R^n)$, and*

$$e^{i(t+q\cdot X+p\cdot D)}u(x) = e^{i(t+q\cdot x+q\cdot p/2)}u(x+p), \tag{1.17}$$

or equivalently

$$e^{i(q\cdot X+p\cdot D)} = e^{iq\cdot p/2}e^{iq\cdot X}e^{ip\cdot D}. \tag{1.18}$$

PROOF. We mainly need to derive the formula (1.17), from which all the technical results of the proposition will be formal consequences. Let $A = t + q \cdot X + p \cdot D$. We compute the action of what should be the one parameter group e^{isA}. Note that if $v(s,x) = e^{isA}u(x)$, then

$$\partial v/\partial s = iAv = \left(\sum p_j \partial/\partial x_j + i(t + q \cdot x)\right) u, \tag{1.19}$$

or

$$\left(\partial/\partial s - \sum p_j \partial/\partial x_j\right) v = i(t + q \cdot x)v, \tag{1.20}$$

while

$$v(0,x) = u(x). \tag{1.21}$$

Solving (1.20)–(1.21) is accomplished by integrating over the straight lines which are the integral curves of $\partial/\partial s - \sum p_j\partial/\partial x_j$, and we obtain

$$v(s,x) = e^{i(st+sq\cdot x+s^2 q\cdot p/2)}u(x+sp). \tag{1.22}$$

Setting $s = 1$ gives (1.17).

It is clear that the family of operators defined by (1.17) leaves $C_0^\infty(R^n)$ invariant. Thus, by Proposition 2.2 of Chapter 0, the proof will be complete if we check the strong continuity of e^{isA}, defined by (1.22). Thus, to finish the proof of Proposition 1.1, it suffices to prove

LEMMA 1.2. *For each $u \in L^2(R^n)$, the function*

$$F(t,q,p) = e^{i(t+q\cdot x+q\cdot p/2)}u(x+p) \tag{1.23}$$

is a continuous function of (t,q,p) with values in $L^2(R^n)$.

PROOF. We have

$$F(t,q,p) = V(t,q,p)u \tag{1.24}$$

where $V(t,q,p)$ is clearly a unitary operator on $L^2(R^n)$ for each $(t,q,p) \in R^{2n+1}$. By Lemma 1.1 of Chapter 0, it suffices to check continuity in (t,q,p) for u belonging to some dense subspace $\mathcal{L}$ of $L^2(R^n)$. If we take $\mathcal{L} = C_0^\infty(R^n)$, the continuity is apparent, and the proof is complete.

Repeated use of (1.18) together with the commutator formula (1.13) gives

$$\begin{aligned} &e^{i(t_1+q_1\cdot X+p_1\cdot D)}e^{i(t_2+q_2\cdot X+p_2\cdot D)} \\ &\quad = \exp i[(t_1+t_2+q_2\cdot p_1/2 - q_1\cdot p_2/2) + (q_1+q_2)\cdot X + (p_1+p_2)\cdot D]. \end{aligned} \tag{1.25}$$

We use this to define the Heisenberg group $\mathbf{H}^n$. As a C^∞ manifold, $\mathbf{H}^n = R^{2n+1}$. If we denote points in $\mathbf{H}^n$ by (t_j, q_j, p_j), with $t_j \in R$, $q_j, p_j \in R^n$, we define the group operation by

$$(1.26)\quad (t_1, q_1, p_1)\cdot(t_2, q_2, p_2) = (t_1 + t_2 + \tfrac{1}{2}p_1\cdot q_2 - \tfrac{1}{2}p_2\cdot q_1, q_1 + q_2, p_1 + p_2).$$

It is straightforward to verify that this is a group operation, with the origin $0 = (0,0,0)$ as the identity element, which makes $\mathbf{H}^n$ a Lie group. Note that the inverse of (t,q,p) is given by $(-t,-q,-p)$. It is also easy to show that Lebesgue measure on $R^{2n+1} = \mathbf{H}^n$ is left invariant and right invariant under the group action defined by (1.26). Thus Lebesgue measure gives the Haar measure on $\mathbf{H}^n$, and this group is unimodular.

The Lie algebra of left invariant vector fields on $\mathbf{H}^n$ is spanned by the vector fields

$$(1.27)\quad T = \partial/\partial t,\quad L_j = \partial/\partial q_j - \tfrac{1}{2}p_j\partial/\partial t,\quad M_j = \partial/\partial p_j + \tfrac{1}{2}q_j\partial/\partial t,\quad 1 \le j \le n.$$

Note that

$$(1.28)\qquad [L_j, M_j] = -[M_j, L_j] = -T,$$

all other commutators being zero, where the commutator $[X,Y]$ of two vector fields is $XY - YX$.

We could use a parametrization of $\mathbf{H}^n$ suggested by the representation (1.15), instead of (1.16). Indeed, replacing (1.25) by

$$(1.29)\qquad \begin{aligned} &e^{it_1}e^{iq_1\cdot X}e^{ip_1\cdot D}e^{it_2}e^{iq_2\cdot X}e^{ip_2\cdot D} \\ &\quad = e^{i(t_1+t_2+q_2\cdot p_1)}e^{i(q_1+q_2)\cdot X}e^{i(p_1+p_2)\cdot D}, \end{aligned}$$

which follows from (1.13), we see we get a group $\tilde{\mathbf{H}}^n$ isomorphic to $\mathbf{H}^n$ if the group law is

$$(1.30)\qquad (t_1, q_1, p_1)\odot(t_2, q_2, p_2) = (t_1 + t_2 + p_1\cdot q_2, q_1 + q_2, p_1 + p_2).$$

The isomorphism from $\mathbf{H}^n$ to $\tilde{\mathbf{H}}^n$ is given by

$$(1.31)\qquad (t,q,p) \mapsto (t + q\cdot p/2, q, p).$$

Note that matrix multiplication for $(n+2)\times(n+2)$ matrices of the form

$$(1.32)\qquad \begin{pmatrix} 1 & 0 & 0 \\ q^t & I & 0 \\ t & p & 1 \end{pmatrix},\qquad q, p \in R^n,$$

gives the group law (1.30). The parametrization of $\mathbf{H}^n$ giving the group law (1.26) has the advantage of being more symmetrical, which will make more apparent the action of the automorphisms of $\mathbf{H}^n$, as we will see in §4.

2. Representations of $\mathbf{H}^n$. If $\mathbf{H}^n$ is the Heisenberg group constructed in §1, the operators (1.17) in Proposition 1.1 give a unitary representation of $\mathbf{H}^n$ on $L^2(R^n)$, which we will denote by π_1,

$$\pi_1(t,q,p) = e^{i(t+q\cdot X+p\cdot D)}, \tag{2.1}$$

or equivalently

$$\pi_1(t,q,p)u(x) = e^{i(t+q\cdot x+q\cdot p/2)}u(x+p). \tag{2.2}$$

The group homomorphism property

$$\pi_1(gg_1) = \pi_1(g)\pi_1(g_1) \tag{2.3}$$

follows from the identity (1.25) and the definition (1.26) of the group operation on $\mathbf{H}^n$. The strong continuity property

$$g_j \to g \quad \text{in } \mathbf{H}^n, \qquad u \in L^2(R^n) \Rightarrow \pi_1(g_j)u \to \pi_1(g)u \quad \text{in } L^2(R^n) \tag{2.4}$$

follows from Lemma 1.2.

The building blocks for representations of a Lie group are the irreducible unitary representations. Recall that a representation π of G on H is irreducible if and only if the only closed linear subspaces of H invariant under $\pi(g)$ for all $g \in G$ are $V = H$ and $V = 0$. We now show that irreducibility is a property of π_1.

THEOREM 2.1. *π_1, given by (2.2), is an irreducible unitary representation of $\mathbf{H}^n$.*

PROOF. By the easy part of Schur's lemma, to prove π_1 is irreducible it suffices to show that, if A is a bounded operator which commutes with $\pi_1(g)$ for all $g \in \mathbf{H}^n$, then it is a scalar. So let A be such an operator. In particular,

$$Ae^{iq\cdot X} = e^{iq\cdot X}A \quad \text{for all } q \in R^n. \tag{2.5}$$

Using the Fourier decomposition

$$b(x) = (2\pi)^{-n/2}\int \hat{b}(q)e^{iq\cdot x}\,dq,$$

we deduce that, for all $b \in \mathcal{S}(R^n)$,

$$A(b(x)u(x)) = b(x)Au(x) \quad \text{in } L^2(R^n), \text{ for all } u \in L^2(R^n). \tag{2.6}$$

We claim this implies

$$Au(x) = a(x)u(x) \tag{2.7}$$

for some $a \in L^\infty(R^n)$. Indeed, let $\chi_\nu(x)$ be the characteristic function of $B_\nu = \{x \in R^n\colon |x| < \nu\}$, and let $a_\nu(x) = A\chi_\nu(x)$. Note that, if $\mu \geq \nu$ and $f \in C_0^\infty(B_\nu)$, then applying (2.6) to $b = f$, $u = \chi_\mu$ gives

$$Af = A(f\chi_\mu) = a_\mu(x)f(x).$$

This implies $a_\mu(x) = a_\nu(x)$ a.e. for $x \in B_\nu$, if $\mu \geq \nu$, so $a(x)$, set equal to $a_\nu(x)$ on B_ν, is well defined a.e., and (2.7) holds for all $u \in C_0^\infty(R^n)$. It follows that $a \in L^\infty(R^n)$, with $\|a\|_{L^\infty} = \|A\|$, and (2.7) holds for all $u \in L^2(R^n)$.

Now if A commutes with $\pi_1(g)$ for all g, we also have

$$Ae^{ip\cdot D} = e^{ip\cdot D}A \quad \text{for all } p \in R^n, \tag{2.8}$$

which implies

$$a(x)u(x+p) = a(x+p)u(x+p) \quad \text{for all } p \in R^n,\ u \in L^2(R^n). \tag{2.9}$$

This implies $a(x)$ is constant a.e., and the proof of Theorem 2.1 is complete.

We sketch a second proof of Theorem 2.1. Suppose $\pi_1 = \pi_{\mathrm{I}} \oplus \pi_{\mathrm{II}}$ is an orthogonal decomposition on $L^2(R^n) = H_{\mathrm{I}} \oplus H_{\mathrm{II}}$. Pick $v_{\mathrm{I}} \in H_{\mathrm{I}}$, $v_{\mathrm{II}} \in H_{\mathrm{II}}$, unit vectors. Then $\pi_1(g)v_{\mathrm{I}}$ is always orthogonal to v_{II}, so we have

$$\int e^{iq\cdot x} v_{\mathrm{I}}(x+p)\overline{v_{\mathrm{II}}(x)}\, dx = 0$$

for all $q, p \in R^n$. In particular, for each $p \in R^n$, the Fourier transform of $v_{\mathrm{I}}(x+p)\overline{v_{\mathrm{II}}(x)}$, an element of $L^1(R^n)$, vanishes identically. This implies $v_{\mathrm{I}}(x+p)v_{\mathrm{II}}(x) = 0$ a.e., for each p, which in turn implies $v_{\mathrm{I}} = 0$ a.e. if $v_{\mathrm{II}}(x) \neq 0$ on a set with positive measure. This contradiction proves the irreducibility of π_1. A variant of this second argument, with some needed additional technical details, will appear in the proof of Lemma 5.1.

Let us identify the space $C^\infty(\pi_1)$ of smooth vectors for the representation π_1. Recall that to say $u \in L^2(R^n)$ belongs to $C^\infty(\pi_1)$ is to say that

$$F(t,q,p) = \pi_1(t,q,p)u \tag{2.10}$$

is a C^∞ function of $(t,q,p) \in \mathbf{H}^n$ with values in $L^2(R^n)$. From the explicit formula (2.2), it is clear that every $u \in \mathcal{S}(R^n)$, defined by (1.8), belongs to $C^\infty(\pi_1)$. In fact, this gives all smooth vectors:

PROPOSITION 2.2. *We have*

$$C^\infty(\pi_1) = \mathcal{S}(R^n). \tag{2.11}$$

In order to prove that $C^\infty(\pi_1) \subset \mathcal{S}(R^n)$, it is convenient to have the following analysis of smooth vectors for the subgroup of operators $\{\tau_p : p \in R^n\}$.

PROPOSITION 2.3. *Suppose $u \in L^2(R^n)$ is a C^k vector for the group $\{\tau_p : p \in R^n\}$. If $k > n/2$, then u is bounded and continuous. If $k = k_0 + l$, $k_0 > n/2$, $l \geq 0$, then $D^\alpha u$ is bounded and continuous for all $|\alpha| \leq l$.*

PROOF. By Proposition 2.5 of Chapter 0, u is a C^k vector for τ_p if and only if

$$D^\alpha u \in L^2(R^n) \quad \text{for all } |\alpha| \leq k. \tag{2.12}$$

Thus $\xi^\alpha \hat{u}(\xi) \in L^2(R^n)$ for $|\alpha| \leq k$, i.e.,

$$(1+|\xi|)^k \hat{u} \in L^2(R^n). \tag{2.13}$$

But for $k > n/2$, using Cauchy's inequality, we have

$$\begin{aligned}\int |\hat{u}(\xi)|\,d\xi &= \int (1+|\xi|)^{-k}(1+|\xi|)^{k}|\hat{u}(\xi)|\,d\xi \\ &\leq \left[\int (1+|\xi|)^{-2k}\,d\xi\right]^{1/2}\cdot\left[\int (1+|\xi|)^{2k}|\hat{u}(\xi)|^2\,d\xi\right]^{1/2} \\ &< \infty,\end{aligned} \tag{2.14}$$

i.e., $\hat{u} \in L^1(R^n)$. The Fourier inversion formula

$$u(x) = (2\pi)^{-n/2}\int \hat{u}(\xi)e^{ix\cdot\xi}\,d\xi \tag{2.15}$$

then implies u is bounded and continuous. This proves the first part of Proposition 2.3, and the rest follows easily.

Proposition 2.3 is a special case of a Sobolev imbedding theorem. For further studies of Sobolev spaces, see [**121, 234, 239, 267**]. We now show how it applies to prove Proposition 2.2.

If $u \in C^\infty(\pi_1)$, then, by Proposition 2.5 of Chapter 0, we know that

$$u \in \mathcal{D}(D^\alpha M_N) \quad \text{for all } \alpha, N, \tag{2.16}$$

where

$$M_N u(x) = (1+|x|^2)^N u(x). \tag{2.17}$$

In view of Proposition 2.3, this implies

$$D^\beta((1+|x|^2)^N u(x)) \text{ is bounded and continuous on } R^n \tag{2.18}$$

for all β, N. This coincides with the definition of $\mathcal{S}(R^n)$ given in (1.8), so the reverse inclusion $C^\infty(\pi_1) \subset \mathcal{S}(R^n)$ is established, and Proposition 2.2 is proved.

We construct other representations of $\mathbf{H}^n$ on $L^2(R^n)$ by a variant of (2.1)–(2.2), utilizing the observation that, for any $\lambda > 0$,

$$\delta_{\pm\lambda}\colon \mathbf{H}^n \to \mathbf{H}^n \tag{2.19}$$

defined by

$$\delta_{\pm\lambda}(t,q,p) = (\pm\lambda t, \pm\lambda^{1/2}q, \lambda^{1/2}p) \tag{2.20}$$

is a group homomorphism. Thus each $\delta_{\pm\lambda}$ is an automorphism of $\mathbf{H}^n$. Other automorphisms are studied in §4. It follows that, for each such λ,

$$\pi_{\pm\lambda}(g) = \pi_1(\delta_{\pm\lambda}g) \tag{2.21}$$

defines a representation of $\mathbf{H}^n$ on $L^2(R^n)$. Since (2.20) is continuous, strong continuity of $\pi_{\pm\lambda}$ follows from that of π_1. Also, since for each $\pm\lambda$ the set $\{\pi_{\pm\lambda}(g)\colon g \in \mathbf{H}^n\}$ coincides with the set $\{\pi_1(g)\colon g \in \mathbf{H}^n\}$, it is clear that each representation $\pi_{\pm\lambda}$ is irreducible. Note that

$$\pi_{\pm\lambda}(t,q,p) = e^{i(\pm\lambda t \pm\lambda^{1/2}q\cdot X + \lambda^{1/2}p\cdot D)} \tag{2.22}$$

is given explicitly on $L^2(R^n)$ by

$$\pi_{\pm\lambda}(t,q,p)u(x) = e^{i(\pm\lambda t \pm \lambda^{1/2} q\cdot x + \lambda q\cdot p/2)} u(x + \lambda^{1/2} p). \tag{2.23}$$

We remark that π_λ is unitarily equivalent to $\pi_\lambda^\#$ ($\lambda \in R \setminus 0$), defined by

$$\pi_\lambda^\#(t,q,p) = e^{i(\lambda t + \lambda q\cdot X + p\cdot D)}, \tag{2.24}$$

and given explicitly on $L^2(R^n)$ by

$$\pi_\lambda^\#(t,q,p)u(x) = e^{i\lambda(t+q\cdot x+q\cdot p/2)} u(x+p), \qquad \lambda \in R \setminus 0. \tag{2.25}$$

There are also the following one-dimensional representations of $\mathbf{H}^n$, which of course are irreducible. For $(y,\eta) \in R^{2n}$, set

$$\pi_{(y,\eta)}(t,q,p) = e^{i(y\cdot q + \eta\cdot p)}. \tag{2.26}$$

Two unitary representations π and ρ of a Lie group G on Hilbert spaces H and H' are unitarily equivalent provided there is a unitary transformation $U\colon H \to H'$ such that

$$U\pi(g)U^{-1} = \rho(g) \quad \text{for all } g \in G. \tag{2.27}$$

Such an operator U is called an intertwining operator between π and ρ.

PROPOSITION 2.4. *No two different representations of* $\mathbf{H}^n$ *given by* (2.22) *and* (2.26) *are unitarily equivalent.*

PROOF. The only point with is not completely trivial is the impossibility of

$$U\pi_\lambda(g)U^{-1} = \pi_{\lambda'}(g) \quad \text{for all } g \in \mathbf{H}^n, \tag{2.28}$$

if $\lambda \neq \lambda'$. Indeed, letting $g = (t,0,0)$, we see that (2.28) implies

$$Ue^{i\lambda t}U^{-1} = e^{i\lambda' t} \quad \text{for all } t \in R,$$

and since $e^{i\lambda t}$ is scalar, this implies $e^{i\lambda t} = e^{i\lambda' t}$ for all $t \in R$, which implies $\lambda = \lambda'$.

We will call the representations $\pi_{\pm\lambda}$ the Schrödinger representations of $\mathbf{H}^n$.

The following important result is due to Stone and von Neumann. A proof will be given in Chapter 5.

THEOREM 2.5. *Every irreducible unitary representation of* $\mathbf{H}^n$ *is unitarily equivalent to one of the form* (2.22) *or* (2.26).

In analogy with the study of the exponential functions $e^{ix\cdot\xi}$, which give the irreducible unitary representations of Euclidean space R^n, we associate to a function f on G a "Fourier transform"

$$\pi(f) = \int_G f(g)\pi(g)\,dg, \tag{2.29}$$

which will be discussed further in the next section. In analogy with the Plancherel theorem for Euclidean space, there is the following result for $\mathbf{H}^n$.

THEOREM 2.6 (PLANCHEREL THEOREM FOR THE HEISENBERG GROUP).

$$\int_{\mathbf{H}^n} |f(z)|^2\,dz = c_n \int_{-\infty}^{\infty} \|\pi_\lambda(f)\|_{\mathrm{HS}}^2 |\lambda|^n\,d\lambda. \tag{2.30}$$

Here $\|T\|_{\mathrm{HS}}^2 = \mathrm{tr}(T^*T)$ is the squared Hilbert-Schmidt norm. This result will be proved in §3. Note that polarization of (2.30) gives

$$\int_{\mathbf{H}^n} f(z)\overline{g(z)}\,dz = c_n \int_{-\infty}^{\infty} \mathrm{tr}(\pi_\lambda(g)^*\pi_\lambda(f))|\lambda|^n\,d\lambda. \tag{2.31}$$

If we replace g by a sequence in $C_0^\infty(\mathbf{H}^n)$ tending to the delta function and pass to the limit, we get

THEOREM 2.7 (INVERSION FORMULA FOR THE HEISENBERG GROUP).

$$f(z) = c_n \int_{-\infty}^{\infty} \mathrm{tr}(\pi_\lambda(z)^*\pi_\lambda(f))|\lambda|^n\,d\lambda. \tag{2.32}$$

Note that the representations (2.26) do not appear in (2.30)–(2.32). One says this set of representations has zero Plancherel measure. Formula (2.30) can be stated as saying that $c_n|\lambda|^n\,d\lambda$ on $R\setminus 0$ is the Plancherel measure on the set of equivalence classes of irreducible unitary representations of $\mathbf{H}^n$.

The way the representations $\pi_{\pm\lambda}$ and $\pi_{(y,\eta)}$ act on the Lie algebra of $\mathbf{H}^n$ is easily read off from (2.22) and (2.26). We have

$$\pi_{\pm\lambda}(T) = \pm i\lambda, \qquad \pi_{\pm\lambda}(L_j) = \pm i\lambda^{1/2}x_j, \qquad \pi_{\pm\lambda}(M_j) = \lambda^{1/2}\partial/\partial x_j, \tag{2.33}$$

and

$$\pi_{(y,\eta)}(T) = 0, \qquad \pi_{(y,\eta)}(L_j) = iy_j, \qquad \pi_{(y,\eta)}(M_j) = i\eta_j. \tag{2.34}$$

3. Convolution operators on $\mathbf{H}^n$ and the Weyl calculus. Recall from Chapter 0 that, if f and g are two functions on a Lie group G, the convolution $f * g$ is defined by

$$f * g(z) = \int_G f(w)g(w^{-1}z)\,dw. \tag{3.1}$$

Here dw is Haar measure on G; as mentioned, Haar measure on $\mathbf{H}^n$ is Lebesgue measure on R^{2n+1}. This convolution is defined on various classes of functions and distributions, and we recall that $(f,g)\mapsto f*g$ defines various bilinear maps, including

$$\begin{gathered} L^1(G)\times L^1(G)\to L^1(G), \qquad C_0^\infty(G)\times C_0^\infty(G)\to C_0^\infty(G), \\ \mathcal{E}'(G)\times\mathcal{E}'(G)\to\mathcal{E}'(G). \end{gathered} \tag{3.2}$$

Given a function or distribution k on G, the operator K_L defined by

$$K_L f = f * k \tag{3.3}$$

is left invariant, i.e., commutes with the operators L_w, $w\in G$, defined by

$$L_w f(z) = f(w^{-1}z), \tag{3.4}$$

and the operator K_R defined by

$$K_R f = k * f \tag{3.5}$$

is right invariant, i.e., commutes with the operators R_w, $w \in G$, defined by

$$R_w f(z) = f(zw). \tag{3.6}$$

If π is a unitary representation of G on a Hilbert space H, then, as seen in Chapter 0, we can define $\pi(f) = \int_G f(z)\pi(z)\,dz$ for $f \in L^1(G)$. Also, in Chapter 0, $\pi(k)$ is defined on $C^\infty(\pi)$ for any compactly supported distribution, $k \in \mathcal{E}'(G)$. We have seen that, generally, $\pi(k_1 * k_2) = \pi(k_1)\pi(k_2)$, given, e.g., $k_j \in \mathcal{E}'(G)$. Thus we expect knowledge of $\pi(k)$ for all irreducible unitary representations of G to give a great deal of information about the operators K_L and K_R defined by (3.3) and (3.5). Here we look into this for $G = \mathbf{H}^n$.

Recall from §2 the representations

$$\pi_{\pm\lambda}(t,q,p) = e^{i(\pm\lambda t \pm \lambda^{1/2} q\cdot X + \lambda^{1/2} p\cdot D)} \tag{3.7}$$

and

$$\pi_{(y,\eta)}(t,q,p) = e^{i(y\cdot q + \eta\cdot p)}. \tag{3.8}$$

We get

$$\begin{aligned} \pi_{\pm\lambda}(k) &= \int k(t,q,p) e^{i(\pm\lambda t \pm \lambda^{1/2} q\cdot X + \lambda^{1/2} p\cdot D)}\,dt\,dq\,dp \\ &= \tilde{k}(\pm\lambda, \pm\lambda^{1/2} X, \lambda^{1/2} D). \end{aligned} \tag{3.9}$$

Here $\tilde{k}(\tau, y, \eta)$ denotes the Euclidean inverse Fourier transform

$$\tilde{k}(\tau,y,\eta) = (2\pi)^{-(2n+1)/2} \int k(t,q,p) e^{i(t\tau + q\cdot y + p\cdot\eta)}\,dt\,dq\,dp, \tag{3.10}$$

and the operator $\tilde{k}(\pm\lambda, \pm\lambda^{1/2}X, \lambda^{1/2}D)$, of the form $a(X,D)$, with $\pm\lambda$ as a parameter, is defined by the *Weyl functional calculus*, as

$$a(X,D) = (2\pi)^{-n} \int \hat{a}(q,p) e^{i(q\cdot X + p\cdot D)}\,dq\,dp, \tag{3.11}$$

where $\hat{a}(q,p) = (2\pi)^{-n} \int a(x,\xi) e^{-i(x\cdot q + \xi\cdot p)}\,dx\,d\xi$ denotes the Fourier transform of $a(x,\xi)$. The following gives a constant alternate formula for the Weyl calculus.

PROPOSITION 3.1. *We have, for well-behaved* $a(x,\xi)$,

$$a(X,D)u(x) = (2\pi)^{-n} \int a(\tfrac{1}{2}(x+y), \xi) e^{i(x-y)\cdot\xi} u(y)\,dy\,d\xi. \tag{3.12}$$

PROOF. By (1.17), we have

$$\begin{aligned} a(X,D)u(x) &= (2\pi)^{-n} \int \hat{a}(q,p) e^{i(q\cdot X + p\cdot D)} u(x)\,dq\,dp \\ &= (2\pi)^{-n} \int \hat{a}(q,p) e^{iq\cdot p/2} e^{iq\cdot x} u(x+p)\,dq\,dp \\ &= (2\pi)^{-2n} \int a(y,\xi) e^{-iy\cdot q - i\xi\cdot p} e^{iq\cdot p/2} e^{iq\cdot x} u(x+p)\,dq\,dp\,dy\,d\xi. \end{aligned}$$

First doing the integral with respect to q gives

$$
\begin{aligned}
(3.13)\qquad &(2\pi)^{-n}\int a(y,\xi)\delta(x-y+\tfrac{1}{2}p)e^{-i\xi\cdot p}u(x+p)\,dp\,dx\,d\xi \\
&= (2\pi)^{-n}2^{-n}\int a(y,\xi)e^{2i(x-y)\cdot\xi}u(x+2(y-x))\,dy\,d\xi,
\end{aligned}
$$

and a simple change of variables gives (3.12).

The operator calculus was introduced by Weyl [**262**], and studied by a number of people. Notable investigations include [**86**] and [**119**], on the application to pseudodifferential operators. The pseudodifferential aspects of the Weyl calculus will not concern us here, though they play an important role in the more advanced treatment of harmonic analysis on the Heisenberg group, given in Chapter II of Taylor [**235**].

We note that, for any $a(x,\xi)\in S'(R^{2n})$, the space of tempered distributions, described in Appendix A, we have $a(X,D)$ defined as a continuous linear operator from $S(R^n)$ to $S'(R^n)$. In particular, if $a(x,\xi)$ is a polynomial in (x,ξ), then $a(X,D)$ can be seen to be a differential operator, whose nature we leave as an exercise for the reader.

Returning to (3.9), we can state that result as

$$(3.14)\qquad \pi_{\pm\lambda}(k) = \sigma_K(\pm\lambda)(X,D)$$

where

$$(3.15)\qquad \sigma_K(\pm\lambda)(x,\xi) = \tilde{k}(\pm\lambda,\pm\lambda^{1/2}x,\lambda^{1/2}\xi),$$

or equivalently

$$(3.16)\qquad \tilde{k}(\pm\tau,y,\eta) = \sigma_K(\pm\tau)(\pm\tau^{-1/2}y,\tau^{-1/2}\eta),\qquad \tau>0.$$

The behavior of $\pi_{(y,\eta)}(k)$ is given simply by

$$
\begin{aligned}
(3.17)\qquad \pi_{(y,\eta)}(k) &= \int k(t,q,p)e^{i(y\cdot q+\eta\cdot p)}\,dt\,dq\,dp \\
&= \tilde{k}(0,y,\eta)\cdot(2\pi)^n.
\end{aligned}
$$

Formulas (3.14)–(3.16) reduce the study of harmonic analysis on $\mathbf{H}^n$ to a study of one parameter families of operators on R^n. As a simple application of these formulas, we give here a proof of the Plancherel formula stated in §2, as formula (2.30):

$$(3.18)\qquad \int_{\mathbf{H}^n}|f(z)|^2\,dz = c_n\int_{-\infty}^{\infty}\|\pi_\lambda(f)\|_{\mathrm{HS}}^2|\lambda|^n\,d\lambda.$$

In general the squared Hilbert-Schmidt norm of an operator

$$Au(x) = \int A(x,y)u(y)\,dy$$

is $\int|A(x,y)|^2\,dx\,dy$. By (3.12), we have $a(X,D)u(x)=\int A(x,y)u(y)\,dy$ with

$$(3.19)\qquad A(x,y) = (2\pi)^{-n}\int a(\tfrac{1}{2}(x+y),\xi)e^{i(x-y)\cdot\xi}\,d\xi.$$

Hence

$$\|a(X,D)\|_{\mathrm{HS}}^2 = c_n' \int |a(x,\xi)|^2\,dx\,d\xi, \tag{3.20}$$

so (3.9) implies

$$\begin{aligned} c_n\|\pi_{\pm\lambda}(f)\|_{\mathrm{HS}}^2 &= \int_{R^{2n}} |\tilde{f}(\pm\lambda, \pm\lambda^{1/2}x, \lambda^{1/2}\xi)|^2\,dx\,d\xi \\ &= |\lambda|^{-n}\int_{R^{2n}} |\tilde{f}(\pm\lambda, y, \eta)|^2\,dy\,d\eta, \end{aligned} \tag{3.21}$$

and hence

$$c_n\int_{-\infty}^{\infty} \|\pi_\lambda(f)\|_{\mathrm{HS}}^2 |\lambda|^n\,d\lambda = \int_{R^{2n+1}} |\tilde{f}(\lambda, y, \eta)|^2\,d\lambda\,dy\,d\eta,$$

so (3.18) is seen to be equivalent to the ordinary Euclidean space Plancherel theorem

$$\int_{\mathbf{H}^n} |f(z)|^2\,dz = \int_{R^{2n+1}} |\tilde{f}(\lambda, y, \eta)|^2\,d\lambda\,dy\,d\eta.$$

This completes the proof of (3.18).

We take a paragraph to describe convolution operators on the group $\tilde{\mathbf{H}}^n$ (isomorphic to $\mathbf{H}^n$), with the group law given by (1.30), and representations $\pi'_{\pm\lambda}$ given by

$$\pi'_{\pm\lambda}(t,q,p) = e^{\pm i\lambda t} e^{\pm i\lambda^{1/2} q\cdot X} e^{i\lambda^{1/2} p\cdot D}. \tag{3.22}$$

We have

$$\pi'_\lambda(k) = \int k(t,q,p) e^{\pm i\lambda t} e^{\pm i\lambda^{1/2} q\cdot X} e^{i\lambda^{1/2} p\cdot D}\,dt\,dq\,dp. \tag{3.23}$$

This is seen to give a representation slightly different from the Weyl calculus, defined by (3.11), (3.12). Indeed, for an operator of the form

$$Au(x) = \int \hat{a}(q,p) e^{iq\cdot X} e^{ip\cdot D} u(x)\,dq\,dp, \tag{3.24}$$

we obtain

$$\begin{aligned} Au(x) &= \int \hat{a}(q,p) e^{iq\cdot x} u(x+p)\,dq\,dp \\ &= (2\pi)^{-n}\int a(y,\xi) e^{-iq\cdot y - ip\cdot\xi} e^{iq\cdot x} u(x+p)\,dq\,dp\,dy\,d\xi \\ &= (2\pi)^{-n}\int a(x,\xi) e^{-iz\cdot\xi + ix\cdot\xi} u(z)\,dz\,d\xi \\ &= (2\pi)^{-n/2}\int a(x,\xi)\hat{u}(\xi) e^{ix\cdot\xi}\,d\xi, \end{aligned} \tag{3.25}$$

which is the *Kohn-Nirenberg* representation, given in [**141**], and in many subsequent publications on pseudodifferential operators; we write

$$Au(x) = a(x,D)u. \tag{3.26}$$

Consequently we have

$$\pi'_{\pm\lambda}(k) = \tilde{k}(\pm\lambda, \pm\lambda^{1/2}x, \lambda^{1/2}D), \tag{3.27}$$

i.e., the operator $\pi'_{\pm\lambda}(k)$ is naturally expressed in terms of the Kohn-Nirenberg formalism rather than the Weyl calculus. A comparison of (3.25) with (3.11), using the identity (1.18), shows that

$$a(X, D) = b(x, D), \tag{3.28}$$

with

$$\hat{b}(q, p) = e^{iq\cdot p/2}\hat{a}(q, p), \tag{3.29}$$

which is the same as saying

$$b(x, \xi) = e^{(i/2)D_x\cdot D_\xi}a(x, \xi). \tag{3.30}$$

For more on the relation between the Weyl calculus and the Kohn-Nirenberg calculus, see Hörmander [**119**].

The advantage of using the Weyl calculus over the Kohn-Nirenberg calculus in developing harmonic analysis on the Heisenberg group arises partly for the same reason that the group law (1.26) manifests symmetries more transparently than the group law (1.30).

4. Automorphisms of $\mathbf{H}^n$; the symplectic groups. The formula (1.26) for the group action on $\mathbf{H}^n$ can be written

$$(t_1, w_1) \cdot (t_2, w_2) = (t_1 + t_2 + \tfrac{1}{2}\sigma(w_1, w_2), w_1 + w_2), \tag{4.1}$$

where, if $w_j = (q_j, p_j) \in R^{2n}$, we set

$$\sigma(w_1, w_2) = p_1 \cdot q_2 - q_1 \cdot p_2. \tag{4.2}$$

Thus σ is a nondegenerate, antisymmetric bilinear form, called the *symplectic form.* We have written, set theoretically,

$$\mathbf{H}^n = R \times R^{2n}. \tag{4.3}$$

It follows that, for any linear map T on R^{2n}, preserving the symplectic form σ given by (4.2), the map

$$(t, w) \mapsto (t, Tw) \tag{4.4}$$

is an automorphism of $\mathbf{H}^n$. We say T belongs to $\mathrm{Sp}(n, R)$, the symplectic group.

The symplectic group will get a fuller treatment in Chapter 11, but we need to introduce some concepts here, making some use of the theory of Hamiltonian vector fields as presented in a number of basic texts, such as Arnold [**4**], or [**1**, **239**].

A linear symplectic map on R^{2n} is a special case of a canonical transformation, which is a map $\varphi\colon R^{2n} \to R^{2n}$ preserving the differential form $\omega = \sum_{j=1}^n dq_j \wedge dp_j$, i.e., $\varphi^*\omega = \omega$. Examples of such canonical transformations are elements of flows generated by Hamiltonian vector fields, defined as follows. If $f(q,p)$ is a smooth function on (some domain in) R^{2n}, the Hamiltonian vector field H_f is given as

$$H_f = \sum_j (\partial f/\partial q_j)\partial/\partial p_j - (\partial f/\partial p_j)\partial/\partial q_j. \tag{4.5}$$

As proved in the references above, the flow generated by such a vector field preserves the form ω. If such a flow consists of linear transformations, then of course the bilinear form σ is also preserved. Given $f, g \in C^\infty(R^{2n})$,

$$H_f g = \{f, g\} \tag{4.6}$$

is called the *Poisson bracket.* Note that $\{f,g\} = -\{g,f\}$. Also the Jacobi identity holds, so $C^\infty(R^{2n})$ is an infinite-dimensional Lie algebra under the Poisson bracket. The set $\mathcal{P}$ of polynomials in (q,p) is a Lie subalgebra.

A further basic result in Hamiltonian mechanics is that any vector field X defined on a region $\Omega \subset R^{2n}$ whose flow preserves ω is locally of the form (4.5), with $f(q,p)$ uniquely determined up to an additive constant. If Ω is simply connected, in particular if $\Omega = R^{2n}$, then X is globally of the form (4.5).

Now if Q is a second order homogeneous polynomial on R^{2n}, then H_Q is a vector field with linear coefficients; hence it generates a flow consisting of linear transformations preserving ω, and hence σ, i.e., a one parameter subgroup of $\mathrm{Sp}(n,R)$. We claim this gives an isomorphism of Lie algebras:

$$\mathrm{sp}(n,R) = \{Q\colon Q \text{ is a second order homogeneous polynomial on } R^{2n}\}. \tag{4.7}$$

To establish (4.7), it remains to consider the generator of a general one parameter subgroup of $\mathrm{Sp}(n,R)$. This is a vector field X on R^{2n}, with coefficients which are linear in (q,p), and the flow generated by X preserves ω. As mentioned, this implies $X = H_Q$ for some $Q \in C^\infty(R^{2n})$, uniquely determined up to an additive constant. Since all the first order partial derivatives of $Q(q,p)$ are functions which are linear in (q,p), we see that Q must be a polynomial in (q,p) of degree ≤ 2. Any linear term in Q would lead to unwanted constant terms among the coefficients of X, so Q must be the sum of a homogeneous second order term and a constant. The constant makes no contribution to H_Q, so we can drop it. Since generally $[H_Q, H_P] = H_{\{Q,P\}}$, the Lie bracket is preserved in the identification (4.7), which is now established. Similarly, for the Lie algebra $\mathfrak{h}^n$ of $\mathbf{H}^n$, we have

$$\mathfrak{h}^n = \{l\colon l \text{ polynomial of degree } \leq 1 \text{ on } R^{2n}\}. \tag{4.8}$$

The Poisson bracket (4.6) gives (4.8) as a module over (4.7) and this is the infinitesimal action of $\mathrm{Sp}(n,R)$ as a group of automorphisms of $\mathfrak{h}^n$, induced by its action as a group of automorphisms of $\mathbf{H}^n$, given by (4.4). Note that the direct sum of (4.7) and (4.8), $\mathcal{P}_{(2)}$, the space of polynomials in (x,ξ) of degree ≤ 2, is also a Lie algebra.

The representations $\pi_{\pm\lambda}$ of $\mathbf{H}^n$ given by (2.6) give rise to representations of the Lie algebra $\mathfrak{h}^n$ of $\mathbf{H}^n$, by the usual formula

$$\pi_{\pm\lambda}(X) = \lim_{t\to 0} t^{-1}(\pi_{\pm\lambda}(\exp tX) - I). \tag{4.9}$$

We see that, if the description (4.8) of $\mathfrak{h}^n$ is used, we have

$$\begin{aligned}\pi_{\pm\lambda}(1) &= \pi_{\pm\lambda}(T) = \pm i\lambda,\\ \pi_{\pm\lambda}(q_j) &= \pi_{\pm\lambda}(L_j) = \pm i\lambda^{1/2}x_j,\\ \pi_{\pm\lambda}(p_j) &= \pi_{\pm\lambda}(M_j) = \lambda^{1/2}\partial/\partial x_j.\end{aligned} \tag{4.10}$$

We define a representation ω of the Lie algebra $\mathrm{sp}(n, R)$, given by (4.7), as

$$\begin{aligned}\omega(p_jp_k) &= (1/i)(\partial/\partial x_j)(\partial/\partial x_k),\\ \omega(q_jq_k) &= ix_jx_k,\\ \omega(p_jp_k) &= \tfrac{1}{2}[x_k(\partial/\partial x_j) + (\partial/\partial x_j)x_k].\end{aligned} \tag{4.11}$$

A unified definition of (4.11) is

$$\omega(Q) = iQ(X, D), \tag{4.12}$$

where $Q(X, D)$ is defined by the Weyl calculus, extended to polynomials, as indicated in the remark preceding (3.22). Note that π_1 and ω fit together to give a representation of the larger Lie algebra $\mathcal{P}_{(2)}$.

For each (real valued) second order polynomial Q, the symmetric operator $Q(X, D) = A$ is essentially selfadjoint, so the exponential $e^{iQ(X,D)}$ is well defined. One way to prove this is to show that the linear space

$$\mathfrak{A} = \{p(x)e^{-|x|^2} : p \text{ polynomial on } R^n\}, \tag{4.13}$$

which is clearly dense in $L^2(R^n)$, is a space of *analytic vectors* for A, i.e.,

$$\|A^k u\|_{L^2} \le C(C_1 k)^k \quad \text{for } u \in \mathfrak{A}. \tag{4.14}$$

See Appendix D for a discussion of analytic vectors. The proof that (4.14) holds for $A = Q(X, D)$, Q a second order polynomial, can be reduced to an estimate of high order derivatives of $e^{-|x|^2}$, which in turn follows from the estimate

$$|(d/dx)^k e^{-x^2}| \le C(8^k k!)^{1/2} e^{-x^2/2}, \tag{4.15}$$

which is proved in Appendix D. We leave the details as an exercise, noting that the square root in (4.15) explains why $\mathfrak{A}$ is analytic for second order $Q(X, D)$, not merely first order. On the other hand, $\mathfrak{A}$ is not analytic for higher order $Q(X, D)$, and there exist such operators which are symmetric but not essentially selfadjoint.

General results about analytic vectors, described in Appendix D, imply that the representation ω of the Lie algebra $\mathrm{sp}(n, R)$ generates a unitary representation

$$\tilde{\omega}: \widetilde{\mathrm{Sp}(n, R)} \to \mathrm{U}(L^2(R^n)) \tag{4.16}$$

of the universal covering group $\widetilde{\mathrm{Sp}(n, R)}$ of $\mathrm{Sp}(n, R)$. Actually, ω can be exponentiated to a representation of the two-fold cover of $\mathrm{Sp}(n, R)$, commonly denoted $\mathrm{Mp}(n, R)$, which in turn is covered by $\widetilde{\mathrm{Sp}(n, R)}$. This representation is called the metaplectic representation, and will be studied in Chapter 11. We remark that $\tilde{\omega}$ has two irreducible components, respectively the even and odd parts of $L^2(R^n)$. The proof is similar to that of Theorem 2.2.

It is useful to know that the Schwartz space $\mathcal{S}(R^n)$, consisting of C^∞ functions which, together with all their derivatives, decrease more rapidly than any negative power of $|x|$ at infinity, is invariant under the action of $\tilde{\omega}$. We have

PROPOSITION 4.1. *For all* $g \in \widetilde{\mathrm{Sp}(n, R)}$,

$$\tilde{\omega}(g) \colon \mathcal{S}(R^n) \to \mathcal{S}(R^n). \tag{4.17}$$

PROOF. The group $\widetilde{\mathrm{Sp}(n, R)}$ is connected, so it suffices to show (4.17) for g in a small neighborhood of the identity element. Thus one need only show

$$e^{itQ(X,D)} \colon \mathcal{S}(R^n) \to \mathcal{S}(R^n), \tag{4.18}$$

in the special case when $iQ(X, D)$ is one of the operators (4.11), since clearly finite products of corresponding one parameter subgroups of $\widetilde{\mathrm{Sp}(n, R)}$ fill out a neighborhood of the identity in $\mathrm{Sp}(n, R)$. The case $iQ(X, D) = ix_jx_k$ is trivial, and the case $iQ(X, D) = (1/i)(\partial/\partial x_j)(\partial/\partial x_k)$ follows by taking the Fourier transform. It remains to investigate the case

$$\begin{aligned} iQ(X, D) &= \tfrac{1}{2}[x_k(\partial/\partial x_j) + (\partial/\partial x_j)x_k] \\ &= x_k(\partial/\partial x_j) + \tfrac{1}{2}\delta_{jk}. \end{aligned} \tag{4.19}$$

In this case, integrating an ODE shows

$$e^{itQ(X,D)}u(x) = e^{it\delta_{jk}/2}u(x + tx_ke_j), \tag{4.20}$$

where e_j is the jth coordinate vector. This makes (4.18) clear for this case too. The proposition is proved.

It is clear that each $u \in \mathcal{S}(R^n)$ is a smooth vector for $\tilde{\omega}$. By considering the special case $Q(X, D) = -\Delta + |x|^2$, it can be deduced that conversely each smooth vector belongs to $\mathcal{S}(R^n)$, i.e., $\mathcal{S}(R^n)$ is precisely the space of smooth vectors for $\tilde{\omega}$. It is easy to show that each $\pi_{\pm\lambda}$ representing $\mathbf{H}^n$ leaves $\mathcal{S}(R^n)$ invariant, and that $\mathcal{S}(R^n)$ is precisely the space of smooth vectors for each such representation.

The following important result relates the representations $\pi_{\pm\lambda}$ and ω.

THEOREM 4.2. *Let* $T = \exp H_Q \in \mathrm{Sp}(n, R)$. *Then*

$$\pi_{\pm\lambda}(t, Tw) = e^{-iQ(X,D)}\pi_{\pm\lambda}(t, w)e^{iQ(X,D)}. \tag{4.21}$$

PROOF. If we set $w = (q, p) \in R^{2n}$, then

$$w(X, D) = q \cdot X + p \cdot D, \tag{4.22}$$

and (4.21) follows from

$$(Tw)(X, D) = e^{-iQ(X,D)}w(X, D)e^{iQ(X,D)}, \tag{4.23}$$

which we restate as

$$[(\exp sH_Q)w](X, D) = e^{-isQ(X,D)}w(X, D)e^{isQ(X,D)}. \tag{4.24}$$

The identity (4.24) is equivalent to the operator differential equation

$$(4.25)\qquad (d/ds)[(\exp sH_Q)w](X,D) = -i[Q(X,D),((\exp sH_Q)w)(X,D)].$$

Now the left side of (4.25) is

$$[H_Q(\exp sH_Q)w](X,D),$$

so to prove (4.25), it suffices to show

$$(4.26)\qquad \{Q,v\}(X,D) = -i[Q(X,D),v(X,D)]$$

for any linear $v(x,\xi)$ and any second order polynomial $Q(x,\xi)$. This is readily verified, and the proof of the theorem is complete.

If the covering homomorphism is denoted

$$(4.27)\qquad j\colon \widetilde{\mathrm{Sp}(n,R)} \to \mathrm{Sp}(n,R),$$

and if, for $T \in \mathrm{Sp}(n,R)$, the automorphism (4.4) of $\mathbf{H}^n$ is also denoted T, then we can restate Theorem 4.2 as

$$(4.28)\qquad \pi_{\pm\lambda}(j(g)z) = \tilde{\omega}(g)^{-1}\pi_{\pm\lambda}(z)\tilde{\omega}(g), \qquad z \in \mathbf{H}^n,\ g \in \widetilde{\mathrm{Sp}(n,R)}.$$

The formula

$$(4.29)\qquad a(X,D) = (2\pi)^{-n}\int \hat{a}(q,p)e^{i(q\cdot X+p\cdot D)}\,dq\,dp$$

then yields the following result, known as metaplectic covariance of the Weyl calculus.

PROPOSITION 4.3. *For $g \in \widetilde{\mathrm{Sp}(n,R)}$, let*

$$(4.30)\qquad a_g(x,\xi) = a(j(g)(x,\xi)).$$

Then

$$(4.31)\qquad a_g(X,D) = \tilde{\omega}(g)^{-1}a(X,D)\tilde{\omega}(g).$$

In addition to the automorphisms of $\mathbf{H}^n$ given by (4.4), there is also the family of "dilations," given by

$$(4.32)\qquad \delta_{\pm\lambda}(t,q,p) = (\pm\lambda t, \pm\lambda^{1/2}q, \lambda^{1/2}p), \qquad \lambda > 0,$$

which was introduced in §2. As pointed out there, we have

$$(4.33)\qquad \pi_1(\delta_{\pm\lambda}z) = \pi_{\pm\lambda}(z), \qquad z \in \mathbf{H}^n.$$

5. The Bargmann-Fok representation. There is a representation of $\mathbf{H}^n$ (unitarily equivalent to π_1) on a Hilbert space of entire functions on $\mathbf{C}^n$, introduced by Bargmann and Fok, which provides a useful alternative perspective on the representation theory. Consider the Hilbert space

$$(5.1)\qquad \mathcal{H} = \left\{u(\varsigma) \text{ holomorphic on } \mathbf{C}^n : \int_{\mathbf{C}^n} |u(\varsigma)|^2 e^{-|\varsigma|^2/2}\,d\varsigma < \infty\right\}$$

with inner product

$$(u,v)=\int_{\mathbf{C}^n} u(\varsigma)\overline{v(\varsigma)}e^{-|\varsigma|^2/2}\,d\varsigma. \tag{5.2}$$

The operators $u \mapsto \varsigma_j u$ and $u \mapsto \partial u/\partial\varsigma_j$ are closed linear operators on $\mathcal{H}$ satisfying the commutation relation $[\partial/\partial\varsigma_j, \varsigma_j] = 1$. These operators are not skew adjoint. In fact, integration by parts and use of the Cauchy-Riemann equations give

$$(\partial u/\partial\varsigma_j, v) = (u, \varsigma_j v). \tag{5.3}$$

It follows that

(5.4)

$$\beta_1(T) = iI,\quad \beta_1(L_j) = (i/\sqrt{2})(\partial/\partial\varsigma_j + \varsigma_j),\quad \beta_1(M_j) = (1/\sqrt{2})(\partial/\partial\varsigma_j - \varsigma_j),$$

defines a skew-adjoint representation of $\mathfrak{h}^n$. It generates a unitary representation β_1 of $\mathbf{H}^n$, defined as follows. For $(t,q,p) \in \mathbf{H}^n$, let $z = q + ip$, so we write $(t,q,p) = (t,z)$, $z \in \mathbf{C}^n$. Then $\beta_1(t,z)$ acts on $\mathcal{H}$ by

$$\beta_1(t,z)u(\varsigma) = e^{it+(i/\sqrt{2})\varsigma\cdot z-|z|^2}u(\varsigma + i\bar{z}/\sqrt{2}). \tag{5.5}$$

We claim β_1 is unitarily equivalent to π_1. First we give a direct proof of its irreducibility.

LEMMA 5.1. *The representation β_1 of $\mathbf{H}^n$ on $\mathcal{H}$ is irreducible.*

PROOF. Suppose we have an orthogonal decomposition $\mathcal{H} = \mathcal{H}_{\mathrm{I}} \oplus \mathcal{H}_{\mathrm{II}}$, with $\mathcal{H}_{\mathrm{I}}$ and $\mathcal{H}_{\mathrm{II}}$ closed and each invariant under β_1, so $\beta_1 = \beta_{\mathrm{I}} \oplus \beta_{\mathrm{II}}$. Pick $v_{\mathrm{I}} \in \mathcal{H}_{\mathrm{I}}$, $v_{\mathrm{II}} \in \mathcal{H}_{\mathrm{II}}$, both unit vectors. We can assume v_{I} and v_{II} are *analytic vectors*, since analytic vectors are dense (see Appendix D). Such functions are in particular analytic vectors for the operations of multiplication by ς_j and of $\partial/\partial\varsigma_j$; hence, for some $\varepsilon > 0$, $|v_{\mathrm{I}}(\varsigma)|e^{\varepsilon|\varsigma|}$ and $|v_{\mathrm{II}}(\varsigma)|e^{\varepsilon|\varsigma|}$ are square integrable with respect to the measure $e^{-|\varsigma|^2/2}\,d\varsigma$, and, for $|c|$ small,

$$(v_{\mathrm{I}}(\varsigma + c), v_{\mathrm{II}}(\varsigma + c)) = 0. \tag{5.6}$$

Making a change of variable in the integral (5.2) for (5.6) gives

$$\int v_{\mathrm{I}}(\varsigma)\overline{v_{\mathrm{II}}(\varsigma)}e^{-\mathrm{Re}(c\cdot\varsigma)}e^{-|\varsigma|^2/2}\,d\varsigma = 0 \tag{5.7}$$

for $c \in \mathbf{C}^n$, $|c|$ small. This implies that

$$v_{\mathrm{I}}(\varsigma)\overline{v_{\mathrm{II}}(\varsigma)}e^{-|\varsigma|^2/2} \quad \text{on } R^{2n} = \mathbf{C}^n \tag{5.8}$$

has a Fourier transform which is holomorphic in a tube and vanishes for small purely imaginary values. Hence the Fourier transform of (5.8) vanishes identically, which implies (5.8) vanishes identically. Since $v_{\mathrm{I}}(\varsigma)$ and $v_{\mathrm{II}}(\varsigma)$ are holomorphic, this forces either v_{I} or v_{II} to be identically zero. This contradiction proves irreducibility of β_1.

We remark that the delicate argument of Appendix D showing that analytic vectors are dense is not needed here. Indeed, it is easy to show that the space $\mathcal{P}$

of polynomials in ς is a dense linear subspace of $\mathcal{H}$, consisting of analytic vectors for β_1. If $\beta_1 = \beta_{\mathrm{I}} \oplus \beta_{\mathrm{II}}$, then the projections of $\mathcal{P}$ onto $\mathcal{H}_{\mathrm{I}}$ and $\mathcal{H}_{\mathrm{II}}$ clearly would provide dense linear subspaces of these Hilbert spaces, consisting of analytic vectors.

Since we have

$$\beta_1(t,0) = e^{it}, \tag{5.9}$$

the Stone-von Neumann theorem implies β_1 must be unitarily equivalent to π_1. In fact, we can produce a unitary intertwining operator

$$K\colon L^2(R^n) \to \mathcal{H} \tag{5.10}$$

such that

$$\pi_1 = K^{-1}\beta_1 K, \tag{5.11}$$

in the form

$$Kf(\varsigma) = \int_{R^n} K(x,\varsigma)f(x)\,dx, \tag{5.12}$$

with $K(x,\varsigma)$ constructed below. We remark that, once the formula (5.12) for K is obtained, one can verify directly that it defines a unitary transformation satisfying (5.10)–(5.11), so in the end we do not need to depend on the Stone-von Neumann theorem.

Since we have for the Schrödinger representation

$$\pi_1(T) = iI, \qquad \pi_1(L_j) = ix_j, \qquad \pi_1(M_j) = \partial/\partial x_j, \tag{5.13}$$

comparison with (5.4) shows the kernel $K(x,\varsigma)$ must satisfy the differential equations

$$\begin{aligned} x_j K(x,\varsigma) &= (1/\sqrt{2})(\partial/\partial\varsigma_j + \varsigma_j)K(x,\varsigma), \\ -(\partial/\partial x_j)K(x,\varsigma) &= (1/\sqrt{2})(\partial/\partial\varsigma_j - \varsigma_j)K(x,\varsigma). \end{aligned} \tag{5.14}$$

as well as being holomorphic in ς; thus $K(x,\varsigma)$ is determined by its values for $\varsigma \in R^n$. If we first consider the case $n = 1$, we see that, given $K(0,0) = C_1$, the first equation specifies $K(0,\varsigma)$ uniquely as

$$K(0,\varsigma) = C_1 e^{-\varsigma^2/2}. \tag{5.15}$$

Subtracting the second equation in (5.14) from the first gives

$$(\partial/\partial x_j + x_j)K(x,\varsigma) = \sqrt{2}\varsigma_j K(x,\varsigma),$$

so we can integrate in the x variable from (5.15) to get

$$K(x,\varsigma) = C_1 \exp(\sqrt{2}\varsigma x - \tfrac{1}{2}(\varsigma^2 + x^2)), \tag{5.16}$$

in case $n = 1$. We can take products to get for general n.

$$K(x,\varsigma) = C_1^n \exp(\sqrt{2}\varsigma\cdot x - \tfrac{1}{2}(\varsigma\cdot\varsigma + |x|^2)). \tag{5.17}$$

Here, as in (5.5), our dot product is bilinear. Computing K^*K reduces to computing Gaussian integrals; for K to be unitary, one takes $C_1^n = \pi^{-n/4}$. We omit the details.

If we use the complex vector fields on $\mathbf{H}^n$,

$$Z_j = L_j + iM_j, \qquad \overline{Z}_j = L_j - iM_j, \tag{5.18}$$

then (5.4) implies

$$\beta_1(Z_j) = i\sqrt{2}\partial/\partial\varsigma_j, \qquad \beta_1(\overline{Z}_j) = i\sqrt{2}\varsigma_j. \tag{5.19}$$

Considering both the Bargmann-Fok and the Schrödinger representations of $\mathbf{H}^n$ leads to many useful insights, as we will see in the following two sections, and also in Chapter 11.

6. (Sub)Laplacians on $\mathbf{H}^n$ and harmonic oscillators. A central object in the application of harmonic analysis on $\mathbf{H}^n$ to partial differential equations is the "Heisenberg Laplacian"

$$\mathcal{L}_0 = \sum_{j=1}^{n}(L_j^2 + M_j^2). \tag{6.1}$$

This second order differential operator is not elliptic; it is associated to a degenerate metric on $\mathbf{H}^n$. We also study certain other sub-Laplacians in this section. In particular, we need to understand the spectrum of the operator $\pi_{\pm\lambda}(\mathcal{L}_0)$ for each $\pm\lambda$. Note that

$$\pi_{(y,\eta)}(\mathcal{L}_0) = \sum_{j=1}^{n}(y_j^2 + \eta_j^2), \tag{6.2}$$

and, by (2.14),

$$\pi_{\pm\lambda}(\mathcal{L}_0) = \sum_{j=1}^{n}(\partial^2/\partial x_j^2 - x_j^2) = -\lambda(-\Delta + |x|^2). \tag{6.3}$$

Thus understanding the spectrum of $\pi_{\pm\lambda}(\mathcal{L}_0)$ is equivalent to understanding the spectrum of the operator

$$H = -\Delta + |x|^2. \tag{6.4}$$

This operator is called the *harmonic oscillator Hamiltonian*, as it arises in the Schrödinger equation $\partial u/\partial t = iHu$ for the quantum mechanical harmonic oscillator.

To analyze the spectrum of (6.4), it suffices to treat the one-dimensional case

$$H = -d^2/dx^2 + x^2. \tag{6.5}$$

We have seen in §4 that e^{itH} preserves the Schwartz space $\mathcal{S}(R)$. It follows that H is essentially selfadjoint on $\mathcal{S}(R)$ (see Appendix A). In this case, it is easy to see that the domain of H is precisely

$$\mathcal{D}(H) = \{u \in L^2(R) \colon u''(x) \in L^2(R) \text{ and } x^2u(x) \in L^2(R)\}. \tag{6.6}$$

We can also deduce that a bounded subset $\mathcal{B}$ of $\mathcal{D}(H)$ is relatively compact in $L^2(R)$. In fact, from $u_j''(x)$ bounded in $L^2(R)$ it follows that u_j' is locally uniformly bounded and hence u_j is locally uniformly Lipschitz, so a subsequence can be selected, converging locally uniformly, $u_{j_k} \to u$. On the other hand, if $(1+x^2)u_j$ is bounded in $L^2(R)$, then for any $\varepsilon > 0$ there is an interval $[-N, N] = I$ such that $\int_{R\setminus I} |u_j(x)|^2\,dx < \varepsilon$ for all j. Hence $u_{j_k} \to u$ in the $L^2(R)$ norm. It follows that

$$(6.7) \qquad (I+H)^{-1} \text{ is a compact operator on } L^2(R).$$

Thus $L^2(R)$ has an orthonormal basis consisting of eigenfunctions for H. Each eigenspace has finite dimension (dimension one for (6.5), as we will see in a moment), and the eigenvalues of H must tend to $+\infty$. In a moment we will see exactly what they are.

Generalizing (6.6), we find

$$(6.8) \qquad \mathcal{D}(H^k) = \{u \in L^2(R) \colon x^j D_x^l u \in L^2(R) \text{ for } j+l \le 2k\}.$$

In particular,

$$(6.9) \qquad \bigcap_k \mathcal{D}(H^k) = \mathcal{S}(R),$$

so all the eigenfunctions of H belong to $\mathcal{S}(R)$.

As is emphasized in quantum mechanics texts, one easy way to analyze the spectrum of (6.5) is to use the operators

$$(6.10) \qquad A = -d/dx - x, \qquad A^* = d/dx - x.$$

A calculation gives

$$(6.11) \qquad A^*A = H - 1, \qquad AA^* = H + 1.$$

The first identity shows 1 is the smallest possible eigenvalue of H. These identities imply

$$(6.12) \qquad AH = AA^*A + A, \qquad HA = AA^*A - A,$$

and hence

$$(6.13) \qquad [A, H] = 2A.$$

Similarly,

$$(6.14) \qquad [A^*, H] = -2A^*.$$

The identities (6.13)–(6.14) are equivalent to

$$(6.15) \qquad HA = A(H-2), \qquad HA^* = A^*(H+2).$$

Hence, if the eigenspace decomposition of H is

$$(6.16) \qquad L^2(R) = \bigoplus_{\mu \in \operatorname{spec} H} V_\mu,$$

then

$$A\colon V_\mu \to V_{\mu-2} \tag{6.17}$$

and

$$A^*: V \to V_{\mu+2}. \tag{6.18}$$

In particular, if $\mu \in \operatorname{spec} H$, then either $\mu - 2 \in \operatorname{spec} H$ or A annihilates V_μ. From (6.10) we see A annihilates only the linear span of

$$h_0(x) = \pi^{-1/4} e^{-x^2/2}. \tag{6.19}$$

The factor $\pi^{-1/4}$ has been thrown in to make h_0 a unit vector in $L^2(R)$. Thus we see that V_1 is the linear span of h_0 and that

$$\operatorname{spec} H = \{1, 3, 5, 7, \ldots\}. \tag{6.20}$$

By (6.11), $A\colon V_\mu \to V_{\mu-2}$ is an isomorphism for $\mu \geq 3$ an odd integer, so each V_{2k+1} is one-dimensional and is the linear span of

$$h_k(x) = c_k (d/dx - x)^k e^{-x^2/2} = c_k H_k(x) e^{-x^2/2}, \tag{6.21}$$

where $H_k(x)$ is the *Hermite polynomial*

$$\begin{aligned} H_k(x) &= (-1)^k e^{x^2} (d/dx)^k e^{-x^2} \\ &= \sum_{j=0}^{[k/2]} (-1)^j [k!/j!(k-2j)!] (2x)^{k-2j}. \end{aligned} \tag{6.22}$$

The constants c_k are chosen so $\|h_k\| = 1$; they will be specified shortly. Since the eigenspaces V_μ are mutually orthogonal, we know the $h_k(x)$ are mutually orthogonal, i.e.,

$$\int_{-\infty}^{\infty} H_j(x) H_k(x) e^{-x^2}\, dx = 0 \quad \text{if } j \neq k. \tag{6.23}$$

We can evaluate the c_k in (6.21) by noting that, with $h_k \in V_{2k+1}$,

$$\|A^* h_k\|^2 = (AA^* h_k, h_k) = 2(k+1)\|h_k\|^2. \tag{6.24}$$

Consequently, if $\|h_k\| = 1$, in order for

$$h_{k+1} = \gamma_{k+1} A^* h_k \tag{6.25}$$

to have unit norm, we need

$$\gamma_{k+1} = (2k+2)^{-1/2}. \tag{6.26}$$

Thus the constants c_k in (6.21) are given by

$$c_k = [\pi^{1/2} 2^k (k!)]^{-1/2}. \tag{6.27}$$

Another approach to (6.27) is to use the recursion formula

$$H_{k+1}(x) - 2x H_k(x) + 2k H_{k-1}(x) = 0 \quad (\text{convention: } H_{-1}(x) = 0), \tag{6.28}$$

which can be deduced from the generating function identity

$$\sum_{k=0}^{\infty} H_k(x)t^k/k! = e^{2xt-t^2}. \tag{6.29}$$

See Lebedev [**154**] for details on this. Note that (6.29) follows immediately from (6.22).

Having shown that the spectrum of $-d^2/dx^2 + x^2$ consists of the positive odd integers, all simple eigenvalues, we deduce that the spectrum of $-\Delta + |x|^2$ on $L^2(R^n)$ consists of all the positive integers of the form $n + 2j$, $j = 0, 1, 2, \ldots$:

$$\operatorname{spec} H = \{n, n+2, n+4, n+6, \ldots\}. \tag{6.30}$$

Note that, for $H = -\Delta + |x|^2$ on R^n, an orthonormal basis of the $n + 2j$ eigenspace of H is given by

$$\{h_\alpha(x) = h_{\alpha_1}(x_1)\cdots h_{\alpha_n}(x_n)\colon |\alpha| = j\}. \tag{6.31}$$

The combinatorially rather complicated form of the Hermite functions given by (6.21)–(6.22) induces one to seek a simpler approach. In fact, use of the Bargmann-Fok representation provides a cleaner route to understanding the spectrum of H. Note, from formula (5.4),

$$\begin{aligned}\beta_1(\mathcal{L}_0) &= \sum_{j=1}^{n}(\varsigma_j\partial/\partial\varsigma_j + \partial/\partial\varsigma_j\varsigma_j)\\ &= 2\sum\varsigma_j\partial/\partial\varsigma_j + n = W,\end{aligned} \tag{6.32}$$

which is equivalent to saying H and W are intertwined by the unitary operator K defined by (5.6)–(5.14).

$$KHK^{-1} = W. \tag{6.33}$$

Now it is very easy to verify that

$$w_\alpha(\varsigma) = (2/\alpha!)^{1/2}\varsigma^\alpha, \qquad \alpha \geq 0,$$

is an orthonormal basis of the Hilbert space $\mathcal{H}$ of entire functions on $\mathbf{C}^n$ defined by (5.1), and, by (6.32),

$$Ww_\alpha = \sum_{j=1}^{n}(2\alpha_j + 1)w_\alpha = (2|\alpha| + n)w_\alpha, \tag{6.34}$$

so the spectrum of W is precisely $\{n + 2j\colon j = 0, 1, 2, \ldots\}$, and an orthonormal basis for the $2j + n$ eigenspace of W is given by $\{w_\alpha\colon |\alpha| = j\}$. The formula (6.32) also implies that the unitary group e^{itW} generated by the skew adjoint operator iW is given by

$$e^{itW}f(\varsigma) = e^{int}f(e^{2it}\varsigma), \qquad f \in \mathcal{H}. \tag{6.35}$$

This is a neat formula. The formula for e^{itH}, which we will consider in the next section, is more complicated.

We can use this result on the spectrum of the harmonic oscillator Hamiltonian H to determine invertibility of the operators $\pi_{\pm\lambda}(\mathcal{L}_\alpha)$, where

$$\mathcal{L}_\alpha = \mathcal{L}_0 + i\alpha T. \tag{6.36}$$

Note, from (2.14) and (6.3),

$$\pi_{\pm\lambda}(\mathcal{L}_\alpha) = -\lambda H \mp \lambda\alpha = -\lambda(H \pm \alpha). \tag{6.37}$$

From the analysis of the spectrum of H, we have

PROPOSITION 6.1. *$\pi_{\pm\lambda}(\mathcal{L}_\alpha)$ is invertible, for all $\lambda \in (0,\infty)$, if and only if*

$$\mp\alpha \text{ avoids the set } \{n+2j\colon j = 0,1,2,\dots\}. \tag{6.38}$$

As will be seen later, this condition is equivalent to hypoellipticity of the operator $\mathcal{L}_\alpha$.

We now consider more general second order differential operators on $\mathbf{H}^n$, namely operators of the form

$$P_0 = \sum_{j,k=1}^{2n} a_{jk} Y_j Y_k \tag{6.39}$$

where

$$Y_j = L_j, \quad Y_{n+j} = M_j, \qquad 1 \le j \le n, \tag{6.40}$$

and (a_{jk}) is a symmetric, positive definite matrix of real numbers. By (2.14) we have

(6.41)

$$\pi_{\pm\lambda}(Y_j) = \pm i\lambda^{1/2} x_j, \quad \pi_{\pm\lambda}(Y_{n+j}) = \lambda^{1/2}\partial/\partial x_j = i\lambda^{1/2} D_j, \qquad 1 \le j \le n,$$

so if P_0 is given by (6.39),

$$\pi_{\pm\lambda}(P_0) = -\lambda Q(\pm X, D), \tag{6.42}$$

where

$$Q(x,\xi) = \sum_{j,k=1}^{2n} a_{jk}\chi_j\chi_k$$

with

$$\chi_j = x_j, \quad \chi_{n+j} = \xi_j, \qquad 1 \le j \le n.$$

The operator $Q(\pm X, D)$ appearing in (6.42) is a positive selfadjoint second order differential operator, and we want to find its spectrum. This is gotten from studying the interplay between the quadratic form $Q(x,\xi)$ on R^{2n} and the symplectic form

$$\sigma((x,\xi),(x',\xi')) = x\cdot\xi' - x'\cdot\xi \quad \text{on } R^{2n}. \tag{6.43}$$

We define the Hamilton map of $Q(x,\xi)$ to be the linear map F on R^{2n} given by

$$\sigma(u, Fv) = Q(u,v), \qquad u,v \in R^{2n}, \tag{6.44}$$

where $Q(u,v)$ is the symmetric bilinear form on R^{2n} polarizing the quadratic form $Q(u)$, i.e., $Q(u) = Q(u,u)$. We are assuming Q is positive definite. So we see that F is skew symmetric and invertible, so its eigenvalues must all be pure imaginary, nonzero, and occur in complex conjugate pairs, i.e., be of the form

$$\pm i\mu_j, \qquad 1 \le j \le n, \quad \mu_j > 0. \tag{6.45}$$

It turns out that we can pick a symplectic basis of R^{2n}, diagonalizing Q, as follows.

LEMMA 6.2. *If Q is positive definite, there is a symplectic basis of R^{2n}, $\{e_j, f_j : 1 \le j \le n\}$, i.e., a basis satisfying*

$$\sigma(e_j, e_k) = 0 = \sigma(f_j, f_k), \qquad \sigma(e_j, f_k) = \delta_{jk}, \tag{6.46}$$

such that, if

$$u = \sum_{j=1}^{n} \alpha_j e_j + \beta_j f_j, \tag{6.47}$$

then

$$Q(u,u) = \sum_{j=1}^{n} \mu_j(\alpha_j^2 + \beta_j^2), \tag{6.48}$$

with μ_j given by (6.45).

PROOF. F^{-1} is characterized by $\sigma(u,v) = Q(u, F^{-1}v)$, so F^{-1} is skew symmetric with respect to the inner product Q. Thus there is a basis $\{E_j, F_j : 1 \le j \le n\}$ of R^{2n}, orthonormal with respect to Q, such that

$$F^{-1}E_j = -\lambda_j F_j, \quad F^{-1}F_j = \lambda_j E_j, \qquad \lambda_j > 0.$$

Set $e_j = \lambda_j^{-1/2} E_j$, $f_j = \lambda_j^{-1/2} F_j$. Then $\{e_j, f_j : 1 \le j \le n\}$ is a symplectic basis of R^{2n}, since $\sigma(\lambda_j^{-1}E_j, F_k) = Q(\lambda_j^{-1}E_j, \lambda_k F_k) = \delta_{jk}$, etc. Furthermore, with $\mu_j = \lambda_j^{-1}$, we immediately have (6.48). This completes the proof of the lemma.

Granted this lemma, we can use Proposition 4.2 to unitarily conjugate $Q(X,D)$ to a sum of one-dimensional harmonic oscillator Hamiltonians. In fact, if $g \in \widetilde{\mathrm{Sp}(n,R)}$ is such that $j(g) \in \mathrm{Sp}(n,R)$ effects the change of basis of R^{2n} to $\{e_j, f_j\}$, then

$$\tilde{\omega}(g)^{-1} Q(X,D) \tilde{\omega}(g) = Q_g(X,D) \tag{6.49}$$

with

$$Q_g(x,\xi) = \sum_{j=1}^{n} \mu_j(x_j^2 + \xi_j^2), \tag{6.50}$$

so $Q(X,D)$ is unitarily equivalent to

$$\sum_{j=1}^{n} \mu_j(-\partial^2/\partial x_j^2 + x_j^2). \tag{6.51}$$

Our analysis of the spectrum of the one-dimensional harmonic oscillator then yields the following result.

PROPOSITION 6.3. *If $Q(x,\xi)$ is a second order homogeneous positive definite polynomial, then the spectrum of $Q(X,D)$ is*

$$\left\{ \sum_{j=1}^{n} (2k_j+1)\mu_j : k_j \in Z^+ \cup \{0\} \right\}, \tag{6.52}$$

where $\pm i\mu_j$, $1 \le j \le n$, are the eigenvalues of the Hamilton map F associated with Q by (6.44), $\mu_j > 0$.

Note that changing (x,ξ) to $(-x,\xi)$ changes the sign of the symplectic form on R^n, so the Hamilton map of $Q(-x,\xi)$ is similar to the negative of that of $Q(x,\xi)$. Thus the two Hamilton maps of $Q(\pm x,\xi)$ have the same set of eigenvalues, so $Q(-X,D)$ has the same spectrum as $Q(X,D)$; the two operators are unitarily equivalent.

Generalizing Proposition 6.1, we can determine invertibility of the operators $\pi_{\pm\lambda}(P_\alpha)$, where

$$P_\alpha = P_0 + i\alpha T, \quad P_0 \text{ given by (6.39)}. \tag{6.53}$$

Note that

$$\pi_{\pm\lambda}(P_\alpha) = -\lambda[Q(\pm X,D) \pm \alpha]. \tag{6.54}$$

We have

PROPOSITION 6.4. *$\pi_{\pm\lambda}(P_\alpha)$ is invertible for all $\lambda \in (0,\infty)$, if and only if*

$$\mp\alpha \text{ avoids the set (6.52)}. \tag{6.55}$$

This condition turns out to be equivalent to *hypoellipticity* of P_α, and also implies hypoellipticity of any operator of the form

$$P = P_0 + i\alpha T + \sum_{j=1}^{2n} a_j Y_j, \qquad a_j \in \mathbf{C}. \tag{6.56}$$

An operator P is said to be hypoelliptic if, for any $u \in \mathcal{D}'$, u is smooth wherever Pu is. The connection between Proposition 6.4 and hypoellipticity is exploited in many places; we mention [**64**, **108**, **204**, **235**, **173**]. We will say a little more about this in the next section.

It is worth pointing out that an alternative approach to Proposition 6.4 is to choose an automorphism of $\mathbf{H}^n$ which puts P_α in the form

$$\sum \mu_j(L_j^2 + M_j^2) + i\alpha T. \tag{6.57}$$

That this can be arranged is also a consequence of Lemma 6.2.

7. Functional calculus for Heisenberg Laplacians and for harmonic oscillator Hamiltonians. We would like to understand the behavior of functions $f(-\mathcal{L}_\alpha)$, $f(-P_\alpha)$ of the selfadjoint operators $\mathcal{L}_\alpha$, P_α considered in §6. Recall that

$$\mathcal{L}_\alpha = \sum_{j=1}^{2n} Y_j^2 + i\alpha T, \tag{7.1}$$

and, more generally,

$$P_\alpha = \sum_{j,k=1}^{2n} a_{jk} Y_j Y_k + i\alpha T, \tag{7.2}$$

where

$$Y_j = L_j, \quad Y_{j+n} = M_j, \qquad 1 \leq j \leq n,$$

and (a_{jk}) is a positive definite symmetric matrix of real numbers. The operators $f(-\mathcal{L}_\alpha)$ and $f(-P_\alpha)$ are defined by the spectral theorem. Recall that

$$\sigma_{\mathcal{L}_\alpha}(\pm\lambda)(X, D) = -\lambda(-\Delta + |x|^2 \pm \alpha), \tag{7.3}$$

$$\sigma_{P_\alpha}(\pm\lambda)(X, D) = -\lambda[Q(\pm X, D) \pm \alpha]. \tag{7.4}$$

We have also set

$$\pi_{\pm\lambda}(k) = \sigma_K(\pm\lambda)(X, D), \tag{7.5}$$

for $Ku = u * k$, as defined in (3.14).

It follows that

$$\sigma_{f(\mathcal{L}_\alpha)}(\pm\lambda)(X, D) = f(-\lambda(-\Delta + |x|^2 \pm \alpha)), \tag{7.7}$$

and more generally,

$$\sigma_{f(P_\alpha)}(\pm\lambda)(X, D) = f(-\lambda[Q(\pm X, D) \pm \alpha]). \tag{7.8}$$

Thus we need to understand

$$f(H), \qquad H = -\Delta + |x|^2, \tag{7.9}$$

and more generally to understand $f(Q(X, D))$, hopefully with a parameter λ thrown in.

The first case of (7.9) we study is $f(H) = e^{-tH}$. We produce the Weyl symbol of such operators, as follows.

PROPOSITION 7.1. *We have*

$$e^{-tH} = h_t(X, D) \tag{7.10}$$

with

$$h_t(x, \xi) = (\cosh t)^{-n} e^{-(\tanh t)(|x|^2 + |\xi|^2)}. \tag{7.11}$$

Equivalently,

$$\hat{h}_t(q, p) = (2 \sinh t)^{-n} e^{-(1/4)(\coth t)(|q|^2 + |p|^2)}. \tag{7.12}$$

This result goes back to Mehler [**154**]. As it is a central result, we will give several proofs.

First note that, by commutativity, $e^{-tH} = e^{-tH_1} \cdots e^{-tH_n}$ where $H_j = -\partial^2/\partial x_j^2 + x_j^2$. Thus $h_t(x, \xi) = h_t(x_1, \xi_1) \cdots h_t(x_n, \xi_n)$, and $\hat{h}_t(q, p)$ satisfies the analogous multiplicativity conditions. Hence it suffices to prove the proposition for $H = -d^2/dx^2 + x^2$, acting on functions of one variable.

The Weyl symbol $h_t(x,\xi)$ is related to the kernel of the operator e^{-tH}, defined by

$$e^{-tH}u(x) = \int K_t(x,y)u(y)\,dy, \tag{7.13}$$

by

$$K_t(x,y) = (2\pi)^{-n}\int h_t(\tfrac{1}{2}(x+y),\xi)e^{i(x-y)\cdot\xi}\,d\xi. \tag{7.14}$$

Consequently, the identity (7.11) (for $n=1$) is equivalent to

(7.15)
$$K_t(x,y) = (2\pi)^{-1/2}(\sinh 2t)^{-1/2}\exp\{[-\tfrac{1}{2}(\coth 2t)(x^2+y^2) - xy]/\sinh 2t\}.$$

This is Mehler's formula for the fundamental solution to $\partial u/\partial t = Hu$. By virtue of the analysis of spec H and the formula (6.21) of §6, this is in turn equivalent to the generating function identity

$$\begin{aligned}\sum_{j=0}^{\infty} h_j(x)h_j(y)t^j &= \pi^{-1/2}(1-t^2)^{-1/2} \\ &\quad\cdot \exp\{[2xyt + (x^2+y^2)t]/(1-t^2)\}e^{-(x^2+y^2)/2}\end{aligned} \tag{7.16}$$

for products of Hermite functions. A direct proof of (7.16) is given in Lebedev [**154**], pages 61–63.

The next proof we give of Proposition 7.1 is independent of the above, and this provides a proof of (7.16), as remarked by Peetre [**194**], whose derivation we follow in (7.21)–(7.24). First we derive a general formula for (the Fourier transform of) the Weyl symbol of an operator.

LEMMA 7.2. *We have*

$$\hat{F}(q,p) = c_n\,\mathrm{tr}(e^{-i(q\cdot X+p\cdot D)}F(X,D)). \tag{7.17}$$

PROOF. We will apply the inversion formula on the Heisenberg group, which reads

$$g(t,q,p) = c_n\int_{-\infty}^{\infty}\mathrm{tr}(\pi_\lambda(-t,-q,-p)\pi_\lambda(g))|\lambda|^n\,d\lambda. \tag{7.18}$$

Now, if $(\mathcal{F}_1 g)(\lambda,q,p)$ denotes the Fourier transform with respect to the first variable, we have, setting $\lambda=1$,

$$(\mathcal{F}_1 g)(1,q,p) = c_n\,\mathrm{tr}(e^{-i(q\cdot X+p\cdot D)}\pi_1(g)). \tag{7.19}$$

Now we can apply this to any case where $\pi_1(g) = F(X,D)$, and given any $F(X,D)$ there exist many such g. Since according to (3.16) we have $F(y,\eta) = \tilde{g}(1,y,\eta)$, we get (7.17), upon taking Fourier transforms.

We now turn to a second proof of Proposition 7.1. We have already reduced to the case $n=1$. Now (7.17) implies

$$\hat{h}_t(q,p) = c_n\,\mathrm{tr}(\pi_1(0,q,p)^* e^{-tH}). \tag{7.20}$$

In order to calculate this trace, one would naturally choose a basis for $L^2(R)$ diagonalizing H, i.e., the basis $h_j(x)$ given by (6.21). We can avoid a lot of combinatorial work in evaluating the result if we instead use the Bargmann-Fok representation β_1, defined by (5.5). Note that

$$\pi_1(0,q,p)f(H) = K^{-1}\beta_1(0,q,p)f(W)K, \tag{7.21}$$

where K is the unitary operator given by (5.8) and W is given by (6.32), i.e.,

$$W = 2\varsigma\partial/\partial\varsigma + 1 \quad (\text{for } n = 1). \tag{7.22}$$

In this case, recall the orthonormal basis of $\mathcal{H}$ diagonalizing W is

$$w_j(\varsigma) = (2/j!)^{1/2}\varsigma^j. \tag{7.23}$$

Hence, if (q,p) is identified with $z = q + ip \in \mathbf{C}$,

$$\begin{aligned} &c_n^{-1}\hat{h}_t(z) \\ &= \operatorname{tr}(\beta_1(0,-z)e^{-tW}) \\ &= \sum_{j=0}^{\infty} e^{-(2j+1)t}(\beta_1(0,-z)w_j, w_j) \\ &= (\pi i/2)\iint_{\mathbf{C}} e^{-|z|^2/2 - i\varsigma z\sqrt{2}} \\ &\qquad \cdot \left(\sum_{j=0}^{\infty} e^{-(2j+1)t}\bar{\varsigma}^j(\varsigma - i\bar{z}/\sqrt{2})^j/j!\right) e^{-|\varsigma|^2/2}\, d\varsigma \\ &= (\pi i/2)\iint_{\mathbf{C}} e^{-|z|^2/2}e^{-t}\exp[e^{-2t}\bar{\varsigma}(\varsigma - i\bar{z}/\sqrt{2}) - i\varsigma z/\sqrt{2} - |\varsigma|^2/2]\, d\varsigma \\ &= \pi e^{-|z|^2/2}e^{-t}\exp[-e^{-2t}|z|^2/2(1-e^{-2t})]. \end{aligned} \tag{7.24}$$

This proves (7.12), and (7.11) follows by taking Fourier transforms.

A third proof of Proposition 7.1 will be given after the proof of Lemma 7.8.

We can use the same multiplicativity argument as above to deduce that, if

$$Q(X,D) = \sum_{j=1}^{n} \mu_j(-\partial^2/\partial x_j^2 + x_j^2), \qquad \mu_j > 0, \tag{7.25}$$

then

$$e^{-tQ(X,D)} = h_t^Q(X,D), \tag{7.26}$$

with

$$h_t^Q(x,\xi) = \prod_{j=1}^{n}(\cosh t\mu_j)^{-1}\exp\left[-\sum_{j=1}^{n}(x_j^2+\xi_j^2)(\tanh t\mu_j)\right] \tag{7.27}$$

and

$$\hat{h}_t^Q(q,p) = c_n'\prod_{j=1}^{n}(\sinh t\mu_j)^{-1}\exp\left[-\sum_{j=1}^{n}(q_j^2+p_j^2)(\coth t\mu_j)/4\right]. \tag{7.28}$$

We can use Proposition 7.2 to analyze the solution operator

$$e^{s\mathcal{L}_0} \tag{7.29}$$

to the "Heisenberg group heat equation"

$$\partial u/\partial s = \mathcal{L}_0 u. \tag{7.30}$$

In fact, by (3.16), we have

$$e^{s\mathcal{L}_0}\delta_0(t,q,p) = k_s(t,q,p) \tag{7.31}$$

with

$$\begin{aligned}\tilde{k}_s(\pm\tau,y,\eta) &= \sigma_{e^s\mathcal{L}_0}(\pm\tau)(\pm\tau^{-1/2}y,\tau^{-1/2}\eta)\\ &= h_{s\tau}(\pm\tau^{-1/2}y,\tau^{-1/2}\eta).\end{aligned} \tag{7.32}$$

Hence, if

$$(\mathcal{F}_1 k_s)(\lambda,q,p) = \int_{-\infty}^{\infty} e^{-it\lambda}k_s(t,q,p)\,dt, \tag{7.33}$$

we have, by (7.12),

$$(\mathcal{F}_1 k_s)(\tau,q,p) = c_n\tau^n(\sinh s\tau)^{-n}\exp[-(\tau\coth s\tau)(|q|^2+|p|^2)/4], \tag{7.34}$$

which implies

(7.35)

$$\begin{aligned}k_s(t,q,p) &= c_n\int_{-\infty}^{\infty} e^{i\tau t}\tau^n(\sinh s\tau)^{-n}\exp[-(\tau\coth s\tau)(|q|^2+|p|^2)/4]\,d\tau\\ &= c_n s^{-n-1}\int_{-\infty}^{\infty} e^{i\tau(t/s)}(\tau/\sinh\tau)^n\exp[-(\tau\coth\tau)(|q|^2+|p|^2)/4s]\,d\tau.\end{aligned}$$

Note that the right side of (7.35) is equal to $s^{-n-1}k_1(t/s,q/\sqrt{s},p/\sqrt{s})$, with

$$k_1(t,q,p) = c_n\int_{-\infty}^{\infty} e^{it\tau}(\tau/\sinh\tau)^n\exp[-(\tau\coth\tau)(|q|^2+|p|^2)/4]\,d\tau, \tag{7.36}$$

a smooth rapidly decreasing function on R^{2n+1}, i.e., an element of the Schwartz space $\mathcal{S}(R^{2n+1})$. This gives the following conclusions for the "heat kernel."

PROPOSITION 7.3. *We have, for $s>0$,*

$$e^{s\mathcal{L}_0}\delta_0(t,q,p) = k_s(t,q,p), \tag{7.37}$$

with

$$k_s(t,q,p) = s^{-n-1}k_1(t/s,q/\sqrt{s},p/\sqrt{s}), \tag{7.38}$$

and $k_1\in\mathcal{S}(R^{2n+1})$ given by (7.36).

This result was obtained by B. Gaveau [**70**], by a different method, utilizing a diffusion process construction. See also Hulanicki [**127**].

We can more generally construct the kernel of the "heat semigroup" e^{-tP_α} for an operator P_α of the form (7.2), with $|\mathrm{Re}\,\alpha|$ sufficiently small, as follows.

First, as noted in §6, by choosing an appropriate automorphism of $\mathbf{H}^n$, we can suppose

$$P_0 = \sum_{j=1}^{n} \mu_j (L_j^2 + M_j^2), \tag{7.39}$$

and then (7.4) holds with $Q(X, D)$ given by (7.25). Thus we have

$$e^{sP_0}\delta_0(t,q,p) = k_s^Q(t,q,p), \tag{7.40}$$

with

$$\begin{aligned} \hat{k}_s^Q(\pm\tau, y, \eta) &= \sigma_{e^{sP_0}}(\pm\tau)(\pm\tau^{-1/2}y, \tau^{-1/2}\eta) \\ &= h_{s\tau}^Q(\pm\tau^{-1/2}y, \tau^{-1/2}\eta), \end{aligned} \tag{7.41}$$

where $h_{s\tau}^Q(x,\xi)$ is given by (7.27). Hence if $(\mathcal{F}_1 k_s^Q)(\tau, q, p)$ denotes the partial Fourier transform of k_s^Q with respect to the first variable, we have, by (7.28),

$$(\mathcal{F}_1 k_s^Q)(\tau,q,p) = c_n \prod_{j=1}^{n} (\tau/\sinh s\tau\mu_j) \exp\left[-\sum_{j=1}^{n} (\tau \coth s\tau\mu_j)(q_j^2 + p_j^2)/4\right]. \tag{7.42}$$

As before, we obtain

PROPOSITION 7.4. *With P_0 given by (7.39), we have, for $s > 0$,*

$$e^{sP_0}\delta_0(t,q,p) = k_s^Q(t,q,p), \tag{7.43}$$

with

$$k_s^Q(t,q,p) = s^{-n-1} k_1^Q(t/s, q/\sqrt{s}, p/\sqrt{s}), \tag{7.44}$$

and $k_1^Q \in \mathcal{S}(R^{2n+1})$ given by

$$\begin{aligned} k_1^Q(t,q,p) = c_n \int_{-\infty}^{\infty} & e^{it\tau} \prod_{j=1}^{n} (\tau/\sinh \mu_j\tau) \\ & \cdot \exp\left[-\sum_{j=1}^{n} (\tau\coth\mu_j\tau)(q_j^2 + p_j^2)/4\right] d\tau. \end{aligned} \tag{7.45}$$

It is desirable to express this kernel in invariant form, not depending on the coordinate system chosen to implement (7.39). We do this using the Hamilton map F_Q associated with the quadratic form Q, as defined in §6, i.e.,

$$\sigma(u, F_Q v) = Q(u,v), \qquad u, v \in R^{2n}. \tag{7.46}$$

From the discussion of F_Q in the proof of Lemma 6.2, it follows that

$$\det \sinh(\tau/i) F_Q = -\left(\prod_{j=1}^{n} \sinh \mu_j \tau\right)^2, \tag{7.47}$$

so

$$\prod_{j=1}^{n} (\tau / \sinh \mu_j \tau) = (-\tau^{-2n} \det \sinh(\tau/i) F_Q)^{-1/2}. \tag{7.48}$$

Now let

$$A_Q = (-F_Q^2)^{1/2} \tag{7.49}$$

be the unique square root of the matrix $-F_Q^2$ with positive spectrum, and define a quadratic form γ on R^{2n} by

$$\gamma(A_Q z, z) = Q(z,z).$$

We see that, in the symplectic coordinate system on R^{2n} such that $Q(z,z) = \sum \mu_j(q_j^2 + p_j^2)$, $z = (q,p)$, we have $\gamma(z,z) = \sum(q_j^2+p_j^2)$, and thus $\gamma(f(A_Q)z,z) = \sum f(\mu_j)(q_j^2+p_j^2)$. Consequently,

$$\begin{aligned} -(\tau/2)\sum_{j=1}^{n}(\coth \mu_j\tau)(q_j^2+p_j^2) &= -(\tau/2)\gamma(\coth \tau A_Q z, z) \\ &= -(\tau/2) Q(A_Q^{-1} \coth \tau A_Q z, z). \end{aligned} \tag{7.50}$$

Thus we can rewrite the heat kernel (7.45) invariantly as

$$k_1^Q(t,z) = c_n \int_{-\infty}^{\infty} e^{it\tau} \Phi_Q(\tau, z)\, d\tau \tag{7.51}$$

with

(7.52)

$$\Phi_Q(\tau,z) = (-\tau^{-2n} \det \sinh(\tau/i)F_Q)^{-1/2} \exp[-(\tau/4)Q(A_Q^{-1}\coth \tau A_Q z, z)].$$

Note that, if $P_\alpha = P_0 + i\alpha T$,

$$\begin{aligned} e^{sP_\alpha}\delta_0(t,q,p) &= e^{sP_0}\delta_0(t+is\alpha, q, p) \\ &= s^{-n-1} k_1^Q((t/s)+i\alpha, q/\sqrt{s}, p/\sqrt{s}) \end{aligned} \tag{7.53}$$

where $k_1^Q(t+i\alpha,q,p)$ is defined from (7.51) by analytic continuation, as long as

$$|\operatorname{Re}\alpha| < \sum_{j=1}^{n} \mu_j. \tag{7.54}$$

We proceed now to some general observations about functions of the harmonic oscillator Hamiltonian $H = -\Delta + |x|^2$. It is no accident that the symbol $h_t(x,\xi)$ of e^{-tH}, given by (7.11), is a function of $|x|^2 + |\xi|^2$. This result, for general $f(H)$, follows by symmetry, using the metaplectic representation discussed in §4. Recall the formula

$$a_g(x,\xi) = a(j(g)(x,\xi)) \Rightarrow a_g(X,D) = \tilde{\omega}(g)^{-1} a(X,D)\tilde{\omega}(g), \tag{7.55}$$

with $g \in \widetilde{\mathrm{Sp}(n,R)}$ and $j\colon \widetilde{\mathrm{Sp}(n,R)} \to \mathrm{Sp}(n,R)$ the covering homomorphism, where the symplectic group acts on R^{2n}, preserving the symplectic form

$$\sigma((x,\xi),(x',\xi')) = x\cdot\xi' - x'\cdot\xi.$$

Note that we can write (noncanonically), $R^{2n} = \mathbf{C}^n$ with $z = x + i\xi$, and then $\sigma(z, z') = \operatorname{Im} z \cdot \bar{z}'$. This makes it clear that the unitary group $\mathrm{U}(n)$ on $\mathbf{C}^n$ preserves the symplectic form. Since the unitary group acts transitively on the unit sphere in $\mathbf{C}^n$, we have immediately

PROPOSITION 7.5. *The symbol $a(x, \xi)$ is a function of $|x|^2 + |\xi|^2$ alone if and only if $a(X, D)$ commutes with $\tilde{\omega}(g)$ for all $g \in j^{-1}\mathrm{U}(n)$.*

We will suppose the amplitude $a(x, \xi)$ is sufficiently well behaved. One tacit restriction we make is $a(X, D)$ maps S to S and S' to S'. We define Rad to consist of $a(x, \xi)$ of the form $b(|x|^2 + |\xi|^2)$ and OPRad to consist of the associated $a(X, D)$. Proposition 7.5 yields

PROPOSITION 7.6. *If $f\colon R^+ \to R$ is polynomially bounded, then $f(H) \in$ OPRad.*

In fact, the converse is true.

PROPOSITION 7.7. *If $A \in$ OPRad, then $A = f(H)$ for some (polynomially bounded) $f(\lambda)$.*

PROOF. Note that under the representation $\tilde{\omega}$ of $j^{-1}\mathrm{U}(n)$ on $L^2(R^n)$, each eigenspace of H is invariant. We claim that any $A \in$ OPRad acts as a scalar on each eigenspace of H, which, by Proposition 7.5, is equivalent to the assertion that

$$\tilde{\omega} j^{-1}\mathrm{U}(n) \text{ acts irreducibly on each eigenspace of } H. \tag{7.56}$$

In order to establish (7.56), it is convenient to work with the form of the metaplectic representation associated with the Bargmann-Fok representation of $\mathbf{H}^n$, namely, let

$$\omega'(g) = K\tilde{\omega}(g)K^{-1}. \tag{7.57}$$

Then $\omega' j^{-1}\mathrm{U}(n)$ acts on each eigenspace of $W = KHK^{-1}$, and we have seen (recall (6.16)), that these eigenspaces (associated to eigenvalue $2j + 1$) consist precisely of the polynomials in ς, homogeneous of degree j.

LEMMA 7.8. *The action of $\omega' j^{-1}\mathrm{U}(n)$ on $\mathcal{H}$ is given up to scalar multiplication by*

$$\{\rho(g)\colon g \in \mathrm{U}(n)\}, \tag{7.58}$$

where

$$\rho(g)f(\varsigma) = f(g^{-1}\varsigma), \qquad f \in \mathcal{H}. \tag{7.59}$$

PROOF. The infinitesimal generator of any one parameter group $f(\varsigma) \mapsto f(e^{itA}\varsigma)$ is a first order differential operator in $\partial/\partial\varsigma_j$, $1 \le j \le n$, with coefficients linear in ς_k, which is intertwined by K with some operator $Q(X, D)$, of the form

(4.12). In fact, the Lie algebra of such infinitesimal generators (with A selfadjoint on $\mathbf{C}^n$) has dimension n^2 and is spanned by

$$L'_{jk} = i(\varsigma_j\partial/\partial\varsigma_k + \varsigma_k\partial/\partial\varsigma_j), \qquad M'_{jk} = \varsigma_j\partial/\partial\varsigma_k - \varsigma_k\partial/\partial\varsigma_j. \tag{7.60}$$

These operators are intertwined by K to

$$L_{jk} = i\lambda_{jk}(X,D), \qquad M_{jk} = i\mu_{jk}(X,D), \tag{7.61}$$

where

$$\lambda_{jk}(x,\xi) = x_jx_k + \xi_j\xi_k, \qquad \mu_{jk}(x,\xi) = x_j\xi_k - x_k\xi_j. \tag{7.62}$$

In turn it is easy to verify that the vector fields $H_{\lambda_{jk}}$ and $H_{\mu_{jk}}$ generate $\mathrm{U}(n)$ in $\mathrm{Sp}(n,R)$. All the functions (7.62) have zero Poisson bracket with $|x|^2+|\xi|^2$, and $\mathrm{U}(n)$ is precisely the subgroup of $\mathrm{Sp}(n,R)$ leaving invariant the quadratic form $|x|^2+|\xi|^2$. This proves that every element of (7.60) comes from some element of the Lie algebra of $j^{-1}\mathrm{U}(n)$ via the homomorphism ω', which is enough to establish the lemma.

Now the unitary representation ρ of $\mathrm{U}(n)$ on $\mathcal{H}$, given by (7.59), restricts to a unitary representation on $\mathcal{P}_j$, the space of polynomials in ς homogeneous of degree j, which we shall label D_j. Then (7.56) is equivalent to the irreducibility of D_j on $\mathcal{P}_j$. This, finally, is a classical result in the representation theory of the unitary group $\mathrm{U}(n)$, which will be proved in Chapter 2, so Proposition 7.7 is proved.

We now give a third proof of Proposition 7.1, which is taken from Appendix A of [**235**]. We can deduce from Proposition 7.6 that $h_t(x,\xi)$ is a function of $|x|^2+|\xi|^2$. If we set

$$b_t(X,D) = Hh_t(X,D), \tag{7.63}$$

then we can deduce a formula for $b_t(x,\xi)$. In fact, for a general quadratic $Q(X,D)$, and a general Weyl operator $a(X,D)$, a straightforward calculation shows

$$Q(X,D)a(X,D) = b(X,D) \tag{7.64}$$

with

$$\begin{aligned} b(x,\xi) &= Q(x,\xi)a(x,\xi) - \tfrac{i}{2}\{Q,a\}(x,\xi) \\ &\quad - \tfrac{1}{8}\textstyle\sum_k(\partial^2/\partial y_k\partial\xi_k - \partial^2/\partial x_k\partial\eta_k)^2Q(x,\xi)a(y,\eta) \end{aligned} \tag{7.65}$$

where the last term is evaluated at $y=x$, $\eta=\xi$. In our case, where $Q(x,\xi) = |x|^2+|\xi|^2$, we have $\{Q,h_t\}=0$, so

$$b_t(x,\xi) = Q(x,\xi)h_t(x,\xi) - \tfrac{1}{4}\textstyle\sum_k(\partial^2/\partial x_k^2 + \partial^2/\partial\xi_k^2)h_t(x,\xi). \tag{7.66}$$

Hence $h_t(x,\xi)$ must satisfy the equation

$$\begin{aligned} (\partial/\partial t)h_t(x,\xi) = &- (|x|^2+|\xi|^2)h_t(x,\xi) \\ &+ \tfrac{1}{4}\textstyle\sum_k(\partial^2/\partial x_k^2 + \partial^2/\partial\xi_k^2)h_t(x,\xi). \end{aligned} \tag{7.67}$$

If we write

$$h_t(x,\xi) = g(t,Q), \qquad Q = |x|^2 + |\xi|^2, \tag{7.68}$$

then (7.67) becomes

$$\partial g/\partial t = -Qg + Q\partial^2 g/\partial Q^2 + n\partial g/\partial Q. \tag{7.69}$$

Now, having grown accustomed to (7.11), we make the "inspired guess" that

$$h_t(x,\xi) = a(t)e^{-b(t)(|x|^2+|\xi|^2)}, \quad \text{i.e.,} \quad g(t,Q) = a(t)e^{-b(t)Q}. \tag{7.70}$$

Then the left side of (7.69) is $(a'/a - b'Q)g$ and the right side is $(-Q + Qb^2 - nb)g$, so the identity (7.69) is equivalent to

$$a'(t)/a(t) = -nb(t) \quad \text{and} \quad b'(t) = 1 - b(t)^2. \tag{7.71}$$

We can solve the second equation for $b(t)$ by separation of variables. Since $h_0(x,\xi) = 1$, we need $b(0) = 0$, so the unique solution is seen to be $b(t) = \tanh t$. Then the equation $a'/a = -n\tanh t$, with $a(0) = 1$, gives $a(t) = (\cosh t)^{-n}$. This completes the third proof of the identity (7.11).

We now obtain the following general formula for an operator $f(H)$. An equivalent result was obtained by Peetre [**194**]; see also Miller [**175**], Geller [**75**], and Nachman [**181**].

PROPOSITION 7.9. *We have*

$$f(H) = c_n \sum_{j=0}^{\infty} f(2j+1)\tilde{\psi}_j(X,D) \tag{7.72}$$

with

$$\tilde{\psi}_j(x,\xi) = (-1)^j L_j^{n-1}(\tfrac{1}{2}|x|^2 + \tfrac{1}{2}|x|^2)e^{-(1/4)(|x|^2+|\xi|^2)}. \tag{7.73}$$

Here L_j^{n-1} is the Laguerre polynomial:

$$L_j^m(x) = e^x(x^{-m}/j!)(d^j/dx^j)(e^{-x}x^{m+j}). \tag{7.74}$$

For properties of Laguerre polynomials, see Lebedev [**154**], page 76. The one property we use here is the generating function identity

$$\sum_{j=0}^{\infty} L_j^m(x^2+\xi^2)\theta^j = (1-\theta)^{-m-1}\exp[-\theta(x^2+\xi^2)/(1-\theta)]. \tag{7.75}$$

In fact, this identity makes (7.73) an immediate consequence of (7.11). Similarly, (7.12) and the same generating function identity yield

$$f(H) = F(X,D) \tag{7.76}$$

with

$$\hat{F}(q,p) = c_n \sum_{j=0}^{\infty} f(2j+1)\psi_j(q,p), \tag{7.77}$$

where

$$\psi_j(q,p) = L_j^{n-1}(2|q|^2 + 2|p|^2)e^{-|q|^2-|p|^2}. \tag{7.78}$$

It seems that any concrete information on $f(H)$ obtainable from (7.72), (7.73), could be obtained just as easily from the formula (7.11) for e^{-tH}, by some kind of synthesis. For example, if we denote the resolvent of H by

$$(H+\alpha)^{-1} = r_\alpha(X,D), \tag{7.79}$$

then

(7.80)

$$\begin{aligned} r_\alpha(x,\xi) &= \int_0^\infty e^{-\alpha t} h_t(x,\xi)\,dt \\ &= \int_0^\infty e^{-\alpha t}(e^{-t}/(1+e^{-2t}))^n \\ &\qquad \cdot \exp[-(1-e^{-2t})(|x|^2+|\xi|^2)/2(1+e^{-2t})]\,dt \\ &= \frac{1}{2}\int_0^1 y^{-1+(\alpha+n)/2}(1+y)^{-n}\exp[-(1-y)(|x|^2+|\xi|^2)/2(1+y)]\,dy \\ &= 2^{-n}\int_0^1 (1-s)^{-1+(\alpha+n)/2}(1+s)^{-n-1-(\alpha+n)/2}e^{-s(|x|^2+|\xi|^2)/2}\,ds, \end{aligned}$$

where the identity is first seen to be valid for $\operatorname{Re}\alpha > 0$. We leave it as an exercise to the reader to analytically continue (7.80) to all complex α not an integer of the form $-2j-n$, i.e., not in the spectrum of $-H$.

One can also analytically continue the formulas associated with e^{-tH} to analyze e^{itH}, $t \in R$, and synthesize operators from this unitary group. This is discussed in Taylor [**235**].

Formula (7.80) could be used to get a formula for

$$\mathcal{L}_\alpha^{-1}\delta_0(t,q,p) = l_\alpha(t,q,p). \tag{7.81}$$

Alternatively, as pointed out by Gaveau [**70**], we can get a formula for l_α from the identity

$$\mathcal{L}_\alpha^{-1} = \int_0^\infty e^{s\mathcal{L}_\alpha}\,ds, \qquad |\operatorname{Re}\alpha| < n, \tag{7.82}$$

as follows. The formulas (7.36)–(7.38) give

$$\begin{aligned} l_\alpha(t,q,p) &= \int_0^\infty s^{-n-1}k_1((t/s)+i\alpha, q/\sqrt{s}, p/\sqrt{s})\,ds \\ &= \int_0^\infty \int_{-\infty}^\infty s^{-n-1}e^{i\tau[(t/s)+i\alpha]}(\tau/\sinh\tau)^n \\ &\qquad \cdot e^{-(\tau\coth\tau)(|q|^2+|p|^2)/2s}\,d\tau\,ds. \end{aligned} \tag{7.83}$$

If we first integrate with respect to s and use

$$\begin{aligned} \int_0^\infty s^{-n-1}e^{-\beta/s}\,ds &= -\int_0^\infty x^{n-1}e^{-\beta x}\,dx \\ &= -(n-1)!\beta^{-n}, \end{aligned} \tag{7.84}$$

we get, for $|\mathrm{Re}\,\alpha| < n$,
(7.85)
$$\begin{aligned} l_\alpha(t,q,p) &= c_n' \int_{-\infty}^{\infty} e^{-\alpha\tau}(\tau/\sinh\tau)^n[(\tau\coth\tau)(|q|^2+|p|^2)/2 - it\tau]^{-n}\,d\tau \\ &= c_n' \int_{-\infty}^{\infty} e^{-\alpha\tau}[(\cosh\tau)(|q|^2+|p|^2)/2 - i(\sinh\tau)t]^{-n}\,d\tau. \end{aligned}$$

Note the mixed homogeneity:

$$l_\alpha(rt, r^{1/2}q, r^{1/2}p) = r^{-n} l_\alpha(t,q,p). \tag{7.86}$$

We leave it as an exercise to the reader to verify that (7.85) defines a function smooth for $(t,q,p) \neq (0,0,0)$, smooth even near the rays $q = p = 0$, $t \neq 0$. We also leave it as an exercise to analytically continue (7.85) to $\alpha \in \mathbf{C}$ such that $\pm\alpha$ avoids the set $\{n+2j : j = 0,1,2,\ldots\}$. Integration by parts is effective. It is perhaps convenient to write (7.85) as

$$\begin{aligned} l_\alpha(t,q,p) = c_n' \int_{-\infty}^{\infty} &[(x^2+1)^{1/2} - x]^\alpha[(x^2+1)^{1/2}(|q|^2+|p|^2)/2 - ixt]^{-n} \\ &\cdot (x^2+1)^{-1/2}\,dx. \end{aligned} \tag{7.87}$$

Equivalent formulas were given in Folland and Stein [**64**]. The smoothness of (7.85) away from $(t,q,p) = (0,0,0)$ implies that, if $f \in \mathcal{E}'(\mathbf{H}^n)$, then $\mathcal{L}_\alpha^{-1} f$ is smooth where f is. Hence if $u \in \mathcal{E}'$ and $\mathcal{L}_\alpha u = f$, then u is smooth where f is, i.e., $\mathcal{L}_\alpha$ is *hypoelliptic*.

We can also calculate the kernel

$$(-\mathcal{L}_0)^{-1/2} e^{-s(-\mathcal{L}_0)^{1/2}} \delta_0(t,q,p) = P_s(t,q,p) \tag{7.88}$$

from the heat kernel (7.35), using the *subordination identity* (which follows by taking the A-derivative of the formula (I.11) in the Introduction):

$$A^{-1}e^{-yA} = \pi^{-1/2}\int_0^\infty e^{-y^2/4t} e^{-tA^2} t^{-1/2}\,dt, \tag{7.89}$$

with $y = s$, $A = (-\mathcal{L}_0)^{1/2}$. We get

$$\begin{aligned} P_s(t,q,p) &= \pi^{-1/2}\int_0^\infty e^{-s^2/4v} v^{-1/2} k_v(t,q,p)\,dv \\ &= c_n \int_0^\infty \int_{-\infty}^{\infty} v^{-n-3/2} e^{-s^2/4v} e^{it\tau/v} (\tau/\sinh\tau)^n \\ &\qquad \cdot e^{-(\tau\coth\tau)(|q|^2+|p|^2)/2v}\,d\tau\,dv, \end{aligned} \tag{7.90}$$

and if we first do the v integral, using (7.84), we get
(7.91)
$$\begin{aligned} &P_s(t,q,p) \\ &\quad = c_n' \int_{-\infty}^{\infty} (\tau/\sinh\tau)^n\left[s^2/4 + (\tau\coth\tau)(|q|^2+|p|^2)/2 - it\tau\right]^{-n-1/2}\,d\tau. \end{aligned}$$

We see that P_s is smooth in (t, q, p), for $s > 0$. Note the homogeneity

$$P_s(t,q,p) = s^{-2n-1} P_1(t/s^2, q/s, p/s). \tag{7.92}$$

Note that (7.91) continues naturally for complex s such that $|\arg s| < \pi/4$. Note that $e^{-s(-\mathcal{L}_0)^{1/2}}$ is a holomorphic semigroup for $\operatorname{Re} s \geq 0$, and it is desirable to understand its kernel, particularly for $s = i\sigma$, pure imaginary. Analytic continuation of (7.91) and application to the study of the wave equation on the Heisenberg group will be made in the next section.

We end this section with a brief look at spectral asymptotics for the Heisenberg Laplacian $\mathcal{L}_0$ and certain other operators, on compact quotients of $\mathbf{H}^n$. Let $\Gamma \subset \mathbf{H}^n$ be a discrete subgroup such that $\mathbf{H}^n/\Gamma$ is compact. Such co-compact groups appear naturally if one considers $\tilde{\mathbf{H}}^n$, isomorphic to $\mathbf{H}^n$, as the group of matrices (1.32). One can take $\tilde{\Gamma}$ to be the set of such matrices such that t, q, p have all integer entries, for example, in which case $\tilde{\Gamma}$ is a discrete subgroup of $\tilde{\mathbf{H}}^n$ such that $\tilde{\mathbf{H}}^n/\tilde{\Gamma}$ is diffeomorphic to a circle bundle over $\mathbf{T}^{2n}$. The formula for $e^{s\mathcal{L}_0}$ on $\mathbf{H}^n/\Gamma$ is obtained from that on $\mathbf{H}^n$ by the standard method of images (also called Poisson summation). If $k_s^\#(t,q,p)$ denotes $e^{s\mathcal{L}_0}\delta_0(t,q,p)$ on $\mathbf{H}^n/\Gamma$, we have

$$k_s^\#(t,q,p) = \sum_{\gamma \in \Gamma} k_s(\gamma \cdot (t,q,p)) \tag{7.93}$$

where $k_s(t, q, p)$ is given by (7.35). The rapid decrease of this function implies the sum in (7.93) is convergent, to a smooth function, for $s > 0$. Thus, for $s > 0$, $e^{s\mathcal{L}_0}$ is a smoothing operator. In particular it is a compact selfadjoint operator in $L^2(\mathbf{H}^n/\Gamma)$.

We deduce that $\mathcal{L}_0$ has a discrete spectrum tending to $-\infty$, on $L^2(\mathbf{H}^n/\Gamma)$. We can study the asymptotic distribution of the eigenvalues of $\mathcal{L}_0$ by examining the asymptotic behavior as $s \downarrow 0$ of

$$\operatorname{tr} e^{s\mathcal{L}_0} = \sum_j e^{-\mu_j s} \tag{7.94}$$

where $\{-\mu_j\}$ is the set of eigenvalues of $\mathcal{L}_0$, repeated according to multiplicity. Since $\mathbf{H}^n/\Gamma$ is homogeneous, we have

$$\operatorname{tr} e^{s\mathcal{L}_0} = (\operatorname{vol} \mathbf{H}^n/\Gamma) k_s^\#(0,0,0). \tag{7.95}$$

In light of (7.93) and the behavior of $k_s(t, q, p)$, we deduce

$$\operatorname{tr} e^{s\mathcal{L}_0} = (\operatorname{vol} \mathbf{H}^n/\Gamma) k_s(0,0,0) + O(s^\infty), \qquad s \downarrow 0. \tag{7.96}$$

Thus, from (7.39), we have

$$\operatorname{tr} e^{s\mathcal{L}_0} = \mathfrak{x}_n (\operatorname{vol} \mathbf{H}^n/\Gamma) s^{-n-1} + O(s^\infty), \qquad s \downarrow 0, \tag{7.97}$$

as a complete asymptotic expansion for the trace, where

$$\mathfrak{x}_n = k_1(0,0,0) = c_n \int_{-\infty}^{\infty} (\tau/\sinh \tau)^n \, d\tau. \tag{7.98}$$

This is a special case of "heat asymptotics" for some general classes of subelliptic operators, studied in greater generality in [**171**, **15**, **235**]. More generally, one gets an infinite sum involving increasing powers of s, rather than just one term. This is analogous to the complete asymptotic expansion for the trace of the ordinary heat kernel on a general compact Riemannian manifold (see [**166**]), compared to that for the torus $\mathbf{T}^n$.

A standard Tauberian argument, due to Karamata (see, e.g., [**234**], Chapter XII), shows that (7.97) has the following implication on the counting function

$$N(\mu) = \text{cardinality of } \{-\mu_j \in \operatorname{spec} \mathcal{L}_0 \colon \mu_j < \mu\}. \tag{7.99}$$

Namely,

$$\lim_{\mu \to \infty} \mu^{-n-1} N(\mu) = \mathfrak{r}_n (\operatorname{vol} \mathbf{H}^n/\Gamma)/\Gamma(n+2). \tag{7.100}$$

On $\mathbf{H}^n/\Gamma$ is also a family of Riemannian metrics, depending on a parameter ε, degenerating (or rather exploding) as $\varepsilon \downarrow 0$, characterized by having associated Laplace operators

$$L_\varepsilon = \mathcal{L}_0 + \varepsilon T^2. \tag{7.101}$$

It is desirable to consider the spectral asymptotics of L_ε as $\varepsilon \downarrow 0$. The analysis given here arose in conversations between the author and Jeff Cheeger. We will produce a uniform analysis of

$$\operatorname{tr} e^{sL_\varepsilon} \tag{7.102}$$

for $\varepsilon \in [0, E]$, a compact interval in $\overline{R^+}$.

The same computations giving the kernel of $e^{s\mathcal{L}_0}$ on $\mathbf{H}^n$ in Proposition 7.3 show that, on $\mathbf{H}^n$,

$$e^{sL_\varepsilon} \delta_0(t, q, p) = k_{\varepsilon,s}(t, q, p) \tag{7.103}$$

with

(7.104)

$$\begin{aligned} k_{\varepsilon,s}(t,q,p) &= c_n s^{-n-1} \int_{-\infty}^{\infty} e^{i\tau(t/s)} (\tau/\sinh\tau)^n e^{-(\tau\coth\tau)(|q|^2+|p|^2)/s} e^{-\varepsilon\tau^2/s}\, d\tau \\ &= s^{-n-1} \kappa(t/s, q/\sqrt{s}, p/\sqrt{s}, \varepsilon/s), \end{aligned}$$

where

$$\kappa(t,q,p,\eta) = c_n \int_{-\infty}^{\infty} e^{i\tau t} (\tau/\sinh\tau)^n e^{-(\tau\coth\tau)(|q|^2+|p|^2)} e^{-\eta\tau^2}\, d\tau. \tag{7.105}$$

Now, if $\mathbf{H}^n/\Gamma$ is compact, the same Poisson summation arguments used above show that

$$\operatorname{tr} e^{sL_\varepsilon} = (\operatorname{vol} \mathbf{H}^n/\Gamma) k_{\varepsilon,s}(0,0,0) + O(s^\infty), \qquad s \downarrow 0, \tag{7.106}$$

uniformly for $\varepsilon \in [0, E]$. Here, the volume element on $\mathbf{H}^n/\Gamma$ is induced from a fixed Haar measure on $\mathbf{H}^n$. The degenerating metrics give a family of volume elements proportional to this, with a scaling by a factor of $\varepsilon^{-1/2}$. In view of (7.104), we have

$$k_{\varepsilon,s}(0,0,0) = s^{-n-1}\kappa(0,0,0,\varepsilon/s), \tag{7.107}$$

so we have

$$\operatorname{tr} e^{sL_\varepsilon} = c_n(\operatorname{vol}\mathbf{H}^n/\Gamma)s^{-n-1}\psi(\varepsilon/s) + O(s^\infty), \qquad s \downarrow 0, \tag{7.108}$$

where

$$\psi(\eta) = \int_{-\infty}^{\infty} (\tau/\sinh\tau)^n e^{-\eta\tau^2}\, d\tau, \qquad 0 \le \eta < \infty. \tag{7.109}$$

The behavior of $\psi(\eta)$ as $\eta \to \infty$ follows simply from Laplace's asymptotic method (see, e.g., [**187**]). We have

$$\psi(\eta) \sim \sum_{j\ge 1} c_j' \eta^{-j/2} \quad \text{as } \eta \to +\infty. \tag{7.110}$$

Of course, ψ is C^∞ on $[0,\infty)$. Again, our fundamental formula (7.108) is uniformly valid for $0 \le \varepsilon \le E$. For any $\varepsilon > 0$, the leading term is

$$c_1' c_n(\operatorname{vol}\mathbf{H}^n/\Gamma)\varepsilon^{-1/2}s^{-n-1/2}, \tag{7.111}$$

the usual leading term for the trace of the heat kernel on a compact Riemannian manifold of dimension $2n+1$ and Riemannian volume $\varepsilon^{-1/2}(\operatorname{vol}\mathbf{H}^n/\Gamma)$. The formula (7.108) shows precisely the transition between the behavior (7.111) and (7.97), as $\varepsilon \downarrow 0$.

8. The wave equation on the Heisenberg group. We discuss here solutions to the initial value problem for $u(s,g)$, $g \in \mathbf{H}^n$:

$$(\partial^2/\partial s^2 - \mathcal{L}_0)u = 0, \tag{8.1}$$

$$u(0,g) = f_1(g), \qquad (\partial/\partial s)u(0,g) = f_2(g). \tag{8.2}$$

If $f_1, f_2 \in \mathcal{E}'(\mathbf{H}^n)$, the solution is given by

$$u(s) = \cos s(-\mathcal{L}_0)^{1/2} f_1 + (-\mathcal{L}_0)^{-1/2}\sin s(-\mathcal{L}_0)^{1/2} f_2. \tag{8.3}$$

Note that

$$\begin{aligned} 2i(-\mathcal{L}_0)^{-1/2}\sin s(-\mathcal{L}_0)^{1/2} = (-\mathcal{L}_0)^{-1/2}&e^{is(-\mathcal{L}_0)^{1/2}} \\ &- (-\mathcal{L}_0)^{-1/2}e^{-is(-\mathcal{L}_0)^{1/2}}. \end{aligned} \tag{8.4}$$

Since $\cos s(-\mathcal{L}_0)^{1/2} = (\partial/\partial s)(-\mathcal{L}_0)^{-1/2}\sin s(-\mathcal{L}_0)^{1/2}$, we should get a good analysis of (8.3) knowing $(-\mathcal{L}_0)^{-1/2}e^{is(-\mathcal{L}_0)^{1/2}}$, $s \in R$. This we propose to obtain from the analysis of $(-\mathcal{L}_0)^{-1/2}e^{-s(-\mathcal{L}_0)^{1/2}}$ given in §7, by analytic continuation. An analysis of the Heisenberg group wave equation, from another point of view, was given by Nachman [**181**].

Recall formulas (7.79), (7.82); for $g = (t,z) = (t,q,p) \in \mathbf{H}^n$, we have

$$(-\mathcal{L}_0)^{-1/2}e^{-s(-\mathcal{L}_0)^{1/2}}\delta_0(t,z) = P_s(t,z) \tag{8.5}$$

with

$$P_s(t,z) = c_n'' \int_{-\infty}^{\infty} (\tau/\sinh\tau)^n [s^2 + 2|z|^2 \tau \coth\tau - 4it]^{-n-1/2}\, d\tau. \tag{8.6}$$

Let us set $\varsigma = -i\tau$ and write this as a complex contour integral

$$P_s(t,z) = c_n \int_\gamma f(\varsigma)[s^2 + g_{A,B}(\varsigma)]^{-n-1/2}\, d\varsigma \tag{8.7}$$

where the path γ is the pure imaginary axis, from $-i\infty$ to $+i\infty$,

$$f(\varsigma) = (\varsigma/\sin\varsigma)^n, \tag{8.8}$$

and

$$g_{A,B}(\varsigma) = B\varsigma\cot\varsigma + A\varsigma, \tag{8.9}$$

with

$$A = 4t, \qquad B = 2|z|^2. \tag{8.10}$$

We will drop the subscripts from $g_{A,B}(\varsigma)$, writing $g(\varsigma)$.

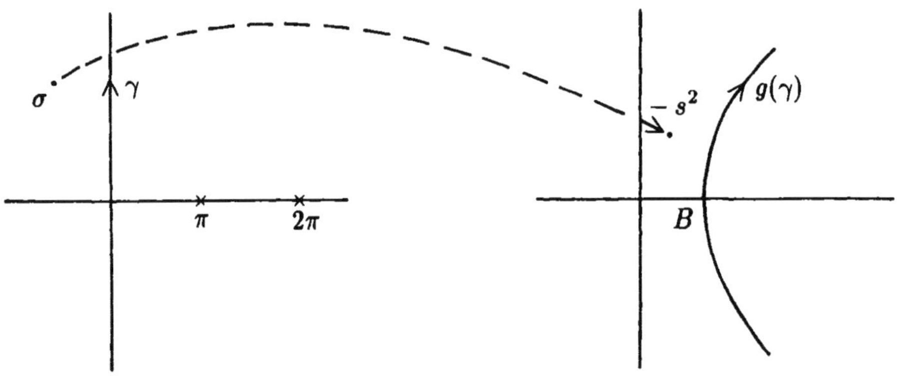

FIGURE 8.1

It is clear that (8.7) continues analytically in a straightforward manner from positive real s to all complex s such that $-s^2$ lies to the *left* of the image curve $g(\varsigma)$. (Note that, for $B = 0, g(\varsigma)$ is also the vertical axis, so the common domain of such straightforward analytic continuation is $\{s \in \mathbf{C}\colon |\arg s| < \pi/4\}$.) To continue analytically beyond this domain, we need to deform the contour γ.

As a simple case of this, consider the case $B = 0$, i.e., $z = 0$. Then we are considering

$$P_s(t,0) = c_n \int_\gamma f(\varsigma)(s^2 + A\varsigma)^{-n-1/2}\, d\varsigma. \tag{8.11}$$

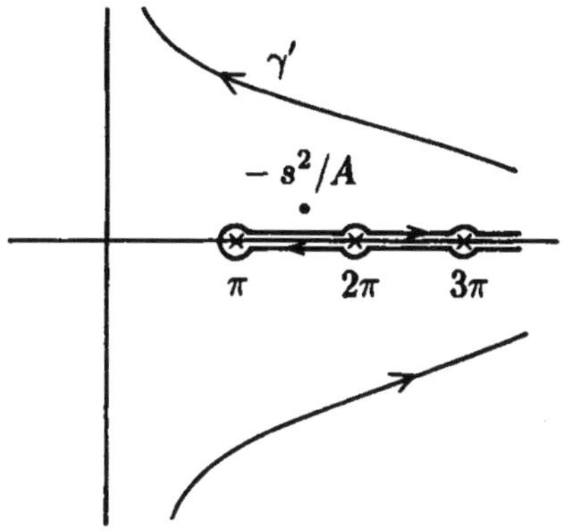

FIGURE 8.2

Now $f(\varsigma)$ has poles at $\varsigma = \pi, 2\pi, 3\pi, \ldots$, so we can deform γ as indicated in Figure 8.2.

It is clear that we can continue $P_s(t,0)$ to the entire half plane $\{s \in \mathbf{C} \colon \operatorname{Re} s > 0\}$, and, on the boundary of this region, i.e., the imaginary axis, $P_s(t,0)$ continues analytically as long as $-s^2/A$ avoids the points $j\pi$, $j = 1,2,3,\ldots$. Thus $P_{is}(t,0) = (-\mathcal{L}_0)^{-1/2} e^{is(-\mathcal{L}_0)^{1/2}} \delta_0(t,0)$ is analytic as long as $s^2 \neq 4|t|j\pi$, $j = 1,2,3,\ldots$. Note also that $P_{is}(t,0)$ and $P_{-is}(t,0)$ agree for $s^2/4|t| < \pi$, so $(-\mathcal{L}_0)^{-1/2} \sin s(-\mathcal{L}_0)^{1/2} \delta_0(t,0)$ vanishes for $|t| > s^2/4\pi$. This is a special case of the finite propagation speed which we will derive.

We want in general to deform the contour γ so that its image $g(\gamma')$ will hug a segment of the positive real axis. Let us note that, for $x \in \mathbf{R}$, $x \cot x$ is monotonically decreasing from 1 to $-\infty$ for $x \in [0,\pi)$, with derivative zero at $x = 0$, going to $-\infty$ as $x \to \pi$. Thus, assuming $A > 0$, for $x \in [0,\pi)$, $g(x)$ increases from $g(0) = B$ to a maximum at $x = x_0$ and then decreases to $-\infty$ as $x \to \pi$. The point $x_0 = x_0(A,B) \in [0,\pi)$ is defined by $g'(x_0) = 0$, i.e.,

$$A \sin^2 x_0 = B(x_0 - \sin x_0 \cos x_0), \tag{8.12}$$

since

$$g'(\varsigma) = [A \sin^2 \varsigma - B(\varsigma - \sin \varsigma \cos \varsigma)] / \sin^2 \varsigma. \tag{8.13}$$

Thus deforming the path γ so it crosses the real axis at x_0 and proceeds for a while along the curve orthogonal to the real axis through x_0 along which $g(\varsigma)$ is real-valued, effects a deformation of the contour $g(\gamma)$ to a curve which hugs a segment of the real axis $[E, E+\varepsilon)$. See Figure 8.3.

Here

$$E = E(A,B) = \max\{g_{A,B}(x) \colon 0 \leq x < \pi\}. \tag{8.14}$$

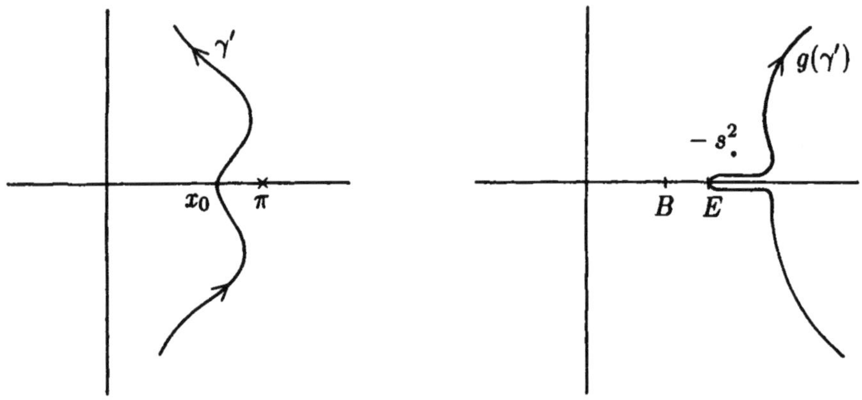

FIGURE 8.3

We see that $P_{is}(t,z)$ and $P_{-is}(t,z)$ agree for $s^2 < E$, which gives us our result on finite propagation speed:

PROPOSITION 8.1. *The fundamental solution of the wave equation*

$$(-\mathcal{L}_0)^{-1/2} \sin s(-\mathcal{L}_0)^{1/2}\delta_0(t,z)$$

vanishes for $E(4|t|, 2|z|^2) > s^2$, *where* $E(A,B)$ *is defined by* (8.14).

Note the simple estimate

$$E(A,B) \geq B + c_0 A \tag{8.15}$$

for some positive c_0.

We proceed to make a more global deformation of the path γ, to a path γ' on which the function $g(\varsigma)$ is real. Start at x_0 and go along the path γ' orthogonal to the real axis at x_0, on which $g(\varsigma)$ is real. Continue until you hit a point ς where $g'(\varsigma) = 0$. Then make a turn of $\pi/2$ counterclockwise (if $g'' \neq 0$ there) and continue, still keeping $g(\varsigma)$ real. In order to analyze what sort of path we get, it is useful to have the following result.

LEMMA 8.2. *If* $g(\varsigma)$ *is real and* $g'(\varsigma) = 0$, *then* ς *is real.*

PROOF. Note that

$$g(\varsigma) = (\varsigma/\sin\varsigma)(B\cos\varsigma + A\sin\varsigma) \tag{8.16}$$

and

$$g'(\varsigma) = (B\cos\varsigma + A\sin\varsigma - B\varsigma/\sin\varsigma)/\sin\varsigma. \tag{8.17}$$

Thus, if $g'(\varsigma) = 0$,

$$g(\varsigma) = B(\varsigma/\sin\varsigma)^2 = B^{-1}(B\cos\varsigma + A\sin\varsigma)^2. \tag{8.18}$$

In particular, if $g(\varsigma)$ is real, then $B\cos\varsigma + A\sin\varsigma$ is either real or pure imaginary. It follows that, if

$$\varsigma = u + iv, \qquad u, v \text{ real}, \tag{8.19}$$

then either ς is real, or

$$B\cos u + A\sin u = 0 \quad \text{or} \quad -B\sin u + A\cos u = 0. \tag{8.20}$$

Now, write

$$\begin{aligned} \cos\varsigma &= \cos u\cosh v - i\sin u\sinh v, \\ \sin\varsigma &= \sin u\cosh v + i\cos u\sinh v. \end{aligned} \tag{8.21}$$

If $g'(\varsigma) = 0$, we must have, by (8.17),

$$\varsigma = \cos\varsigma\sin\varsigma + (A/B)\sin^2\varsigma, \tag{8.22}$$

and substituting in (8.21) and equating real and imaginary parts, respectively, gives

$$\begin{aligned} u &= (\sin u\cos u + (A/B)\sin^2 u)\cosh^2 v \\ &\qquad + (\sin u\cos u - (A/B)\cos^2 u)\sinh^2 v, \\ v &= (\cos^2 u - \sin^2 u + 2(A/B)\sin u\cos u)\sinh v\cosh v. \end{aligned} \tag{8.23}$$

Now, the first possibility in (8.20) yields

$$u = -(A/B)[1 + (A/B)^2]\sin^2 u\sinh^2 v, \tag{8.24}$$

$$v = -[1 + (A/B)^2]\sin^2 u\sinh v\cosh v, \tag{8.25}$$

and dividing these equations gives

$$u = (A/B)v\tanh v. \tag{8.26}$$

Note that (8.24) implies $(B/A)u \leq 0$ and (8.26) implies $(B/A)u \geq 0$. Thus we see that the first possibility in (8.20) does not allow for a nonreal ς satisfying the hypotheses of the lemma.

On the other hand, the second possibility in (8.20) yields

$$u = (A/B)[1 + (B/A)^2]\sin^2 u\cosh^2 v, \tag{8.27}$$

$$v = [1 + (B/A)^2]\sin^2 u\sinh v\cosh v. \tag{8.28}$$

Also $\cos u = (B/A)\sin u$ implies

$$1 = [1 + (B/A)^2]\sin^2 u,$$

which, together with (8.28), yields

$$v/\sinh v = \cosh v. \tag{8.29}$$

But the left side of (8.29) is ≤ 1 and the right side is ≥ 1, with equality only at $v = 0$. Again we get no nonreal ς satisfying the hypotheses of the lemma, so the proof is complete.

It remains to study the real numbers x such that $g'_{A,B}(x) = 0$. We have already discussed the unique such point in the interval $[0, \pi)$, denoted $x_0(A, B)$. From the way $\varsigma \cot \varsigma$ decreases from $+\infty$ to $-\infty$ on any interval $(j\pi, (j+1)\pi)$, $j \geq 1$, one can see that on each such interval, $g'_{A,B}(x)$ vanishes either nowhere, or on a pair of points $x_j(A, B) < y_j(A, B)$, which for certain values of A and B coalesce to a double root $x_j(A, B) = y_j(A, B)$. For given A, B, there are only finitely many such points. These points approach the poles $\{k\pi\}$ as $B \to 0$.

The deformation of γ to take is now clear. Starting on the path described above through x_0, proceed into the complex domain on a path on which $g(\varsigma)$ is real, which either remains away from the real axis indefinitely and on which $g'(\varsigma)$ is never zero, or which returns to the real axis, say at $x_j(A, B)$. Proceed from $x_j(A, B)$ to $y_j(A, B)$ along the real axis, and then take off into the complex domain, along a path orthogonal to the real axis at $y_j(A, B)$ on which $g(\varsigma)$ is real. Continue in this fashion, to effect an analytic continuation of $P_s(t, z)$ to $\{s \in \mathbf{C} \colon \operatorname{Re} s > 0\}$. At regular points of this curve, it is possible to continue a little further, so we obtain our result on the singularities of the kernel $P_{is}(t, z)$.

THEOREM 8.3. *The fundamental solution of the wave equation,*

$$(-\mathcal{L}_0)^{-1/2} \sin s(-\mathcal{L}_0)^{1/2} \delta_0(t, z)$$

has singularities only where, for some j,

$$s^2 = g_{A,B}(x_j(A, B)) \quad \text{or} \quad g_{A,B}(y_j(A, B)), \tag{8.30}$$

where $g_{A,B}(\varsigma)$ *is given by* (8.9), $A = 4|t|$, $B = 2|z|^2$, *and the points* x_j, y_j *are the real zeros of* $g'_{A,B}(x)$.

We can also describe the fundamental solution to the wave equation

$$(\partial^2/\partial s^2 - \mathcal{L}_\alpha)u = 0, \tag{8.31}$$

for $|\operatorname{Re} \alpha| < n$. Indeed, from the formula (7.53) for the heat kernel $e^{s\mathcal{L}_\alpha}$, it is clear that

$$(-\mathcal{L}_\alpha)^{-1/2} e^{-s(-\mathcal{L}_\alpha)^{1/2}} \delta_0(t, z) = P_{\alpha,s}(t, z) \tag{8.32}$$

has the same representation as (8.7), with $f(\varsigma) = (\varsigma / \sin \varsigma)^n$ replaced by

$$f(\varsigma) = e^{i\alpha\varsigma}(\varsigma / \sin \varsigma)^n. \tag{8.33}$$

Consequently all the arguments above apply without change to this more general case, yielding the results given above on finite propagation speed and location of singularities.

CHAPTER 2

The Unitary Group

The unitary groups $\mathrm{U}(n)$ have the simplest structure of all the compact Lie groups (other than the tori T^k). We will give a full account in this chapter of the representations of SU(2), and also of SU(3), granted some arguments to be presented in Chapter 3. This chapter also presents the basics of the representation theory of $\mathrm{U}(n)$ for general n. A complete parametrization of the irreducible representations of $\mathrm{SU}(n)$ for general n will be given in Chapter 3, as an application of some general results about compact Lie groups, which will be seen to generalize the development of §2 of this chapter. More detailed accounts of $\mathrm{SU}(n)$ and $\mathrm{U}(n)$ can be found in Zelobenko [**268**] and Weyl [**261**]. We also consider a subelliptic operator on SU(2), whose behavior is analogous to the behavior of the subelliptic operator $\mathcal{L}_0$ studied in Chapter 1, but here the analysis is not pursued in nearly as much detail.

In considering irreducible unitary representations of $\mathrm{U}(n)$, we will automatically restrict attention to finite-dimensional representations. As shown in the next chapter, any irreducible unitary representation of a compact Lie group is finite-dimensional.

1. Representation theory for SU(2), SO(3), **and some variants.** The group SU(2) is the group of 2×2 complex unitary matrices of determinant 1. Recall that a $k \times k$ matrix A is unitary provided $A^*A = I$. Then SU(2) is the set of matrices

$$\left\{ \begin{pmatrix} z_1 & z_2 \\ -\bar{z}_2 & \bar{z}_1 \end{pmatrix} : |z_1|^2 + |z_2|^2 = 1, z_j \in \mathbf{C} \right\}. \tag{1.1}$$

Note that, as a set, SU(2) is naturally identified with the unit sphere S^3 in $\mathbf{C}^2$. Its Lie algebra su(2) consists of 2×2 complex skew adjoint matrices of trace 0. A basis of su(2) consists of

$$X_1 = \frac{1}{2}\begin{pmatrix} i & 0 \\ 0 & -i \end{pmatrix}, \qquad X_2 = \frac{1}{2}\begin{pmatrix} 0 & 1 \\ -1 & 0 \end{pmatrix}, \qquad X_3 = \frac{1}{2}\begin{pmatrix} 0 & i \\ i & 0 \end{pmatrix}. \tag{1.2}$$

Note the commutation relations

$$[X_1, X_2] = X_3, \qquad [X_2, X_3] = X_1, \qquad [X_3, X_1] = X_2. \tag{1.3}$$

These are the same as the relations $i \times j = k$, $j \times k = i$, $k \times i = j$ so su(2) is seen to be isomorphic to R^3 with the cross product.

Note that the following generators of the Lie algebra so(3) of SO(3) have the same commutation relations as (1.3). Indeed, so(3) is spanned by

(1.4)

$$J_1 = \begin{pmatrix} 1 & & \\ & 0 & 1 \\ & -1 & 0 \end{pmatrix}, \quad J_2 = \begin{pmatrix} 0 & & -1 \\ & 1 & \\ 1 & & 0 \end{pmatrix}, \quad J_3 = \begin{pmatrix} 0 & 1 & \\ -1 & 0 & \\ & & 1 \end{pmatrix},$$

generating rotations about the x_1, x_2, and x_3 axis, respectively.

Thus SU(2) and SO(3) have isomorphic Lie algebras. In fact, there is an explicit homomorphism

$$p\colon \mathrm{SU}(2) \to \mathrm{SO}(3) \tag{1.5}$$

which exhibits SU(2) as a double cover of SO(3). One way to construct p is the following. The linear span $\mathfrak{g}$ of (1.2) is a three-dimensional real vector space, with an inner product given by

$$(X, Y) = -\operatorname{tr} XY. \tag{1.6}$$

It is clear that the representation p of SU(2) as a group of linear transformations on $\mathfrak{g}$ given by

$$p(g)X = g^{-1}Xg \tag{1.7}$$

preserves the inner product (1.6) and gives (1.5). Note that $\ker p = \{I, -I\}$.

We now tackle the problem of classifying all the irreducible unitary representations of SU(2). We will work with the complexified Lie algebra, and with polynomials in the X_j, i.e., with the universal enveloping algebra. Indeed, let

$$\Delta = X_1^2 + X_2^2 + X_3^2. \tag{1.8}$$

Here we are identifying X_j with a vector field on SU(2), and Δ is a left invariant differential operator on $C^\infty(\mathrm{SU}(2))$. One easily verifies using (1.3) that X_j and Δ commute:

$$\Delta X_j = X_j \Delta, \qquad 1 \le j \le 3. \tag{1.9}$$

Suppose π is an irreducible unitary representation of SU(2) on V. Then π induces a skew adjoint representation of the Lie algebra su(2), and an algebraic representation of the universal enveloping algebra, which we will also denote π. Note that, by (1.9),

$$\pi(\Delta)\pi(X_j) = \pi(X_j)\pi(\Delta), \qquad j = 1, 2, 3. \tag{1.10}$$

Thus, if π is irreducible, Schur's lemma implies

$$\pi(\Delta) = -\lambda^2 I \tag{1.11}$$

for some $\lambda \in R$ (since $\pi(\Delta)$ is a sum of squares of skew adjoint operators, it must be negative). Let

$$L_j = \pi(X_j), \qquad C = \pi(\Delta). \tag{1.12}$$

Now we will diagonalize L_1 on V. Say

$$V_\mu = \{v \in V : L_1 v = i\mu v\}, \tag{1.13}$$

so

$$V = \bigoplus_{i\mu \in \operatorname{spec} L_1} V_\mu. \tag{1.14}$$

The structure of π is defined by how L_2 and L_3 behave on V_μ. This is equivalent to how $L_\pm$ behave, where we set

$$L_\pm = L_2 \mp iL_3. \tag{1.15}$$

Note that the commutation relations (1.3) yield $[X_1, X_\pm] = \pm iX_\pm$ if $X_\pm = X_2 \mp iX_3$, so

$$[L_1, L_\pm] = \pm iL_\pm. \tag{1.16}$$

The identity (1.16) is the key to the structure of $L_\pm$. It yields the following result.

PROPOSITION 1.1. *We have*

$$L_\pm : V_\mu \to V_{\mu\pm 1}. \tag{1.17}$$

In particular, if $i\mu \in \operatorname{spec} L_1$, *then either* $L_+ = 0$ *on* V_μ (*resp.* $L_- = 0$ *on* V_μ), *or* $\mu + 1 \in \operatorname{spec} L_1$ (*resp.* $\mu - 1 \in \operatorname{spec} L_1$).

PROOF. Let $\xi \in V_\mu$. By (1.16), we have

$$L_1 L_\pm \xi = L_\pm L_1 \xi \pm iL_\pm \xi = i(\mu \pm 1)L_\pm \xi,$$

which establishes the proposition.

If π is irreducible on V, we claim that $\operatorname{spec} L_1$ must consist of a sequence

$$\operatorname{spec} L_1 = \{\mu_0, \mu_0 + 1, \dots, \mu_0 + k = \mu_1\}, \tag{1.18}$$

with

$$L_+ : V_{\mu_0+j} \to V_{\mu_0+j+1} \quad \text{isomorphism for } 0 \le j \le k-1, \tag{1.19}$$

and

$$L_- : V_{\mu_1-j} \to V_{\mu_1-j-1} \quad \text{isomorphism for } 0 \le j \le k-1. \tag{1.20}$$

In fact, we can compute

$$\begin{aligned} L_- L_+ &= L_2^2 + L_3^2 + i[L_3, L_2] \\ &= C - L_1^2 - iL_1 \\ &= -\lambda^2 - L_1^2 - iL_1 \quad \text{on } V, \end{aligned}$$

and

$$L_+L_- = C - L_1^2 + iL_1$$
$$= -\lambda^2 - L_1^2 + iL_1 \quad \text{on } V.$$

Thus

$$L_-L_+ = \mu(\mu+1) - \lambda^2 \quad \text{on } V_\mu, \tag{1.21}$$

and

$$L_+L_- = \mu(\mu-1) - \lambda^2 \quad \text{on } V_\mu. \tag{1.22}$$

Note that, since L_1 and L_2 are skew adjoint, we have

$$L_+L_- = -(L_2 + iL_3)^*(L_2 + iL_3),$$

negative selfadjoint, and also L_-L_+ is negative selfadjoint. We see that $\ker L_+ = \ker L_-L_+$; $\ker L_- = \ker L_+L_-$. These observations make the assertions (1.18)–(1.20) fairly transparent. We indicate the situation pictorially:

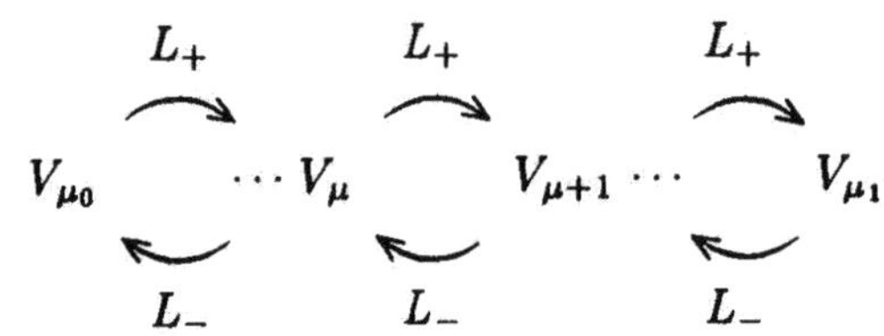

FIGURE 1.1.

From (1.21) and (1.22) we see that $\mu_1(\mu_1+1) = \lambda^2 = \mu_0(\mu_0 - 1)$. In particular, if

$$k = \dim V - 1 = \mu_1 - \mu_0, \tag{1.23}$$

then

$$\mu_0 = -k/2, \qquad \mu_1 = k/2, \tag{1.24}$$

and

$$\lambda^2 = k(k+2)/4 = (\dim V^2 - 1)/4. \tag{1.25}$$

Since L_+ and L_- are inverses of each other, up to a well-defined scalar multiple, on each segment of Figure 1.1, and each V_μ is one-dimensional (by (1.24), or directly from irreducibility), we see that an irreducible representation π of SU(2) on V is determined uniquely up to equivalence by $\dim V$. Thus there is precisely one equivalence class of irreducible representations of SU(2) on $\mathbf{C}^{k+1}$ for $k = 0, 1, 2, \ldots$. A convenient model for this is

$$\mathcal{P}_k = \{p(z)\colon p \text{ homogeneous polynomial of degree } k \text{ on } \mathbf{C}^2\}, \tag{1.26}$$

with SU(2) acting on $\mathcal{P}_k$ by

$$\pi_k(g)f(z) = f(g^{-1}z), \qquad g \in \mathrm{SU}(2), z \in \mathbf{C}^2. \tag{1.27}$$

We leave it to the reader to check that this has the same structure as indicated in Figure 1.1. Note that $\mu_1 = k/2$, and if the complexification of su(2) is identified with the set of 2×2 complex matrices of trace zero in such a way that

$$(1.28)\quad L_+ = \begin{pmatrix} 0 & 1 \\ 0 & 0 \end{pmatrix}, \qquad L_- = \begin{pmatrix} 0 & 0 \\ -1 & 0 \end{pmatrix}, \qquad L_1 = (i/2)\begin{pmatrix} 1 & 0 \\ 0 & -1 \end{pmatrix},$$

we have, for $0 \le j \le k$,

$$(1.29)\qquad V_{-k/2+j} = \text{linear span of } z_1^{k-j} z_2^j.$$

We can deduce the classification of irreducible unitary representations of SO(3) from the discussion above as follows. We have a double covering homomorphism p: SU(2) $\to$ SO(3), with

$$(1.30)\qquad \ker p = \{\pm I\}.$$

Now each irreducible representation d_j of SO(3) defines an irreducible representation $d_j \circ p$ of SU(2), which must be equivalent to one of the π_k given by (1.26), (1.27). We see that π_k factors through to a representation of SO(3) if and only if π_k is the identity on $\ker p$, i.e., if and only if $\pi_k(-I) = I$. Clearly this holds if and only if k is even. Thus all the irreducible unitary representations of SO(3) are given by representations d_j on $\mathcal{P}_{2j}$, uniquely defined by

$$(1.31)\qquad d_j(pg) = \pi_{2j}(g), \qquad g \in \mathrm{SU}(2).$$

In other words, half of the representations (1.26), (1.27) of SU(2) give rise to representations of SO(3), namely those on vector spaces of odd dimension. Note that in this case the operator Δ is represented by

$$(1.32)\qquad d_j(\Delta) = -j(j+1).$$

We note that the irreducible representations of U(2) can be classified, using the results of SU(2). Indeed, we have the exact sequence

$$(1.33)\qquad 0 \to K \to S^1 \times \mathrm{SU}(2) \to \mathrm{U}(2) \to 0$$

where

$$(1.34)\qquad \begin{aligned} K &= \{(\omega, g) \in S^1 \times \mathrm{SU}(2) \colon g = \omega^{-1} I, \omega^2 = 1\} \\ &= \{(1, I), (-1, -I)\}. \end{aligned}$$

The irreducible representations of $S^1 \times$ SU(2) are given by

$$(1.35)\qquad \pi_{mk}(\omega, g) = \omega^m \pi_k(g) \quad \text{on } \mathcal{P}_k,$$

with $m \in \mathbf{Z}$, $k \in \mathbf{Z}^+ \cup \{0\}$. Those giving the complete set of irreducible representations of U(2) are those for which $\pi_{mk}(K) = I$, i.e., for which $(-1)^m \pi_k(-I) = I$. Since $\pi_k(-I) = (-1)^k I$, we see the condition is that $m + k$ be an even integer.

Finally let us note that SO(4) is covered by SU(2)×SU(2). To see this, equate the unit sphere $S^3 \subset R^4$, with its standard metric, to SU(2), with a bi-invariant metric. Then SO(4) is the connected component of the identity in the isometry group of S^3. Meanwhile, SU(2) × SU(2) acts as a group of isometries, by

$$(g_1, g_2) \cdot x = g_1^{-1} x g_2, \qquad g_j, x \in \mathrm{SU}(2). \tag{1.36}$$

Thus we have a map

$$\tau\colon \mathrm{SU}(2) \times \mathrm{SU}(2) \to \mathrm{SO}(4). \tag{1.37}$$

This is a group homomorphism. Note that $(g_1, g_2) \in \ker \tau$ implies $g_1 = g_2 = \pm I$, so

$$\mathrm{SO}(4) \approx \mathrm{SU}(2) \times \mathrm{SU}(2)/\{\pm(I, I)\}, \tag{1.38}$$

since a dimension count shows τ must be surjective.

In the next chapter we will prove the following proposition. Let G_1, G_2 be compact Lie groups, $G = G_1 \times G_2$. Then the set of all irreducible unitary representations of G, up to unitary equivalence, is given by

$$\{\pi(g) = \pi_1(g_1) \otimes \pi_2(g_2)\colon \pi_j \in \hat{G}_j\},$$

where $g = (g_1, g_2) \in G$ and $\pi_j \in \hat{G}_j$ is a general irreducible unitary representation of G_j. In particular, the irreducible unitary representations of SU(2)×SU(2), up to equivalence, are precisely the representations of the form

$$\delta_{kl}(g) = \pi_k(g_1) \otimes \pi_l(g_2), \qquad k, l \in \mathbf{Z}^+ \cup \{0\},$$

acting on $\mathbf{C}^{k+1} \otimes \mathbf{C}^{l+1}$, where π_k is given by (1.26)–(1.27). By (1.38), the irreducible unitary representations of SO(4) are given by all δ_{kl} such that $k + l$ is even, since, for $p_0 = (-I, -I) \in \mathrm{SU}(2) \times \mathrm{SU}(2)$, $\delta_{kl}(p_0) = (-1)^{k+l} I$.

2. Representation theory for U(n). Let π be a finite-dimensional unitary representation of U(n), on a vector space V. For the moment we will not assume π is irreducible. Then there is induced a skew adjoint representation (also denoted π) of the Lie algebra u(n) of skew adjoint $n \times n$ complex matrices, and a complex linear representation of its complexification, which is naturally isomorphic to $\mathrm{gl}(n, \mathbf{C}) = M(n, \mathbf{C})$, the algebra of $n \times n$ complex matrices, since any element of $M(n, \mathbf{C})$ can be uniquely written as $A + iB$, with A and B skew adjoint. The complexified representation on $M(n, \mathbf{C})$ is also denoted π. $M(n, \mathbf{C})$ has a natural structure as a complex Lie algebra, and is the Lie algebra of $\mathrm{GL}(n, \mathbf{C})$, the group of invertible $n \times n$ complex matrices, which has a natural structure as a complex Lie group. This complex linear representation π of $M(n, \mathbf{C})$ on V exponentiates to a representation (also denoted π) of $\mathrm{GL}(n, \mathbf{C})$ on V which is holomorphic, in the sense that, with respect to a basis of V, the matrix entries of such a representation are holomorphic functions on $\mathrm{GL}(n, \mathbf{C})$.

A useful tool in the study of representations of U(n), in addition to their analytic continuations to $\mathrm{GL}(n, \mathbf{C})$ defined above, is the Gauss decomposition, which is the following. If D denotes the group of diagonal matrices in $\mathrm{GL}(n, \mathbf{C})$,

Z_+ (resp., Z_-) the group of upper (resp., lower) triangular matrices in $\mathrm{GL}(n, \mathbf{C})$ with ones on the diagonal, then

$$Z_- D Z_+ = G_{\mathrm{reg}}$$

is a dense subset of $\mathrm{GL}(n, \mathbf{C})$. For a proof, see Zelobenko [**268**], page 28, or the reader can try it as an exercise. We note that a weaker result is easy to prove. Namely, the Lie algebras of D, Z_+, and Z_- span $M(n, \mathbf{C})$, and hence, by the inverse function theorem, $Z_- D Z_+$ contains a neighborhood of the identity in $\mathrm{GL}(n, \mathbf{C})$. For a study of holomorphic functions on $\mathrm{GL}(n, \mathbf{C})$, this weaker result can be just as effective.

Let T^n be the group of diagonal elements of $\mathrm{U}(n)$; it is an n-torus. Let its Lie algebra be denoted by H. Then the action of π on T^n and on H can be simultaneously diagonalized. We can set, for an element $\lambda \in H'$, the dual space to H,

$$V_\lambda = \{v \in V : \pi(h)v = i\lambda(h)v, \ \forall h \in H\}, \tag{2.1}$$

and we have a finite direct sum

$$V = \bigoplus V_\lambda. \tag{2.2}$$

If $\lambda \in H'$ is such that $V_\lambda \neq 0$, we call λ a *weight*, and any nonzero $v \in V_\lambda$ a *weight vector*. We will see that, for certain $e_{ij} \in \mathrm{gl}(n, \mathbf{C})$, $\pi(e_{ij})$ acts on the V_λ's in a revealing fashion, similar to the operators $L_\pm$ of §1.

Let e_{ij} be the $n \times n$ matrix with 1 at row i, column j, and 0 in all other spots. We have the commutation relations

$$[e_{ij}, e_{kl}] = \delta_{jk} e_{il} - \delta_{il} e_{kj}. \tag{2.3}$$

Note that, if $i < j$ and $k < l$, the nontrivial commutation relations among e_{ij} and e_{kl} reduce to $j = k$ and

$$[e_{ij}, e_{jk}] = e_{ik} \quad \text{if } i < j < k. \tag{2.4}$$

Let

$$e_j = i e_{jj}. \tag{2.5}$$

Then $\{e_j : 1 \leq j \leq n\}$ spans H. If $h \in H$ is of the form

$$h = \sum t_j e_j, \tag{2.6}$$

then

$$[h, e_{ij}] = i(t_i - t_j) e_{ij}. \tag{2.7}$$

Thus the element ω_{jk} of H' is defined by

$$\omega_{jk}(h) = t_j - t_k \tag{2.8}$$

is a weight for the adjoint representation of U(n); we call ω_{jk} a *root.* Similarly we call e_{jk} a *root vector* in gl$(n, \mathbf{C})$.

Now let

$$E_{ij} = \pi(e_{ij}). \tag{2.9}$$

The commutation relation (2.7) implies

$$\pi(h)E_{jk} = E_{jk}(\pi(h) + i\omega_{jk}(h)I). \tag{2.10}$$

We can utilize this identity in a fashion parallel to our use of (1.10) of the last section.

PROPOSITION 2.1. *We have*

$$E_{jk}\colon V_\lambda \to V_{\lambda+\omega_{jk}}. \tag{2.11}$$

In particular, if $\lambda \in H'$ is a weight for the representation π, then either E_{jk} annihilates V_λ or $\lambda + \omega_{jk}$ is a weight.

PROOF. Let $\xi \in V_\lambda$. By (2.10), we have

$$\begin{aligned} \pi(h)E_{jk}\xi &= E_{jk}\pi(h)\xi + iE_{jk}\omega_{jk}(h)\xi \\ &= i(\lambda(h) + \omega_{jk}(h))E_{jk}\xi, \end{aligned} \tag{2.12}$$

which establishes the proposition.

Note that, if we factor i out of the diagonal skew adjoint matrices making up the Lie algebra H of T^n, we have, by reading down the diagonal, a natural isomorphism of H with R^n. If $\alpha, \beta \in R^n$, we say $\alpha < \beta$, or $\beta - \alpha > 0$, if the first nonzero coordinate of $\beta - \alpha$ is positive. Thus there is a natural ordering of the weights. For a given finite-dimensional representation π, with respect to this ordering there will be a *highest weight* λ_m (also called a maximal weight), and also a *lowest weight* λ_s. From Proposition 2.1 and (2.12), we see that

$$E_{jk} = 0 \text{ on } V_{\lambda_m} \quad \text{if } j < k, \tag{2.13}$$

$$E_{jk} = 0 \text{ on } V_{\lambda_s} \quad \text{if } j > k. \tag{2.14}$$

We call E_{jk} a *raising operator* if $j < k$ and a *lowering operator* if $j > k$. Thus all raising operators annihilate V_{λ_m} and all lowering operators annihilate V_{λ_s}. More generally, we say a weight λ is nonraisable if all raising operators annihilate V_λ and nonlowerable if all lowering operators annihilate V_λ. In a little while we will show that, if π is irreducible, then the only nonraisable weight is maximal. At present, we record the following progress.

PROPOSITION 2.2. *If π is a unitary representation of* U(n) *on V, $\dim V < \infty$, then there exists a nonzero highest weight vector ξ, and in particular there exists a nonzero $\xi \in V$ annihilated by all raising operators, which is also a weight vector.*

It is very interesting that this can be used as a *tool* for establishing that certain representations of $\mathrm{U}(n)$ are irreducible. In fact, we have the following

PROPOSITION 2.3. *Let π be a unitary representation of* $\mathrm{U}(n)$ *on* V, $\dim V < \infty$. *Suppose the set of vectors $\xi \in V$ annihilated by all raising operators which are also weight vectors, is equal to the set of nonzero multiples of a single element. Then π is irreducible.*

PROOF. Otherwise, $V = V_1 \oplus V_2$ with π acting on each factor, and Proposition 2.2 produces two linearly independent weight vectors $\xi_j \in V_j$ annihilated by all raising operators.

As an example, consider the natural action of $\mathrm{U}(n)$ on $S^k\mathbf{C}^n$, the k-fold symmetric tensor product of $\mathbf{C}^n$. We can identify this with

$$\mathcal{P}_k = \{p(z)\colon p \text{ homogeneous polynomial of degree } k \text{ in } z \in \mathbf{C}^n\}, \tag{2.15}$$

with the action given by

$$\pi_k(g)f(z) = f(g^{-1}z). \tag{2.16}$$

We see that, for $|\alpha| = k$, z^α is a weight vector, with weight $\alpha \in R^n = H'$. The action of the operators E_{ij} on $\mathcal{P}_k$ is defined by

$$\begin{aligned} E_{ij}p(z) &= (d/dt)p(z_1, \dots, z_{j-1}, z_j + tz_i, z_{j+1}, \dots, z_n)|_{t=0} \\ &= z_i(\partial/\partial z_j)p(z). \end{aligned} \tag{2.17}$$

In particular, the only weight vector z^α annihilated by all raising operators is z_1^k. We see that π_k is an irreducible representation of $\mathrm{U}(n)$ on $\mathcal{P}_k$.

As a second example, consider the natural action of $\mathrm{U}(n)$ on $\bigwedge^k \mathbf{C}^n$. We denote this representation d_k,

$$d_k(g)(v_1 \wedge \cdots \wedge v_k) = gv_1 \wedge \cdots \wedge gv_k, \qquad g \in \mathrm{U}(n), v_j \in \mathbf{C}^n. \tag{2.18}$$

If $u_1, \dots, u_n$ denotes the standard orthonormal basis of $\mathbf{C}^n$, we see that

$$u_{j_1} \wedge \cdots \wedge u_{j_k} = u_\gamma \qquad (j_1 < \cdots < j_k) \tag{2.19}$$

is a weight vector, with weight $\gamma = (\gamma_1, \dots, \gamma_n) \in R^n = H'$, where $\gamma_j = 1$ if j is some j_k occurring in (2.19), $\gamma_j = 0$ otherwise. The action of the operators E_{ij} on $\bigwedge^k \mathbf{C}^n$ is defined by

$$E_{ij}u_{j_1} \wedge \cdots \wedge u_{j_k} = \begin{cases} u_{j_1} \wedge \cdots \wedge u_{j_{\nu-1}} \wedge u_i \wedge u_{j_{\nu+1}} \wedge \cdots \wedge u_{j_k} & \text{if } j = j_\nu, \\ 0 \quad \text{if } j \notin \{j_1, \dots, j_k\}. \end{cases} \tag{2.20}$$

In particular, the only weight vector u_γ annihilated by all raising operators is $u_1 \wedge \cdots \wedge u_k$. We see that d_k is an irreducible representation of $\mathrm{U}(n)$ on $\bigwedge^k \mathbf{C}^n$.

Note that, in each of the two examples above, $V = S^k\mathbf{C}^n$ and $V = \bigwedge^k \mathbf{C}^n$, the action of $\mathrm{U}(n)$ on V is generated by the action of $\mathrm{SU}(n)$ on V and the action of scalars $e^{i\alpha}I$ on V, $\alpha \in R$. Thus, all these representations restrict to representations of $\mathrm{SU}(n)$ which are irreducible.

Another class of representations of $\mathrm{U}(n)$ (and of $\mathrm{SU}(n)$) is on the subspace (of codimension 1) of $S^k\mathbf{C}^n \otimes S^l(\mathbf{C}^n)^*$ consisting of tensors $A^{i_1,\ldots,i_l}_{j_1,\ldots,j_k}$ which are annihilated by the contraction operator

$$\sum_{\nu=1}^{n} A^{\nu,i_2,\ldots,i_l}_{\nu,j_2,\ldots,j_k} = 0. \tag{2.21}$$

In this case, one can verify the weight vectors are given by monomials and the unique weight vector annihilated by all raising operators is defined by

$$A^{i_1,\ldots,i_l}_{j_1,\ldots,j_k} = \begin{cases} 1 & \text{if } (i_1,\ldots,i_l) = (0,\ldots,0,1) \text{ and } (j_1,\ldots,j_k) = (1,0,\ldots), \\ 0 & \text{otherwise.} \end{cases} \tag{2.22}$$

Thus we obtain a two parameter family of irreducible unitary representations of $\mathrm{SU}(n)$. For $n = 3$, this list turns out to be exhaustive.

We return to our general study. Let π be an irreducible representation of $\mathrm{U}(n)$ on a complex vector space, and retain the conventions of the first paragraph of this section. We define the representation $\overline{\pi}$, contragredient to π, of $\mathrm{U}(n)$ on V^*, by

$$\langle \xi, \overline{\pi}(g)\eta\rangle = \langle \pi(g^{-1})\xi, \eta\rangle. \tag{2.23}$$

The pairing in (2.23) is taken to be bilinear. Now suppose $\xi_0 \in V$ is a nonraisable weight vector for π, with weight $\lambda \in H'$, and suppose $\eta_0 \in V^*$ is a nonlowerable weight vector for $\overline{\pi}$, with weight $-\mu \in H'$. We will study the function

$$\alpha(g) = \langle \pi(g)\xi_0, \eta_0\rangle. \tag{2.24}$$

We can take the extension of π to a representation of $\mathrm{Gl}(n,\mathbf{C})$, as mentioned in the first paragraph, so $\alpha(g)$ is defined for $g \in \mathrm{Gl}(n,\mathbf{C})$, the complexification of $\mathrm{U}(n)$, and holomorphic. Let Z_+ be the subgroup of $\mathrm{Gl}(n,\mathbf{C})$ whose Lie algebra $\mathfrak{z}_+$ is generated by the raising operators E_{ij}, $i < j$, and let Z_- be the subgroup of $\mathrm{Gl}(n,\mathbf{C})$ whose Lie algebra $\mathfrak{z}_-$ is generated by the lowering operators E_{ij}, $i > j$. Then Z_+ (resp. Z_-) consists of matrices which are upper triangular (resp. lower triangular), with ones along the diagonal. Let D denote the group of diagonal matrices in $\mathrm{Gl}(n,\mathbf{C})$. There is the Gauss decomposition, mentioned above, which says

$$Z_- D Z_+ = G_{\mathrm{reg}} \tag{2.25}$$

is a dense subset of $\mathrm{Gl}(n,\mathbf{C})$, or (in a weaker form) contains a neighborhood of the identity. Let

$$g = \varsigma\delta z, \qquad \varsigma \in Z_-, \delta \in D, z \in Z_+. \tag{2.26}$$

We see that, if $\delta = \exp(h_1 + ih_2)$, $h_j \in H$,

$$\alpha(\varsigma g) = \alpha(g), \qquad \alpha(\delta g) = e^{i(\lambda(h_1)+i\lambda(h_2))}\alpha(g) \tag{2.27}$$

and

$$\alpha(gz) = \alpha(g), \qquad \alpha(g\delta) = e^{i(\mu(h_1)+i\mu(h_2))}\alpha(g). \tag{2.28}$$

Thus

$$\alpha(\varsigma\delta z) = \alpha(\delta) = \exp i(\lambda(h_1) + i\lambda(h_2))\alpha(e)$$
$$= \exp i(\mu(h_1) + i\mu(h_2))\alpha(e). \tag{2.29}$$

Now we claim $\alpha(e) \neq 0$. Otherwise $\alpha(g) \equiv 0$. But if

$$V_0 = \{\xi \in V : \langle \pi(g)\xi, \eta_0 \rangle = 0 \ \forall g \in \mathrm{Gl}(n, \mathbf{C})\} \tag{2.30}$$

then V_0 is invariant, and since clearly $V_0 \neq V$, we have $V_0 = 0$ if π is irreducible. This shows $\alpha(e) \neq 0$. From (2.29) we can deduce the following important result.

PROPOSITION 2.4. *If π is irreducible on V, then there is only one weight λ which is nonraisable, namely the highest weight. Furthermore, the highest weight vector is unique, up to scalar multiple. Finally, if π and π' are irreducible and have the same highest weight, they are unitarily equivalent.*

PROOF. The identity $\lambda = \mu$ proves the uniqueness of λ. Note that if we normalize the weight vectors so $\alpha(e) = 1$, the function $\alpha(g)$ is uniquely characterized by the following three properties:

$$\alpha(g) \in C^\infty(\mathrm{Gl}(n, \mathbf{C})) \quad \text{(in fact, holomorphic)}, \tag{2.31}$$

$$\alpha(\varsigma g z) = \alpha(g), \qquad \varsigma \in Z_-, z \in Z_+, \tag{2.32}$$

$$\alpha(\delta) = \exp i(\lambda(h_1) + i\lambda(h_2)) \quad \text{for } \delta = \exp(h_1 + ih_2) \in D. \tag{2.33}$$

Thus, if ξ_1 were another highest weight vector, also normalized so $\langle \xi_1, \eta_0 \rangle = 1$, we would have $\langle \pi(g)\xi_1, \eta_0 \rangle = \alpha(g)$, so

$$\langle \pi(g)(\xi_1 - \xi_0), \eta_0 \rangle = 0 \quad \forall g,$$

or

$$\langle \xi_1 - \xi_0, \overline{\pi}(g)\eta_0 \rangle = 0 \quad \forall g.$$

Since $\overline{\pi}$ is irreducible, this implies $\xi_1 = \xi_0$.

As for the final assertion of the proposition, let π' be an irreducible representation on V'. Pick a maximal weight vector ξ_0' for π' and a minimal weight vector η_0' for its contragredient representation $\overline{\pi}'$, normalized so $\langle \xi_0', \eta_0' \rangle = 1$ (which is possible by the argument above implying the inner product is nonzero), and form

$$\alpha'(g) = \langle \pi'(g)\xi_0', \eta_0' \rangle.$$

Since $\alpha'(g)$ satisfies the conditions (2.31)–(2.33) provided λ is also the highest weight for π', we see that $\alpha(g) = \alpha'(g)$. But if π' and π are not equivalent, the Weyl orthogonality relations (see Chapter 3) imply the restrictions of α and α' to $\mathrm{U}(n)$ must be orthogonal, and certainly cannot coincide. This proves that two irreducible representations π and π' must be equivalent if their maximal weights coincide.

The classification of the irreducible unitary representations of $\mathrm{U}(n)$ requires us to specify which elements $\lambda \in \mathfrak{h}'$ can be highest weights. We will wait until

Chapter 3 to discuss the general case; see the end of §2, Chapter 3 for this, including some details for the closely related group SU(n).

3. The subelliptic operator $X_2^2 + X_3^2$ on SU(2). The Laplace operator on SU(2)

$$\Delta = X_1^2 + X_2^2 + X_3^2 \tag{3.1}$$

is elliptic. We want to study the nonelliptic operator

$$\mathcal{L} = X_2^2 + X_3^2. \tag{3.2}$$

Note that $\mathcal{L}$ and Δ commute, so $\mathcal{L}$ acts on each eigenspace of Δ. In particular, the spectrum of $\mathcal{L}$ is discrete. We can examine the spectrum of $\mathcal{L}$ by decomposing $L^2(\mathrm{SU}(2))$ into eigenspaces of Δ, which is equivalent to decomposing it into subspaces irreducible for the regular action of $\mathrm{SU}(2) \times \mathrm{SU}(2)$ given by

$$(g_1, g_2) \cdot f(x) = f(g_1^{-1} x g_2), \qquad g_j, x \in \mathrm{SU}(2). \tag{3.3}$$

Note, by the Peter-Weyl theorem (discussed in Chapter 3), the irreducible subspaces of $L^2(\mathrm{SU}(2))$ are precisely the spaces of the form

$$V_k = \text{linear span of } \pi_k^{ij}(x), \qquad 1 \le i, j \le k+1, \tag{3.4}$$

where π_k is the representation of SU(2) on $\mathbf{C}^{k+1}$ described by (1.26), (1.27), with matrix representation $\pi_k^{ij}(x)$. Of course, each space V_k is an eigenspace of the Laplace operator Δ; by (1.25) the associated eigenvalue is $-k(k+2)/4$.

If we consider the left regular representation

$$g \cdot f(x) = f(g^{-1} x), \tag{3.5}$$

then V_k is a direct sum of $k+1$ representations of SU(2), each equivalent to π_k. One decomposition of V_k into irreducible subspaces for (3.5) is

$$V_k = \bigoplus_{l=1}^{k+1} V_{kl} \tag{3.6}$$

where V_{kl} = linear span of $\pi_k^{il}(x)$, $1 \le i \le k+1$. Each V_{kl} breaks up into eigenspaces for the operator X_1, as discussed in §1:

$$V_{kl} = \bigoplus_{\mu} V_{kl\mu} \tag{3.7}$$

where

$$\mu \in \{-k/2, -k/2+1, \dots, k/2\} \tag{3.8}$$

and

$$X_1 = i\mu \quad \text{on } V_{kl\mu}. \tag{3.9}$$

Since $\mathcal{L} = \Delta - X_1^2$, we have

$$\begin{aligned} \mathcal{L} &= -(\lambda^2 - \mu^2) \quad \text{on } V_{kl\mu} \\ &= -(k^2/4 + k/2 - \mu^2) \quad \text{on } V_{kl\mu}. \end{aligned} \tag{3.10}$$

By (3.8), we deduce

PROPOSITION 3.1. *On the* $-k(k+2)/4$ *eigenspace* V_k *of* Δ, *the operator* $-\mathcal{L}$ *has* $k+1$ *eigenvalues, ranging from a minimum of* $k/2$ *to a maximum of* $k(k+2)/4$ *or* $k(k+2)/4-1/4$, *for* k *even or odd, respectively.*

Suppose $u \in \mathcal{D}'(\mathrm{SU}(2))$ satisfies the equation

$$\mathcal{L}u = f. \tag{3.11}$$

We want to examine smoothness of u given smoothness of f. Note that Proposition 3.1 implies $\mathcal{L}$ is invertible on $\{f \in L^2\colon \int f = 0\}$, and

$$\|(-\Delta)^{1/2} f\|_{L^2} \le C\|\mathcal{L}f\|_{L^2}. \tag{3.12}$$

We can take, for example, $C = \frac{1}{2}$. More generally,

$$\|(-\Delta)^{k/2} f\|_{L^2} \le C^k\|\mathcal{L}^k f\|_{L^2}. \tag{3.13}$$

Given $f \in L^2(\mathrm{SU}(2))$, write

$$f = \sum_k f_k, \qquad f_k \in V_k. \tag{3.14}$$

It is well known, and easy to prove, that

$$f \in C^\infty(\mathrm{SU}(2)) \Leftrightarrow \|f_k\|_{L^2} \le C_N k^{-N} \quad \text{for all } N, \tag{3.15}$$

and

$$f \text{ is real analytic } \Leftrightarrow \|f_k\|_{L^2} \le C_\varepsilon e^{-\varepsilon k} \quad \text{for some } \varepsilon > 0. \tag{3.16}$$

Note that the solution u to (3.11) satisfies

$$u = \sum u_k, \qquad u_k \in V_k, \tag{3.17}$$

with $u_k = \mathcal{L}^{-1} f_k$ and, by Proposition 3.1,

$$\|u_k\|_{L^2} \le Ck^{-1}\|f_k\|_{L^2}. \tag{3.18}$$

Thus, from (3.15) and (3.16), it is clear that $u \in C^\infty(\mathrm{SU}(2))$ if $f \in C^\infty$ and u is real analytic if f is. We say $\mathcal{L}$ is globally C^∞ hypoelliptic and globally analytic hypoelliptic. It is a special case of general theorems (see Hörmander [**120**], Boutet de Monvel et al. [**25**]) that $\mathcal{L}$ is actually locally C^∞ hypoelliptic and in fact (see Tartakoff [**233**], Treves [**240**]), even locally analytic hypoelliptic.

The C^∞ and analytic hypoellipticity of $\mathcal{L}$ are properties shared with the Laplace operator Δ. Since Δ is elliptic, these properties are classical in this case. There is another property of Δ which is not shared by $\mathcal{L}$, namely the Kotake-Narasimhan theorem says

$$\|\Delta^k u\|_{L^2} \le C(Ck)^{2k} \quad \forall k \Rightarrow u \text{ is real analytic}. \tag{3.19}$$

Note that the hypothesis of (3.19) implies the function

$$v(y,x) = \sum_{k=0}^{\infty} [y^{2k}/(2k)!](-\Delta)^k u(x), \qquad y \in (-\varepsilon, \varepsilon), x \in \mathrm{SU}(2),$$

is a convergent series for y in some interval $(-\varepsilon, \varepsilon)$, and we see v satisfies the equation

$$(\partial^2/\partial y^2 + \Delta)v = 0. \tag{3.20}$$

Since $\partial^2/\partial y^2 + \Delta$ is elliptic, local analytic hypoellipticity implies v is analytic, so $u(x) = v(0,x)$ must be analytic. Now suppose the hypothesis in (3.19) is replaced by

$$\|\mathcal{L}^k u\|_{L^2} \le C(Ck)^{2k}, \quad \text{all } k \in \mathbf{Z}^+. \tag{3.21}$$

We can form

$$v(y,x) = \sum_{k=0}^{\infty} [y^{2k}/(2k)!](-\mathcal{L})^k u(x), \tag{3.22}$$

and also get convergence for y in some interval $(-\varepsilon, \varepsilon)$, and

$$(\partial^2/\partial y^2 + \mathcal{L})v = 0. \tag{3.23}$$

Now the operator $\partial^2/\partial y^2 + \mathcal{L}$, which has double characteristics, is locally C^∞ hypoelliptic. But its characteristic set is not a symplectic manifold, so general results on analytic hypoellipticity do not apply. In fact, we will show that (3.21) does not imply u is analytic, and hence $\partial^2/\partial y^2 + \mathcal{L}$ is not analytic hypoelliptic.

Indeed, pick $w_k \in V_k$ such that

$$w_k \in V_{k,1,k/2}, \qquad \|w_k\|_{L^2} = 1, \tag{3.24}$$

and set

$$w = \sum_{k=1}^{\infty} e^{-\sqrt{k}} w_k. \tag{3.25}$$

By (3.16), we see that w is definitely not real analytic. Note that

$$\mathcal{L}^k w = \sum_{j=1}^{\infty} e^{-\sqrt{j}} (j/2)^k w_j, \tag{3.26}$$

so

$$\|\mathcal{L}^k w\|_{L^2}^2 = \sum_{j=1}^{\infty} e^{-2\sqrt{j}} (j/2)^{2k}. \tag{3.27}$$

Thus

$$\begin{aligned} \sum_k [y^{4k}/(4k)!] \|\mathcal{L}^k w\|_{L^2}^2 &\le \sum_j \sum_k e^{-2\sqrt{j}} [1/(4k)!] (y\sqrt{j}/\sqrt{2})^{4k} \\ &\le \sum_j e^{-2\sqrt{j}} \cdot e^{y\sqrt{j}/\sqrt{2}}, \end{aligned} \tag{3.28}$$

which is finite if $|y| < 2\sqrt{2}$. This shows that the hypothesis (3.21) is satisfied. Hence the hypothesis (3.21) does not imply u is analytic; the Kotake-Narasimhan theorem fails for $\mathcal{L}$, and in particular $\partial^2/\partial y^2 + \mathcal{L}$ cannot be analytic hypoelliptic.

There is a natural equivalence between the set of operators on $C^\infty(\mathrm{SU}(2))$ which are left invariant and the set of operators on $S(R^2)$ which commute with the harmonic oscillator Hamiltonian $H = -\Delta + |x|^2$, or equivalently with the set of operators on the Bargmann-Fok space $\mathcal{H}$ which commute with the operator $W = 2(z_1\partial/\partial z_1 + z_2\partial/\partial z_2) + 2$ (described in Chapter 1, §6), which we now define. Let $\tilde{\pi}_k$ denote the natural representation of SU(2) on $\mathcal{H}$ (given by $\tilde{\pi}_k(g)u(z) = u(g^{-1}z)$, $g \in \mathrm{SU}(2)$, $z \in \mathbf{C}^2$), restricted to the $2k+2$ eigenspace of W, which consists of polynomials in $z \in \mathbf{C}^2$ homogeneous of degree k. Let π_k denote the associated representation of SU(2) on the $2k+2$ eigenspace of H, i.e.,

$$\pi_k(g) = K^{-1}\tilde{\pi}_k(g)K, \tag{3.29}$$

where $K\colon L^2(R^2) \to \mathcal{H}$ is the unitary operator intertwining the Schrödinger and Bargmann-Fok representations of the Heisenberg group, defined by (5.8), (5.9) of Chaper 1. If $f \in C^\infty(\mathrm{SU}(2))$ and $\pi(f) = \int_{\mathrm{SU}(2)} f(g)\pi(g)\,dg$, we set

$$Tf = \sum_k \int_{\mathrm{SU}(2)} f(g)\pi_k(g)\,dg, \tag{3.30}$$

and

$$\tilde{T}f = \sum_k \int_{\mathrm{SU}(2)} f(g)\tilde{\pi}_k(g)\,dg, \tag{3.31}$$

so Tf is an operator on $L^2(R^2)$ commuting with H and $\tilde{T}f$ is an operator on $\mathcal{H}$ commuting with W.

It follows from Lemma 7.8 of Chapter 1, and the formulas (7.60)–(7.62) in the proof of that lemma, the generators X_1, X_2, X_3 of su(2) are taken by T to

$$T(X_j) = ia_j(X, D) \tag{3.32}$$

where

$$\begin{aligned} a_1(x,\xi) &= \tfrac{1}{4}(x_1^2 + \xi_1^2 - x_2^2 - \xi_2^2), \\ a_2(x,\xi) &= \tfrac{1}{2}(x_1x_2 + \xi_1\xi_2), \\ a_3(x,\xi) &= \tfrac{1}{2}(x_1\xi_2 - x_2\xi_1). \end{aligned} \tag{3.33}$$

In particular, for the Laplace operator (3.1) on SU(2), we have

$$T(-\Delta) = \tfrac{1}{4}(-\partial^2/\partial x_1^2 - \partial^2/\partial x_2^2 + x_1^2 + x_2^2) - 1 = \tfrac{1}{4}(H_1 + H_2) - 1, \tag{3.34}$$

and for the subelliptic operator $\mathcal{L} = X_2^2 + X_3^2$, we have

$$T(-\mathcal{L}) = (-\partial^2/\partial x_1^2 + x_1^2)(-\partial^2/\partial x_2^2 + x_2^2) - 1 = H_1H_2 - 1. \tag{3.35}$$

Once again our analysis of subelliptic operators leads us to harmonic oscillator Hamiltonians. Compare with Chapter 1, §7.

We now obtain an inversion formula for the transformation T, so that we could analyze functions of the operator $\mathcal{L}$ in terms of functions of the operator $H_1H_2 - 1$, which appears on the right side of (3.35). First we impose a Hilbert space structure on some subspace of $\mathcal{D}'(R^2 \times R^2)$ containing $C_0^\infty(R^2 \times R^2)$ so that T maps $L^2(\mathrm{SU}(2))$ isometrically into this Hilbert space. Then T^* will map the range of T isometrically onto $L^2(\mathrm{SU}(2))$ and will provide the inverse. The orthogonality relations, which will be proved in Chapter 3, imply

$$\|f\|^2_{L^2(\mathrm{SU}(2))} = \sum_k (k+1)\|\pi_k(f)\|^2_{\mathrm{HS}}. \tag{3.36}$$

Thus the Hilbert space norm $|||\ |||$ on $C_0^\infty(R^2 \times R^2)$ should satisfy the identity

$$|||Tf|||^2 = \sum_k (k+1)\|\pi_k(f)\|^2_{L^2(R^2\times R^2)}, \tag{3.37}$$

and since the harmonic oscillator Hamiltonian H on R^2 is equal to $2(k+1)$ on V_{π_k}, we have

$$\begin{aligned} |||Tf|||^2 &= \tfrac{1}{2}(H_1 Tf, Tf)_{L^2(R^2\times R^2)} \\ &= \tfrac{1}{4}((H_1+H_2)Tf, Tf)_{L^2(R^4)} \end{aligned} \tag{3.38}$$

where $H_1 + H_2$ is the harmonic oscillator Hamiltonian on R^4. Thus we pick our pre-Hilbert space norm on $C_0^\infty(R^2 \times R^2)$ to be

$$|||v|||^2 = \tfrac{1}{4}((H_1+H_2)v, v)_{L^2(R^4)}. \tag{3.39}$$

Hence

$$(T^* v, f)_{L^2(\mathrm{SU}(2))} = \tfrac{1}{4}((H_1+H_2)v, Tf)_{L^2(R^4)}. \tag{3.40}$$

To get a more explicit formula for T^*v, v an operator on $L^2(R^2)$ commuting with H, we want to compute Tf in the limit when f is the delta function δ_g, $g \in \mathrm{SU}(2)$. Note that the operator $\tilde{T}\delta_g$ is defined by

$$(\tilde{T}\delta_g)u(z) = u(g^{-1}z), \qquad g \in \mathrm{SU}(2), z \in \mathbf{C}^2. \tag{3.41}$$

Since

$$Tf = K^*(\tilde{T}f)K \tag{3.42}$$

with

$$Kw(z) = \int_{R^2} K(x,z)w(x)\,dx \tag{3.43}$$

and

$$K^*u(x) = \int_{\mathbf{C}^2} \overline{K(x,z)}u(z)e^{-|z|^2/2}\,d\,\mathrm{vol}(z), \tag{3.44}$$

where, as given in Chapter 1, §5, we have

$$K(x,z) = \exp[\sqrt{2}z\cdot x - (z\cdot z + |x|^2)/2], \tag{3.45}$$

we obtain the formula for the operator $T\delta_g\colon C_0^\infty(R^2) \to C^\infty(R^2)$:

$$(T\delta_g)w(x) = \int_{R^2}\int_{\mathbf{C}^2} K(y, g^{-1}z)\overline{K(x,z)}e^{-|z|^2/2}w(y)\,d\,\mathrm{vol}(z)\,dy. \tag{3.46}$$

Thus we have the inversion formula

$$\begin{aligned} T^*v(g) = \frac{1}{4}\iint_{R^2\times R^2}\int_{\mathbf{C}^2} & (H_1+H_2)v(x,y)K(y,g^{-1}z) \\ & \cdot \overline{K(x,z)}e^{-|z|^2/2}w(y)\,d\,\mathrm{vol}(z)\,dx\,dy. \end{aligned} \tag{3.47}$$

Returning to our subelliptic operator $\mathcal{L}$, we have

PROPOSITION 3.2. *Let $f(-\mathcal{L})\delta_e = k_1(g) \in \mathcal{D}'(\mathrm{SU}(2))$ and let $f(H_1H_2 - 1)$ have kernel $k_2(x,y) \in \mathcal{D}'(R^2 \times R^2)$. Then*

$$
\begin{aligned}
k_1(g) = \frac{1}{4} \iint_{R^2 \times R^2} \int_{\mathbf{C}^2} & (H_1 + H_2) k_2(x,y) K(y, g^{-1}z) \\
& \cdot \overline{K(x,z)} e^{-|z|^2/2} \, d\,\mathrm{vol}(z)\, dx\, dy.
\end{aligned} \tag{3.48}
$$

CHAPTER 3

Compact Lie Groups

Compactness of a Lie group G leads to a much simpler theory of its unitary representations than the general case, due to the fact that the regular representation of G on $L^2(G)$, $\rho(g)u(x) = u(g^{-1}x)$, breaks up into mutually orthogonal finite-dimensional irreducible representations of G, and all irreducible unitary representations of G are contained there. One can analyze these representations in terms of their restrictions to a maximal torus in G. Most of this theory is due to E. Cartan and H. Weyl. Much more extensive discussions of the theory of representations of compact Lie groups are given in Zelobenko [**268**] and Weyl [**261**]; see also Wallach [**253**].

The "classical" compact Lie groups include U(n), SU(n), O(n), SO(n), the double cover Spin(n) of SO(n), discussed further in Chapter 12, and one more family which has been neglected in these notes, the group Sp(n), consisting of $n \times n$ matrices A of *quaternions*, satisfying $A^*A = I$. These groups seem to play a smaller role in analysis than their more powerful noncompact cousins Sp(n, R), introduced in Chapter 1 and studied in more detail in Chapter 11. There are also five "exceptional" (simply connected) compact Lie groups, whose Lie algebras have complexifications denoted $\mathfrak{g}_2$, $\mathfrak{f}_4$, $\mathfrak{e}_6$, $\mathfrak{e}_7$, and $\mathfrak{e}_8$. For more on these, see [**129, 159, 101**].

1. Weyl orthogonality relations and the Peter-Weyl theorem. Here we collect some basic results on finite-dimensional representations of a compact Lie group G. If π is a representation of G on a finite-dimensional complex vector space V, first note we can put an inner product on V so π is unitary. Indeed, let $((u, v))$ be any Hermitian inner product on V, and set

$$(u, v) = \int_G ((\pi(g)u, \pi(g)v))\, dg. \tag{1.1}$$

Note that, if V_1 is a subspace of V invariant under $\pi(g)$ for all $g \in G$, then its orthogonal complement with respect to such an inner product is also invariant. Thus, if π is not irreducible on V, we can decompose it, and we can obviously continue this process only a finite number of times, if $\dim V$ is finite. Thus π breaks up into a direct sum of irreducible unitary representations of G.

Let π and λ be two representations of G, on V and W, respectively. We say they are equivalent if there is $A \in \mathcal{L}(V,W)$, invertible, such that

$$\pi(g) = A^{-1}\lambda(g)A, \quad \text{for all } g \in G. \tag{1.2}$$

If these representations are unitary, we say they are unitarily equivalent if A can be taken unitary. Suppose that π and λ are irreducible and unitary, and (1.2) holds. Then

$$\pi(g)^* = A^*\lambda(g)^*(A^{-1})^*, \qquad g \in G, \tag{1.3}$$

and since $\pi(g)^* = \pi(g)^{-1}$, we also have

$$\pi(g) = A^*\lambda(g)(A^*)^{-1} \quad \text{for all } g \in G. \tag{1.4}$$

Hence

$$\pi(g) = A^*A\pi(g)(A^*A)^{-1}. \tag{1.5}$$

By Schur's lemma (for a proof of which, see Appendix B), A^*A must be a (positive) scalar, say λ^2. Replacing A by $\lambda^{-1}A$ makes it unitary. Breaking a general π up into irreducible representations, we deduce that, whenever π and λ are finite-dimensional unitary representations of G, if they are equivalent, then they are unitarily equivalent.

We now derive the Weyl orthogonality relations and the Peter-Weyl theorem. As one preliminary to the Weyl orthogonality relations, let π and λ be two irreducible representations of a compact group G, on finite-dimensional spaces V and W, respectively. Consider the representation

$$\nu = \pi \otimes \overline{\lambda} \quad \text{on } V \otimes W' \approx \mathcal{L}(W,V). \tag{1.6}$$

This can be defined by

$$\nu(g)(A) = \pi(g)A\lambda(g)^{-1}, \qquad g \in G, \ A \in \mathcal{L}(W,V). \tag{1.7}$$

Let Z be the linear subspace of $\mathcal{L}(W,V)$ on which ν acts trivially. We want to specify Z. Note that $A_0 \in Z$ if and only if

$$\pi(g)A_0 = A_0\lambda(g), \quad \text{all } g \in G. \tag{1.8}$$

Since this implies range A_0 is invariant under π and $\ker A_0$ is invariant under $\lambda(g)$, we see that either $A_0 = 0$ or A_0 is an isomorphism from W to V. In the latter case we have

$$\pi(g) = A_0\lambda(g)A_0^{-1}, \tag{1.9}$$

so the representations π and λ would have to be equivalent. In this case, for arbitrary $A \in Z$, we would have $\pi(g)A = A\lambda(g) = AA_0^{-1}\pi(g)A_0$, or $\pi(g)AA_0^{-1} = AA_0^{-1}\pi(g)$, so Schur's lemma implies AA_0^{-1} is a scalar. Thus we have proved

PROPOSITION 1.1. *If π and λ are finite-dimensional irreducible representations of G, and if $\nu = \pi \otimes \overline{\lambda}$, then the trivial representation occurs not at all in ν if π and λ are not equivalent, and it occurs acting on a one-dimensional subspace of $V \otimes W'$ if π and λ are equivalent.*

Our next ingredient for the orthogonality relation is the study of the operator

$$P = \int_G \pi(g)\, dg. \tag{1.10}$$

Here π is a finite-dimensional representation of the compact group G on V, not necessarily irreducible, and dg denotes Haar measure, normalized so meas$(G) = 1$. Note that

$$\pi(y)P = \int_G \pi(yg)\, dg = P = P\pi(y) \tag{1.11}$$

for all $y \in G$, and hence

$$P^2 = P\int_G \pi(g)\, dg = \int_G P\pi(g)\, dg = P, \tag{1.12}$$

so P is a projection. Also, if π is unitary, a simple calculation gives

$$P^* = P. \tag{1.13}$$

We see that, if π is unitary, π gives a representation on both the range $\mathcal{R}(P)$ and the kernel $\ker P$. It is clear from (1.10) that, given $v \in V$, $\|Pv\| < \|v\|$ unless $\pi(g)v = v$ for all $g \in G$. Consequently, π operates like the identity on $\mathcal{R}(P)$, but $\pi(g)v = v$ for all $g \in G$ does not hold for any nonzero $v \in \ker P$. We have proved

PROPOSITION 1.2. *If π is a unitary representation of G on V, then P, given by* (1.10), *is the orthogonal projection onto the subspace of V on which π acts trivially.*

We remark that the proof of this also holds for infinite-dimensional unitary representations of G. A special case of Proposition 1.2 is

COROLLARY 1.3. *If π is a nontrivial irreducible unitary representation, P given by* (1.10), *then $P = 0$.*

We apply Proposition 1.2 to

$$P_1 = \int_G \pi(g) \otimes \overline{\lambda}(g)\, dg \tag{1.14}$$

with π and λ irreducible. By Proposition 1.1, we see that

$$P_1 = 0 \quad \text{if } \pi \text{ and } \lambda \text{ are not equivalent.} \tag{1.15}$$

On the other hand, if we set $\lambda = \pi$, P_1 has as its range the set of scalar multiples of the identity operator on V (if π acts on V). Note that $\pi \otimes \overline{\pi}$ leaves invariant the space of elements $A \in \mathcal{L}(V,V)$ of trace zero, which is complementary to the scalars, so P_1 must annihilate this space. Thus, P_1 is uniquely characterized by

$$P_1(A) = (d^{-1}\operatorname{tr} A)I, \quad \text{if } \pi = \lambda,\ d = \dim V. \tag{1.16}$$

The identities (1.15) and (1.16) are equivalent to the Weyl orthogonality relations. If we express π and λ as matrices, with respect to some orthonormal bases, we get

THEOREM 1.4. *Let π and λ be inequivalent irreducible unitary representations of G, on V and W, with matrix entries π_{ij}, λ_{kl}. Then*

$$\int_G \pi_{ij}(g)\overline{\lambda_{kl}(g)}\,dg = 0. \tag{1.17}$$

Also

$$\int_G \pi_{ij}(g)\overline{\pi_{kl}(g)}\,dg = 0 \quad \textit{unless } i = k \textit{ and } j = l, \tag{1.18}$$

and finally

$$\int_G |\pi_{ij}(g)|^2\,dg = d^{-1}. \tag{1.19}$$

where $d = \dim V = \operatorname{tr}\pi(e)$.

Thus if $\{\pi^k\}$ is a complete set of inequivalent representations of G on spaces V_k, of dimension d_k, then

$$d_k^{1/2}\pi_{ij}^k(g) \tag{1.20}$$

forms an orthonormal set in $L^2(G)$. The content of the Peter-Weyl theorem is

THEOREM 1.5. *The orthonormal set* (1.20) *is complete.*

In other words, the linear span of (1.20) is dense. If G is given as a group acting on some vector space $\mathbf{C}^N$, this result is elementary. In fact, the linear span of (1.20) is an *algebra* (take tensor products of π^k and π^l and decompose into irreducibles), closed under complex conjugates (take the contragredient representations), so if we know it separates points (which is clear if $G \subset \mathrm{U}(N)$), the Stone-Weierstrass theorem applies.

If we do not know a priori that G is faithfully represented on some $\mathbf{C}^N$, we can recover this result by considering the regular representation of G on $L^2(G)$, acting on the eigenspaces of the Laplace operator Δ on G (endowed with a bi-invariant Riemannian metric). The proof is as follows. Since Δ is an elliptic selfadjoint operator on G, the resolvent $(I-\Delta)^{-1}$ is compact, and hence Δ has a sequence E_j of eigenspaces, which span $L^2(G)$. We claim that each linear space V_k, the linear span of $\pi_{ij}^k(g)$ for $1 \le i,j \le d_k$, is contained in some eigenspace E_l of Δ. Indeed, Δ leaves V_k invariant and commutes with the action $\pi^{(2)}$ of $G \times G$ on V_k given by $\pi^{(2)}(g_1,g_2)v(x) = v(g_1^{-1}xg_2)$, which is easily seen to be irreducible, so Schur's lemma implies Δ is scalar on V_k. Now, if Theorem 1.5 were false, then the representation B_{l_0} of $G \times G$ on some eigenspace E_{l_0} defined by $B_{l_0}(g_1, g_2)u(x) = u(g_1^{-1}xg_2)$ can be split into irreducibles, which would include each V_k contained in E_{l_0} (if any), and then at least one irreducible piece F left over, orthogonal to each of the V_k contained in E_{l_0}, and indeed to all of the spaces V_k which are linear spans of the functions (1.20). We will show by contradiction that such F cannot exist.

Pick an orthonormal basis $f_1, \dots, f_K$ for F. Then, for all $g_j \in G$, $x \in G$, $f_j(g_1^{-1}xg_2) = \sum_k B_{jk}(g_1, g_2) f_k(x)$. The matrix $(B_{jk}(g_1, g_2))$ is the matrix of the representation B_{l_0} of $G \times G$, restricted to F, with respect to this basis. If we set $g_1 = e$, then we get

$$f_j(xg) = \sum b_{jk}(g) f_k(x), \tag{1.21}$$

and $(b_{jk}(g))$ is the matrix of a representation of G. This representation breaks down into irreducibles, and hence each function $b_{jk}(g)$ on G must be a finite linear combination of functions of the form (1.20). However, (1.21) implies $b_{jk}(g) = \int_G f_j(xg)\overline{f_k(x)}\, dx$, and since $f_j(g)$ belongs to the space F, invariant under the action of B_{l_0}, it follows that each function $b_{jk}(g)$ belongs to F. This contradicts the orthogonality of F with each V_k, and completes the proof of the Peter-Weyl theorem.

In this proof we found it convenient to exploit the Laplace operator, or equivalently its resolvent $(1 - \Delta)^{-1}$. This could be replaced by any bi-invariant compact operator K on $L^2(G)$, which is selfadjoint and positive, with only a finite-dimensional kernel. Here is another way to construct such an operator. Consider the convolution operator

$$K_f u = \check{f} * f * u \tag{1.22}$$

where $\check{f}(g) = \overline{f(g^{-1})}$. Note that

$$(K_f u, u) = \|f * u\|_{L^2}^2. \tag{1.23}$$

Now, let f_j be a countable dense subset of the unit ball of $C(G)$, and form

$$K_0 = \sum 2^{-j} K_{f_j}. \tag{1.24}$$

If G has unit Haar measure, it is easy to see this is a norm convergent series, so K_0 is a compact operator. Note that

$$K_0 u = \left(\sum 2^{-j} \check{f}_j * f_j\right) * u. \tag{1.25}$$

K_0 is selfadjoint and positive, with zero kernel. The operator K_0 is right invariant, but it may not be left invariant. To fix this up, we need merely form

$$K = \int_G L(g)^{-1} K_0 L(g)\, dg \tag{1.26}$$

where

$$L(g)u(x) = u(g^{-1}x), \qquad g, x \in G. \tag{1.27}$$

It is clear that this proof works not only for a compact Lie group, but for any compact metrizable group. The Peter-Weyl theorem can even be extended to any compact topological group, not necessarily metrizable; see [**111**].

Throughout this section we have considered only finite-dimensional irreducible unitary representations of G. In fact, all irreducible unitary representations of a compact Lie group are finite-dimensional. We end this section with a proof of this fact, which is a simple consequence of the Peter-Weyl theorem.

LEMMA 1.6. *Let π and λ be two inequivalent irreducible unitary representations of G, with matrix entries $(\pi_{ij}(g))$ and $(\lambda_{kl}(g))$ with respect to some orthonormal bases of their representation spaces. Assume λ is finite-dimensional, but not necessarily π. Then the conclusion* (1.17) *in Theorem* 1.4 *remains valid, i.e.,*

$$\int_G \pi_{ij}(g)\overline{\lambda_{kl}(g)}\,dg = 0.$$

PROOF. All the details of the proof of (1.17) in Theorem 1.4 continue to apply, from (1.6), on, including the identification $V \otimes W' \approx \mathcal{L}(V, W)$, provided V is finite-dimensional.

PROPOSITION 1.7. *If π is an irreducible unitary representation of the compact Lie group G, then π is finite-dimensional.*

PROOF. If π is infinite-dimensional, it is not equivalent to any finite-dimensional representation, so $(\pi_{ij}, \lambda_{kl})_{L^2(G)} = 0$ for all finite-dimensional representations λ. By the Peter-Weyl theorem, this implies $\pi_{ij}(g) = 0$, a contradiction.

We made use of the Peter-Weyl theorem to provide a short proof of Proposition 1.7. For a proof independent of the Peter-Weyl theorem, which also applies to Banach space representations, see Warner [**256**]. Another simple corollary will be drawn from the Peter-Weyl theorem in the following classification of the irreducible unitary representations of a Cartesian product of two compact Lie groups.

PROPOSITION 1.8. *If G_1 and G_2 are two compact Lie groups, then the irreducible unitary representations of $G = G_1 \times G_2$ are, up to unitary equivalence, precisely those of the form*

$$\pi(g) = \pi_1(g_1) \otimes \pi_2(g_2) \tag{1.28}$$

where $g = (g_1, g_2) \in G$, and $\pi_j \in \hat{G}_j$ is a general irreducible unitary representation of G_j.

PROOF. Given π_j irreducible unitary representations of G_j, the irreducibility and unitarity of (1.28) are clear. It remains to prove the completeness of the set of such representations. For this, it suffices to show that the matrix entries of such representations have dense linear span in $L^2(G_1 \times G_2)$. This follows from the general elementary fact that tensor products of bases of $L^2(G_1)$ and $L^2(G_2)$

form a basis of $L^2(G_1 \times G_2)$, since the matrix entries of the irreducible unitary representations of each G_j, normalized, form a basis of $L^2(G_j)$.

2. Roots, weights, and the Borel-Weil theorem. The notions of roots and weights, described for U(n) in §2 of Chapter 2, have natural generalizations for compact Lie groups, which we now discuss.

Let G be a compact connected Lie group, with Lie algebra $\mathfrak{g}$. Endow G with a bi-invariant metric $\langle\ ,\ \rangle$, so for $X \in \mathfrak{g}$, the map $\operatorname{ad} X$ is a skew adjoint operator on $\mathfrak{g}$. Let T^n be a torus in G (hence a commutative subgroup of G) of maximal dimension. We call T^n a *maximal torus.* Denote the Lie algebra of T^n by $\mathfrak{h}$; $\mathfrak{h}$ is a commutative subalgebra of $\mathfrak{g}$, of maximal dimension.

If $h_1, \dots, h_n$ form a basis of $\mathfrak{h}$, we can simultaneously put the skew adjoint operators $\operatorname{ad} h_j$ $(1 \le j \le n)$ acting on $\mathfrak{g}$ in normal form. In fact, for almost all choices $r_j \in R^+$, $h^b = \sum_j r_j h_j \in \mathfrak{h}$ separates out the spectra of $\operatorname{ad} h_j$, $1 \le j \le n$, and it suffices to put $\operatorname{ad} h^b$ in normal form. So there is a set of elements $x_1, \dots, x_k, y_1, \dots, y_k$ $(2k + n = \dim \mathfrak{g})$, such that

$$\{h_1, \dots, h_n, x_1, \dots, x_k, y_1, \dots, y_k\}$$

form a basis of $\mathfrak{g}$ and such that

$$\operatorname{ad} h(x_j \pm i y_j) = \pm i \alpha_j(h)(x_j \pm i y_j), \qquad h \in \mathfrak{h}, \tag{2.1}$$

for certain $\alpha_j \in \mathfrak{h}'$. Put another way, we can write the complexified algebra $\mathbf{C}\mathfrak{g}$ as

$$\mathbf{C}\mathfrak{g} = \mathbf{C}\mathfrak{h} + \bigoplus_{\alpha} \mathfrak{g}_\alpha \tag{2.2}$$

where, given $\alpha \in \mathfrak{h}'$,

$$\mathfrak{g}_\alpha = \{z \in \mathbf{C}\mathfrak{g} \colon [h, z] = i\alpha(h) z, \ \text{all } h \in \mathfrak{h}\}. \tag{2.3}$$

If $\mathfrak{g}_\alpha \ne 0$, we call α a *root*, and nonzero elements of $\mathfrak{g}_\alpha$ are called *root vectors*, provided $\alpha \ne 0$. Note that $\mathfrak{g}_0 = \mathbf{C}\mathfrak{h}$. From the Jacobi identity, which implies

$$\operatorname{ad} h([z_\alpha, z_\beta]) = [\operatorname{ad} h(z_\alpha), z_\beta] + [z_\alpha, \operatorname{ad} h(z_\beta)], \tag{2.4}$$

it follows that

$$[\mathfrak{g}_\alpha, \mathfrak{g}_\beta] \subset \mathfrak{g}_{\alpha+\beta}. \tag{2.5}$$

Note from (2.1) that if α is a root, so is $-\alpha$.

Now introduce an ordering on $\mathfrak{h}'$. Pick a basis of $\mathfrak{h}'$ to make $\mathfrak{h}' \approx R^n$, and then say $\alpha - \beta > 0$ (or $\alpha > \beta$) if the first nonzero coordinate of $\alpha - \beta$ is positive. The root vectors corresponding to positive roots will play the role of raising operators in the representation theory for G. If we consider as a special case the adjoint representation of G on $\mathfrak{g}$, we get information on the structure of $\mathfrak{g}$. This turns out to be the key to the classification of compact Lie algebras. We will not do much on that here, but will present a few facts, including establishing

$$\dim \mathfrak{g}_\alpha = 1 \quad \text{for each root } \alpha. \tag{2.6}$$

Extend the inner product $\langle\ ,\ \rangle$ on $\mathfrak{g}$ to a symmetric *bilinear* form on $\mathbf{C}\mathfrak{g} \times \mathbf{C}\mathfrak{g}$. Also, use the restriction of the metric $\langle\ ,\ \rangle$ to $\mathfrak{h}$ to define an isomorphism of $\mathfrak{h}'$ with $\mathfrak{h}$. Thus, each root α determines an element $H_\alpha \in \mathfrak{h}$ by

$$\alpha(h) = \langle H_\alpha, h\rangle \quad \text{for all } h \in \mathfrak{h}. \tag{2.7}$$

LEMMA 2.1. *If $X \in \mathfrak{g}_\alpha$, $Y \in \mathfrak{g}_{-\alpha}$, then*

$$[X, Y] = i\langle X, Y\rangle H_\alpha. \tag{2.8}$$

PROOF. By (2.5), $[X, Y] \in \mathbf{C}\mathfrak{h} = \mathfrak{g}_0$. Now, for any $H \in \mathfrak{h}$,

$$\langle [X, Y], H\rangle = \langle Y, [H, X]\rangle = i\alpha(H)\langle Y, X\rangle, \tag{2.9}$$

and, by (2.7), this identity is equivalent to (2.8).

Note that

$$\alpha(H_\alpha) = \langle H_\alpha, H_\alpha\rangle > 0. \tag{2.10}$$

We are now in a position to prove

PROPOSITION 2.2.

$$\dim \mathfrak{g}_\alpha = 1 \quad \textit{for each root } \alpha.$$

PROOF. Suppose $\dim \mathfrak{g}_\alpha \geq 2$. Pick linearly independent $X, Z \in \mathfrak{g}_\alpha$, and pick $Y \in \mathfrak{g}_{-\alpha}$. Granted $\mathfrak{g}_\alpha$ and $\mathfrak{g}_{-\alpha}$ are not orthogonal, which we check in a moment, we can suppose normalization made so that $\langle X, Y\rangle = 1$, $\langle Z, Y\rangle = 0$. This implies $[X, Y] = iH_\alpha$, $[Z, Y] = 0$. Now an inductive argument (exercise) shows

$$\operatorname{ad} Y (\operatorname{ad} X)^n Z = -\tfrac{1}{2} n(n+1)\alpha(H_\alpha)(\operatorname{ad} X)^{n-1} Z. \tag{2.11}$$

By (2.10), it follows that all the elements $(\operatorname{ad} X)^n Z \in \mathfrak{g}_{(n+1)\alpha}$ are nonzero. This would imply $\dim \mathfrak{g} = \infty$, a contradiction.

It remains to show $\mathfrak{g}_\alpha$ and $\mathfrak{g}_{-\alpha}$ are not orthogonal with respect to $\langle\ ,\ \rangle$. Indeed, $x + iy \in \mathfrak{g}_\alpha (x, y \in \mathfrak{g}) \Rightarrow x - iy \in \mathfrak{g}_{-\alpha}$. But $\langle x + iy, x - iy\rangle = \langle x, x\rangle + \langle y, y\rangle > 0$. This completes the argument.

For each root α, pick nonzero vectors $e_\alpha \in \mathfrak{g}_\alpha$. We can normalize so $\langle e_\alpha, e_{-\alpha}\rangle = 1$. Thus

$$[e_\alpha, e_{-\alpha}] = iH_\alpha. \tag{2.12}$$

Note that, by (2.1), we can write

$$e_{\pm\alpha} = x_\alpha \pm i y_\alpha, \qquad x_\alpha, y_\alpha \in \mathfrak{g}. \tag{2.13}$$

The commutation relations (2.12) are equivalent to

$$[x_\alpha, y_\alpha] = -\tfrac{1}{2} H_\alpha. \tag{2.14}$$

We turn now to the representation theory of G. Let π be a unitary representation of G on a complex vector space V, $\dim V < \infty$. Then π gives a skew adjoint representation of the Lie algebra $\mathfrak{g}$, which in turn extends to a complex linear representation (also denoted π) of the complexified Lie algebra $\mathbf{C}\mathfrak{g}$. As in

Chapter 2, we also find it convenient to analytically continue π on G. Let $\tilde{G}_{\mathbf{C}}$ denote the simply connected Lie group with Lie algebra $\mathbf{C}\mathfrak{g}$. $\tilde{G}_{\mathbf{C}}$ has a natural complex manifold structure. Let G_1 be the Lie subgroup of $\tilde{G}_{\mathbf{C}}$ with Lie algebra $\mathfrak{g}$. We claim G_1 covers G. To prove this we can suppose that G, being compact, is faithfully represented on a finite-dimensional space W, $\rho\colon G \to \operatorname{End}(W)$. Then ρ defines a representation of $\mathfrak{g}$, hence of $\mathbf{C}\mathfrak{g}$, and hence a representation $\tilde{\rho}$ of $\tilde{G}_{\mathbf{C}}$, which restricts to the desired covering $\tilde{\rho}\colon G_1 \to G$. The kernel of this covering homomorphism is a discrete subgroup Γ of the center of G_1. Analytic continuation implies it belongs to the center of $\tilde{G}_{\mathbf{C}}$, so $G_{\mathbf{C}} = \tilde{G}_{\mathbf{C}}/\Gamma$ is a group which is the complexification of G. A complex linear representation π of the complex Lie algebra $\mathbf{C}\mathfrak{g}$ of $G_{\mathbf{C}}$ (whose restriction to $\mathfrak{g}$ exponentiates to G) exponentiates to a holomorphic representation (also denoted π) of $G_{\mathbf{C}}$; i.e., with respect to a basis of V, the matrix entries of π are holomorphic functions on $G_{\mathbf{C}}$.

To pursue our analysis of a representation π of G on V, let T^n denote a maximal torus of G. Thus the action of T^n on V can be simultaneously diagonalized. Thus, given $\lambda \in \mathfrak{h}'$, if we set

$$V_\lambda = \{v \in V : \pi(h)v = i\lambda(h)v, \text{ for all } h \in \mathfrak{h}\}, \tag{2.15}$$

we have

$$V = \bigoplus V_\lambda. \tag{2.16}$$

If $\lambda \in \mathfrak{h}'$, and $V_\lambda \neq 0$, we call λ a *weight*, and any nonzero $v \in V_\lambda$ a *weight vector*. For e_α the root vectors in $\mathbf{C}\mathfrak{g}$ defined above, set

$$E_\alpha = \pi(e_\alpha). \tag{2.17}$$

We call E_α a *raising operator* if $\alpha > 0$ and a lowering operator if $\alpha < 0$.

We want to study the action of the operators E_α on the V_λ's. To do this, we use the commutation relations

$$[h, e_\alpha] = i\alpha(h)e_\alpha, \qquad h \in \mathfrak{h}, \tag{2.18}$$

to get

$$\pi(h)E_\alpha = E_\alpha\pi(h) + i\alpha(h)E_\alpha. \tag{2.19}$$

This implies

PROPOSITION 2.3. *For each root α, we have*

$$E_\alpha : V_\lambda \to V_{\lambda+\alpha}. \tag{2.20}$$

In particular, if λ is a weight, and α is a root, then either E_α annihilates V_λ, or $\lambda + \alpha$ is a weight.

PROOF. If $\xi \in V_\lambda$, we have

$$\begin{aligned} \pi(h)(E_\alpha\xi) &= E_\alpha(\pi(h)\xi) + i\alpha(h)E_\alpha\xi \\ &= i(\lambda(h) + \alpha(h))E_\alpha\xi, \end{aligned} \tag{2.21}$$

which proves the proposition.

The ordering we have put on $\mathfrak{h}'$ induces an ordering on the weights. For a given finite-dimensional representation π, with respect to this ordering there will be a *highest weight* λ_m and also a lowest weight λ_s. From Proposition 2.3 we see that

$$\begin{aligned} &E_\alpha = 0 \text{ on } V_{\lambda_m} \text{ for all raising operators } E_\alpha, \\ &E_\alpha = 0 \text{ on } V_{\lambda_s} \text{ for all lowering operators } E_\alpha. \end{aligned} \tag{2.22}$$

In general, call a weight λ nonraisable if V_λ is annihilated by all raising operators and call it nonlowerable if V_λ is annihilated by all lowering operators. Shortly we will show that, if π is irreducible, then the only nonraisable weight is maximal. At present, we record the following progress.

PROPOSITION 2.4. *If π is a unitary representation of the compact Lie group G on V, $\dim V < \infty$, then there exists a highest weight vector ξ, and in particular there exists a nonzero weight vector $\xi \in V$ annihilated by all raising operators.*

This result gives a tool for showing that certain representations of G are irreducible, namely,

COROLLARY 2.5. *Let π be a unitary representation of G on V, $\dim V < \infty$. Suppose the set of weight vectors $\xi \in V$ annihilated by all raising operators is equal to the set of nonzero multiples of a single element. Then π is irreducible.*

PROOF. Otherwise, $V = V_1 \oplus V_2$ with π acting on each factor, and Proposition 2.4 produces two linearly independent weight vectors $\xi_j \in V_j$ annihilated by all raising operators.

We now head for a circle of results which includes a converse for this corollary, which will closely parallel the treatment of Proposition 2.4 in Chapter 2. With π representing G on V as above, let $\overline{\pi}$ be the contragredient representation, of G on V', defined by

$$\langle \xi, \overline{\pi}(g)\eta \rangle = \langle \pi(g^{-1})\xi, \eta \rangle, \qquad \xi \in V, \eta \in V'. \tag{2.23}$$

Suppose $\xi_0 \in V$ is a nonraisable weight vector for π, with weight $\lambda \in \mathfrak{h}'$, and suppose $\eta_0 \in V'$ is a nonlowerable weight vector for $\overline{\pi}$, with weight $-\mu \in \mathfrak{h}'$. As in the section on the unitary group $\mathrm{U}(n)$, we will study the function

$$\varphi(g) = \langle \pi(g)\xi_0, \eta_0 \rangle. \tag{2.24}$$

We can take the extension of π to a representation of the complexified group $G_{\mathbf{C}}$, so $\varphi(g)$ is defined and holomorphic on $G_{\mathbf{C}}$. Let N_+ be the subgroup of $G_{\mathbf{C}}$ whose Lie algebra $\mathcal{N}_+$ is generated by the raising operators e_α, $\alpha > 0$, and let N_- be the subgroup of $G_{\mathbf{C}}$ whose Lie algebra $\mathcal{N}_-$ is generated by the lowering operators e_α, $\alpha < 0$. Let D denote the complexification of the maximal torus T^n, which is isomorphic to $(\mathbf{C}^*)^n$, $\mathbf{C}^* = \{z \in \mathbf{C}: z \neq 0\}$. Since $\mathbf{C}\mathfrak{h}$, $\mathcal{N}_+$, and $\mathcal{N}_-$ clearly span $\mathbf{C}\mathfrak{g}$, it follows from the inverse function theorem that

$$N_- D N_+ = G_{\text{reg}} \tag{2.25}$$

is a subset of $G_{\mathbb{C}}$ which contains an open neighborhood of the identity element e. Let

$$g = \varsigma\delta z, \qquad \varsigma \in N_-,\ \delta \in D,\ z \in N_+. \tag{2.26}$$

We see that, if $\delta = \exp(h_1 + ih_2)$, $h_j \in \mathfrak{h}$, then

$$\varphi(\varsigma g) = \varphi(g), \qquad \varphi(\delta g) = e^{i(\mu(h_1)+i\mu(h_2))}\varphi(g), \tag{2.27}$$

and

$$\varphi(gz) = \varphi(g), \qquad \varphi(g\delta) = e^{i(\lambda(h_1)+i\lambda(h_2))}\varphi(g). \tag{2.28}$$

Thus

$$\begin{aligned} \varphi(\varsigma\delta z) = \varphi(\delta) &= \exp i(\lambda(h_1) + i\lambda(h_2))\varphi(e) \\ &= \exp i(\mu(h_1) + i\mu(h_2))\varphi(e). \end{aligned} \tag{2.29}$$

LEMMA 2.6. $\varphi(e) \neq 0$.

PROOF. If $\varphi(e) = 0$, then (2.29) implies $\varphi(g) = 0$ on G_{reg}. Since φ is holomorphic and G_{reg} contains a neighborhood of e, it follows that $\varphi(g) = 0$ on G. But if

$$V_0 = \{\xi \in V : \langle \pi(g)\xi, \eta_0 \rangle = 0 \text{ for all } g \in G\} \tag{2.30}$$

then V_0 is invariant, and since clearly $V_0 \neq V$, we have $V_0 = 0$ if π is irreducible. This proves the lemma.

From (2.29), we can deduce the following important result.

THEOREM 2.7. *If π is irreducible on V, then the only weight λ which is non-raisable is the highest weight. Furthermore, the highest weight vector is unique, up to scalar multiple. Finally, if π and π_2 have the same highest weight, they are unitarily equivalent.*

PROOF. The identity $\lambda = \mu$ proves the uniqueness of λ. Note that if we normalize the weight vectors so $\varphi(e) = 1$, the function $\varphi(g)$ is uniquely characterized by the following three properties:

$$\varphi(g) \text{ holomorphic on } G_{\mathbb{C}}, \tag{2.31}$$

$$\varphi(\varsigma g z) = \varphi(g), \qquad \varsigma \in N_-,\ z \in N_+, \tag{2.32}$$

$$\varphi(\delta) = \exp(i(\lambda(h_1) + i\lambda(h_2))) \quad \text{for } \delta = \exp(h_1 + ih_2) \in D. \tag{2.33}$$

Thus, if ξ_1 were another highest weight vector, also normalized so $\langle \xi_1, \eta_0 \rangle = 1$, we would have $\langle \pi(g)\xi_1, \eta_0 \rangle = \varphi(g)$, so $\langle \pi(g)(\xi_1 - \xi_0), \eta_0 \rangle = 0$ for all g, or $\langle \xi_1 - \xi_0, \overline{\pi}(g)\eta_0 \rangle = 0$ for all g. Since $\overline{\pi}$ is irreducible, this implies $\xi_1 = \xi_0$.

As for the final assertion of the proposition, let π_2 be an irreducible representation on V_2. Pick a maximal weight vector ξ_2 for π_2 and a minimal weight vector η_2 for its contragredient representation $\overline{\pi}_2$, normalized so $\langle \xi_2, \eta_2 \rangle = 1$, and form $\varphi_2(g) = \langle \pi_2(g)\xi_2, \eta_2 \rangle$. Since $\varphi_2(g)$ satisfies the conditions (2.31)–(2.33) provided λ is also the highest weight for π_2, we see that $\varphi(g) = \varphi_2(g)$.

But if π and π_2 are not equivalent, the Weyl orthogonality relations imply the restrictions of π and π_2 to G must be orthogonal, and certainly cannot coincide. This proves that the two irreducible representations π and π_2 must be equivalent if their maximal weights coincide. This completes the proof of the theorem.

To each $\xi \in V$, associate the function

$$f_\xi(g) = \langle \pi(g)\xi, \eta_0 \rangle. \tag{2.34}$$

As shown in the proof that $V_0 = 0$, with V_0 given by (2.30), if π is irreducible, this map provides an isomorphism of V onto a certain space of functions in $C^\infty(G)$, namely the space

$$W_\lambda = \{f_\xi : \xi \in V\}. \tag{2.35}$$

The action of π on G is given in this realization by the right regular representation

$$\rho(g)f_\xi(x) = f_\xi(xg), \qquad f_\xi \in W_\lambda. \tag{2.36}$$

Note that every $f \in W_\lambda$ extends uniquely to a holomorphic function on $G_{\mathbf{C}}$. Since η_0 is a lowest weight vector for $\overline{\pi}$, we have, for $f \in W_\lambda$,

$$f(\varsigma x) = f(x), \qquad x \in G_{\mathbf{C}},\ \varsigma \in N_-. \tag{2.37}$$

We also have

$$f(\delta x) = e^{i(\lambda(h_1)+i\lambda(h_2))} f(x), \qquad \delta = \exp(h_1 + ih_2) \in D. \tag{2.38}$$

Note that the left regular representation l of G on $L^2(G)$:

$$l(g)u(x) = u(g^{-1} \cdot x) \tag{2.39}$$

represents a vector field $X \in \mathfrak{g}$ by

$$l(X)u = (d/dt)u((\exp tX)^{-1} \cdot g)|_{t=0} = -\hat{X}u, \tag{2.40}$$

where $\hat{X}$ is the *right invariant* vector field on G, equal to X at e. We see that, if $e_\alpha = x_\alpha + iy_\alpha$, α is a negative root, (2.37) gives

$$(\hat{x}_\alpha + i\hat{y}_\alpha)f = 0, \quad \text{all } f \in W_\lambda. \tag{2.41}$$

Let us denote by $\overline{Z}_\alpha$ the right invariant vector field $\hat{x}_\alpha + i\hat{y}_\alpha$, for any negative root α. We have the following useful description of the space W_λ on which the representation π is realized.

THEOREM 2.8. *The space W_λ is equal to the set of $f \in C^\infty(G)$ such that*

$$\overline{Z}_\alpha f = 0 \quad \textit{for each negative root } \alpha, \textit{ and} \tag{2.42}$$

$$f(\delta x) = e^{i\lambda(h)} f(x), \quad \textit{if } \delta = \exp h \in T^n. \tag{2.43}$$

PROOF. For the moment, denote the space of functions satisfying these conditions by $W^\#$. We have seen that $W_\lambda \subset W^\#$. To prove they are identical, it suffices to show the right regular representation of G leaves $W^\#$ invariant and is irreducible on $W^\#$. Since $\overline{Z}_\alpha$ commute with right translations, it is clear that $f(x) \in W^\#$ implies $\rho(g)f(x)$ belongs to $W^\#$ for all $g \in G$. To check irreducibility, let $f^\# \in W^\#$ be a nonraisable weight vector. If we can prove $f^\#$ is a multiple of $\varphi(x)$, defined by (2.24), we will be done.

Since the system of equations (2.42) satisfies the condition of transversal ellipticity with respect to the cosets gT^n, results from PDE guarantee that $\dim W^\# < \infty$, and each element of $W^\#$ is real analytic. Thus $f^\#$ extends uniquely to be holomorphic in a neighborhood Ω of G in $G_{\mathbf{C}}$, and (2.42) implies

$$f^\#(\varsigma x) = f^\#(x), \qquad \varsigma \in N_-, \ x \in \Omega. \tag{2.44}$$

The hypothesis that $f^\#$ is a nonraisable weight vector implies

$$\rho(e_\alpha)f^\# = 0, \quad \text{each positive root } \alpha, \tag{2.45}$$

which implies, for its holomorphic extension,

$$f^\#(xz) = f(x), \qquad x \in \Omega, \ z \in N_+. \tag{2.46}$$

Thus, by the analysis of the conditions (2.31)–(2.33) characterizing φ, we see that $f^\#(x) = f^\#(e)\varphi(x)$, $x \in G$. Thus G acts irreducibly on $W^\#$, so the proof that $W^\# = W_\lambda$ is complete.

We can rephrase the result of Theorem 2.8 in an elegant fashion, as observed by Borel and Weil. The identity (2.43) makes f naturally equivalent to a section $\tilde{f}$ of the line bundle E_λ over $M = G/T^n$ obtained from the principal bundle $G \to G/T^n$ with fiber T^n via the homomorphism $l\colon T^n \to \mathbf{C}^*$ defined by $l(\exp h) = e^{i\lambda(h)}$, $h \in \mathfrak{h}$. The right invariant vector fields $\overline{Z}_\alpha$ on G give rise to linearly independent vector fields, call than $\overline{L}_\alpha$, on M. Since $\overline{Z}_\alpha$ span a complex Lie algebra, so do the $\overline{L}_\alpha$, as α runs over the negative roots, so by the Newlander-Nirenberg theorem (see Hörmander [**122**]) they define an integrable almost complex structure on M. Thus M is a complex manifold in a natural manner. Using the yoga of vector bundles, one sees that E_λ is a holomorphic line bundle over M, and the equations (2.42) are equivalent to saying $\tilde{f}$, the image of f, is a holomorphic section of E_λ over M. Thus Theorem 2.8 is equivalent to the following result, known as the Borel-Weil theorem.

THEOREM 2.9. *Let π be an irreducible unitary representation of the compact Lie group G on V. Let $\lambda \in \mathfrak{h}'$ be its highest weight. Then V is isomorphic to the space of holomorphic sections of the line bundle E_λ over $M = G/T^n$, with its natural complex structure and its natural G action.*

To complete the classification of the irreducible unitary representations of G, it remains to specify which $\lambda \in \mathfrak{h}'$ can occur as highest weights. It is clear that only countably many can occur, by the Weyl orthogonality relations, since $L^2(G)$ is separable. We will now derive the appropriate restriction which $\lambda \in \mathfrak{h}'$ must satisfy in order for it to be a maximal weight.

Recall that, for any positive root α, we can pick $x_\alpha, y_\alpha \in \mathfrak{g}$ such that (2.12)–(2.14) hold. In other words,

$$[x_\alpha, y_\alpha] = -\tfrac{1}{2}H_\alpha, \quad [H_\alpha, x_\alpha] = -\alpha(H_\alpha)y_\alpha, \quad [H_\alpha, y_\alpha] = \alpha(H_\alpha)x_\alpha. \tag{2.47}$$

Now, if we let

$$(2.48)\quad X_1 = \alpha(H_\alpha)^{-1}H_\alpha, \quad X_2 = (2/\alpha(H_\alpha))^{1/2}y_\alpha, \quad X_3 = (2/\alpha(H_\alpha))^{1/2}x_\alpha,$$

recalling from (2.10) that $\alpha(H_\alpha) > 0$, we see that

$$(2.49)\qquad [X_1, X_2] = X_3, \qquad [X_1, X_3] = -X_2, \qquad [X_2, X_3] = X_1,$$

i.e., we have the generators of the Lie algebra of SU(2), studied in §2. The associated Lie algebra injection su(2) $\hookrightarrow \mathfrak{g}$ induces a homomorphism of Lie groups SU(2) $\to G$. Thus an irreducible unitary representation π of G on V composes to give a (possibly reducible) representation $\pi_1 = \pi \circ \rho$ of SU(2) on V.

Let λ be the highest weight of π. Say $V_\lambda \subset V$ is the weight space for λ. Then V_λ is also a space of nonraisable weight vectors for π_1. By the classification of representations of SU(2) achieved in Chapter 2, §1, we see that, for some nonnegative integer n,

$$(2.50)\qquad \pi_1(X_1) = in/2$$

on V_λ. From (2.48) it follows that

$$(2.51)\qquad \pi(H_\alpha) = \alpha(H_\alpha)\pi_1(X_1) = i\alpha(H_\alpha)n/2,$$

on V_λ. Since $\pi(H_\alpha) = i\lambda(H_\alpha)$ on V_λ, we deduce $\lambda(H_\alpha) = \alpha(H_\alpha)n/2$, or

$$(2.52)\qquad 2\lambda(H_\alpha)/\alpha(H_\alpha) = n.$$

In other words, the left side of (2.52) must be a nonnegative integer for each positive root α. Such an element of $\mathfrak{h}'$ is called a *dominant integral weight.*

If G is a compact semisimple Lie group, this "quantization condition" turns out to be all that is needed, at least to produce a finite-dimensional representation of $\mathfrak{g}$ and hence of the simply connected cover of G. This fundamental result is known as the Theorem of the Highest Weight. For a proof, we refer to Chapter 4 of Wallach [**253**]. But we will note that this result can be established simply for the classical groups, and sketch details for $G = \mathrm{SU}(n)$. Our reasoning will use the following simple result.

LEMMA 2.10. *If λ and μ are highest weights for irreducible representations of G, so is $\lambda + \mu$.*

PROOF. If $v_\lambda \in W_\lambda$ and $v_\mu \in W_\mu$ are highest weight vectors for representations π_λ and π_μ of G, on spaces W_λ and W_μ, respectively, then $v_\lambda \otimes v_\mu \in W_\lambda \otimes W_\mu$ is a nonraisable weight vector for $\pi_\lambda \otimes \pi_\mu$, with weight $\lambda + \mu$, the unique such up to a scalar, so a copy of $\pi_{\lambda+\mu}$ is contained in $\pi_\lambda \otimes \pi_\mu$.

From the description of dominant integral weights given above, it is easy to see there exist *fundamental weights* $e_1, \dots, e_k$ ($k = \operatorname{rank} G$) such that each dominant integral weight λ can be written uniquely in the form $\lambda = \sum_{j=1}^k n_j e_j$, n_j a nonnegative integer. In view of Lemma 2.10, we see that to establish the Theorem of the Highest Weight for G, it suffices to show each fundamental weight e_j is a highest weight for some irreducible representation of G.

Let us see how this works for $G = \mathrm{SU}(n)$. A maximal torus consists of T^{n-1}, the set of diagonal matrices whose diagonal elements are unit complex numbers, the product of which is 1. Thus $\mathfrak{h}$ is the set of diagonal skew adjoint matrices of trace 0. In this case, the fundamental weights can be taken to be

$$e_1 = \omega_1, e_2 = \omega_1 + \omega_2, \dots, e_{n-1} = \omega_1 + \omega_2 + \cdots + \omega_{n-1}, \tag{2.53}$$

where $\omega_j \in \mathfrak{h}'$ selects the jth diagonal entry of $\mathfrak{h}$ (and divides by i). The dominant integral weights are of the form

$$\lambda = \sum_{j=1}^{n-1} m_j \omega_j, \qquad m_1 \geq m_2 \geq \cdots \geq m_{n-1}, \tag{2.54}$$

where m_j are nonnegative integers.

It follows from the computations (2.18)–(2.20) of Chapter 2 that the natural representation of $\mathrm{SU}(n)$ on $\bigwedge^k \mathbf{C}^n$ is irreducible with weight e_k, for $1 \leq k \leq n-1$. This establishes the Theorem of the Highest Weight for $\mathrm{SU}(n)$.

The computation (2.17) of Chapter 2 shows that the natural representation of $\mathrm{SU}(n)$ of $S^k\mathbf{C}^n$, also known to be irreducible, has highest weight $k\omega_1 = ke_1$. The natural representation of $\mathrm{SU}(n)$ on $\bigwedge^{n-1} \mathbf{C}^n$ can be seen to be contragredient to that of $\mathrm{SU}(n)$ on $\mathbf{C}^n$, so the contragredient representation of $\mathrm{SU}(n)$ on $S^k(\mathbf{C}^n)'$ has highest weight ke_{n-1}. The irreducible representation of $\mathrm{SU}(n)$ on the trace free part of $S^k(\mathbf{C}^n) \otimes S^l(\mathbf{C}^n)'$, discussed in §2 of Chapter 2, therefore has highest weight $ke_1 + le_{n-1}$. In particular, we see this provides a concrete realization of a complete set of irreducible representations of $\mathrm{SU}(3)$.

Generally, the irreducible representation of $\mathrm{SU}(n)$ with highest weight $\lambda = \sum_{j=1}^{n-1} \nu_j e_j$, where e_j are given by (2.53), is realized on the cyclic subspace of

$$V^{(\nu_1)} \otimes \left(\textstyle\bigwedge^2 V\right)^{(\nu_2)} \otimes \cdots \otimes \left(\textstyle\bigwedge^{n-1} V\right)^{(\nu_{n-1})} \tag{2.55}$$

where $V = \mathbf{C}^n$, $W^{(\nu)} = W \otimes \cdots \otimes W$ (ν factors), generated by the highest weight vector

$$v_\lambda = u_1^{\nu_1} \otimes (u_1 \wedge u_2)^{\nu_2} \otimes \cdots \otimes (u_1 \wedge u_2 \wedge \cdots \wedge u_{n-1})^{\nu_{n-1}} \tag{2.56}$$

where $u_1, \dots, u_n$ denotes an orthonormal basis of $\mathbf{C}^n$, and, for $w \in W$, $w^\nu = w \otimes \cdots \otimes w$ (ν factors) is defined as an element of $W^{(\nu)}$. This cyclic subspace can be described as a linear subspace of

$$\bigotimes^N \mathbf{C}^n, \qquad N = \nu_1 + 2\nu_2 + \cdots + (n-1)\nu_{n-1},$$

as follows. Identifying $\bigotimes^N \mathbf{C}^n$ with $(\bigotimes^N (\mathbf{C}^n)')'$, we regard a tensor in $\bigotimes^N \mathbf{C}^n$ as an N-linear form on $(\mathbf{C}^n)'$. Rather than writing the arguments of $\psi \in (\bigotimes^N (\mathbf{C}^n)')'$ in a line, as $\psi(v_1, \dots, v_N)$, $v_j \in \mathbf{C}^{n\prime}$, we will arrange the arguments in a two-dimensional pattern. Put $v_{1j1}, \dots, v_{1,j,n-1}$ on the first ν_{n-1} columns, $j = 1, \dots, \nu_{n-1}$ (starting with $v_{111}, \dots, v_{1,1,n-1}$ down the first column, if $\nu_{n-1} > 0$), then $v_{2j1}, \dots, v_{2,j,n-1}$ on the next ν_{n-2} columns, etc., and

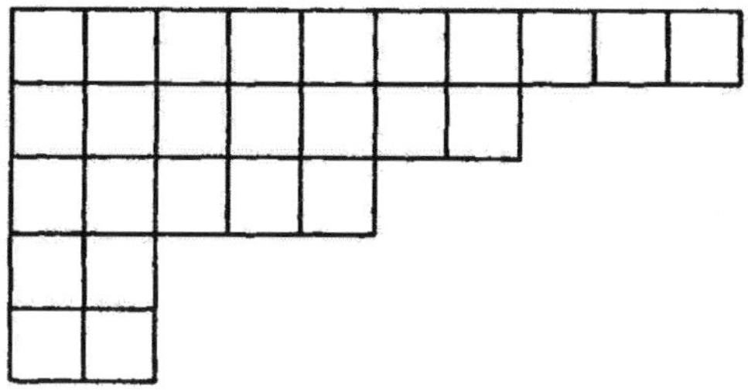

FIGURE 2.1

write the action of ψ as

$$\psi\begin{pmatrix} v_{111}\cdots v_{n-1,1,1}\cdots v_{n-1,\nu_1,1} \\ \vdots \\ v_{1,1,n-1} \end{pmatrix}. \tag{2.57}$$

Many columns will be shorter than the first column, and the pattern of the spots occupied by the v_{jkl} will look like Figure 2.1, called a Young tableau. This particular example is for SU(6), with $\nu_5 = 2$, $\nu_4 = 0$, $\nu_3 = 3$, $\nu_2 = 2$, and $\nu_1 = 3$.

The representation space W_λ associated to the highest weight $\lambda = \sum_{j=1}^{n-1} \nu_j e_j$ is contained in (2.55), so all vectors in $W_\lambda \subset \bigotimes^N \mathbf{C}^n = (\bigotimes^N (C^n)')'$ are antisymmetric under the interchange of any two vectors in any one column of (2.57). An additional symmetry requirement is suggested by our analysis of the representation with highest weight in $(\mathbf{C}^n)^{(k)}$, which is on the space $S^k\mathbf{C}^n$. We are motivated to look at the following two projections in End($\bigotimes^N \mathbf{C}^n$). Let Q be the projection obtained by antisymmetrization with respect to elements in any common column, and let P be the projection obtained by symmetrization with respect to the elements in any common row. P and Q do not commute, and it is a remarkable, though elementary, fact that $C = QP \Rightarrow C^2 = \mu C$ for some $\mu > 0$, so $\mu^{-1}C$ is a projection. It is easy to see that the highest weight vector (2.56) is in the range of this projection. Since the SU(n) action on $\bigotimes^N \mathbf{C}^n$ commutes with the action of the permutation group S_N, it follows that the range of the projection $\mu^{-1}C$ is invariant under SU(n). It can be shown that the action of SU(n) on the range of $\mu^{-1}C$ is irreducible, so the irreducible representation of SU(n) with highest weight $\lambda = \sum_{j=1}^{n-1} \nu_j e_j$ is identified. For the details on this last argument and further use of Young tableaux, we refer to [**261**, **262**], or [**268**].

3. Representations of compact groups on eigenspaces of Laplace operators. Suppose a compact Lie group G acts transitively on a compact Riemannian manifold M as a group of isometries. Then if we set

$$l(g)f(x) = f(g^{-1}x), \qquad g \in G,\ x \in M \tag{3.1}$$

we have a unitary representation of G on $L^2(M)$. Note that $l(g)\colon C^\infty(M) \to C^\infty(M)$. Now if Δ is the Laplace operator, we have

$$l(g)\Delta = \Delta l(g), \qquad g \in G. \tag{3.2}$$

Thus if the eigenspaces of Δ are denoted

$$V_\lambda = \{u \in C^\infty(M)\colon \Delta u = -\lambda^2 u\}, \tag{3.3}$$

we have

$$l(g)\colon V_\lambda \to V_\lambda, \tag{3.4}$$

for each λ such that $-\lambda^2 \in \operatorname{spec} \Delta$.

We introduce the notion of zonal function. Fix $p_0 \in M$, and let

$$K = \{g \in G\colon g \cdot p_0 = p_0\}. \tag{3.5}$$

Let K_0 be the connected component of the identity in K. Let V be a finite-dimensional subspace of $C^\infty(M)$ invariant under $l(g)$. We define

$$\mathfrak{z}(V) = \{v \in V\colon l(k)v = v \text{ for all } k \in K_0\}, \tag{3.6}$$

and call this the space of *zonal functions* in V.

PROPOSITION 3.1. *If $V \neq 0$, there is a nonzero element of $\mathfrak{z}(V)$.*

PROOF. Define $\xi \in V'$ by

$$\xi(f) = f(p_0), \qquad f \in V. \tag{3.7}$$

Then $\xi \neq 0$ since V contains a function not identically zero and G is transitive on M. Thus $\ker \xi \subset V$ is a hyperplane; it is invariant under K_0. Thus $L = (\ker \xi)^\perp \subset V$ is also invariant under K_0. We have $\dim_{\mathbf{C}} L = 1$. It follows from (3.1) that $L = \mathbf{C} \otimes L_R$, $\dim_R L_R = 1$. Since K_0 is connected, and acts on L_R, it must act trivially, so the proof is complete.

COROLLARY 3.2. *If $\dim \mathfrak{z}(V) = 1$, then $l(g)$ is irreducible on V.*

PROOF. If $l(g)$ acts on each factor of $V_1 \oplus V_2 = V$, we get two linearly independent zonal functions $v_j \in V_j$, by Proposition 3.1.

We can use this to show that, under certain circumstances, G acts irreducibly on all the eigenspaces V_λ of Δ.

DEFINITION. We say M is a rank one symmetric space if K_0 acts transitively on the set of unit tangent vectors to M at p_0.

PROPOSITION 3.2. *If M is a compact rank one symmetric space, then G acts irreducibly on each eigenspace V_λ of Δ on M.*

PROOF. It suffices to show that

$$\dim \mathfrak{z}(V_\lambda) = 1. \tag{3.8}$$

Suppose v and w are two linearly independent elements of $\mathfrak{z}(V_\lambda)$. Pick a sequence $r_j \to 0$ such that $v(x)$ and $w(x)$ are nonzero for $\operatorname{dist}(p_0, x) = r_j$. (Note that if

M is a rank 1 symmetric space, then elements $f(x)$ of $\mathfrak{z}(V_\lambda)$ depend only on $\operatorname{dist}(p_0, x)$, for x near p_0.) Since elements of V_λ are real analytic, this can be arranged. Let $\Omega_j = \{x \in M : \operatorname{dist}(p_0, x) < r_j\}$. Say $u(x) = a_j$, $v(x) = b_j$ on $\partial\Omega_j$. Let $w_j(x) = b_j u(x) - a_j v(x)$. Then $w_j \in V_\lambda \subset C^\infty(M)$ and $(\Delta + \lambda^2)w_j = 0$ on Ω_j, $w_j = 0$ on $\partial\Omega_j$. However, it is clear that, for j sufficiently large, the smallest eigenvalue of $-\Delta$ on Ω_j, with Dirichlet boundary condition on $\partial\Omega_j$, must exceed λ^2, so this contradiction establishes the proposition.

The following are examples of compact rank one symmetric spaces, where this last proposition applies.

$$G = \mathrm{SO}(n+1), \qquad M = S^n, \tag{3.9}$$

$$G = \mathrm{SU}(n+1), \qquad M = \mathbf{C}P^n, \tag{3.10}$$

$$G = \mathrm{Sp}(n+1), \qquad M = \mathbf{H}P^n, \qquad K = \mathrm{Sp}(n) \times \mathrm{Sp}(1), \tag{3.11}$$

$$G = F_4, \quad M = \mathbf{Ca}P^2 \text{ (projective Cayley 2-space)}, \quad K = \mathrm{Spin}(9). \tag{3.12}$$

We remark that the generator of $\mathfrak{z}(V_\lambda)$ in general will not coincide with the maximal weight vector. Indeed, if $G = \mathrm{SO}(3)$, then $K = \mathrm{SO}(2)$ is a maximal torus, and a zonal function is then a weight vector with weight 0, certainly not maximal. In other situations, the zonal function need not even be a weight vector.

Considering the study of f_ξ in §2, one would expect there could be something new to be learned by studying the transformation

$$f_v(x) = (v, \pi_\lambda(g)z_\lambda), \qquad x = gK \in M, \tag{3.13}$$

given $v \in V_\lambda$, $\pi_\lambda(g) = l(g)$ restricted to V_λ. Assume K is connected, so $K_0 = K$. We pick $z_\lambda \in \mathfrak{z}(V_\lambda)$ such that

$$\|z_\lambda\|_{L^2} = 1. \tag{3.14}$$

The irreducibility of π_λ on V_λ shows that $v \mapsto f_v$ is injective. Clearly $f_v \in V_\lambda$. Furthermore, the representation π_λ on v is carried to the regular representation $l(g)$ on f_v. By Schur's lemma, we have

$$f_v = c_\lambda v, \tag{3.15}$$

for some $c_\lambda \in \mathbf{C}$, and setting $v = z_\lambda$ and $x = p_0$ gives

$$c_\lambda = z_\lambda(p_0)^{-1}. \tag{3.16}$$

Thus the transformation

$$P_\lambda v(x) = c_\lambda^{-1} \int_M v(y) z_\lambda(g^{-1} \cdot y)\, dy, \qquad x = gK \in M, \tag{3.17}$$

defines the orthogonal projection of $L^2(M)$ onto V_λ. Furthermore, if (π, V) is any irreducible unitary representation of G on V with a K-invariant vector z, the map $(Av)(x) = (v, \pi(g)z)$, $x = gK \in M$, maps V isomorphically to some irreducible subspace of $L^2(M)$, which hence must be some eigenspace V_λ. Such a representation is called a *class one* representation, and we have just seen that, if $M = G/K$ is a rank one symmetric space, with K connected, then $L^2(M)$ contains each class one representation, exactly once, as an eigenspace of Δ.

The identity (3.17) for $v \in V_\lambda$ allows us to obtain the following integral identity.

PROPOSITION 3.3. *Let $M = G/K$ be a compact rank one symmetric space, K connected. Let φ be continuous on $M \times M$ and suppose*

$$\varphi(g \cdot x, g \cdot y) = \varphi(x, y), \quad \text{all } g \in G. \tag{3.18}$$

Then, for all $v \in V_\lambda$,

$$\int_M \varphi(x, y) v(y)\, dy = c_\lambda v(x) \int_M \varphi(p_0, y) z_\lambda(y)\, dy. \tag{3.19}$$

The usefulness of this lies in the fact that the integral on the right side of (3.19) can be reduced to a one-dimensional integral, and then often recognized as a special function. Note that (3.18) is equivalent to saying $\varphi(x, y)$ depends only on the distance from x to y in M.

PROOF. Let

$$F(x) = \int_G \overline{\varphi(x, g \cdot p_0)} \pi_\lambda(g) z_\lambda\, dg, \tag{3.20}$$

so $F(x) \in V_\lambda$ for each $x \in M$. We see that, if (3.18) holds,

$$F(g' \cdot x) = \pi_\lambda(g') F(x). \tag{3.21}$$

In particular,

$$F(p_0) \in \mathfrak{z}(V_\lambda), \tag{3.22}$$

so, since $\dim \mathfrak{z}(V_\lambda) = 1$,

$$F(p_0) = (F(p_0), z_\lambda) z_\lambda. \tag{3.23}$$

Thus, if $x = g' \cdot p_0$, the left side of (3.19), multiplied by c_λ, is equal to

$$\begin{aligned} &c_\lambda \int_G \varphi(x, g \cdot p_0) v(g \cdot p_0)\, dg \\ &\quad = (v, F(g' \cdot p_0)) = (v, \pi_\lambda(g') F(p_0)) \\ &\quad = (v, \pi_\lambda(g') z_\lambda) \overline{(F(p_0), z_\lambda)} \\ &\quad = (v, \pi_\lambda(g') z_\lambda) \int_G \varphi(p_0, g \cdot p_0)(z_\lambda, \pi_\lambda(g) z_\lambda)\, dg, \end{aligned} \tag{3.24}$$

and the last formula is seen to coincide with c_λ times the right side of (3.19), by (3.15). This completes the proof.

SECOND PROOF. If $\varphi(x, y)$ is a function of $\text{dist}(x, y)$, then the operator $T_\varphi \colon L^2(M) \to L^2(M)$ defined by

$$T_\varphi v(x) = \int_M \varphi(x, y) f(y)\, dy \tag{3.25}$$

commutes with the G action on $L^2(M)$, so it commutes with the Laplace operator Δ and with all its spectral projections, for any compact homogeneous space M. Thus

$$T_\varphi v(x) = A_\lambda v(x) \quad \text{on } V_\lambda \text{ (since } V_\lambda \text{ is irreducible)} \tag{3.26}$$

for some $A_\lambda \in \mathbf{C}$. To find A_λ, let $v = z_\lambda$ and $x = p_0$ in (3.26). This result is equivalent to (3.19).

In considering zonal functions, both $\|z_\lambda\|_{L^2}$ (which we have normalized to be 1) and $z_\lambda(p_0)$ play a role. It is useful to know they have a simple relationship. Indeed, we have

PROPOSITION 3.4. *We have*

$$z_\lambda(p_0) = (\dim V_\lambda)^{1/2}\|z_\lambda\|_{L^2}. \tag{3.27}$$

PROOF. The Weyl identity (1.14) implies

$$\int_G |(z_\lambda, \pi_\lambda(g)z_\lambda)|^2\,dg = (\dim V_\lambda)^{-1}\|z_\lambda\|_{L^2}^2. \tag{3.28}$$

In view of (3.13)–(3.16) we see that the left side of (3.28) is equal to

$$c_\lambda^2 \int_{G/K} |z_\lambda|^2\,dx$$

and comparing this with the right side of (3.28) gives (3.27).

Leaving the study of zonal functions, let us recall that, in the proof of the Peter-Weyl theorem given in §1, we considered the Laplace operator on a compact group G itself, endowed with a bi-invariant Riemannian metric. Suppose G is semisimple; we can take the metric on $T_eG \approx \mathfrak{g}$ to be given by the negative of the Killing form: $(X,Y) = -\operatorname{tr}(\operatorname{ad} X \operatorname{ad} Y)$. As we have seen, Δ is a scalar on the linear span V_λ of the matrix entries of any irreducible representation π_λ. There is the following pretty formula for the action of Δ on each space V_λ.

THEOREM 3.5. *If π_λ is the irreducible representation of G with highest weight λ, then Δ on V_λ is given by*

$$-\Delta v = (\|\lambda+\delta\|^2 - \|\delta\|^2)v, \qquad v \in V_\lambda, \tag{3.29}$$

where δ is equal to half the sum of the positive roots.

PROOF. Use the decomposition (2.2) of the complexification of $\mathfrak{g}$. Pick an orthonormal basis H_j of $\mathfrak{h}$ and, for each positive root $\alpha \in \Delta^+$, we can pick

$$E_{\pm\alpha} = X_\alpha \pm iY_\alpha \in \mathfrak{g}_\alpha, \quad \|X_\alpha\|^2 = \|Y_\alpha\|^2 = \tfrac{1}{2}, \quad (X_\alpha, Y_\alpha) = 0. \tag{3.30}$$

Then

$$\begin{aligned}\Delta &= \sum H_j^2 + \sum X_\alpha^2 + \sum Y_\alpha^2\\ &= \sum H_j^2 + \sum_{\alpha\in\Delta^+} (E_\alpha E_{-\alpha} + E_{-\alpha}E_\alpha)\\ &= \sum H_j^2 + i\sum_{\alpha\in\Delta^+} H_\alpha + 2\sum_{\alpha\in\Delta^+} E_{-\alpha}E_\alpha.\end{aligned} \tag{3.31}$$

To prove (3.29), it suffices to take v to be the highest weight vector. Then $E_\alpha v = 0$ for all $\alpha \in \Delta^+$, so from (3.31) we deduce

$$
\begin{aligned}
\Delta v &= \sum \pi_\lambda(H_j)^2 v + i \sum_{\alpha\in\Delta^+} \pi_\lambda(H_\alpha) v \\
&= -\langle\lambda,\lambda\rangle v - \sum_{\alpha\in\Delta^+} \langle\lambda,\alpha\rangle v \qquad (3.32)\\
&= -(\langle\lambda,\lambda\rangle + 2\langle\lambda,\delta\rangle)v,
\end{aligned}
$$

which implies (3.29).

We can generalize Theorem 3.5, partially, to say something about the action of a general bi-invariant differential operator on G, also called a Casimir operator. Of course, any Casimir operator is scalar on each V_λ. A general left invariant differential operator P of order m on G can be written uniquely as a linear combination of operators of the form

$$P_{\gamma\sigma\delta} = E_{-\alpha_1}^{\gamma_1} \cdots E_{-\alpha_K}^{\gamma_K} H_1^{\sigma_1} \cdots H_n^{\sigma_n} E_{\alpha_1}^{\delta_1} \cdots E_{\alpha_K}^{\delta_K} \tag{3.33}$$

where $\{\alpha_1, \ldots, \alpha_K\}$ is the set of positive roots, and $E_{\pm\alpha}, H_j$ are as in the proof of Theorem 3.5, as follows from the Poincaré-Birkhoff-Witt theorem. Note that

$$[H_j, P_{\gamma\sigma\delta}] = \{(\delta_1 - \gamma_1)\alpha_1(H_j) + \cdots + (\delta_K - \gamma_K)\alpha_K(H_j)\} P_{\gamma\sigma\delta}. \tag{3.34}$$

It is not hard to show all monomials of the form (3.33) are linearly independent over $\mathbf{C}$, so a necessary condition for a sum

$$P = \sum_{|\gamma|+|\sigma|+|\delta|\le m} b_{\gamma\sigma\delta} P_{\gamma\sigma\delta} \tag{3.35}$$

to be bi-invariant, so that $[P, H_j] = 0$ for all $H_j \in \mathfrak{h}$, is that

$$\gamma_1 + \cdots + \gamma_K = \delta_1 + \cdots + \delta_K. \tag{3.36}$$

With this in mind, we can apply P to a highest weight vector and prove the following partial generalization of Theorem 3.5.

PROPOSITION 3.6. *Let P be a bi-invariant differential operator of order m on G. Then we have*

$$Pv = p^\#(\lambda) v \quad \text{on } V_\lambda, \tag{3.37}$$

where $p^\#(\lambda)$ is a polynomial of degree m in λ. The part of $p^\#(\lambda)$ homogeneous of degree m coincides with the principal symbol of P, restricted to $\mathfrak{h}' \subset \mathfrak{g}' = T_e^*G$.

PROOF. Write P in the form (3.35), with (3.36) holding. Since P is scalar on each V_λ, it suffices to check (3.37) when v is a highest weight vector, i.e., $E_\alpha v = 0$ for each positive root α. In that case, all terms (3.33) annihilate v unless $|\delta| = 0$; by (3.36) this requires $|\gamma| = 0$, so the only terms that contribute to Pv are

$$\sum b_{0\sigma 0} H_1^{\sigma_1} \cdots H_n^{\sigma_n} v = \sum b_{0\sigma 0} i^{|\sigma|} \lambda(H_1)^{\sigma_1} \cdots \lambda(H_n)^{\sigma_n} v. \tag{3.38}$$

Consequently we have (3.37) with

$$p^{\#}(\lambda) = \sum b_{0\sigma 0} i^{|\sigma|} \lambda(H_1)^{\sigma_1} \cdots \lambda(H_n)^{\sigma_n}. \tag{3.39}$$

It is clear that the top order term in (3.39) coincides with the principal symbol of P restricted to $\mathfrak{h}' \subset T_e^* G$.

Proposition 3.6 has a very important refinement, which we state here without proof; see Zelobenko [**268**], pages 365–370.

THEOREM 3.7. *Let P be a bi-invariant differential operator on G of order m. Then we have*

$$Pv = p(\lambda + \delta)v \quad \text{on } V_\lambda, \tag{3.40}$$

where $p(\lambda)$ is a polynomial of degree m in λ, invariant under the Weyl group. The part of $p(\lambda)$ homogeneous of degree m coincides with the principal symbol of P, restricted to $\mathfrak{h}' \subset T_e^* G$. Conversely, for any polynomial $p(\lambda)$ of order m which is invariant under the Weyl group, (3.40) defines a bi-invariant differential operator, of order m.

The *Weyl group* is the group of linear transformations on $\mathfrak{g}'$ induced by inner automorphisms of G leaving the maximal torus T^k invariant. This is a finite group with many interesting properties, important for a complete appreciation of the structure and representation theory of compact Lie groups. We will not develop this study in this monograph, but instead refer the reader to the treatments of Wallach [**253**] and Zelobenko [**268**]. The converse part of the theorem rests upon the following result of Chevalley: restriction from $\mathfrak{g}'$ to $\mathfrak{h}'$ produces an *isomorphism* of the space of Ad^*-invariant polynomials on $\mathfrak{g}'$ (homogeneous of order m) onto the space of Weyl group-invariant polynomials on $\mathfrak{h}'$ (homogeneous of order m). There is a generalization of Theorem 3.7 to pseudodifferential operators, which provides a tool to generalize a good deal of classical harmonic analysis from the torus to general compact Lie groups. This is given in Chapter 12, §6, of the book [**234**]. This theory has been used to study asymptotic properties of representations of Lie groups, in [**32**] and [**31**]. Furthermore it has been effective in studying some properties of quantum Hamiltonians associated with Yang-Mills potentials, in [**208**].

We conclude this section with a discussion of a certain "generating function" associated to the representations π_λ of G, and some applications, including a second derivation of Proposition 3.6. Our generating function is essentially (2.24), but we will use slightly different notation, emphasizing the dependence on λ. If π_λ is an irreducible unitary representation with highest weight λ, let v_λ be a highest weight vector; by Theorem 2.7 it is unique, up to a complex scalar; normalize so $\|v_\lambda\| = 1$. Form

$$\Psi_\lambda(g) = (\pi_\lambda(g)v_\lambda, v_\lambda). \tag{3.41}$$

Note that, although v_λ is defined only up to a phase $e^{i\theta}$, $\Psi_\lambda(g)$ is independent of this phase factor. There is the following simple but remarkable identity, proved in Zelobenko [**268**], and also in Gillmore [**79**]; see also Simon [**217**].

LEMMA 3.8. *We have*

$$\Psi_\lambda(g)\Psi_\mu(g) = \Psi_{\lambda+\mu}(g). \tag{3.42}$$

PROOF. If λ and μ are highest weights for π_λ and π_μ, then $\lambda+\mu$ is the highest weight for the representation $\pi_\lambda \otimes \pi_\mu$, and the unique highest weight vector is $v_\lambda \otimes v_\mu$; this vector is contained in a copy of $\pi_{\lambda+\mu}$ and (3.42) is an immediate consequence.

This result can be rephrased as follows. Among the dominant integral weights, there exist a finite number of *fundamental weights* $e_1, \dots, e_K$, with the property that any dominant integral weight can be written uniquely in the form

$$\lambda = \sum_{j=1}^{K} l_j e_j, \qquad l_j \geq 0 \text{ integer}. \tag{3.43}$$

Then

$$\Psi_\lambda(g) = \prod_{j=1}^{K} \Psi_j(g)^{l_j}, \tag{3.44}$$

where

$$\Psi_j(g) = \Psi_{e_j}(g). \tag{3.45}$$

This reduces the problem of determining $\Psi_\lambda(g)$ for all representations π_λ to the finite problem of determining the cases (3.45).

The role of $\Psi_\lambda(g)$ in the analysis of the behavior of Casimir operators arises as follows. For any left invariant differential operator $P \in \mathfrak{U}(\mathfrak{g})$, we have

$$(\pi_\lambda(P)v_\lambda, v_\lambda) = P\Psi_\lambda(g)\big|_{g=e}. \tag{3.46}$$

Given (3.44), it is an obvious consequence of Leibniz' identity that $P\Psi_\lambda(g)|_{g=e}$ is a polynomial in $(l_1, \dots, l_K)$ of degree m (i.e., a polynomial in λ of degree m) if the operator P has order m. In order to see what this polynomial is, we first note that

$$(\pi_\lambda(X)v_\lambda, v_\lambda) = i\lambda(X), \qquad X \in \mathfrak{g}. \tag{3.47}$$

In order to see this, recall that $\pi_\lambda(X)v_\lambda = i\lambda(X)v_\lambda$ for $X \in \mathfrak{h}$ since v_λ is a weight vector. Since it is the highest weight vector, $\pi_\lambda(Z)v_\lambda = 0$ when $Z \in \mathbf{C}\mathfrak{g}$ is any root vector in $\mathfrak{g}_\mu$, for any positive root μ. By duality, if Z is a root vector for a negative root, $\pi_\lambda(Z)v_\lambda$ is orthogonal to v_λ. This proves (3.47). It is hence clear that the principal part of the polynomial given by the right side of (3.46) is

$$\sum b_{j_1\cdots j_m}\lambda(X_{j_1})\cdots\lambda(X_{j_m}) \tag{3.48}$$

provided

$$P = \sum b_{j_1\cdots j_m}X_{j_1}\cdots X_{j_m}, \qquad X_{j_k} \in \mathfrak{g}. \tag{3.49}$$

Thus, in case P is bi-invariant, we obtain another proof of Proposition 3.6.

The function $\Psi_\lambda(g)$ played a role in the investigation in [**208**] of quantum partition functions for Hamiltonians associated with an external Yang-Mills potential, following previous work in [**79, 217**]. We mention one further property of $\Psi_\lambda(g)$, namely its connection with the *character*

$$\chi_\lambda(g) = \operatorname{tr} \pi_\lambda(g). \tag{3.50}$$

It is easy to show that $\chi_\lambda(g)$ is the unique element of V_λ, up to a scalar multiple, which is invariant under conjugation. Thus, averaging conjugates of $\Psi_\lambda(g)$ over G produces a scalar multiple of $\chi_\lambda(g)$; since $\Psi_\lambda(e) = 1$ while $\chi_\lambda(e) = d_\lambda$, the dimension of the representation space of π_λ, we have

$$\begin{aligned} \chi_\lambda(x) &= d_\lambda \int_G \Psi_\lambda(g^{-1}xg)\, dg \\ &= d_\lambda \int_G \Psi_1(g^{-1}xg)^{l_1} \cdots \Psi_K(g^{-1}xg)^{l_K}\, dg \end{aligned} \tag{3.51}$$

for $x, g \in G$. Furthermore, since $\chi_\lambda(g)$ has L^2 norm on G equal to 1, as a consequence of Weyl's identities (1.18)–(1.19), we deduce that

$$d_\lambda^{-2} = \int_G \left| \int_G \Psi_\lambda(g^{-1}xg)\, dg \right|^2 dx. \tag{3.52}$$

The classical formula of Weyl for d_λ is

$$d_\lambda = \prod_{\alpha\in\Delta^+} \langle \lambda + \delta, \alpha\rangle / \langle \delta, \alpha\rangle, \tag{3.53}$$

where as before Δ^+ is the set of positive roots and δ is half the sum of the positive roots. The inner product is induced by the Killing form. The author has not seen a direct demonstration of the equivalence of (3.52) with (3.53), nor of (3.51) with Weyl's celebrated character formula, given and proved, along with (3.53), in Chapter 4 of Wallach [**253**].

CHAPTER 4

Harmonic Analysis on Spheres

Harmonic analysis on S^2 was initiated over two hundred years ago by Legendre, Laplace, and others. It arose from studying the Laplace operator in polar coordinates, and spherical harmonics became an important tool for treating the Laplace operator and its kin on balls and complements of balls. Conversely, direct attacks on various PDEs on S^n yield valuable information on harmonic analysis on spheres. We mention the connection between the solution to the wave equation on $R \times S^2$ and the Mehler-Dirichlet formula for the Legendre functions, discussed in §3, as an example of this.

1. The Laplace operator on polar coordinates. If we denote a point $x \in R^n$ by $x = r\omega$, $r = |x|$, $\omega \in S^{n-1}$, then the Laplace operator Δ on R^n is given by

$$\Delta = \partial^2/\partial r^2 + (n-1)r^{-1}\partial/\partial r + r^{-2}\Delta_S \tag{1.1}$$

where Δ_S is the Laplace operator on the unit sphere S^{n-1}. In fact, (1.1) gives the Laplace operator on the cone $C(S)$ over any Riemannian manifold S of dimension $n-1$, which is $R^+ \times S$ with the metric

$$g = dr^2 + r^2 g_S. \tag{1.2}$$

Implications of this are pursued further in Chapter 7. The formula (1.1) is the source of much interest in the Laplace operator Δ_S on S^{n-1}, and also the source of much information on it, as we shall see.

We take a look at the Dirichlet problem for the unit ball in R^n

$$B = \{x \in R^n : |x| < 1\}, \tag{1.3}$$

which is

$$\Delta u = 0 \quad \text{in } B, \qquad u = f \quad \text{on } S^{n-1} = \partial B, \tag{1.4}$$

given $f \in \mathcal{D}'(S^{n-1})$. The Poisson integral formula for the solution is

$$u(x) = V_{n-1}^{-1}(1 - |x|^2) \int_{S^{n-1}} \|x - \omega'\|^{-n} f(\omega')\, d\omega' \tag{1.5}$$

where V_{n-1} is the volume of S^{n-1},

$$V_{n-1} = 2\pi^{n/2}/\Gamma(n/2). \tag{1.6}$$

Equivalently, if we set $x = r\omega$ as above, we have

$$u(r\omega) = V_{n-1}^{-1}(1 - r^2) \int_{S^{n-1}} (1 - 2r\omega \cdot \omega' + r^2)^{-n/2} f(\omega')\, d\omega'. \tag{1.7}$$

This classical formula is given in many books on potential theory, such as [**109**]. Details on one derivation are presented in this monograph, in Chapter 10, §2. We briefly discribe the method of Chapter 10 here.

First, the validity of (1.5) at the origin $x = 0$ is clear, by the mean value theorem for harmonic functions in B:

$$u(0) = V_{n-1}^{-1} \int_{S^{n-1}} f(\omega')\, d\omega'. \tag{1.8}$$

The mean value theorem is a consequence of rotational invariance of Δ. Indeed, if one averages $u(g \cdot x)$ over $g \in \mathrm{SO}(n)$, one gets a harmonic function $v(x)$ such that $v(0) = u(0)$ and $v|_{S^{n-1}}$ is a constant C, namely the mean value of f. Now $v_1(x) = C$ is a solution to such a Dirichlet problem, and granted that such a solution is unique, we have $v(x) = C$. To deduce (1.5) for general $x = r\omega \in B$, we can move x to the origin by a *conformal* transformation, obtain a transformation on functions on B which preserves the property of being harmonic, and deduce the general case from the mean value theorem. See Chapter 10 for details.

We can derive an alternative formula for the solution of (1.4) if we use (1.1) and regard $\Delta u = 0$ as an operator valued ODE in r; it is an Euler equation, with solution

$$u(r\omega) = r^{-(n-2)/2+\nu} f(\omega), \qquad r \le 1, \tag{1.9}$$

where ν is an operator on $\mathcal{D}'(S^{n-1})$, defined by

$$\nu = (-\Delta_S + (n-2)^2/4)^{1/2}. \tag{1.10}$$

If we set $r = e^{-t}$ and compare (1.9) and (1.7), we obtain a formula for the semigroup $e^{-t\nu}$, as follows. Let $\theta(\omega, \omega')$ denote the geodesic distance on S^{n-1} from ω to ω', so $\cos\theta(\omega,\omega') = \omega \cdot \omega'$. We can rewrite (1.7) as

$$\begin{aligned} u(r\omega) &= 2V_{n-1}^{-1} \sinh(\log r^{-1}) r^{-(n-2)/2} \\ &\quad \cdot \int_{S^{n-1}} [2\cosh(\log r^{-1}) - 2\cos\theta(\omega,\omega')]^{-n/2} f(\omega')\, d\omega'. \end{aligned} \tag{1.11}$$

In other words, by (1.9),

$$e^{-t\nu} f(\omega) = 2V_{n-1}^{-1} \sinh t \int_{S^{n-1}} (2\cosh t - 2\cos\theta(\omega,\omega'))^{-n/2} f(\omega')\, d\omega'. \tag{1.12}$$

Identifying an operator on $\mathcal{D}'(S^{n-1})$ with its Schwartz kernel in $\mathcal{D}'(S^{n-1} \times S^{n-1})$, we write

$$e^{-t\nu} = 2V_{n-1}^{-1}(\sinh t)(2\cosh t - 2\cos\theta)^{-n/2}, \qquad t > 0. \tag{1.13}$$

Note that integration of (1.13) from t to ∞ produces the formula

$$\nu^{-1}e^{-t\nu} = 2A_n(2\cosh t - 2\cos\theta)^{-(n-2)/2}, \qquad t > 0, \tag{1.14}$$

where

$$A_n = [(n-2)V_{n-1}]^{-1} = \tfrac{1}{4}\pi^{-n/2}\Gamma(\tfrac{1}{2}n - 1). \tag{1.15}$$

2. Classical PDE on spheres. We will examine the basic Laplace, wave, and heat equations on the sphere S^{n-1}, using the analysis of the last section. The particular equations we have in mind are

$$(\partial^2/\partial y^2 + L)u = 0 \quad \text{(Laplace equation)}, \tag{2.1}$$

$$(\partial/\partial t)u = Lu \quad \text{(heat equation)}, \tag{2.2}$$

$$(\partial^2/\partial t^2 - L)u = 0 \quad \text{(wave equation)}, \tag{2.3}$$

where

$$L = \Delta_S - (n-2)^2/4 = -\nu^2. \tag{2.4}$$

Note that if we specify $u(0) = f \in \mathcal{D}'(S^{n-1})$, then the solutions to (2.1), (2.2) which do not blow up as y or t tend to $+\infty$ are, respectively,

$$u(y) = e^{-y\nu}f, \tag{2.5}$$

$$u(t) = e^{-t\nu^2}f, \tag{2.6}$$

and if we prescribe Cauchy data $u(0) = f$, $(\partial/\partial t)u(0) = g$ for (2.3), the solution is

$$u(t) = (\cos t\nu)f + \nu^{-1}(\sin t\nu)g. \tag{2.7}$$

The method of analysis here follows Cheeger and Taylor [**37**].

We will be able to obtain formulas, in these cases, from the formulas (1.13), (1.14) derived in §1, which we record here:

$$\nu^{-1}e^{-t\nu} = -2A_n(2\cosh t - 2\cos\theta)^{-(n-2)/2}, \qquad t > 0, \tag{2.8}$$

$$e^{-t\nu} = 2V_{n-1}^{-1}\sinh t(2\cosh t - 2\cos\theta)^{-n/2}, \qquad t > 0. \tag{2.9}$$

We suppose $n - 1 \geq 2$. In fact, (2.9) already furnishes the formula for (2.5), the solution to (2.1). Note the formal equivalence therefore between the Laplace equation (2.1) on $R^+ \times S^{n-1}$ and the Laplace equation (1.4) on the unit ball in R^n, with solution (1.9).

Our next goal is to analyze the wave equation (2.3). By (2.7), it will be desirable to analyze $e^{it\nu}$ and $\nu^{-1}e^{it\nu}$, which we will obtain from (2.8), (2.9) by analytic continuation. In fact, both sides of (2.8), (2.9) are holomorphic in $\{t \in \mathbf{C}\colon \operatorname{Re} t > 0\}$, so, for any $\varepsilon > 0$, we can write

$$\nu^{-1}e^{(it-\varepsilon)\nu} = -2A_n[2\cosh(it-\varepsilon) - 2\cos\theta]^{-(n-2)/2}, \tag{2.10}$$

$$e^{(it-\varepsilon)\nu} = 2V_{n-1}^{-1}\sinh(it-\varepsilon)[2\cosh(it-\varepsilon) - 2\cos\theta]^{-n/2}. \tag{2.11}$$

Now we can pass to the limit $\varepsilon \downarrow 0$, to obtain

(2.12) $\nu^{-1}e^{it\nu} = \lim_{\varepsilon\downarrow 0} -2A_n(2\cosh\varepsilon\cos t - 2i\sinh\varepsilon\sin t - 2\cos\theta)^{-(n-2)/2},$

(2.13) $e^{it\nu} = \lim_{\varepsilon\downarrow 0} 2V_{n-1}^{-1} i \sin t(2\cosh\varepsilon\cos t - 2i\sinh\varepsilon\sin t - 2\cos\theta)^{-n/2}.$

Now we have

$$\cos t\nu = \operatorname{Re} e^{it\nu}, \qquad \nu^{-1}\sin t\nu = \operatorname{Im}(\nu^{-1}e^{it\nu}), \tag{2.14}$$

so (2.12), (2.13) provide formulas for these kernels.

For example, on S^2 $(n = 3)$, we have

$$\nu^{-1}e^{it\nu} = -2A_3(2\cos t - 2\cos\theta)^{-1/2}, \tag{2.15}$$

which implies that, for $|t| \le \pi$,

$$\nu^{-1}\sin t\nu = \begin{cases} -2A_3(2\cos\theta - 2\cos t)^{-1/2}, & |\theta| < |t|, \\ 0, & |\theta| > |t|, \end{cases} \tag{2.16}$$

with an analogous expression for general t, determined by the identity

$$\nu^{-1}\sin(t + 2\pi)\nu = -\nu^{-1}\sin t\nu \quad \text{on } S^2.$$

The last line on the right in (2.16) reflects the well-known finite propagation speed for solutions to the hyperbolic equation (2.3).

On the other hand, on odd dimensional spheres (where n is even), the exponent on the right side of (2.12) is an integer, so the distribution kernel for $\nu^{-1}\sin t\nu$ must *vanish* if $|t| \neq \theta$. In other words, the kernel is supported on the shell $\theta = |t|$. This is the strict Huygens principle, well known for the wave equation on Euclidean space (for odd space dimensions). In case $n - 1$ is odd, one obtains, from (2.12), (2.13),

$$\nu^{-1}\sin t\nu f(x) = \frac{1}{2}(2\pi)^{-(n-2)/2}\left(\frac{1}{\sin s}\frac{\partial}{\partial s}\right)^{(n-4)/2} (\sin^{n-3} s\overline{f}(x,s))|_{s=t} \tag{2.17}$$

and

(2.18)
$$\cos t\nu f(x) = \frac{1}{2}(2\pi)^{-(n-2)/2}\sin s\left(\frac{1}{\sin s}\frac{\partial}{\partial s}\right)^{(n-2)/2} (\sin^{n-3} s\overline{f}(x,s))|_{s=t},$$

where

$$\overline{f}(x,s) = \text{mean value of } f \text{ on } \textstyle\sum_s(x) = \{y \in S^{n-1} \colon \theta(x,y) = |s|\}. \tag{2.19}$$

We can examine general functions of the operator ν by the functional calculus

$$\begin{aligned} f(\nu) &= (2\pi)^{-1/2}\int_{-\infty}^{\infty} \hat{f}(t)e^{it\nu}\,dt \\ &= (2\pi)^{-1/2}\int_{-\infty}^{\infty} \hat{f}(t)\cos t\nu\,dt, \end{aligned} \tag{2.20}$$

the last identity holding provided f is an *even* function. We can rewrite this, using the fact that, for n even, $\cos t\nu$ has period 2π in t, while, for n odd, $\cos t\nu$

has period 4π in t. This periodicity follows from (2.13), and is equivalent to the assertion that the spectrum of ν consists of integers for n even and half integers for n odd. For more on this, see the first paragraph of §3. In the case of odd-dimensional spheres S^{n-1} this fact, together with (2.18), shows that the kernel of $f(\nu)$ (for f even) is given by

$$f(\nu) = (2\pi)^{-1/2}\left(-\frac{1}{2\pi}\frac{1}{\sin\theta}\frac{\partial}{\partial\theta}\right)^{(n-2)/2}\sum_{k=-\infty}^{\infty}\hat{f}(\theta+2k\pi). \tag{2.21}$$

As an example, we calculate the heat kernel on odd-dimensional spheres. Take $f(\lambda) = e^{-t\lambda^2}$. Then $\hat{f}(s) = (2\pi t)^{-1/2}e^{-s^2/4t}$, and

$$\begin{aligned}(2\pi)^{-1/2}\sum_k \hat{f}(s+2\pi k) &= (4\pi t)^{-1/2}\sum_k e^{-(s+2\pi k)^2/4t}\\ &= \vartheta(s,t)\end{aligned} \tag{2.22}$$

where $\vartheta(s,t)$ is the theta function. Thus the kernel of $e^{-t\nu^2}$ on S^{n-1} ($n-1$ odd) is given by

$$e^{-t\nu^2} = \left(-\frac{1}{2\pi}\frac{1}{\sin\theta}\frac{\partial}{\partial\theta}\right)^{(n-2)/2}\vartheta(\theta,t). \tag{2.23}$$

A similar analysis on S^{n-1} for $n-1$ even gives an integral, with the theta function appearing in the integrand. We omit the details.

We end this section with a remark on how to get the fundamental solution to the wave equation on *hyperbolic space* $\mathcal{H}^{n-1}$, the simply connected Riemannian manifold of constant curvature -1. This can be achieved by analytic continuation of the metric tensor. In fact, let p_0 be the origin (north pole) of S^{n-1} in geodesic polar coordinates, and consider the one parameter family of metrics giving the spaces of constant curvature K. The unit sphere corresponds to $K=1$; $K\in(0,1)$ corresponds to dilated spheres, and the fundamental kernel for $\nu^{-1}\sin t\nu$ with

$$\nu = (-\Delta + K(n-2)^2/4)^{1/2} \tag{2.24}$$

can be obtained explicitly from that on the unit sphere by a change of scale. The explicit representation so obtained analytically continues to all real values of K, and at $K=-1$ gives the following formula for the wave kernel

$$\nu^{-1}\sin t\nu = \lim_{\varepsilon\downarrow 0} -2A_n \operatorname{Im}[2\cos(it-\varepsilon) - 2\cosh r]^{-(n-2)/2}, \tag{2.25}$$

where r denotes the distance between the source point and the observation point. Here

$$\nu = (-\Delta - (n-2)^2/4)^{1/2}. \tag{2.26}$$

Thus for odd-dimensional hyperbolic space we also have the strict Huygens principle, i.e., $\nu^{-1}\sin t\nu$ and $\cos t\nu$ are supported on the shell $r=|t|$. Compare with the derivation of Lax and Phillips **[151]**, which involves an "inspired guess" when

$n - 1 = 3$ (which becomes plausible once one suspects the strict Huygens principle should hold), and a variant of the method of descent to handle $n - 1 = 2$. That the solution to

$$\partial^2 R/\partial t^2 - (\Delta - K(n-2)^2/4)R = 0,$$
$$R(0,x) = 0, \qquad (\partial/\partial t)R(0,x) = \delta_{p_0}(x)$$

on the space of curvature K depends analytically on K, in an appropriate sense (which can be deduced from the Cauchy-Kowalewski theorem), implies that the formula for $\nu^{-1} \sin t\nu \delta_{p_0}(x)$ depends analytically on K. It follows that (2.25) is valid.

In connection with the appearance of the operator (2.26), it is useful to note that the Laplace operator on $\mathcal{H}^n$, of constant curvature -1, has the following bound on its spectrum:

$$\operatorname{spec}(-\Delta) \subset [(n-1)^2/4, \infty). \tag{2.27}$$

We give a proof of this. $\mathcal{H}^n$ will be identified with the upper half space in R^n, $x_n \geq 0$, with metric

$$x_n^{-2} \sum_1^n dx_j^2. \tag{2.28}$$

A derivation of this metric can be found in Chapter 10, §2. Then the Laplace operator, given on a Riemannian manifold by $\Delta = g^{-1/2}\partial_j g^{jk} g^{1/2} \partial_k$, is of the form

$$\Delta u = \sum_j x_n^n \partial_j x_n^{2-n} \partial_j u. \tag{2.29}$$

The volume element on $\mathcal{H}^n$ is $x_n^{-n} dx_1 \cdots dx_n$, so, for $u \in C_0^\infty(\mathcal{H}^n)$, we have

$$\begin{aligned}((-\Delta - (n-1)^2/4)u, u) = &\int [(\partial_n u)^2 - ((n-1)u/2x_n)^2] x_n^{2-n}\, dx_1 \cdots dx_n \\ &+ \sum_1^{n-1} \int (\partial_j u)^2 x_n^{2-n}\, dx_1 \cdots dx_n.\end{aligned} \tag{2.30}$$

On the other hand, integration by parts implies

$$\begin{aligned}&\int_{\mathcal{H}^n} [(\partial_n u)^2 - ((n-1)u/2x_n)^2] x_n^{2-n}\, dx_1 \cdots dx_n \\ &\quad = \int_{\mathcal{H}^n} [\partial_n (x_n^{-(n-1)/2} u)]^2 x_n\, dx_1 \cdots dx_n,\end{aligned} \tag{2.31}$$

so, for $u \in C_0^\infty(\mathcal{H}^n)$, the expression (2.30) is ≥ 0. This proves (2.27).

3. Spherical harmonics. We want to determine the spectral projections E_k of ν onto its eigenspaces V_k, which of course are the eigenspaces of the Laplace operator Δ_S. Suppose $\nu_0 < \nu_1 < \cdots$ give the spectrum of the operator ν. If

$\nu = \nu_k$ on V_k, then $-\Delta_S = \nu_k^2 - (n-2)^2/4$ on V_k. Let us first make a simple remark about the spectrum of ν. Since the function $u(x)$ given by

$$u(r\omega) = r^{-(n-2)/2+\nu} f(\omega) \tag{3.1}$$

solving the Dirichlet problem (1.3) must be smooth at $x = 0$, we see that the exponent of r must take on only integral values. Thus

(3.2) ν_k are all integers if n is even and all half-integers if n is odd.

We will see that all ν_k in the appropriate category which are $\geq (n-2)/2$ actually occur as eigenvalues of ν. First, note that, if $f \in V_k$, then $u(x) = r^{\nu_k-(n-2)/2} f(\omega)$ must be a polynomial in $x \in R^n$, because u is smooth at the origin and homogeneous of degree $d_k = \nu_k - (n-2)/2$. Such $u(x)$ is a homogeneous harmonic polynomial. Conversely, of course, if $u(x)$ is a harmonic polynomial, homogeneous of degree d_k, then $u|_{S^{n-1}} \in V_k$. In other words, V_k can be identified with the space H_k of harmonic polynomials, homogeneous of degree d_k. Now we will exhibit harmonic polynomials homogeneous of any integral degree k, so we can conclude

$$d_k = \nu_k - (n-2)/2 = k, \qquad k = 0, 1, 2, \ldots. \tag{3.3}$$

Indeed, consider, for $c = (c_1, \ldots, c_n) \in \mathbf{C}^n$,

$$p_c(x) = (c_1 x_1 + \cdots + c_n x_n)^k. \tag{3.4}$$

It is easy to see that $\Delta p_c = 0$ whenever

$$\sum c_j^2 = \langle c, c \rangle = 0. \tag{3.5}$$

This proves (3.3). In fact, we have

PROPOSITION 3.1. *H_k is the linear span of polynomials of the form* (3.4), (3.5).

PROOF. Since S^{n-1} is a rank one symmetric space, we know SO(n) acts irreducibly on V_k (see Chapter 3), hence on H_k, so it suffices to show that SO(n) leaves invariant this linear span. Indeed, if $g \in \mathrm{SO}(n)$, we have $p_c(g^{-1} \cdot x) = p_{g \cdot c}(x)$, where g is extended to act on $\mathbf{C}^n$ by complexification. Since $\langle g \cdot c, g \cdot c \rangle = \langle c, c \rangle$, this completes the proof.

We have identified V_k, but we still want to specify the orthogonal projections E_k of $L^2(S^{n-1})$ on V_k. Recall from Chapter 3 that, if z_k denotes the zonal function in V_k, then

$$E_k f(x) = \int_{S^{n-1}} E_k(x, y) f(y)\, dy, \qquad x \in S^{n-1}, \tag{3.6}$$

with

$$E_k(x, y) = c_k^{-1} z_k(g^{-1} \cdot y), \qquad x = g^{-1} \cdot p_0 \in S^{n-1}, \tag{3.7}$$

where $p_0 = e \cdot \mathrm{SO}(n-1)$ is the "north pole" and $c_k^{-1} = z_k(p_0)$, so computing the kernel of E_k is equivalent to computing the zonal function $z_k(x)$, $x \in S^{n-1}$. One formula for E_k is

$$\begin{aligned} E_k &= (2T)^{-1} \int_{-T}^{T} e^{-i\nu_k t} e^{it\nu}\, dt \\ &= (2T)^{-1} \int_{-T}^{T} \cos t(\nu - \nu_k)\, dt \\ &= (2T)^{-1} \int_{-T}^{T} e^{-i\nu_k t} \cos t\nu\, dt, \end{aligned} \tag{3.8}$$

where $T = \pi$ or 2π depending on whether $n-1$ is odd or even. (In either case, one could take $T = 2\pi$.) The last identity in (3.8) follows from the fact that $\operatorname{spec} \nu \subset R^+$. Substituting the formula (2.13) for the kernel of $e^{it\nu}$ will produce an integral formula for $E_k(x,y)$. We will return to this point later in this section.

First, we will obtain from formula (1.8), which, we recall, is

$$\nu^{-1} e^{-t\nu} = -2A_n(2\cosh t - 2\cos\theta)^{-(n-2)/2}, \tag{3.9}$$

an alternative approach to the kernel $E_k(x,y)$. In fact, we have

$$\sum_{k=0}^{\infty} \nu_k^{-1} e^{-t\nu_k} E_k(x,y) = -2A_n(2\cosh t - 2\cos\theta)^{-(n-2)/2}, \tag{3.10}$$

where $\theta = \theta(x,y)$ is the geodesic distance from x to y in S^{n-1}. If we set $r = e^{-t}$ and use (3.3), which says $\nu_k = k + (n-2)/2$, we get the generating function identity

$$\begin{aligned} B_n(1 - 2r\cos\theta + r^2)^{-(n-2)/2} &= \sum_{k=0}^{\infty} r^k \nu_k^{-1} E_k(x,y) \\ &= \sum_{k=0}^{\infty} r^k p_k(\cos\theta), \end{aligned} \tag{3.11}$$

where we have set

$$p_k(\cos\theta) = \nu_k^{-1} E_k(x,y), \qquad B_n = 2A_n. \tag{3.12}$$

Note that the formula (3.7) implies that the zonal functions z_k on S^{n-1} are given by

$$z_k(x) = (\nu_k / z_k(p_0)) p_k(x \cdot p_0), \tag{3.13}$$

or, solving for $z_k(p_0)$,

$$z_k(x) = (\nu_k / p_k(1))^{1/2} p_k(x \cdot p_0). \tag{3.14}$$

The generating function identity (3.11) can be used to produce formulas for $p_k(t)$, $-1 \le t \le 1$. Indeed, Cauchy's integral formula gives

$$p_k(t) = B_n \frac{1}{2\pi i} \int_\gamma (1 - 2zt + z^2)^{-(n-2)/2} z^{-k-1}\, dz, \tag{3.15}$$

where γ can be taken to be a small circle about the origin. We can make a change of variable, setting

$$1 - uz = (1 - 2tz + z^2)^{1/2}, \tag{3.16}$$

which is equivalent to setting $z = 2(u - t)/(u^2 - 1)$. We obtain

$$(1 - 2tz + z^2)^{-1/2} z^{-k-1}\, dz = 2^{-k}(u^2 - 1)^k (u - t)^{-k-1}\, du,$$

and more generally

(3.17)

$$(1 - 2tz + z^2)^{-l-1/2} z^{-k-1}\, dz = 2^{-k}(u^2 - 1)^{k+2l}(1 + u^2 - 2tu)^{-2l}(u - t)^{-k-1}\, du.$$

Thus we have

$$p_k(t) = B_n 2^{-k} \frac{1}{2\pi i} \int_{\gamma'} \frac{(u^2 - 1)^{k+n-3}}{(1 + u^2 - 2tu)^{n-3}} \frac{du}{(u - t)^{k+1}}, \tag{3.18}$$

where γ' is a small circle about the point $t \in [-1, 1]$. Again, Cauchy's formula gives

$$p_k(t) = B_n (2^k k!)^{-1} (d/du)^k \left[\frac{(u^2 - 1)^{k+n-3}}{(1 + u^2 - 2tu)^{n-3}} \right] \bigg|_{u=t}. \tag{3.19}$$

In the case $n = 3$, where $S^{n-1} = S^2$, we have

$$B_3^{-1} p_k(t) = (2^k k!)^{-1} (d/dt)^k (t^2 - 1)^k, \tag{3.20}$$

which is the formula of Rodrigues for the *Legendre polynomials*, so

$$B_3^{-1} p_k(t) = P_k(t) \quad \text{for } n = 3. \tag{3.21}$$

If we look at S^3 ($n = 4$), the generating function identity (3.11) is more directly useful than (3.19). Indeed, we have

$$\begin{aligned} \sum_{k=0}^{\infty} r^k p_k(t) &= B_4 (1 - 2rt + r^2)^{-1} \\ &= B_4 \sum_{j=0}^{\infty} r^j (2t - r)^j, \end{aligned} \tag{3.22}$$

and comparing coefficients gives a recursion formula for the polynomials $p_k(t)$ and hence for the zonal harmonics on S^3. Formula (3.11) gives similar recursive formulas for the zonal harmonics on S^{n-1} whenever $n - 1$ is odd.

We remark that, generally, the functions $p_k(t)$ are Gegenbauer polynomials. In fact, the Gegenbauer polynomials $C_k^{\alpha}(x)$ are characterized by the generating function identity

$$(1 - 2tr + r^2)^{-\alpha} = \sum_{k=0}^{\infty} C_k^{\alpha}(t) r^k, \tag{3.23}$$

so, in the case of S^{n-1},

$$p_k(t) = B_n C_k^{(n-2)/2}(t). \tag{3.24}$$

The functions $C_k^\alpha(t)$ are given by

$$C_m^\alpha(t) = \sum_{k=0}^{[m/2]} (-1)^k 2^{m-2k} \frac{\Gamma(\alpha+m-k)}{\Gamma(\alpha)k!(m-2k)!} t^{m-2k}. \tag{3.25}$$

See Lebedev [**154**], page 125.

If we return to the formula (3.8) and substitute in the fundamental solution of the wave equation obtained in §2, in the case of S^2 ($n = 3$), we obtain the integral formula

$$P_k(\cos\theta) = (1/\pi)\int_{-\theta}^{\theta} (2\cos t - 2\cos\theta)^{-1/2} \cos(k+\tfrac{1}{2})t\, dt, \tag{3.26}$$

for the Legendre polynomials, known as the Mehler-Dirichlet formula. Similar calculations for $p_k(t)$ on odd-dimensional spheres (n even) produce expressions in closed form, due to the manifestation of the strict Huygens principle. For example, on S^3, we have

$$\begin{aligned} E_k u(x) &= (2\pi)^{-1}\int_{-\pi}^{\pi} e^{iks}\frac{\partial}{\partial s}(\sin s\overline{u}(x,s))\, ds \\ &= -(ik/2\pi)\int_{-\pi}^{\pi} e^{iks}\sin s\overline{u}(x,s)\, ds \\ &= (2k/\pi)\int_0^{\pi} [\cos(k+1)s - \cos(k-1)s]\overline{u}(x,s)\, ds. \end{aligned} \tag{3.27}$$

Now if $\cos s = t$, then the Tchebecheff polynomials are defined by

$$T_k(t) = \cos ks, \tag{3.28}$$

so we have

$$E_k u(x) = (2k/\pi)\int_0^{\pi} [T_{k+1}(x\cdot y) - T_{k-1}(x\cdot y)]V(s)^{-1}\int_{\sum_s(x)} u(y)\, d_T y\, ds \tag{3.29}$$

where

$$\sum\nolimits_s(x) = \{y \in S^3 : \mathrm{dist}(x,y) = |s|\}, \qquad V(s) = \mathrm{vol}\sum\nolimits_s(x). \tag{3.30}$$

Note that, if $\cos s = t$,

$$V(s) = C(1-t^2), \tag{3.31}$$

so (3.29) becomes

$$E_k u(x) = C(2k/\pi)\int_{S^3} [T_{k+1}(x\cdot y) - T_{k-1}(x\cdot y)](1-(x\cdot y)^2)^{-1}u(y)\, dy. \tag{3.32}$$

Thus, on S^3, the zonal functions are given by (3.14), with

$$p_k(t) = C'k[T_{k+1}(t) - T_{k-1}(t)]/(1-t^2). \tag{3.33}$$

This could be interpreted as an identity between Tchebecheff polynomials and the Gegenbauer polynomials $C_k^1(t)$.

We devote a few words to the exploitation of the highest weight vectors among the spherical harmonics. The highest weight vectors w_k in V_k can be identified as follows. Pick a maximal torus in SO(n) and a basis $e_1, \ldots, e_\mu$ ($\mu = [n/2]$) for its Lie algebra $\mathfrak{h}$ such that e_1 generates a rotation in the x_1x_2 plane. Then, with respect to a natural ordering on $\mathfrak{h}'$, the highest weight vector is

$$w_k(x) = (x_1 + ix_2)^k. \tag{3.34}$$

If we consider SO($n-1$) as acting on the variables $(x_2, \ldots, x_n)$, so

$$p_0 = (1, 0, \ldots, 0),$$

then we can recover the zonal function $z_k(x)$ from (3.34) as

$$z_k(x) = z_k(p_0) \int_{\mathrm{SO}(n-1)} (x_1 + ig^* x_2)^k \, dg. \tag{3.35}$$

We may as well normalize by setting

$$z_k^\#(x) = z_k(p_0)^{-1} z_k(x), \tag{3.36}$$

so $z_k^\#(p_0) = 1$. In particular, on S^2, we have

$$z_k^\#(x) = (2\pi)^{-1} \int_{-\pi}^{\pi} (x_1 + ix_2 \cos s + ix_3 \sin s)^k \, ds. \tag{3.37}$$

This suggests another generating function identity. If we multiply (3.37) by $r^k/k!$ and sum, we get, on S^2,

$$\sum_{k=0}^{\infty} (r^k/k!) z_k^\#(x) = (2\pi)^{-1} e^{rx_1} \int_{-\pi}^{\pi} e^{i(x_2 \cos s + x_3 \sin s)} \, ds. \tag{3.38}$$

The integral on the right is seen to be a special case of the integral characterizing Bessel functions:

$$\begin{aligned} J_\nu(z) &= [\Gamma(\tfrac{1}{2})\Gamma(\nu + \tfrac{1}{2})]^{-1} (z/2)^\nu \int_{-1}^{1} (1 - t^2)^{\nu - 1/2} \cos zt \, dt \\ &= [\Gamma(\tfrac{1}{2})\Gamma(\nu + \tfrac{1}{2})]^{-1} (z/2)^\nu \int_0^{\pi} \cos(z \cos\theta) \sin^{2\nu}\theta \, d\theta. \end{aligned} \tag{3.39}$$

So we have

$$\sum_{k=0}^{\infty} (r^k/k!) z_k^\#(x) = \tfrac{1}{4}\pi^{-1/2} e^{rx_1} J_0(r\sqrt{x_2^2 + x_3^2}). \tag{3.40}$$

If we specialize to $x_1^2 + x_2^2 + x_3^2 = 1$, this becomes

$$\sum_{k=0}^{\infty} (r^k/k!) P_k(t) = \tfrac{1}{4}\pi^{-1/2} e^{rt} J_0(r\sqrt{1 - t^2}). \tag{3.41}$$

If instead we multiply (3.37) by r^k and sum, we can evaluate the integral, via the identity

$$\int_0^{\pi} (a + b\cos s)^{-1} \, ds = \pi(a^2 - b^2)^{-1/2}, \quad \text{if } |b| > |a|, \tag{3.42}$$

and rederive the generating function identity (3.11), for $n = 3$.

We can also derive formulas for an orthogonal basis of V_k (for S^2). If we decompose the action of $S^1 = \mathrm{SO}(2)$ on V_k we get eigenvectors $z_{kj}(x)$ on which the generator of SO(2) acts as multiplication by ij, $j = -k, \dots, k$, and, generalizing (3.37), we can write

$$z_{kj}(x) = (2\pi)^{-1} \int_{-\pi}^{\pi} (x_1 + ix_2 \cos s + ix_3 \sin s)^k e^{ijs}\, ds. \tag{3.43}$$

If we multiply by r^k and sum, we get

$$\sum_{k=0}^{\infty} r^k z_{kj}(x) = (2\pi)^{-1} \int_{-\pi}^{\pi} (1 - x_1 - ix_2 \cos s - ix_3 \sin s)^{-1} e^{ijs}\, ds, \tag{3.44}$$

and if we use the identity

$$\int_{-\pi}^{\pi} (a + b\cos s)^{-1} \cos js\, ds = 2\pi(a^2 - b^2)^{-1/2}[(-a + \sqrt{a^2 - b^2})/b]^j, \quad \text{if } |b| > |a|, \tag{3.45}$$

which of course can be derived from the residue theorem, we get

$$\sum_{k=0}^{\infty} r^k z_{kj}(x) = i^{-j} e^{ij\psi}(1 - 2rx_1 + r^2)^{-1/2} \left[\frac{rx_1 - 1 + \sqrt{1 - 2rx_1 + r^2}}{r\sqrt{1 - x_1^2}}\right]^j \tag{3.46}$$

on S^2, where

$$x_2 + ix_3 = (x_2^2 + x_3^2)^{1/2} e^{i\psi}, \tag{3.47}$$

so ψ denotes the angular coordinate about the x_1-axis in R^3. Comparing with the case $j = 0$, one can deduce

$$z_{kj}(x) = i^{-j}(-1)^j[(k-j)!/k!]e^{ij\psi} P_k^j(x_1) \quad \text{on } S^2, \tag{3.48}$$

with

$$P_k^j(t) = (-1)^j (1 - t^2)^{|j|/2} (d/dt)^{|j|} P_k(t). \tag{3.49}$$

These functions are called associated Legendre functions.

Note that the kernel $E_k(x, y)$ of the projection E_k clearly satisfies

$$E_k(x, y) = \sum_j z_{kj}^b(x) \overline{z_{kj}^b(y)}, \tag{3.50}$$

where we have normalized the basis functions by setting $z_{kj}^b = z_{kj}/\|z_{kj}\|$. If we substitute the result of (3.48) on the right side of (3.50) and the result of (3.12), (3.24) on the left, we get the "addition theorem" for Legendre functions, obtained by Legendre in 1782.

We note the following alternate method of producing an orthonormal basis of V_k, in $L^2(S^2)$. If angular coordinates ψ, θ are chosen on S^2, with $(1, 0, 0) = p_0$ the north pole, θ the geodesic distance from p_0, ψ given by (3.47), then the regular representation of SO(3) on $L^2(S^2)$ defines the operators

$$L_1 = \partial/\partial\psi, \qquad L_\pm = ie^{\pm i\psi}(\pm\partial/\partial\theta + i\cot\theta\,\partial/\partial\psi). \tag{3.51}$$

As we saw in Chapter 2, all the weight vectors, giving an orthonormal basis of V_k, can be obtained by applying repeatedly the lowering operator L_- to the highest weight vector

$$w_k = (x_2 + ix_3)^k = \sin^k \theta e^{ik\psi}. \tag{3.52}$$

Alternatively, one could apply both lowering and raising operators repeatedly to the 0-weight vector $z_k(x) = P_k(\cos\theta)$. In either case, we rederive the formulas (3.48)–(3.49) for the associated Legendre functions, up to normalizing factors. Normalization can be achieved by using the formulas for operator norms of $L_\pm$, acting on eigenspaces of L_1, given by (1.21)–(1.22) of Chapter 2. We leave the details to the reader.

We end this section with one more simple fact about the representation of SO(3) on $L^2(S^2)$, which is special to this case.

PROPOSITION 3.2. *The decomposition*

$$L^2(S^2) = \bigoplus_k V_k \tag{3.53}$$

contains each irreducible unitary representation of SO(3), *exactly once.*

PROOF. As we have seen in Chapter 3, §3, since $S^2 = \mathrm{SO}(3)/\mathrm{SO}(2)$ is a rank one symmetric space, the decomposition (3.53) contains precisely once each irreducible representation of SO(3) with a nonzero vector invariant under SO(2), i.e., for which 0 is a weight. In Chapter 2, we classified all the irreducible representations of SU(2) and saw that, among them, precisely those for which 0 is a weight arise from representations of SO(3), so the proof is complete.

4. The subelliptic operators $L_1^2 + L_2^2 + i\alpha L_3$ on S^2. If L_j generates the group of rotations about the x_j axis in R^3, we can regard L_j as an operator on $\mathcal{D}'(S^2)$, and the Laplace operator on S^2 is

$$\Delta = L_1^2 + L_2^2 + L_3^2. \tag{4.1}$$

We want to study the operator

$$P_0 = L_1^2 + L_2^2, \tag{4.2}$$

and more generally

$$P_\alpha = L_1^2 + L_2^2 + i\alpha L_3 = \Delta - L_3^2 + i\alpha L_3. \tag{4.3}$$

Recall that $[L_1, L_2] = L_3$.

The operators P_α are not elliptic. They have characteristics on the equator $\{x \in S^2 : x_3 = 0\}$. In fact, $\operatorname{char} P_\alpha \subset T^*(S^2) \setminus 0$ consists of elements (x, ξ) such that $x_3 = 0$ and such that ξ_x annihilates any tangent vector $V_x \in T_x(S^2)$ which is orthogonal to the equator $x_3 = 0$, with respect to the natural Riemannian metric on S^2.

We want to show an equivalence between the study of P_α and the study of the operator $\mathcal{L}_\alpha$ on SO(3) defined by

$$\mathcal{L}_\alpha = X_1^2 + X_2^2 + i\alpha X_3, \tag{4.4}$$

where $X_j \in \mathrm{SO}(3)$ is the left invariant vector field on SO(3), of unit length, generating the one parameter subgroup of rotations about the x_j axis in R^3. This operator (on SU(2) rather than SO(3)) was studied briefly in Chapter 2. Of course, the general theory of pseudodifferential operators can be applied to both of these operators, but we aim to keep the analytical techniques elementary.

The correspondence $\mathcal{L}_\alpha \mapsto P_\alpha$ is a special case of the map T from left invariant operators on $C^\infty(\mathrm{SO}(3))$ to operators on $C^\infty(S^2)$ which commute with the Laplace operator, which we now define. We can identify the space of left invariant operators on $C^\infty(\mathrm{SO}(3))$ with $\mathcal{D}'(\mathrm{SO}(3))$, by right convolution. Given $f \in \mathcal{D}'(\mathrm{SO}(3))$, write

$$\pi_j(f) = \int_{\mathrm{SO}(3)} f(g)\pi_j(g)\,dg, \tag{4.5}$$

where $\{\pi_j\}$ runs over all the representations of SO(3), each contained once in $L^2(S^2)$, and set

$$Tf = \sum_j \pi_j(f)P_j, \tag{4.6}$$

where P_j is the orthogonal projection on $L^2(S^2)$ whose range is the space where SO(3) acts like π_j. Thus,

$$Tf = \int_{\mathrm{SO}(3)} f(g)\lambda(g)\,dg \tag{4.7}$$

where λ is the regular representation of SO(3) on $L^2(S^2)$:

$$\lambda(g)\varphi(x) = \varphi(g^{-1}\cdot x), \qquad g \in \mathrm{SO}(3), x \in S^2. \tag{4.8}$$

In other words, $Tf = \lambda(f)$. The action of Tf on $\varphi \in C^\infty(S^2)$ is therefore given by

$$(Tf)\varphi(x) = \int_{\mathrm{SO}(3)} f(x)\varphi(g^{-1}\cdot x)\,dg. \tag{4.9}$$

If we let φ tend to a delta function, we get

$$Tf(x,y) = \int_{\mathrm{SO}(3)} f(g)\delta_y(g^{-1}\cdot x)\,dg = \int_{K(x,y)} f(g)\,dg \tag{4.10}$$

where

$$K(x,y) = \{g \in \mathrm{SO}(3) : g\cdot y = x\} \tag{4.11}$$

is a great circle in SO(3). In (4.10), we are identifying the set of operators on $C^\infty(S^2)$ which commute with Δ with

$$\{u \in \mathcal{D}'(S^2 \times S^2) : \Delta_1 u = \Delta_2 u\}. \tag{4.12}$$

Note that if $f \in C^\infty(\mathrm{SO}(3))$, then $Tf \in C^\infty(S^2 \times S^2)$.

We see that (4.10) is a sort of *Radon transform*. We want to obtain an inversion formula, in order to recover functions of $\mathcal{L}_\alpha$ from functions of the operator P_α. Our first goal will be to impose a Hilbert space structure on some subspace of $\mathcal{D}'(S^2 \times S^2)$ containing the smooth functions, so that T maps $L^2(\mathrm{SO}(3))$ isometrically into this Hilbert space. Then T^* will map the range of T isometrically onto $L^2(\mathrm{SO}(3))$ and will provide the inverse.

Note that the orthogonality relations of Chapter 3 imply

$$\|f\|^2_{L^2(\mathrm{SO}(3))} = \sum_j (2j+1)\|\pi_j(f)\|^2_{\mathrm{HS}}, \tag{4.13}$$

since the dimension of the representation space of π_j is $d_j = 2j+1$. Recall that the Laplace operator Δ on S^2 is equal to $-j(j+1)$ on V_{π_j}. Thus the Hilbert space norm $|||\ |||$ on $C^\infty(S^2 \times S^2)$ should satisfy the identity

$$\begin{aligned} |||Tf|||^2 &= \sum_j d_j \|\pi_j(f)\|^2_{\mathrm{HS}} \\ &= \sum_j (2j+1)\|\pi_j(f)\|^2_{L^2(S^2\times S^2)} \\ &= \|(-2\Delta+1)^{1/4}Tf\|^2_{L^2(S^2\times S^2)} \end{aligned} \tag{4.14}$$

where we denote by Δ the operator $\Delta_1 + \Delta_2$, a Laplace operator on $S^2 \times S^2$. Thus we pick our pre-Hilbert space norm on $C^\infty(S^2 \times S^2)$ to be

$$|||u|||^2 = ((-2\Delta+1)^{1/2}u, u)_{L^2(S^2\times S^2)}. \tag{4.15}$$

We can now construct the adjoint of T. Given $u \in C^\infty(S^2 \times S^2)$, $f \in C^\infty(\mathrm{SO}(3))$, we should have

$$(T^*u, f)_{L^2(\mathrm{SO}(3))} = ((-2\Delta+1)^{1/2}u, Tf)_{L^2(S^2\times S^2)}. \tag{4.16}$$

If we let f tend to the delta function $\delta_g \in \mathcal{D}'(\mathrm{SO}(3))$, we have, by (4.10),

$$\begin{aligned} T^*u(g) &= \iint_{S^2\times S^2} (-2\Delta+1)^{1/2}u(x,y) \int_{K(x,y)} \delta_g(g')\,dg'\,dx\,dy \\ &= \int_{S^2} (-2\Delta+1)^{1/2}u(g\cdot y, y)\,dy. \end{aligned} \tag{4.17}$$

This is our inversion formula.

Returning to our operators $\mathcal{L}_\alpha$ and P_α, we have

PROPOSITION 4.1. *Let $f(\mathcal{L}_\alpha)\delta_e = k_1(g) \in \mathcal{D}'(\mathrm{SO}(3))$ and let $f(P_\alpha)$ have kernel $k_2(x,y) \in \mathcal{D}'(S^2 \times S^2)$. Then*

$$k_2(x,y) = \int_{K(x,y)} k_1(g)\,dg, \qquad K(x,y) = \{g \in \mathrm{SO}(3) : g\cdot y = x\}, \tag{4.18}$$

and

$$k_1(g) = \int_{S^2} (-2\Delta+1)^{1/2}k_2(g\cdot y, y)\,dy. \tag{4.19}$$

CHAPTER 5

Induced Representations, Systems of Imprimitivity, and Semidirect Products

In the preceding chapters, representations of groups have been constructed by various straightforward devices. Here we introduce a more sophisticated technique, both for constructing representations and for demonstrating when one has a complete set. In particular, we will be able to show that the set of irreducible unitary representations of $\mathbf{H}^n$ discussed in Chapter 1 is complete, as well as tackle other groups, such as Euclidean motion groups, which are semidirect products of simpler groups.

1. Induced representations and systems of imprimitivity. Let G be a Lie group, H a closed subgroup, and let

$$M = G/H. \tag{1.1}$$

Suppose for the moment that M has a G-invariant measure $d\mu$. Let L be a unitary representation of H on a Hilbert space V. We construct $\mathcal{H}_L$, a Hilbert space consisting of measurable functions f on G with values in V such that

$$f(gh) = L(h^{-1})f(g), \qquad h \in H, \tag{1.2}$$

and

$$\int_M \|f([g])\|_V^2 \, d\mu([g]) < \infty, \tag{1.3}$$

where $[g]$ denotes the image of g in M, under $G \to G/H = M$. We use the inner product

$$(f, f')_{\mathcal{H}_L} = \int_M (f(x), f'(x))_V \, d\mu(x). \tag{1.4}$$

Then we denote by U^L the representation of G on $\mathcal{H}_L$ defined by

$$U^L(g)f(x) = f(g^{-1}x), \qquad g, x \in G, \ f \in \mathcal{H}_L. \tag{1.5}$$

One verifies this is unitary. It is called the representation of G *induced* by the representation L of H. Sometimes it is convenient to use a more complete notation, such as

$$U^L = \mathrm{Ind}^G_{H,L}. \tag{1.6}$$

There are important cases one wants to consider where G does not leave invariant a smooth measure on M. A weaker condition to impose is that the smooth measure satisfy the following: if $\mu_g(\mathcal{B}) = \mu(g \cdot \mathcal{B})$, then

$$d\mu_g/d\mu(x) = \rho_g(x) \in L^\infty_{\mathrm{loc}}(M) \quad \text{for each } g \in G. \tag{1.7}$$

This always holds in the Lie group case. In this more general case, one can define $\mathcal{H}_L$ as above and set

$$U^L(g)f(x) = \rho_g(x)^{1/2} f(g^{-1}x). \tag{1.8}$$

A function f satisfying (1.2) corresponds naturally to a section of the vector bundle E_L over M, constructed from the principal H-bundle $G \to M$ by the homomorphism $L\colon H \to \mathrm{Aut}(V)$. Thus the fibres of E_L are copies of V. The natural G action on E_L then defines a representation of G equivalent to U^L, via the natural analogue of (1.5) or (1.8). We remark that strong continuity of this representation is easily checked for f a smooth compactly supported section of E_L. Strong continuity on $\mathcal{H}_L$ follows, by Lemma 1.1 of Chapter 0.

If M has no smooth measure invariant under G, we can also define $\mathrm{Ind}^G_{H,L}$, using sections of the vector bundle $E_L \otimes |\Lambda^{1/2}|$ over M, where $|\Lambda^{1/2}|$ is the bundle of half-densities (see Guillemin and Sternberg [**87**]). We will not go into details on this.

We next define a system of imprimitivity. Let G be a Lie group, U a unitary representation of G on a Hilbert space $\mathcal{H}$, M a G-space (possibly not transitive), and P a projection valued measure on the Borel sets of M, $P(\mathcal{B})$ being orthogonal projections on $\mathcal{H}$. We say (G, U, M, P) defines a *system of imprimitivity* if $P(M) = I$ and

$$U(g)^{-1}P(\mathcal{B})U(g) = P(g \cdot \mathcal{B}), \quad \text{all } g \in G, \ \mathcal{B} \subset M \text{ Borel set.} \tag{1.9}$$

If G acts transitively on M, so $M = G/H$, we say we have a *transitive system of imprimitivity*. An important example is provided by $M = G/H$, $U = U^L$ an induced representation as above, on L^2 sections of the vector bundle E_L over M, and P^L defined by

$$P^L(\mathcal{B})f(x) = \chi_{\mathcal{B}}(x)f(x), \quad f \text{ section of } E_L \text{ over } M. \tag{1.10}$$

In fact, in many important cases, this is essentially the only example, as is shown by the following important theorem of Mackey.

IMPRIMITIVITY THEOREM. *Let (G, U, M, P) define a transitive system of imprimitivity. Then there exists a representation L of H and a unitary map*

$$K\colon \mathcal{H} \to \mathcal{H}_L \tag{1.11}$$

such that

$$KU(g)K^{-1} = U^L(g), \qquad KP(\mathcal{B})K^{-1} = P^L(\mathcal{B}), \tag{1.12}$$

for all $g \in G$, $\mathcal{B} \subset M$ Borel.

We give a brief indication of a proof. For more details, see the book [**161**] of Mackey, or [**162**].

First, applying the strong version of the spectral theorem, we can suppose $\mathcal{H} = L^2(M, \mu, W)$ for some auxiliary Hilbert space W, and

$$P(\mathcal{B})f(x) = \chi_{\mathcal{B}}(x)f(x). \tag{1.13}$$

Now define the regular representation R of G on $L^2(M, \mu, W)$ by

$$R(g)f(x) = (d\mu_g/d\mu)^{1/2} f(g^{-1} \cdot x). \tag{1.14}$$

Consider now the operator valued function on G:

$$Q(g) = U(g)R(g)^{-1}. \tag{1.15}$$

Since

$$R(g)^{-1}P(\mathcal{B})R(g) = P(g \cdot \mathcal{B}), \tag{1.16}$$

we see that

$$Q(g)P(\mathcal{B}) = P(\mathcal{B})Q(g) \quad \text{for all } g \in G, \text{ Borel } \mathcal{B} \subset M. \tag{1.17}$$

It follows that the unitary operator $Q(g)$ on $L^2(M, \mu, W)$ is realizable as a multiplication operator, i.e., as the operation of multiplication by a function on M taking values in the set of unitary operators on W:

$$Q(g)f(x) = Q(g, x)f(x); \quad Q(g, x) \text{ unitary on } W. \tag{1.18}$$

Now for $U(g)$ to be a representation, we require $Q(g, x)$ to satisfy the identity

$$Q(g_1 g_2, x) = Q(g_1, x)Q(g_2, g_1 \cdot x). \tag{1.19}$$

Let us extend Q from $G \times M$ to $G \times G$, writing

$$\tilde{Q}(g, g') = Q(g, x), \qquad x = g'H \in M. \tag{1.20}$$

Now define

$$B(g) = \tilde{Q}(g_0^{-1} g, g_0), \tag{1.21}$$

with g_0 fixed. Then

$$\tilde{Q}(g, g') = B(g')^{-1} B(gg'). \tag{1.22}$$

Note that $\tilde{Q}(g, g'h) = \tilde{Q}(g, g')$ for all $h \in H$ implies

$$B(gg'h)B(gg')^{-1} = B(g'h)B(g')^{-1}, \quad \text{for all } g, g' \in G, \ h \in H, \tag{1.23}$$

so if we set

$$L(h) = B(gh)B(g)^{-1} \tag{1.24}$$

we see that $L(h)$ is independent of $g \in G$. This defines a representation L of H on W.

From here one verifies that, if $U = U^M$, $P = P^M$ and the process above is used, then one obtains $L = M$. Next, let (U, P) and (U', P') both lead to the same representation L of H. Then one can set up P' to act on $L^2(M,\mu,W)$ exactly as P, and define the operator valued functions $Q', \tilde{Q}', B'$, in analogy with (1.15), (1.20), (1.23). So we have

$$B(gh)B(g)^{-1} = L(h) = B'(gh)B'(g)^{-1}, \tag{1.25}$$

which implies

$$K(g) = B'(g)B(g) \tag{1.26}$$

is invariant under right multiplication by $h \in H$, so we have defined a unitary operator valued function $K(x) = K(g)$, $x = gH \in M$, and the desired unitary operator K on $L^2(M,\mu,W)$ is given by

$$Kf(x) = K(x)f(x). \tag{1.27}$$

2. The Stone-von Neumann theorem. In this section we will prove that the set of irreducible unitary representations of the Heisenberg group $\mathbf{H}^n$ constructed in Chapter 1 is complete. Recall that the group law on $\mathbf{H}^n$ is given by

$$(t,q,p)\cdot(t',q',p') = (t+t'+\tfrac{1}{2}q\cdot p' - \tfrac{1}{2}p\cdot q', q+q', p+p'), \tag{2.1}$$

where $t \in R$, $q \in R^n$, $p \in R^n$. Denote by C the subgroup

$$C = \{(t,0,0) : t \in R\}. \tag{2.2}$$

C is the center of $\mathbf{H}^n$. Any irreducible unitary representation U of $\mathbf{H}^n$ is scalar on C. If it is trivial on C, it must represent the commutative group $\mathbf{H}^n/C$, and so be one of the one-dimensional representations $\pi_{(y,\eta)}$ given in Chapter 1:

$$\pi_{(y,\eta)}(t,q,p) = e^{iy\cdot q + i\eta\cdot p}. \tag{2.3}$$

So we can restrict attention to U nontrivial on C, say

$$U((t,0,0)) = e^{i\lambda t}I. \tag{2.4}$$

We claim that if U is irreducible, it must be equivalent to π_λ, constructed in Chapter 1, or equivalently, any two irreducible unitary representations of $\mathbf{H}^n$ satisfying (2.4), with the same nonzero λ, must be unitarily equivalent. Using the dilations $\delta_{\pm\lambda}$ of $\mathbf{H}^n$, we can suppose $\lambda = 1$.

Thus, suppose U is any irreducible unitary representation of $\mathbf{H}^n$ satisfying

$$U((t,0,0)) = e^{it}I. \tag{2.5}$$

Let

$$V(q) = U(0,q,0), \qquad W(p) = U(0,0,p), \tag{2.6}$$

so V and W are each unitary representations of the commutative group R^n, and, by (2.1),

$$U(t,q,p) = e^{it+iq\cdot p/2}V(q)W(p). \tag{2.7}$$

Conversely, if V and W are unitary representations of R^n on a Hilbert space $\mathcal{H}$, (2.7) defines a representation of $\mathbf{H}^n$ if and only if

$$W(p)V(q) = e^{iq\cdot p}V(q)W(p). \tag{2.8}$$

Now apply Stone's theorem to the unitary group $V(q)$, to get a projection valued measure P on R^n such that

$$V(q) = \int_{R^n} e^{iq\cdot y}\, dP(y). \tag{2.9}$$

The commutation relation (2.8) is equivalent to

$$W(p)P(\mathcal{B})W(p)^{-1} = P(p+\mathcal{B}), \tag{2.10}$$

for $\mathcal{B} \subset R^n$ a Borel set. Now R^n acts transitively on R^n, by translation, so (R^n, W, R^n, P) is a transitive system of imprimitivity. Note that the subgroup of R^n fixing some given point in $M = R^n$ is the trivial group 0. The imprimitivity theorem implies there is a unitary map $K\colon \mathcal{H} \to L^2(R^n)$ such that

$$KWK^{-1} = \mathrm{Ind}_{0,1}^{R^n} = U^1, \qquad KP(\mathcal{B})K^{-1} = P'(\mathcal{B}), \tag{2.11}$$

where

$$P'(\mathcal{B})f(x) = \chi_{\mathcal{B}}(x)f(x) \tag{2.12}$$

and of course U^1 is the regular representation of R^n on $L^2(R^n)$:

$$U^1(p)f(x) = f(x+p). \tag{2.13}$$

This gives the uniqueness of U up to unitary equivalence, and concludes the proof of the Stone-von Neumann theorem.

3. Semidirect products. Let K be a Lie group and A be abelian. In fact, suppose $A = R^l$. Suppose K acts on A by automorphisms

$$\varphi\colon K \to \mathrm{Aut}(A). \tag{3.1}$$

We then define the *semidirect product*

$$G = A \times_\varphi K \tag{3.2}$$

to be set-theoretically $A \times K$, with the group law

$$(a,k)\cdot(a',k') = (a+\varphi(k)a', kk'). \tag{3.3}$$

We want to analyze the irreducible unitary representations of G in terms of those of K and A.

If U is a unitary representation of G, then, since

$$(a,k) = (a,e)\cdot(0,k), \tag{3.4}$$

we can set

$$V(a) = U(a,e), \qquad W(k) = U(0,k), \tag{3.5}$$

and we have

$$U(a,k) = V(a)W(k). \tag{3.6}$$

Also, V is a unitary representation of A and W is a unitary representation of K. Suppose conversely that V and W are unitary representations of A and K, respectively, on a Hilbert space $\mathcal{H}$. Then (3.6) defines an operator valued function on $G = A \times_\varphi K$, and it is a representation of G provided

$$W(k)^{-1}V(\varphi(k)a)W(k) = V(a), \quad \text{for all } a \in A,\ k \in K. \tag{3.7}$$

We will obtain a system of imprimitivity, as follows. Use Stone's theorem to write the commutative group V as

$$V(a) = \int_{A'} e^{ia\cdot\alpha}\, dP(\alpha), \qquad a \in A. \tag{3.8}$$

We see that (3.7) implies

$$W(k)^{-1}P(\mathcal{B})W(k) = P(\varphi(k)' \cdot \mathcal{B}) \tag{3.9}$$

for any Borel set $\mathcal{B} \subset A'$. We now aim for conditions yielding a transitive system of imprimitivity, so we want to force the projection valued measure P to be supported on a single orbit of K in A'.

For starters, suppose the representation U of G is irreducible. Now let S be any K-invariant Borel set in A'. Then (3.9) shows $P(S)$ commutes with $W(k)$ for all $k \in K$. But certainly $P(S)$ commutes with $V(a)$ for all $a \in A$, so $P(S)$ must commute with $U(g)$ for all $g \in G = A \times_\varphi K$. If U is irreducible, this implies either $P(S) = 0$ or $P(S) = I$. Now suppose the following condition is satisfied:

(3.10) There is a countable family $\mathcal{B}_j$ of K-invariant Borel sets in A' such that if $\mathcal{O}, \mathcal{O}'$ are disjoint orbits in A', then there are disjoint $\mathcal{B}_{j_1}, \mathcal{B}_{j_2}$ such that $\mathcal{O} \subset \mathcal{B}_{j_1}$ and $\mathcal{O}' \subset \mathcal{B}_{j_2}$.

If this condition is satisfied, we say the K-orbits in A' are *countably separated.* It is clear that this condition guarantees that, if

$$J = \{j : P(\mathcal{B}_j) = I\}, \tag{3.11}$$

then

$$\mathcal{O} = \bigcap_{j\in J} \mathcal{B}_j \text{ is an orbit,} \tag{3.12}$$

and

$$P(\mathcal{O}) = I. \tag{3.13}$$

Thus $(K, W, \mathcal{O}, P)$ defines a transitive system of imprimitivity, under the hypotheses above, so by the imprimitivity theorem there is a representation L of the subgroup K_α of K fixing some $\alpha \in \mathcal{O}$, such that

$$W \text{ is equivalent to } U^L = \operatorname{Ind}^K_{K_\alpha, L}. \tag{3.14}$$

Note that

$$(3.15) \qquad \mathcal{O} = K/K_\alpha = G/AK_\alpha,$$

and the identity (3.6) says

$$(3.16) \qquad U \text{ is equivalent to } U^{\chi_\alpha L} = \mathrm{Ind}_{AK_\alpha,\chi_\alpha L}^{G},$$

where the representation $\chi_\alpha L$ of AK_α is defined by

$$(3.17) \qquad \chi_\alpha L(a,k) = e^{ia\cdot\alpha}L(k), \qquad k \in K_\alpha.$$

We summarize the result in

THEOREM 3.1. *Let $G = A \times_\varphi K$ be a semidirect product with $A = R^l$. Assume the K-orbits in A' are countably separated. Let U be an irreducible unitary representation of G. Then there is a K-orbit $\mathcal{O} \subset A'$, and if K_α is the subgroup of K fixing an element $\alpha \in \mathcal{O}$, there is a (necessarily irreducible) unitary representation L of K_α such that U is unitarily equivalent to the induced representation $U^{\chi_\alpha L} = \mathrm{Ind}_{AK_\alpha,\chi_\alpha L}^{G}$.*

We state, without proof, a result which complements Theorem 3.1. See Mackey [**162**].

THEOREM 3.2. *The induced representation $U^{\chi_\alpha L}$ constructed above is irreducible if L is an irreducible representation of K_α. $U^{\chi_\alpha L}$ and $U^{\chi_\alpha L'}$ are equivalent if and only if L and L' are equivalent. Also $U^{\chi_\alpha L_1}$ and $U^{\chi_\beta L_2}$ cannot be equivalent unless χ_α and χ_β are in the same K-orbit in A'. If they are in the same orbit, then every $U^{\chi_\alpha L_1}$ is equivalent to $U^{\chi_\beta L_2}$ for some L_2.*

When applied to specific examples, the assertions of Theorem 3.2 will be fairly straightforward, and so we will mainly depend on Theorem 3.1 to guarantee that the set of irreducible representations we produce is *complete*.

The first example we consider is called the "$ax + b$ group," the group of affine transformations of the line. This group is

$$(3.18) \qquad G_2 = R \times_\varphi R^+$$

where R is additive, R^+ is multiplicative, and

$$(3.19) \qquad \varphi(a)b = ab, \qquad a \in R^+, b \in R.$$

Note that there are three orbits of R^+ in $R' = R$:

$$(3.20) \qquad \mathcal{O}_0 = \{0\}, \qquad \mathcal{O}_+ = (0,\infty), \qquad \mathcal{O}_- = (-\infty, 0).$$

For $\mathcal{O}_0$, the group $K_0 \subset R^+$ fixing 0 is all of R^+, so $AK_0 = RR^+ = G_2$. The representations obtained are hence the one-dimensional representations of G_2. The orbit $\mathcal{O}_+$ contains 1, i.e., the representation χ_1 of R given by $\chi_1(a) = e^{ia}$. In this case, the subgroup of R^+ fixing 1 is trivial, so the only irreducible unitary representation is $L = 1$. Thus $\mathcal{O}_+$ gives the representation

$$(3.21) \qquad U_+ = \mathrm{Ind}_{R,\chi_1\cdot 1}^{G_2}.$$

By the same argument, $\mathcal{O}_-$ gives the representation

$$U_- = \mathrm{Ind}_{R,\chi_{-1}\cdot 1}^{G_2}. \tag{3.22}$$

Thus G_2 has exactly these equivalence classes of irreducible unitary representations: the one-dimensional representations and the two infinite-dimensional representations U_+ and U_-, which can be seen to be contragredient to each other.

We note that the Heisenberg group $\mathbf{H}^n$ is a semidirect product. To see this, we find it convenient to use the group $\tilde{\mathbf{H}}^n$ (isomorphic to $\mathbf{H}^n$), with group law

$$(t,q,p) \odot (t',q',p') = (t+t'+q\cdot p', q+q', p+p'). \tag{3.23}$$

Then we see that

$$\tilde{\mathbf{H}}^n = R^{n+1} \times_\varphi R^n, \tag{3.24}$$

with $\varphi\colon R^n \to \mathrm{Aut}(R^{n+1})$ given by

$$\varphi(p)(t,q) = (t + p\cdot q, q). \tag{3.25}$$

Since we have already analyzed the representations of $\mathbf{H}^n$ in §2, we will not do so again here.

Another important class of semidirect products is furnished by the Euclidean groups, which will be the subject of the next section.

4. The Euclidean group and the Poincaré group. The Euclidean group $\mathrm{E}(n)$ is the group of isometries of R^n. Thus it is a semidirect product

$$\mathrm{E}(n) = R^n \times_\varphi \mathrm{O}(n) \tag{4.1}$$

where $\varphi\colon \mathrm{O}(n) \to \mathrm{Aut}(R^n)$ is given by the standard action of orthogonal matrices on R^n. The component of the identity in $\mathrm{E}(n)$ will be denoted $\mathrm{E}_0(n)$, so

$$\mathrm{E}_0(n) = R^n \times_\varphi \mathrm{SO}(n). \tag{4.2}$$

According to the results of §3, we can classify the irreducible unitary representations of $\mathrm{E}_0(n)$ by looking at $\mathrm{SO}(n)$-orbits in R^n. These orbits consist of $\mathcal{O}_0 = \{0\}$ and $\mathcal{O}_r = rS^{n-1} = \{x \in R^n\colon |x| = r\}$, $r \in (0,\infty)$. It is clear that these orbits are countably separated. The orbit $\mathcal{O}_0$ gives rise to the finite-dimensional representations of $\mathrm{E}_0(n)$, arising from $\mathrm{E}_0(n) \to \mathrm{SO}(n)$. Given $r \in (0,\infty)$, pick $\alpha \in \mathcal{O}_r = rS^{n-1}$ to lie on the x_n-axis. Then the subgroup K_α of $\mathrm{SO}(n)$ fixing α is naturally identified with $\mathrm{SO}(n-1)$. (If $n = 2$, $\mathrm{SO}(1)$ consists of only the identity.) For each irreducible unitary representation π_λ of $\mathrm{SO}(n-1)$, we can form

$$U_{r,\lambda} = \mathrm{Ind}_{R^n\cdot \mathrm{SO}(n-1),\alpha\pi_\lambda}^{\mathrm{E}_0(n)}, \qquad |\alpha| = r, \tag{4.3}$$

and by the results of §3, these exhaust the irreducible unitary representations of $\mathrm{E}_0(n)$.

Note that, for $n = 2$, $\mathrm{SO}(n-1) = \{e\}$, so there is just a one parameter family of infinite-dimensional irreducible unitary representations of $\mathrm{E}_0(2)$:

$$U_r = \mathrm{Ind}_{R^2,\alpha}^{\mathrm{E}_0(2)}, \qquad |\alpha| = r. \tag{4.4}$$

For $n = 3$, $\mathrm{SO}(n-1) = \mathrm{SO}(2) = S^1$. In this case, the representations of S^1 are given by

$$\pi_k(e^{i\theta}) = e^{ik\theta}, \tag{4.5}$$

so the infinite-dimensional irreducible unitary representations of $\mathrm{E}_0(3)$ are given by

$$U_{r,k} = \mathrm{Ind}_{R^3\cdot\mathrm{SO}(2),\chi_\alpha\pi_k}^{\mathrm{E}_0(3)}, \qquad |\alpha| = r, k \in \mathbf{Z}. \tag{4.6}$$

For $n = 4$, $\mathrm{SO}(n-1) = \mathrm{SO}(3)$, whose representations were derived rather explicitly in Chapter 2, §1. For $n = 5$, $\mathrm{SO}(n-1) = \mathrm{SO}(4)$, which is covered by $\mathrm{SU}(2)\times\mathrm{SU}(2)$, so its representations are also explicitly described by the results of Chapter 2.

We can similarly treat the inhomogeneous Lorentz group, or Poincaré group,

$$\mathcal{L}(n+1) = R^{n+1} \times_\varphi \mathrm{O}(n,1), \tag{4.7}$$

of isometries of R^{n+1} with the Lorentz metric $\langle x, x\rangle = x_1^2 + \cdots + x_n^2 - x_{n+1}^2$. In this case the orbits fall into four classes:

(4.8)

$$\mathcal{O}^b = \{0\}; \qquad \mathcal{O}_\lambda = \{x \in R^n \setminus 0 \colon \langle x, x\rangle = \lambda\} \quad \text{in cases } \lambda > 0,\ \lambda = 0,\ \lambda < 0.$$

The orbit $\mathcal{O}^b$ gives rise to the irreducible representations of $\mathcal{L}(n+1)$ arising from those of $\mathrm{O}(n,1)$ via $\mathcal{L}(n+1) \to \mathrm{O}(n,1)$. Given $\lambda \in R$, the subgroup of $\mathrm{O}(n,1)$ stabilizing a given point in the orbit $\mathcal{O}_\lambda$ is isomorphic to $\mathrm{O}(n)$ if $\lambda < 0$, to $\mathrm{O}(n-1,1)$ if $\lambda > 0$, and to a group we denote MN which is a product of $M = \mathrm{O}(n-1)$ and an abelian subgroup N of $\mathrm{O}(n,1)$, if $\lambda = 0$. Thus the representations of the inhomogeneous Lorentz group are analyzed as before, in terms of irreducible representations of the groups $\mathrm{O}(n,1)$, $\mathrm{O}(n)$, $\mathrm{O}(n-1,1)$, and MN. This result was obtained by Wigner in his famous paper [**264**], which pointed out the physical significance of those representations of $\mathcal{L}(n+1)$ arising from representations of the compact group $\mathrm{O}(n)$, whose representations were well understood at that time. In Chapters 8 and 9 we shall discuss the representation theory of $\mathrm{SO}(2,1)$ and $\mathrm{SO}(3,1)$. The group MN is isomorphic to the Euclidean group $\mathrm{E}(n-1)$.

CHAPTER 6

Nilpotent Lie Groups

The Heisenberg group $\mathbf{H}^n$ is the simplest example (other than R^n) of a nilpotent Lie group. An inductive process enables one to reduce the representation theory of a general nilpotent Lie group to that of $\mathbf{H}^n$. This is particularly simple for "step two" nilpotent groups, which we study separately in §2 before considering the general case in §3.

Inductive methods are also effective on solvable Lie groups. This works easiest on "exponential" solvable groups, for which $\exp : \mathfrak{g} \to G$ is a surjective diffeomorphism. More general solvable groups need not be "type I," but the type I cases have been identified and their representations classified. In the previous chapter we considered two solvable Lie groups, the Euclidean group E(2) and the "$ax + b$ group," which will appear again in the next chapter, but we will not go into a general study of solvable Lie groups. See [**9**, **10**, **177**, **200**] for information on this topic.

1. Nilpotent Lie algebras and Lie algebras with dilations. A Lie algebra $\mathfrak{g}$ is said to be nilpotent if, for each $X \in \mathfrak{g}$, the linear transformation $\operatorname{ad} X$ on $\mathfrak{g}$, defined by $\operatorname{ad} X(Y) = [X, Y]$, is nilpotent, i.e., $(\operatorname{ad} X)^k = 0$ for some k. Engel's theorem says that $\mathfrak{g}$ satisfies this condition if and only if, with

$$\mathfrak{g}_{(0)} = \mathfrak{g}, \qquad \mathfrak{g}_{(1)} = [\mathfrak{g}, \mathfrak{g}], \qquad \mathfrak{g}_{(j)} = [\mathfrak{g}, \mathfrak{g}_{(j-1)}], \tag{1.1}$$

we have, for some K,

$$\mathfrak{g}_{(0)} \supset \mathfrak{g}_{(1)} \supset \cdots \supset \mathfrak{g}_{(K)} = 0. \tag{1.2}$$

In fact, something stronger is true. If $\mathfrak{g}$ is nilpotent, there exist ideals $\mathfrak{g}_j$ such that $\dim \mathfrak{g}_j = j$, $j = 0, 1, \ldots, \dim \mathfrak{g}$, and $[\mathfrak{g}, \mathfrak{g}_j] \subset \mathfrak{g}_{j-1}$. This fact in turn is a special case of

PROPOSITION 1.1. *If ρ is a representation of a Lie algebra $\mathfrak{g}$ on a finite-dimensional vector space V such that $\rho(X)$ is nilpotent for each $X \in \mathfrak{g}$, then there are linear subspaces V_j, with*

$$V = V_n \supset V_{n-1} \supset \cdots \supset V_0 = 0, \tag{1.3}$$

such that $\dim V_j = j$ *and* $\rho(\mathfrak{g})V_j \subset V_{j-1}$.

This proposition can be proved using the following simple lemma.

LEMMA 1.2. *Under the hypotheses of Proposition* 1.1, *there is a nonzero* $v \in V$ *such that* $\rho(X)v = 0$ *for all* $X \in \mathfrak{g}$.

To deduce Proposition 1.1 from Lemma 1.2, let V_1 be the linear span of such a v. Then $\mathfrak{g}$ acts on V/V_1. Pick a nonzero $\bar{v}_2 \in V/V_1$ annihilated by $\mathfrak{g}$; say $\bar{v}_2$ is the image of $v_2 \in V$. Let V_2 be the linear span of v_2 and v; $\mathfrak{g}$ acts on V/V_2. Continue in this fashion.

We give a proof of Lemma 1.2, using induction. Let $N = \dim \mathfrak{g}$, and suppose the assertion is true for all Lie algebras of dimension less than N. Let $\mathfrak{a}$ be a proper Lie subalgebra of $\rho(\mathfrak{g})$ of maximal dimension. If we consider the natural representation τ of $\mathfrak{a}$ on $\rho(\mathfrak{g})/\mathfrak{a}$, we see by the induction hypothesis that there is an $X_1 \in \rho(\mathfrak{g})$, $X_1 \notin \mathfrak{a}$, such that $\tau(\mathfrak{a})X_1 \in \mathfrak{a}$. This implies $(X_1) + \mathfrak{a}$ is a subalgebra of $\rho(\mathfrak{g})$, hence is all of $\rho(\mathfrak{g})$. Thus $\mathfrak{a}$ is an ideal in $\rho(\mathfrak{g})$ of codimension 1. Let $W = \{v \in V \colon Yv = 0 \text{ for all } Y \in \mathfrak{a}\}$. Then, by induction, $\dim W \geq 1$. If $v \in W$, $Y \in \mathfrak{a}$, then with X_1 as above $YX_1v = [Y, X_1]v + X_1Yv = 0$, so $X_1(W) \subset W$. Since $X_1^k = 0$ for some k, there exists a nonzero $v \in W$ such that $X_1v = 0$. Hence $\rho(\mathfrak{g})v = 0$, and the lemma is proved.

A connected Lie group G whose Lie algebra $\mathfrak{g}$ is nilpotent, is also called nilpotent. We note without proof the standard fact that, if G is nilpotent and simply connected, then $\exp : \mathfrak{g} \to G$ is a diffeomorphism of $\mathfrak{g}$ onto G. Thus G and $\mathfrak{g}$ are identified as C^∞ manifolds. Furthermore, Lebesgue measure on $\mathfrak{g}$ gives a bi-invariant Haar measure on G. For these facts, see the standard references, e.g., [**100, 254**].

Important examples of nilpotent Lie algebras are given by Lie algebras with dilations. We now introduce this concept. Suppose $\mathfrak{g}$ is a Lie algebra on which a one parameter group of automorphisms $\alpha(t)$ acts. Then

$$\delta(X) = (d/dt)\alpha(t)X|_{t=0}, \qquad X \in \mathfrak{g}, \tag{1.4}$$

defines a derivation on $\mathfrak{g}$. In other words, the identity

$$\alpha(t)([X, Y]) = [\alpha(t)X, \alpha(t)Y] \tag{1.5}$$

implies

$$\delta([X, Y]) = [X, \delta(Y)] + [\delta(X), Y]. \tag{1.6}$$

Suppose all the eigenvalues of the linear transformation δ on $\mathfrak{g}$ are *real*. For $b \in \mathcal{B} = \operatorname{spec} \delta$, set

$$\mathfrak{g}_b = \{X \in \mathfrak{g} \colon X \text{ is a generalized } b\text{-eigenvector of } \delta\}. \tag{1.7}$$

Then we have

$$\mathfrak{g} = \bigoplus_{b \in \mathcal{B}} \mathfrak{g}_b \quad \text{linear (not Lie algebra) direct sum.} \tag{1.8}$$

Note that

$$X \in \mathfrak{g}_b \quad \text{if and only if} \quad \alpha(t)X = e^{tb} \times \text{polynomial in } t, \tag{1.9}$$

the polynomial taking values in $\mathfrak{g}$. If $X \in \mathfrak{g}_a$, $Y \in \mathfrak{g}_b$, we have

$$\begin{aligned}(1.10)\qquad \alpha(t)[X,Y] &= [\alpha(t)X, \alpha(t)Y]\\ &= e^{(a+b)t} \times \text{ polynomial in } t.\end{aligned}$$

It follows that

$$(1.11)\qquad [\mathfrak{g}_a, \mathfrak{g}_b] \subset \mathfrak{g}_{a+b}.$$

Of course, if δ is scalar on $\mathfrak{g}_b$ for all $b \in \mathcal{B}$, this is also an immediate consequence of (1.6). An immediate consequence of (1.11) is

$$(1.12)\qquad \mathfrak{g} \text{ is nilpotent if } \mathcal{B} \subset R^+ \text{ (or if } \mathcal{B} \subset R^-).$$

Generally, we will say a one parameter group $\alpha(t)$ of automorphisms of a Lie algebra $\mathfrak{g}$ (necessarily nilpotent) is a group of *dilations* if $\operatorname{spec}\delta \subset R^+$, where δ is the derivation defined by (1.4). We say it is a group of strict dilations if in addition $\alpha(t)$ acts as a scalar on each $\mathfrak{g}_b$, $b \in \operatorname{spec}\delta$.

The Lie algebra $\mathfrak{h}^n$ of the Heisenberg group is the basic example of a nilpotent Lie algebra. Recall from Chapter 1 that $\mathfrak{h}^n$ has a basis T, L_j, M_j $(1 \le j \le n)$, with

$$(1.13)\qquad [M_j, L_j] = -[L_j, M_j] = T, \quad \text{other commutators zero.}$$

We have

$$(1.14)\qquad [\mathfrak{h}^n, \mathfrak{h}^n] = (T),$$

the linear span of T, and

$$(1.15)\qquad [\mathfrak{h}^n, T] = 0.$$

A dilation group on $\mathfrak{h}^n$, which we introduced in Chapter 1, is given by

$$(1.16)\qquad \alpha(t)(sT + q\cdot L + p\cdot M) = e^{2t}sT + e^t q\cdot L + e^t p\cdot M,$$

where $q \cdot L = \sum q_j L_j$ and $p \cdot M$ is similarly defined. Thus $\mathfrak{h}^n$ has a group of strict dilations. In the notation of (1.7), we have

$$(1.17)\qquad \operatorname{spec}\delta = \{1,2\}, \qquad \mathfrak{g}_1 = (L_j, M_j \colon 1 \le j \le n), \qquad \mathfrak{g}_2 = (T),$$

and in the notation of (1.1) we have

$$(1.18)\qquad \mathfrak{g}_{(0)} = \mathfrak{h}^n, \qquad \mathfrak{g}_{(1)} = (T), \qquad \mathfrak{g}_{(2)} = 0.$$

The simplest sort of generalization of $\mathfrak{h}^n$ is the class of Step 2 nilpotent Lie algebras, which we will study in the next section.

2. Step 2 nilpotent Lie groups. We introduce the following concept.

DEFINITION. A Lie algebra $\mathfrak{g}$ is called nilpotent of Step 2 if we have

$$(2.1)\qquad \mathfrak{g}_{(2)} = 0,$$

using the notation of (1.1).

Such algebras always possess dilations. Indeed, we have

PROPOSITION 2.1. *If* $\mathfrak{g}$ *is nilpotent of Step 2, there is a group of strict dilations whose associated derivation satisfies* $\operatorname{spec}\delta = \{1,2\}$.

PROOF. With $\mathfrak{g}_{(1)} = [\mathfrak{g},\mathfrak{g}]$, let V be any linear subspace of $\mathfrak{g}$ which is complementary to $\mathfrak{g}_{(1)}$. Define $\alpha(t)$ on $\mathfrak{g}$ by

$$\alpha(t)(X+Y) = e^t X + e^{2t} Y, \qquad X \in V,\ Y \in \mathfrak{g}_{(1)}.$$

Since $[V,V] \subset \mathfrak{g}_{(1)}$, it is easy to check $\alpha(t)$ is a group of automorphisms. Clearly $\operatorname{spec}\delta = \{1,2\}$, and $\mathfrak{g}_1 = V$, $\mathfrak{g}_2 = \mathfrak{g}_{(1)}$.

The following are examples of $(2n+2)$-dimensional Step 2 nilpotent Lie algebras. Pick generators $X_1,\dots,X_n,Y_1,\dots,Y_n,T_1,T_2$, and define a Lie bracket so that

$$[X_j,X_k] = [Y_j,Y_k] = [X_j,T_k] = [Y_j,T_k] = 0 \tag{2.2}$$

and

$$[X_j,Y_k] = \delta_{jk}(a_jT_1 + b_jT_2), \tag{2.3}$$

where $a_j, b_j \in R$ are given. The reader will notice quite a bit of structure such Lie algebras share with $\mathfrak{h}^n$. That there is a close relation generally is illustrated by

PROPOSITION 2.2. *Let* $\mathfrak{g}$ *be nilpotent of Step* 2; $\mathfrak{g} = \mathfrak{g}_1 + \mathfrak{g}_2$. *If* V *is a linear subspace of* $\mathfrak{g}_2$ *of codimension* 1, *then, for some* n, k,

$$\mathfrak{g}/V \approx \mathfrak{h}^n \oplus R^k \quad (\textit{Lie algebra direct sum}). \tag{2.4}$$

PROOF. With $\mathfrak{g}' = \mathfrak{g}/V$, we see that $\mathfrak{g}'$ is a nilpotent Lie algebra of Step 2, with $\mathfrak{g}'_1 \approx \mathfrak{g}_1$, $\mathfrak{g}'_2 \approx \mathfrak{g}_2/V$, one-dimensional. Picking a nonzero $T \in \mathfrak{g}'_2$, we see that, for $X, Y \in \mathfrak{g}'_1$, $[X,Y] = \sigma(X,Y)T$, where $\sigma(X,Y)$ is an antisymmetric bilinear form on the vector space $\mathfrak{g}'_1$. Let $E = \{X \in \mathfrak{g}'_1 : \sigma(X,Y) = 0$ for all $Y \in \mathfrak{g}'_1\}$. Let F be any linear subspace of $\mathfrak{g}'_1$ complementary to E. Then $\sigma(X,Y)$ is nondegenerate on F. It follows that $\dim F = 2n$ is even, and choosing a symplectic basis of F shows that

$$\mathfrak{g}' = E \oplus (F + (T)) = E \oplus \mathfrak{h}^n \quad \text{(Lie algebra direct sum)}.$$

Here E is a commutative Lie algebra, so $E = R^k$ with $k = \dim E$. The proof is complete.

We remark that an arbitrary Step 2 Lie algebra is constructed as follows. Let V and W be finite-dimensional vector spaces, and

$$B : V \otimes V \to W \tag{2.5}$$

an antisymmetric bilinear map. Then we define a Lie algebra structure on $\mathfrak{g} = V \oplus W$ by

$$[v_1 + w_1, v_2 + w_2] = B(v_1, v_2), \qquad v_j \in V,\ w_j \in W. \tag{2.6}$$

We have $\mathfrak{g}_1 = V$, $\mathfrak{g}_2 = W$. The reader might consider conditions on a pair B_1, B_2 of bilinear maps which give rise to isomorphic Lie algebras.

We now consider the irreducible unitary representations of the simply connected Lie group with Lie algebra $\mathfrak{g}$, assumed to be nilpotent of Step 2. Let π be any such representation of G, giving rise to a skew adjoint representation of $\mathfrak{g}$, also denoted π. If we write $\mathfrak{g} = \mathfrak{g}_1 + \mathfrak{g}_2$ as above, since $\mathfrak{g}_2$ is contained in the *center* of $\mathfrak{g}$, by Schur's lemma, π must act by scalars on $\mathfrak{g}_2$:

$$(1/i)\pi\colon \mathfrak{g}_2 \to R. \tag{2.7}$$

Now either $\pi|_{\mathfrak{g}_2} = 0$ or $V_\pi = \{X \in \mathfrak{g}_2 : \pi(X) = 0\}$ has codimension 1. In the first case, with G_2 the subgroup of G generated by the Lie algebra $\mathfrak{g}_2$, we have π factoring through:

$$\begin{array}{ccc} G & \xrightarrow{\ \pi\ } & \mathrm{U}(H) \\ \downarrow & \nearrow_{\tilde{\pi}} & \\ G/G_2 & & \end{array}$$

However, clearly $G/G_2 \approx R^m$, i.e., G/G_2 is abelian and simply connected. Thus in this case the Hilbert space H is one-dimensional and the set of irreducible unitary representations of $G/G_2 \approx R^m$ is well known.

In the second case, π factors through

$$\begin{array}{ccc} G & \xrightarrow{\ \pi\ } & \mathrm{U}(H) \\ \downarrow & \nearrow_{\tilde{\pi}} & \\ G/H_\pi & & \end{array}$$

where H_π is the subgroup of G generated by the Lie algebra V_π. According to Proposition 2.2,

$$G/H_\pi \approx \mathbf{H}^n + R^k \quad \text{for some } n, k, \tag{2.8}$$

and again the set of irreducible unitary representations of this group is known, by the Stone-von Neumann theorem.

We take a closer look at a nilpotent Lie group of Step 2 which is of particular interest, namely the group $N_{k,2}$ with Lie algebra

$$\mathcal{N}_{k,2} = V \oplus \bigwedge^2 V, \qquad V = R^k, \tag{2.9}$$

the bracket operation on $\mathcal{N}_{k,2}$ being given by

$$[(v, \alpha), (w, \beta)] = (0, v \wedge w). \tag{2.10}$$

These Lie algebras are universal, in the sense that any nilpotent Lie algebra of Step 2 is a quotient of one of them. One says $N_{k,2}$ is the (simply connected) nilpotent Lie group, *free* of Step 2.

If π is a nonscalar irreducible unitary representation of $N_{k,2}$, annihilating a codimension one subspace H of $\bigwedge^2 V$, then π comes from a representation π^b of $\mathfrak{g}_H = V \oplus \bigwedge^2 V/H = V + W_H$, where we have set $W_H = \bigwedge^2 V/H$; $\dim W_H = 1$. Put a standard euclidean metric on V, $\bigwedge^2 V$, and hence $N_{k,2}$. This determines a linear isomorphism $W_H \approx R$; an element $T \in W_H$ of norm 1 corresponds to $1 \in R$. The Lie algebra structure on $V + W_H$ is characterized by

$$[v, w] = \omega(v, w)T, \qquad v, w \in V. \tag{2.11}$$

The bilinear form ω is a general element of $(\bigwedge^2 V)' \approx \bigwedge^2 V'$, i.e., a general anti-symmetric bilinear form on V, normalized (with norm 1 in $\bigwedge^2 V'$). In the case $\dim V = 2l$ is even, the generic such ω is nondegenerate; when ω is nondegenerate, we have $\mathfrak{g}_H = V + W_H$ isomorphic to $\mathfrak{h}^l$, and we get a family of irreducible representations $\pi_{\lambda,H}$, $\lambda \in R\backslash 0$, characterized to within unitary equivalence by $\pi_{\lambda,H}(T) = i\lambda$. We will work out rather explicitly the Plancherel formula and inversion formula on $N_{2l,2}$.

We will deduce the Plancherel formula for $N_{2l,2}$ from the Plancherel formula for the Heisenberg group $\mathbf{H}^l$ proved in Chapter 1, §3:

$$\|f\|^2_{L^2(\mathbf{H}^l)} = C_l \int_{-\infty}^{\infty} \|\pi_\lambda(f)\|^2_{\mathrm{HS}} |\lambda|^l \, d\lambda. \tag{2.12}$$

Now if we identify $N_{2l,2}$ and $\mathcal{N}_{2l,2}$ as C^∞ manifolds, with Haar measure equal to Lebesgue measure, and if Lebesgue measure is imposed on $\mathfrak{g}_H$, let $\varphi(H)$ denote the factor (for H such that $\mathfrak{g}_H \approx \mathfrak{h}^l$) such that

$$\|g\|^2_{L^2(\mathfrak{g}_H)} = \varphi(H) \int_{-\infty}^{\infty} \|\pi_{\lambda,H}(g)\|^2_{\mathrm{HS}} |\lambda|^l \, d\lambda. \tag{2.13}$$

Now we introduce the partial Radon transform

$$\mathcal{R}f(x, w, H) = \int_H f(x, y + w) \, dy \tag{2.14}$$

where $x \in V, y \in H, w \in W_H$, and $V + W_H$ is identified with a linear subspace of $\mathcal{N}_{2l,2}$ by identifying $T \in W_H$ with a vector in $\bigwedge^2 V$ orthogonal to H, of norm 1. The formula (2.13) gives

$$\int_{V+W_H} |\mathcal{R}f(x, w, H)|^2 \, dx \, dw = \varphi(H) \int_{-\infty}^{\infty} \|\pi_{\lambda,H}(f)\|^2_{\mathrm{HS}} |\lambda|^l \, d\lambda. \tag{2.15}$$

Now the Plancherel formula for the Radon transform (see Appendix C) implies

$$\|f\|^2_{L^2(N_{2l,2})} = c_l \int_{\mathcal{G}} \int_{-\infty}^{\infty} \int_V |D_w^{\alpha/2} \mathcal{R}f(x, w, H)|^2 \, dx \, dw \, d\,\mathrm{vol}(H), \tag{2.16}$$

where $\mathcal{G}$ is the Grassmannian manifold of hyperplanes in $\bigwedge^2 V$, and

$$\alpha = \dim \textstyle\bigwedge^2 V - 1 = \binom{2l}{2} - 1 = 2l^2 - l - 1. \tag{2.17}$$

Since $|D_w|$ is represented as $|\lambda|$ by $\pi_{\lambda,H}$, we deduce from (2.15) the Plancherel formula

$$\|f\|^2_{L^2(N_{2l,2})} = c_l \int_{\mathcal{G}} \int_{-\infty}^{\infty} \|\pi_{\lambda,H}(f)\|^2_{\mathrm{HS}} |\lambda|^{l+\alpha} \varphi(H) \, d\lambda \, d\,\mathrm{vol}(H). \tag{2.18}$$

From here, the inversion formula is a simple consequence:

$$f(\varsigma) = c_l \int_{\mathcal{G}} \int_{-\infty}^{\infty} \operatorname{tr}(\pi_{\lambda,H}(\varsigma)^* \pi_{\lambda,H}(f)) |\lambda|^{l+\alpha} \varphi(H)\, d\lambda\, d\operatorname{vol}(H). \tag{2.19}$$

We leave it to the reader to work out analogous formulas for $N_{2l+1,2}$.

More general results on representations of nilpotent Lie groups are given in the next section, but the discussion here has had the advantage of being very simple. For connections between the study of Step 2 nilpotent Lie groups and hypoelliptic operators, see [**174**, **108**], and Rothschild and Stein [**204**] for further generalizations.

3. Representations of general nilpotent Lie groups. In this section we show how to produce a complete set of irreducible unitary representations of a simply connected nilpotent Lie group G, following the theory of Dixmier [**43**] and Kirillov [**135**]. As in §2, this will be achieved by application of the Stone-von Neumann theorem on the representations of the Heisenberg group, but the argument in this case will be somewhat more elaborate.

Let $\mathfrak{g}$ be the Lie algebra of G, and let $\mathfrak{z}$ be the center of $\mathfrak{g}$. If π is an irreducible unitary representation of G on a Hilbert space H, it induces a skew adjoint representation π of $\mathfrak{g}$. This must represent $\mathfrak{z}$ by scalars

$$\pi(Z) = i\lambda(Z)I, \qquad Z \in \mathfrak{z}, \tag{3.1}$$

where λ is a linear functional on $\mathfrak{z}$. If $\mathfrak{z}_1 = \ker\lambda, \mathfrak{z}_1$ generates the subgroup H_1 of G, then π comes from a unitary representation ρ of G/H_1; $\pi = \rho \circ \mu$, where $\mu\colon G \to G/H_1$ is the natural projection. Let us suppose, therefore, that $\dim \mathfrak{z} = 1$ and $\lambda \neq 0$ on $\mathfrak{z}$. Fix a nonzero $Z \in \mathfrak{z}$, which therefore generates $\mathfrak{z}$.

We begin our analysis by producing a subalgebra of $\mathfrak{g}$ isomorphic to $\mathfrak{h}^1$.

LEMMA 3.1. *There exist $X, Y \in \mathfrak{g}$ such that*

$$[X, Y] = Z. \tag{3.2}$$

PROOF. Using Proposition 1.1, with ρ the adjoint representation, we have ideals $\mathfrak{g}_j$ of $\mathfrak{g}$, $\mathfrak{g} = \mathfrak{g}_n \supset \cdots \supset \mathfrak{g}_1 \supset \mathfrak{g}_0 = 0$, with $\dim \mathfrak{g}_j = j$ and $[\mathfrak{g}, \mathfrak{g}_j] \subset \mathfrak{g}_{j-1}$. Hence $\mathfrak{g}_1 = \mathfrak{z}$. Pick any nonzero $Y \in \mathfrak{g}_2 \backslash \mathfrak{g}_1$. Since $Y \notin \mathfrak{z}$, there exists $X \in \mathfrak{g}$ with $[X, Y] \neq 0$. But $[X, Y] \in \mathfrak{g}_1 = \mathfrak{z}$, so $[X, Y]$ is a multiple of Z. Rescaling X gives (3.2).

With X and Y chosen so (3.2) holds, let

$$\mathcal{L} = \{W \in \mathfrak{g} \colon [W, Y] = 0\}. \tag{3.3}$$

We note that

(3.4) $\qquad \mathcal{L}$ is a Lie algebra, and Y and Z are in the center of $\mathcal{L}$.

It is also useful to note

LEMMA 3.2. *We have* $\mathfrak{g} = (X) + \mathcal{L}$.

PROOF. Since $[\mathfrak{g}, \mathfrak{g}_2] \subset \mathfrak{g}_1 = \mathfrak{z}$ and $\dim \mathfrak{z} = 1$, it is clear that $\operatorname{codim} \mathcal{L} \leq 1$. Since X does not belong to $\mathcal{L}$, by (3.2), the result follows.

Now define $\mathfrak{A}$ to be the linear span of X, Y, and Z. Thus $\mathfrak{A}$ is a Lie subalgebra of $\mathfrak{g}$, isomorphic to $\mathfrak{h}^1$, and π is a skew adjoint representation of $\mathfrak{A}$ which is given by (3.1) on the center of $\mathfrak{A}$. The Stone-von Neumann theorem, in the form established in Chapter 5, §2, implies

LEMMA 3.3. *There is a unitary equivalence of* H *with* $L^2(R, H_1)$, *for some Hilbert space* H_1, *such that, for*

$$f \in L^2(R, H_1), \tag{3.5}$$

we have

$$\begin{aligned} \pi(\exp(tZ + qY + pX))f(s) &= \pi_\lambda^\#(t, q, p)f(s) \\ &= e^{i\lambda(t+qs+qp/2)}f(s+p). \end{aligned} \tag{3.6}$$

Here we identify $\lambda \in \mathfrak{z}'$ *with* $\lambda \in R\backslash 0$.

We next examine the behavior of π on L, the subgroup of G generated by $\mathcal{L}$. First of all, by the defining relations (3.3), we see that each $l \in L$ commutes with $\exp tY$, so

$$\pi(l)\pi(\exp tY) = \pi(\exp tY)\pi(l), \qquad l \in L. \tag{3.7}$$

However, if $f \in H$ is represented by (3.5), since $\pi(\exp tY)$ is given by multiplication by $e^{i\lambda ts}$, if $\pi(l)$ commutes with all these multiplication operators, it must be a multiplication operator, so we must have

$$\pi(l)f(s) = U(l, s)f(s) \tag{3.8}$$

where $U(l, s)$ takes values in the unitary operators on H_1. A priori, $U(l, s)$ is a measurable function of s.

Since $\mathcal{L}$ is a subalgebra of $\mathfrak{g}$ of codimension 1, it must be an ideal. Hence, for $l \in L$, we have $\exp(-t_0X)l\exp(t_0X) \in L$. Since

$$\pi(\exp(-t_0X)l\exp(t_0X)) = \pi(\exp(-t_0X))\pi(l)\pi(\exp(t_0X)),$$

and since, by (3.6), $\pi(\exp(t_0X))f(s) = f(s+t_0)$, we see that, almost everywhere,

$$U(l, t - t_0) = U(\exp(-t_0X)l\exp(t_0X), t), \qquad l \in L. \tag{3.9}$$

It follows that, possibly modifying $U(l, t)$ on a set of measure zero, we obtain $U(l, t)$ strongly continuous in t, and

$$U(l, t) = U(\exp(tX)l\exp(-tX), 0). \tag{3.10}$$

Another simple property of $U(l, t)$, which follows from $\pi(ll') = \pi(l)\pi(l')$, is that $U(ll', s) = U(l, s)U(l', s)$ and in particular

$$U(ll', 0) = U(l, 0)U(l', 0). \tag{3.11}$$

Consequently, if

$$U_0(l) = U(l, 0), \qquad l \in L, \tag{3.12}$$

then U_0 gives a unitary representation of L on H_1. The following result gives a simple but incisive relationship between the representation π of G which we set out to analyze and the representation U_0 of L.

PROPOSITION 3.4. *The representation π of G is unitarily equivalent to*

$$\mathrm{Ind}_{L,U_0}^G. \tag{3.13}$$

PROOF. This follows almost immediately from the definition of an induced representation. Indeed, Ind_{L,U_0}^G acts on the Hilbert space $\tilde{H}$ of measurable functions u from G to H_1 satisfying $u(gl) = U_0(l^{-1})u(g)$, $g \in G$, $l \in L$, such that $\int_{G/L} \|u(gl)\|_{H_1}^2 \, d(gL) < \infty$. Since $G/L \approx R$, we get a unitary mapping B from $\tilde{H}$ to $L^2(R, H_1)$ by setting $(Bu)(t) = u(\exp tX)$. If we denote (3.13) by π^b, we have $B\pi^b(\exp tX)f(s) = f(s-t)$ by the definition of the induced representation. Meanwhile, using (3.10), we see that, for $l \in L$,

$$\begin{aligned} B\pi^b(l)f(s) &= f(l^{-1}\exp sX) \\ &= f(\exp(sX)\exp(-sX)l^{-1}\exp(sX)) \\ &= U_0(\exp(-sX)l\exp(sX))f(\exp sX) \\ &= U(l,s)Bf(s) = \pi(l)Bf(s). \end{aligned}$$

Thus B provides the unitary intertwining of π and Ind_{L,U_0}^G.

Note that, if π is to be irreducible, the unitary representation U_0 of L must be irreducible. We have consequently reduced the problem of obtaining all the irreducible unitary representations of G to the same problem for simply connected nilpotent Lie groups of smaller dimension. There is one further detail to be settled, to make sure that all the representations obtained in the form (3.13) are in fact irreducible. In view of the remark (3.4), the fact that this is so is established by the following result.

PROPOSITION 3.5. *Let G be a simply-connected nilpotent Lie group with one-dimensional center. Let L be a connected subgroup of codimension* 1. *Then L is normal, and if Z is the center of G, then $Z \subset L$. Let U_0 be an irreducible unitary representation of L on a Hilbert space H_1, such that $U_0|_Z$ is not trivial. If the center of L has dimension at least* 2, *then Ind_{L,U_0}^G is irreducible.*

PROOF. If we denote by H the representation space of $\pi^b = \mathrm{Ind}_{L,U_0}^G$, suppose A is a bounded linear operator on H which commutes with $\pi^b(g)$ for all $g \in G$. Our goal is to show A is a multiple of the identity.

Let $\mathfrak{g}$ be the Lie algebra of G, $\mathcal{L}$ that of L. Since $\mathcal{L}$ has codimension 1 in $\mathfrak{g}$, one of the arguments in the proof of Lemma 1.2 shows $\mathcal{L}$ is an ideal; hence L is normal. Since $[\mathfrak{g}, \mathcal{L}] \subset \mathcal{L}$, we can apply Proposition 1.1, to write $\mathcal{L} = \mathcal{L}_{n-1} \supset \mathcal{L}_{n-2} \supset \cdots \supset \mathcal{L}_1 \supset \mathcal{L}_0 = 0$, with $\dim \mathcal{L}_j = j$ and $[\mathfrak{g}, \mathcal{L}_j] \subset \mathcal{L}_{j-1}$.

In particular, $[\mathfrak{g}, \mathcal{L}_1] = 0$, so it follows that $\mathcal{L}_1 = \mathfrak{z}$, the center of G. Let Z generate $\mathfrak{z} = \mathcal{L}_1$. Pick $Y \in \mathcal{L}_2$ such that Z and Y generate $\mathcal{L}_2$. Since the center of $\mathcal{L}$ is assumed to have dimension at least 2, we can arrange that $\mathcal{L}_2$ belong to the center of $\mathcal{L}$, so $[\mathcal{L}, Y] = 0$. Pick $X \in \mathfrak{g} \backslash \mathcal{L}$, so $\mathfrak{g} = (X) + \mathcal{L}$. Hence $[X, Y] \neq 0$, since $Y \notin \mathfrak{z}$; however $[X, Y] \in \mathcal{L}_1 = \mathfrak{z}$, so upon rescaling X we can assume $[X, Y] = Z$.

Recalling the argument in the proof of Proposition 3.4, we can realize Ind_{L,U_0}^G on the space of square integrable functions $f\colon R \to H_1$, by

$$\pi^b(\exp tX) f(s) = f(s-t), \tag{3.14}$$

$$\pi^b(l) f(s) = U_0(\exp(-sX) l \exp(sX)) f(s), \qquad l \in L. \tag{3.15}$$

If A is an operator on $L^2(R, H_1)$ commuting with $\pi^b(g)$ for all $g \in G$, in particular A must commute with the translations (3.14). We next show that A must commute with multiplication operators. Indeed, since $[X, Y] = Z$, we have

$$\exp(-sX)\exp(\sigma Y)\exp(sX) = (\exp \sigma Y)(\exp \sigma s Z) + O(\sigma^2), \tag{3.16}$$

and hence, by (3.15),

$$\pi^b(\exp \sigma Y) f(s) = U_0((\exp \sigma Y)(\exp \sigma s Z)) f(s) + O(\sigma^2),$$

if $f \in S(R, H_1)$. Now since Y and Z belong to the center of $\mathcal{L}$, we have

$$\begin{aligned} U_0(\exp \sigma s Z) &= e^{i\sigma s\mu}, \qquad \mu \in R\backslash 0, \\ U_0(\exp \sigma Y) &= e^{i\sigma\gamma}, \qquad \gamma \in R. \end{aligned} \tag{3.17}$$

Thus,

$$\pi^b(\exp \sigma Y) f(s) = e^{i\sigma(s\mu+\gamma)} f(s) + O(\sigma^2), \tag{3.18}$$

for $f \in S(R, H_1)$, and hence $\pi^b(Y) f(s) = i(s\mu+\gamma) f(s)$, so we can sharpen (3.18) to

$$\pi^b(\exp \sigma Y) f(s) = e^{i\sigma(s\mu+\gamma)} f(s). \tag{3.19}$$

Now A commutes with all operators of this form. Hence A must be an operator of multiplication by a bounded measurable function $a(t)$, taking values in $\mathcal{L}(H_1)$. This fact together with the translation invariance of A implies that $a(t) = a \in \mathcal{L}(H_1)$ is constant:

$$(Af)(s) = af(s), \qquad a \in \mathcal{L}(H_1). \tag{3.20}$$

Finally, since A commutes with all operators of the form (3.15), we see that $a \in \mathcal{L}(H_1)$ must commute with $U_0(l)$ for all $l \in L$. Since U_0 is assumed to be irreducible, this implies a is a multiple of the identity. This proves Proposition 3.5.

The last two propositions in principle allow one to obtain a complete set of irreducible unitary representations for any simply connected nilpotent Lie group. In the theory of Kirillov, a neater description of these representations is available. A one-to-one correspondence is set up between equivalence classes of such representations and orbits in $\mathfrak{g}^*$ under the co-adjoint representation. The association of a representation of G to each orbit in $\mathfrak{g}^*$ can be viewed as a special case of a construction known as "geometric quantization," studied by Kostant [**142**], Souriau [**221**], and others. In this framework, a generalization of the results of Kirillov to the representation theory of type I solvable Lie groups has been achieved by Auslander and Kostant [**9**]; see also Brezin [**28**]. We will not go into any of these matters here, referring the reader to these articles, and also the articles of Blattner [**21**], and Weinstein [**260**], and the book of Wallach [**254**].

CHAPTER 7

Harmonic Analysis on Cones

The Laplace operator Δ on a cone can be analyzed in terms of the Laplace operator on the base of the cone, using the dilations of the cone, whose infinitesimal generator together with Δ spans a two-dimensional Lie algebra, the Lie algebra of the "$ax+b$ group." The Hankel transform H_ν arises as an operator intertwining two irreducible unitary representations of this group. Thus its unitarity is suggested by Schur's lemma, though we give a proof of this unitarity and the resulting Hankel inversion formula along different lines, yielding more detailed information.

To illustrate the study of cones, we derive the solution of the wave equation for the cone over an interval of length 2π, which is equivalent to the wave equation in the plane R^2 with a slit along the positive x-axis.

1. Dilations of cones and the $ax+b$ group. A cone with vertex at the origin in R^n is describable as the cone over a subdomain Ω of the unit sphere S^{n-1} in R^n. More generally, if N is any compact Riemannian manifold of dimension $n-1$, possibly with boundary, the cone over N, denoted $C(N)$, is the space $R^+ \times N$ together with the Riemannian metric

$$g = dr^2 + r^2 g_N \tag{1.1}$$

where g_N is the metric tensor on N. We want to study the spectral theory and harmonic analysis of the Laplace operator Δ on $C(N)$, which can be written

$$\Delta = \partial^2/\partial r^2 + (n-1)r^{-1}\partial/\partial r + r^{-2}\Delta_N, \tag{1.2}$$

where Δ_N is the Laplace operator on the base N.

The group R acts as a group of dilations on $C(N)$, by

$$\delta(t)(r,x) = (e^t r, x). \tag{1.3}$$

This magnifies the metric by a factor of e^t, so it multiplies the Laplace operator by a factor of e^{-2t}. Thus

$$D(t)\Delta D(t)^{-1} = e^{-2t}\Delta, \tag{1.4}$$

if

$$D(t)f(r,x) = e^{nt/2} f(e^t r, x). \tag{1.5}$$

The factor $e^{nt/2}$ is thrown in to make $D(t)$ unitary on

$$L^2(C(N)) = L^2(R^+ \times N, r^{n-1} dr\, d_{\mathrm{vol}}(x)).$$

Note that the infinitesimal generator of $D(t)$ is

$$X = r\partial/\partial r + n/2, \tag{1.6}$$

which is skew adjoint on $L^2(R^+, r^{n-1}dr)$, and differentiating (1.4) gives the commutator relation

$$[X, \Delta] = -2\Delta. \tag{1.7}$$

We see that X and Δ generate a two-dimensional Lie algebra, isomorphic to the Lie algebra $\mathrm{aff}(R^1)$ of the "$ax+b$ group" introduced in Chapter 5, §3, with a basis $\{A, B\}$ satisfying commutator relations $[A, B] = -2B$. We have a skew-symmetric representation π of $\mathrm{aff}(R^1)$, defined by

$$\pi(A) = X, \qquad \pi(B) = i\Delta. \tag{1.8}$$

Using the polar coordinate representation (1.2) of Δ, we can obtain a family of representations of $\mathrm{aff}(R^1)$ on $L^2(R^+, r^{n-1}dr)$, via the eigenfunction decomposition of Δ_N on $L^2(N)$, as follows. Note that, if we set

$$Y = r^{-1}\partial/\partial r + \tfrac{1}{2}(n-2)r^{-2}, \tag{1.9}$$

a skew-symmetric operator on $L^2(R^+, r^{n-1}dr)$, we have

$$L = \tfrac{1}{2}(XY + YX) = \partial^2/\partial r^2 + (n-1)r^{-1}\partial/\partial r + \tfrac{1}{4}(n-2)^2 r^{-2}. \tag{1.10}$$

A simple calculation gives

$$[X, Y] = -2Y. \tag{1.11}$$

Generally

$$[X, XY] = XXY - XYX = X[X, Y], \tag{1.12}$$

and

$$[X, YX] = XYX - YXX = [X, Y]X, \tag{1.13}$$

so (1.11) implies

$$[X, L] = -2L. \tag{1.14}$$

Now, on the eigenspace of $L^2(N)$ where $\Delta_N = -\mu^2$, we have

$$\Delta = L - \nu^2/r^2, \tag{1.15}$$

where $-\mu^2 = -\nu^2 + \frac{1}{4}(n-2)^2$, or

$$\nu = [\mu^2 + \tfrac{1}{4}(n-2)^2]^{1/2}. \tag{1.16}$$

Also a straightforward calculation gives

$$[X, F] = -2F, \tag{1.17}$$

where

$$Ff(r) = r^{-2} f(r), \tag{1.18}$$

so (1.14) and (1.17) lead back to (1.7), given (1.15).

Thus we can define a representation $\pi_{\nu,n}$ of $\mathrm{aff}(R^1)$ on $L^2(R^+, r^{n-1}dr)$ by

$$\begin{aligned}\pi_{\nu,n}(A) &= X = r\partial/\partial r + n/2,\\ \pi_{\nu,n}(B) &= i(L - \nu^2 F) = i(\partial^2/\partial r^2 + (n-1)r^{-1}\partial/\partial r - \mu^2 r^{-2}).\end{aligned} \tag{1.19}$$

We will show that the Lie algebra representation of $\mathrm{aff}(R^1)$ defined by (1.19) exponentiates to a unitary representation of the $ax+b$ group, $\mathrm{Aff}_e(R^1)$. Clearly $\pi_{\nu,n}(A)$ exponentiates; we have

$$\pi_{\nu,n}(\exp tA) f(r) = e^{nt/2} f(e^t r). \tag{1.20}$$

To pursue this analysis, we will rewrite the representation (1.19) in terms of the spectral decomposition of $\pi_{\nu,n}(B)$. The operator

$$(1/i)\pi_{\nu,n}(B) = \partial^2/\partial r^2 + (n-1)r^{-1}\partial/\partial r - \mu^2 r^{-2} = L_{\nu,n} \tag{1.21}$$

is a symmetric, negative semidefinite operator on $L^2(R^+, r^{n-1}dr)$, with domain $C_0^\infty(R^+)$. The Friedrichs extension process extends this operator to a negative semidefinite selfadjoint operator (call it $L''_{\nu,n}$), the unique negative semidefinite selfadjoint extension with the property that $C_0^\infty(R^+)$ is dense in the domain of $(-L''_{\nu,n})^{1/2}$. The spectral decomposition of $L''_{\nu,n}$ is a special case of the Titchmarsh-Weyl-Kodaira theory. The (generalized) eigenvalue equation

$$(\partial^2/\partial r^2 + (n-1)r^{-1}\partial/\partial r - \mu^2 r^{-2})u_{\nu,\lambda}(r) = -\lambda^2 u_{\nu,\lambda}(r) \tag{1.22}$$

can be converted to Bessel's equation, and the solution sufficiently well behaved at $r = 0$ to be a generalized eigenfunction is

$$u_{\nu,\lambda}(r) = c r^{1-n/2} J_\nu(\lambda r), \tag{1.23}$$

where $J_\nu(z)$ is the Bessel function, given by the integral formula

$$J_\nu(z) = [\Gamma(\tfrac{1}{2})\Gamma(\nu + \tfrac{1}{2})]^{-1}(z/2)^\nu \int_{-1}^{1} (1-t^2)^{\nu - 1/2} \cos zt\, dt. \tag{1.24}$$

Note that the action of the operators on the right side of (1.19), *formally*, on $u_{\nu,\lambda}$, is given by

$$Xu_{\nu,\lambda} = (r\partial/\partial r + n)u_{\nu,\lambda} \Leftrightarrow e^{tX}u_{\nu,\lambda} = e^t u_{\nu,e^t\lambda}, \tag{1.25}$$

$$i(L - \nu^2 F)u_{\nu,\lambda} = -i\lambda^2 u_{\nu,\lambda} \Leftrightarrow e^{it(L-\nu^2 F)}u_{\nu,\lambda} = e^{-it\lambda^2}u_{\nu,\lambda}. \tag{1.26}$$

This suggests considering the following unitary representation of $\mathrm{Aff}_e(R^1)$ on $L^2(R^+, \lambda d\lambda)$:

$$\gamma(\exp tA) f(\lambda) = e^{-t} f(e^{-t}\lambda), \tag{1.27}$$

$$\gamma(\exp tB) f(\lambda) = e^{-it\lambda^2} f(\lambda). \tag{1.28}$$

It is easy to see that γ is irreducible; indeed γ is unitarily equivalent to the representation U_+ of $\mathrm{Aff}_e(R^1)$ discussed in Chapter 5, §3. Note that

$$\gamma(A) = -\lambda\partial/\partial\lambda - 1, \qquad \gamma(B) = -i\lambda^2, \tag{1.29}$$

and $[-\lambda\partial/\partial\lambda, \lambda^2] = -2\lambda^2$, reflecting the commutator relations in $\mathrm{aff}(R^1)$.

It turns out that for each $\nu \geq 0$ the exponentiated representation $\pi_{\nu,n}$ of $\mathrm{Aff}_e(R^1)$ exists and is unitarily equivalent to γ. We will define an intertwining of $\pi_{\nu,n}$ and γ by means of the *Hankel transform*,

$$H_\nu f(\lambda) = \int_0^\infty f(r)J_\nu(\lambda r)r\,dr. \tag{1.30}$$

The content of the Hankel inversion formula, which we will prove shortly, is that

$$H_\nu\colon L^2(R^+, r\,dr) \to L^2(R^+, \lambda\,d\lambda) \quad \text{is unitary,} \tag{1.31}$$

and

$$f(r) = H_\nu(H_\nu f)(r). \tag{1.32}$$

Granted this, we define a unitary operator

$$H_{\nu,n}\colon L^2(R^+, r^{n-1}dr) \to L^2(R^+, \lambda\,d\lambda) \tag{1.33}$$

by

$$H_{\nu,n}f = H(r^{(n-2)/2}f), \tag{1.34}$$

i.e.,

$$H_{\nu,n}f(\lambda) = \int_0^\infty f(r)r^{n/2}J_\nu(\lambda r)\,dr. \tag{1.35}$$

We note that all this makes sense for any $n \in R$; n need not be an integer ≥ 2.

PROPOSITION 1.1. *For all $Y \in \mathrm{aff}(R^1)$, $n \in R$,*

$$H_{\nu,n}\pi_{\nu,n}(Y) = \gamma(Y)H_{\nu,n}, \tag{1.36}$$

acting on $C_0^\infty(R^+)$.

PROOF. It suffices to show this for $Y = A$ and for $Y = B$. For $Y = A$, we can even work with the exponentiated version, given by (1.20). We have

$$\begin{aligned} H_{\nu,n}\pi_{\nu,n}(\exp tA)f(\lambda) &= \int e^{nt/2}f(e^t r)r^{n/2}J_\nu(\lambda r)\,dr \\ &= e^{-t}\int f(s)s^{n/2}J_\nu(e^{-t}\lambda s)\,ds \\ &= e^{-t}(H_{\nu,n}f)(e^{-t}\lambda) \\ &= \gamma(\exp tA)H_{\nu,n}f(\lambda). \end{aligned} \tag{1.37}$$

Furthermore, by integration by parts, for $f \in C_0^\infty(R^+)$,

$$
\begin{aligned}
H_{\nu,n}\pi_{\nu,n}(B)f(\lambda) &= \int [i(L-\nu^2F)f(r)]r^{1-n/2}J_\nu(\lambda r)r^{n-1}\,dr \\
&= i\int f(r)(L-\nu^2F)(r^{1-n/2}J_\nu(\lambda r))r^{n-1}\,dr \\
&= -i\lambda^2\int f(r)r^{1-n/2}J_\nu(\lambda r)r^{n-1}\,dr \\
&= -i\lambda^2H_{\nu,n}f(\lambda) = \gamma(B)(H_{\nu,n}f)(\lambda).
\end{aligned} \tag{1.38}
$$

This proves the proposition.

The unitarity of $H_{\nu,n}$ implies that $H_{\nu,n}$ effects a unitary equivalence between the irreducible unitary representations γ and $\pi_{\nu,n}$ of $\mathrm{Aff}_e(R^1)$, the latter exponentiating the Lie algebra representation defined by (1.19).

We want to establish the unitarity (1.31) of the Hankel transform, which implies the unitarity of (1.33), in view of (1.34), and is also a special case, as $H_\nu = H_{\nu,2}$. Since $H_{\nu,n}$ intertwines the Lie algebra representations $\pi_{\nu,n}$ and γ, we are led to believe its unitarity is a consequence of Schur's lemma. Note that, once the unitarity of H_ν is established, its symmetry in (1.30), $H_\nu = H_\nu^*$, clearly leads to $H_\nu^2 = I$, which is the inversion formula (1.32). However, we have not given a direct demonstration, apart from the Hankel inversion formula, that $\pi_{\nu,n}$ exponentiates from a Lie algebra representation to a Lie group representation, nor that such a Lie group representation is irreducible, though PDE theory can be used to prove the exponentiability of (1.8), from which that of (1.9) can be deduced, and granted the exponentiability of (1.19), the irreducibility of the associated unitary representation $\pi_{\nu,n}$ of $\mathrm{Aff}_e(R^1)$ is not hard to deduce from the fact that the set of all operators commuting with the dilations (1.20) has a simple structure. We therefore take Schur's lemma as a good hint that (1.31)–(1.32) is true, but rather than pursue a proof of the Hankel inversion formula via Schur's lemma, we attack it more directly.

Our strategy for studying the Hankel transform (1.30) is the following. We have seen in (1.37) (with $n = 2$) that H_ν intertwines the operators $\pi_{\nu,2}(\exp tA)$ with $\gamma(\exp tA)$, which are given by dilations

$$\pi_{\nu,2}(\exp tA)f(r) = e^tf(e^tr), \qquad \gamma(\exp tA)f(\lambda) = e^{-t}f(e^{-t}\lambda). \tag{1.39}$$

We can conjugate with the unitary operator

$$M_r\colon L^2(R^+, r\,dr) \to L^2(R^+, r^{-1}dr) \tag{1.40}$$

defined by $M_rf(r) = rf(r)$, to simplify the associated dilations to $\delta(t)$ and $\delta(-t)$, where

$$\delta(t)f(r) = f(e^tr) \quad \text{on } L^2(R^+, r^{-1}dr). \tag{1.41}$$

Note that $r^{-1}dr$ is Haar measure on the multiplicative group R^+. Furthermore, composing with the unitary operator J defined by

$$Jf(\lambda) = f(\lambda^{-1}) \quad \text{on } L^2(R^+, r^{-1}dr), \tag{1.42}$$

we obtain an operator, which we want to show is unitary on $L^2(R^+, r^{-1}dr)$, which now commutes with the operators (1.41). Such an operator is of course given by convolution (see formula (1.46)), and it can be studied via Fourier analysis on R^+, i.e., the Mellin transform (1.48).

A short computation for the composed operator

$$H_\nu^\# = JM_r H_\nu M_r^{-1}, \tag{1.43}$$

which we want to act on $L^2(R^+, r^{-1}dr)$, gives

$$(H_\nu^\# f)(\lambda) = \int_0^\infty f(r)(r/\lambda)J_\nu(r/\lambda)r^{-1}\,dr. \tag{1.44}$$

Since the natural convolution on R^+ is given by

$$(f * g)(\lambda) = \int_0^\infty f(r)g(\lambda/r)r^{-1}\,dr, \tag{1.45}$$

we see that (1.44) is equivalent to

$$H_\nu^\# f(\lambda) = (f * \mathfrak{J}_\nu)(\lambda) \tag{1.46}$$

where

$$\mathfrak{J}_\nu(r) = r^{-1}J_\nu(r^{-1}). \tag{1.47}$$

To analyze a convolution operator on R^+, it is natural to use the Fourier transform adapted to R^+, i.e., the Mellin transform

$$\mathcal{M}^\# f(s) = \int_0^\infty f(r)r^{is}r^{-1}\,dr. \tag{1.48}$$

A related Mellin-like transform, denoted $\mathcal{M}$, is considered in Chapter 8, §3. $\mathcal{M}^\#$ is related to the Fourier transform via a change of variable:

$$\mathcal{M}^\# f(s) = \int_{-\infty}^\infty f(e^x)e^{isx}\,dx. \tag{1.49}$$

The Fourier inversion formula and Plancherel theorem imply

$$f(r) = (2\pi)^{-1}\int_{-\infty}^\infty (\mathcal{M}^\# f)(s)r^{-is}\,ds \tag{1.50}$$

and

$$\int_{-\infty}^\infty |\mathcal{M}^\# f(s)|^2\,ds = (2\pi)\int_0^\infty |f(r)|^2 r^{-1}\,dr. \tag{1.51}$$

Furthermore, we have

$$\mathcal{M}^\#(f * g)(s) = (\mathcal{M}^\# f)(s)(\mathcal{M}^\# g)(s). \tag{1.52}$$

Consequently, by (1.46),

$$\mathcal{M}^\# H_\nu^\# f(s) = \Phi(s)(\mathcal{M}^\# f)(s) \tag{1.53}$$

where

$$\Phi(s) = \int_0^\infty J_\nu(r^{-1}) r^{is-2}\, dr \tag{1.54}$$
$$= \int_0^\infty J_\nu(t) t^{-is}\, dt.$$

The statement that $H^{\#}$ is unitary on $L^2(R^+, r^{-1}dr)$ is thus equivalent to the statement that

$$|\Phi(s)| = 1 \quad \text{for all } s \in R. \tag{1.55}$$

Indeed, we have the following pretty identity, due to Sonine, Hankel, and other nineteenth century analysts:

$$\int_0^\infty J_\nu(t) t^{-is}\, dt = 2^{-is}\Gamma(\tfrac{1}{2}(\nu + 1 - is))/\Gamma(\tfrac{1}{2}(\nu + 1 + is)), \tag{1.56}$$

which manifestly implies (1.55). See Watson [**257**], pages 385–393, for several derivations of this identity. The left side of (1.56) is not absolutely integrable; $J_\nu(t)$ is bounded for $t \in [0, \infty)$, provided $\nu \geq 0$, but it decreases only like $t^{-1/2}$ as $t \to \infty$. Thus the left side of (1.56) exists as a tempered distribution on R, and can be evaluated by an approximation procedure, e.g., multiplying the integrand on the left side of (1.56) by e^{-at}, or by $e^{-p^2t^2}$, and passing to the limit, $a \downarrow 0$ or $p \downarrow 0$. In such a case, there are the formulas

$$\int_0^\infty J_\nu(t) e^{-at} t^{-is}\, dt = 2^{-\nu}(\Gamma(\nu + 1 - is)/\Gamma(\nu + 1)) a^{-\nu-1+is} \times {}_2F_1(\tfrac{1}{2}(\nu + 1 - is), \tfrac{1}{2}(\nu + 2 - is); \nu + 1; -1/a^2) \tag{1.57}$$

and

$$\int_0^\infty J_\nu(t) e^{-p^2t^2} t^{-is}\, dt = 2^{-\nu-1}(\Gamma(\tfrac{1}{2}(\nu + 1 - is))/\Gamma(\nu + 1)) p^{-\nu-1+is} \times {}_1F_1(\tfrac{1}{2}(\nu + 1 - is); \nu + 1; -1/4p^2), \tag{1.58}$$

due to Hankel (see [**257**], pages 385 and 393). Here ${}_2F_1$ and ${}_1F_1$ are hypergeometric functions, given by integral formulas

$$ {}_2F_1(\alpha, \beta; \gamma; z) = (\Gamma(\gamma)/\Gamma(\beta)\Gamma(\gamma - \beta)) \int_0^1 t^{\beta-1}(1 - t)^{\gamma-\beta-1}(1 - tz)^{-\alpha}\, dt, \tag{1.59}$$

for $\operatorname{Re}\gamma > \operatorname{Re}\beta > 0$, $z \in \mathbf{C} \setminus [1, \infty)$, and

$$ {}_1F_1(\alpha; \gamma; z) = (\Gamma(\gamma)/\Gamma(\alpha)\Gamma(\gamma - \alpha)) \int_0^1 e^{zt} e^{\alpha-1}(1 - t)^{\gamma-\alpha-1}\, dt, \tag{1.60}$$

for $\operatorname{Re}\gamma > \operatorname{Re}\alpha > 0$. Limiting properties of these hypergeometric functions yield (1.56) from either (1.57) or (1.58), as shown in Watson.

Since H_ν is clearly symmetric, its being unitary implies that its spectrum must consist of $\{-1, 1\}$, so it must be its own inverse. This proves the Hankel inversion formula.

Now the unitary operator $H_{\nu,n}$ of (1.33) intertwines the representation γ with $\pi_{\nu,n}$, making these representations equivalent. Proposition 1.1 implies that the skew adjoint generator of $\pi_{\nu,n}(\exp tB)$ is an extension (call it $iL'_{\nu,n}$) of the skew symmetric operator $iL_{\nu,n}$ on $L^2(R^+, r^{n-1}dr)$, defined by (1.21), with domain $C_0^\infty(R^+)$. It remains to identify this generator (times $1/i$) with the Friedrichs extension of $L_{\nu,n}$, which as mentioned above is the unique negative semidefinite selfadjoint extension $L''_{\nu,n}$ such that $C_0^\infty(R^+) \subset \mathcal{D}(L''_{\nu,n})$ and such that $C_0^\infty(R^+)$ is *dense* in $\mathcal{D}((-L''_{\nu,n})^{1/2})$, in its natural graph topology.

We note that it suffices to give the proof for a single value of n. For example, the unitary operator

$$M_{r^{n/2}}\colon L^2(R^+, r^{n-1}dr) \to L^2(R^+, r^{-1}dr) \tag{1.61}$$

conjugates the operator $L_{\nu,n}$, given by (1.19), to

$$\begin{aligned} L^b_{\nu,n} &= \partial^2/\partial r^2 - r^{-1}\partial/\partial r - [\mu^2 + \tfrac{1}{4}n(n-4)]r^{-2} \\ &= L_{\nu,0}, \quad \text{symmetric on } L^2(R^+, r^{-1}dr), \end{aligned} \tag{1.62}$$

so it suffices to analyze $L_{\nu,0}$. We could choose other standard forms; the unitary operator $M_{r^{(n-1)/2}}\colon L^2(R^+, r^{n-1}dr) \to L^2(R^+, dr)$ conjugates $L_{\nu,n}$ to

$$\begin{aligned} L^0_{\nu,n} &= \partial^2/\partial r^2 - [\mu^2 + \tfrac{1}{4}(n-1)(n-3)]r^{-2} \\ &= L_{\nu,1}, \quad \text{symmetric on } L^2(R^+, dr), \end{aligned} \tag{1.63}$$

and the unitary operator $M_{r^{(n-1)/2}}\colon L^2(R^+, r^{n-1}dr) \to L^2(R^+, r\,dr)$ conjugates $L_{\nu,n}$ to

$$\begin{aligned} L^\#_{\nu,n} &= \partial^2/\partial r^2 + r^{-1}\partial/\partial r - [\mu^2 + \tfrac{1}{4}(n-2)^2]r^{-2} \\ &= \partial^2/\partial r^2 + r^{-1}\partial/\partial r - \nu^2 r^{-2} \\ &= L_{\nu,2}, \quad \text{symmetric on } L^2(R^+, r\,dr). \end{aligned} \tag{1.64}$$

The standard form (1.64) makes the eigenvalue equation (1.22) closest in form to Bessel's equation. The form (1.63) is given a spectral analysis in Dunford-Schwartz [**50**], pages 1532–1535, an analysis somewhat different in detail from that presented here. We choose the form (1.62), because $r^{-1}dr$ is Haar measure on the multiplicative group R^+, and the Mellin transform is implemented with a minimum of further changes of variables. The representation $\pi_{\nu,0}$ is intertwined by the unitary operator $JH^\#_\nu$ on $L^2(R^+, r^{-1}dr)$, described by (1.43)–(1.46), to the unitary representation one gets by conjugating γ in (1.27)–(1.28) by the unitary operator $M_r\colon L^2(R^+, r\,dr) \to L^2(R^+, r^{-1}dr)$, i.e., $\tilde{\gamma}$, given by

$$\tilde{\gamma}(\exp tA)f(r) = f(e^{-t}r), \qquad \tilde{\gamma}(\exp tB)f(r) = e^{-itr^2}f(r). \tag{1.65}$$

In particular, $H^\#$ conjugates $L'_{\nu,0}$ with multiplication by r^2, on $L^2(R^+, r^{-1}dr)$. Note that

$$H^\#_\nu = \mathcal{M}^{\#-1} M_\Phi \mathcal{M}^\#, \tag{1.66}$$

where M_Φ denotes the operator of multiplication by $\Phi(s)$, given by (1.56). We therefore consider the image of $C_0^\infty(R^+)$ under $(JH_\nu^\#)^{-1} = H^{\#-1}J$, i.e., the image of $C_0^\infty(R^+)$ under $H_\nu^{\#-1} = \mathcal{M}^{\#-1}M_{\Phi^{-1}}\mathcal{M}^\#$.

Since the Mellin transform $\mathcal{M}^\#$ is the Fourier transform in exponential coordinates, by (1.49), the image of $C_0^\infty(R^+)$ under $\mathcal{M}^\#$ coincides with the image of $C_0^\infty(R)$ under the ordinary Fourier transform $\mathcal{F}$. This image is known (see, e.g., Yosida [**267**], Chapter VI) to consist precisely of

$$\{f(z) \text{ holomorphic for } z \in \mathbf{C}\colon |f(x+iy)| \le C_N(1+|x|)^{-N}e^{A|y|} \text{ for some } A, C_N, \text{ all } N\}. \tag{1.67}$$

With the idea of meeting halfway in $L^2(R)$, let us consider the image under $\mathcal{M}^\#$ of $\mathcal{D}(M_{r^\alpha})$:

$$\mathcal{M}^\#\mathcal{D}(M_{r^\alpha}) = \left\{f \text{ holomorphic for } -\alpha < \operatorname{Im} z < 0\colon \int_{-\infty}^{\infty} |f(x+iy)|^2\,dx \le C < \infty \text{ for } -\alpha < y < 0\right\}. \tag{1.68}$$

It remains to see how $M_{\Phi^{-1}}$ relates (1.67) and (1.68).

Clearly the poles of $\Gamma(z)$, i.e., the zeros of $1/\Gamma(z)$, at $z = 0, -1, 2, \ldots$ (particularly the one at $z = 0$) play a dominant role in this step of the analysis. Away from such poles, one has simple estimates

$$\begin{aligned} |\Phi(x+ia)| &\le C_a|x|^{2a}, \\ |\Phi(x+ia)|^{-1} &\le C_a|x|^{-2a}, \end{aligned} \qquad |x| \ge 1. \tag{1.69}$$

Thus we see that, for any $\alpha > 0$, the image of $C_0^\infty(R^+)$ under $H^{\#-1}$ is contained in $\mathcal{D}(M_{r^\alpha})$, which affirms that the domain of $L'_{\nu,0}$ contains $C_0^\infty(R^+)$. Furthermore, we see that the image of (1.67) under $M_{\Phi^{-1}}$ is dense in (1.68) precisely when $\alpha \le \nu + 1$. Since, by (1.16), we always have $\nu \ge 0$, it follows that the image of $C_0^\infty(R^+)$ under $H^{\#-1}$ is dense in M_r, so $L'_{\nu,0}$ is indeed the Friedrichs extension of $L_{\nu,0}$; we drop the prime and just denote the operator by $L_{\nu,0}$. Recalling the reduction from general n, we record the more precise result just established.

PROPOSITION 1.2. *We have, for any* $\alpha > 0$,

$$C_0^\infty(R^+) \quad \textit{dense in} \quad \mathcal{D}((-L_{\nu,n})^{\alpha/2}) \tag{1.70}$$

if and only if

$$\alpha \le \nu + 1 = [\mu^2 + \tfrac{1}{4}(n-2)^2]^{1/2} + 1. \tag{1.71}$$

Note that, if $\nu \ge 1$, then (1.70) holds with $\alpha = 2$, i.e.,

(1.72)

$L_{\nu,n}$ is essentially selfadjoint on $C_0^\infty(R^+)$ if and only if $\mu^2 + \frac{1}{4}(n-2)^2 \ge 1$.

In the next section we apply this material to the harmonic analysis of the Laplace operator Δ on a cone $C(N)$.

We remark that the two-dimensional Lie algebra generated by X and $i\Delta$ can be enlarged to a three-dimensional Lie algebra, generated by X, $i\Delta = U$, and $ir^2 = V$. We have the commutation relations

$$[X,U] = -2U, \qquad [X,V] = 2V, \qquad [U,V] = 4X. \tag{1.73}$$

This Lie algebra is isomorphic to $\mathrm{sl}(2,R)$, and we get a representation of the universal covering group of $\mathrm{SL}(2,R)$. In the special case where $C(N)$ is the cone over the unit circle and hence $C(N) = R^2$, we get the holomorphic discrete series of representations of $\mathrm{SL}(2,R)$; see Chapter 8 for details on this. We note here that, unlike the associated representations of the $ax+b$ group, the irreducible unitary representations of $\mathrm{SL}(2,R)$ gotten by decomposing $L^2(C(N))$ as above are not mutually equivalent, so the role of the Hankel transform described above does not extend.

2. Spectral representation and functional calculus for the Laplacian on a cone. Our spectral analysis of the Laplace operator on a cone $C(N)$ will follow the work of Cheeger [**35**] and Cheeger and Taylor [**37**]. It starts with the method of separation of variables, applied to the polar coordinate form of Δ,

$$\Delta = \partial^2/\partial r^2 + (n-1)r^{-1}\partial/\partial r + r^{-2}\Delta_N, \tag{2.1}$$

where Δ_N is the Laplace operator on N. Let μ_j^2, $\varphi_j(x)$ denote the eigenvalues and eigenfunctions of $-\Delta_N$, and set $\nu_j = (\mu_j^2 + \frac{1}{4}(n-2)^2)^{1/2}$. If

$$g(r,x) = \sum_{j\geq 0} g_j(r)\varphi_j(x) \tag{2.2}$$

with $g_j(r)$ well-behaved, then

$$\Delta g(r,x) = \sum (L - \nu_j^2 F) g_j(r)\varphi_j(x) \tag{2.3}$$

where

$$\begin{aligned} L - \nu_j^2 F &= \partial^2/\partial r^2 + (n-1)r^{-1}\partial/\partial r + \tfrac{1}{4}(n-2)^2 r^{-2} - \nu_j^2 r^{-2} \\ &= \partial^2/\partial r^2 + (n-1)r^{-1}\partial/\partial r - \mu_j^2 r^2, \end{aligned} \tag{2.4}$$

as in §1. From our formula (1.19) for the representation π_{ν_j} of $\mathrm{Aff}_e(R^1)$ on $L^2(R^+, r^{n-1}dr)$, and the unitary equivalence with the representation γ of $\mathrm{Aff}_e(R^1)$ on $L^2(R^+, \lambda\, d\lambda)$, defined by (1.27), implemented by the unitary operator $H_{\nu_j,n}$ defined by

$$H_{\nu,n} f(\lambda) = \int_0^\infty f(r) r^{n/2} J_\nu(\lambda r)\, dr, \tag{2.5}$$

in particular from the intertwining identity (1.38), which is equivalent to

$$H_{\nu,n}((L^2 - \nu^2 F)g) = -\lambda^2 H_{\nu,n} g, \tag{2.6}$$

we see that the map

$$\mathcal{H}\colon L^2(C(N)) \to L^2(R^+, \lambda\, d\lambda, l^2) \tag{2.7}$$

defined by

$$\mathcal{H}g = (H_{\nu_0,n}(g_0), H_{\nu_1,n}(g_1), \ldots) \tag{2.8}$$

is a unitary map with the property that Δ is carried into multiplication by $-\lambda^2$. Thus (2.8) provides a spectral representation of Δ.

We can use this spectral representation to define a functional calculus for operators $f(-\Delta)$. In fact, for well-behaved functions f, we have

$$\begin{aligned} &f(-\Delta)g(r,x) \\ &\quad = r^{-(n-2)/2}\sum_j \int_0^\infty f(\lambda^2)J_{\nu_j}(\lambda r)\lambda \int_0^\infty s^{n/2}J_{\nu_j}(\lambda s)g_j(s)\,ds\,d\lambda\varphi_j(x). \end{aligned} \tag{2.9}$$

Now we can interpret this in the following fashion. Define the *operator* ν on N by

$$\nu = (-\Delta_N + \tfrac{1}{4}(n-2)^2)^{1/2}. \tag{2.10}$$

Thus $\nu\varphi_j = \nu_j\varphi_j$. Identifying operators with their distribution kernels, we can describe the kernel of $f(-\Delta)$ as a function on $R^+ \times R^+$ taking values in the space of operators on $C^\infty(N)$, by the formula

$$\begin{aligned} f(-\Delta) &= (r_1r_2)^{-(n-2)/2}\int_0^\infty f(\lambda^2)J_\nu(\lambda r_1)J_\nu(\lambda r_2)\lambda\,d\lambda \\ &= K(r_1,r_2,\nu), \end{aligned} \tag{2.11}$$

since the volume element on $C(N)$ is $r^{n-1}\,dr\,dx$.

We can use this to analyze the solution operator $(-\Delta)^{-1/2}\sin t(-\Delta)^{1/2}$ for the wave equation on the cone $C(N)$. We have

$$\begin{aligned} &(-\Delta)^{-1/2}\sin t(-\Delta)^{1/2} \\ &\quad = -\lim_{\varepsilon\downarrow 0}(r_1r_2)^{-(n-2)/2}\mathrm{Im}\int_0^\infty e^{-(\varepsilon+it)\lambda}J_\nu(\lambda r_1)J_\nu(\lambda r_2)\,d\lambda \\ &\quad = -\mathrm{Im}\,\pi^{-1}(r_1r_2)^{-(n-1)/2}\lim_{\varepsilon\downarrow 0}Q_{\nu-1/2}((r_1^2+r_2^2+(\varepsilon+it)^2)/2r_1r_2), \end{aligned} \tag{2.12}$$

where the last identity is known as the Lipschitz-Hankel integral formula; see Watson [**257**], page 389. The Legendre function $Q_{\nu-1/2}$ is given by the integral formula

$$Q_{\nu-1/2}(\cosh\eta) = \int_\eta^\infty (2\cosh s - 2\cosh\eta)^{-1/2}e^{-s\nu}\,ds, \tag{2.13}$$

where the path of integration is a suitable path from η to $+\infty$ in the complex plane. Thus we see that the kernel of $(-\Delta)^{-1/2}\sin t(-\Delta)^{1/2}$ is equal to

$$0 \quad \text{if } t < |r_1 - r_2|, \tag{2.14}$$

$$(1/\pi)(r_1r_2)^{-(n-2)/2}\int_0^{\beta_1}[t^2-(r_1^2+r_2^2-2r_1r_2\cos s)]^{-1/2}\cos\nu s\,ds \qquad \text{if } |r_1-r_2|<t<r_1+r_2, \tag{2.15}$$

$$(1/\pi)(r_1r_2)^{-(n-2)/2}\cos\pi\nu\int_{\beta_2}^{\infty}[r_1^2+r_2^2+2r_1r_2\cosh s-t^2]^{-1/2}e^{-s\nu}\,ds \qquad \text{if } t>r_1+r_2, \tag{2.16}$$

where

$$\beta_1=\cos^{-1}((r_1^2+r_2^2-t^2)/2r_1r_2), \qquad \beta_2=\cosh^{-1}((t^2-r_1^2-r_2^2)/2r_1r_2). \tag{2.17}$$

These formulas provide a basis for analyzing the diffraction of waves by a cone.

Here, following [**37**], we will show how they lead to an analysis of the classical problem of diffraction of waves by a slit along the positive x-axis on the plane R^2. In fact, if waves propagate in R^2 with this ray removed, on which Dirichlet boundary conditions are placed, we can regard the space as the cone over an interval of length 2π, with Dirichlet boundary conditions at the endpoints. By the method of images, it suffices to analyze the case of the cone over a circle of circumference 4π (twice the circumference of the standard unit circle). Thus $C(N)$ is a double cover of $R^2 \setminus 0$ in this case. We divide up the space-time into regions I, II, and III, respectively, as described by (2.14), (2.15), (2.16). Region I contains only points on $C(N)$ too far away from the source point to be influenced by time t; that the fundamental solution is 0 here is a consequence of finite propagation speed.

Since the circle has dimension $n-1=1$, we see that

$$\nu=(-\Delta_N)^{1/2}=(-d^2/d\theta^2)^{1/2} \tag{2.18}$$

in this case, if $\theta\in R/4\pi\mathbf{Z}$ is the parameter on the circle of circumference 4π. On the line, we have

$$\cos s\nu\delta_{\theta_1}(\theta_2)=\tfrac{1}{2}[\delta(\theta_1-\theta_2+s)+\delta(\theta_1-\theta_2-s)]. \tag{2.19}$$

To get $\cos s\nu$ on $R/4\pi\mathbf{Z}$ we simply make (2.19) periodic by the method of images. Consequently, from (2.15), we have the wave kernel $(-\Delta)^{-1/2}\sin t(-\Delta)^{1/2}$ equal to

$$\begin{array}{ll}(1/2\pi)[t^2-(r_1^2+r_2^2-2r_1r_2\cos(\theta_1-\theta_2))]^{-1/2} & \text{if } |\theta_1-\theta_2|\le\pi,\\ 0 & \text{if } |\theta_1-\theta_2|>\pi,\end{array} \tag{2.20}$$

in region II. Of course, for $|\theta_1-\theta_2|<\pi$ this coincides with the free space fundamental solution, so (2.20) also follows by finite propagation speed. It is only in region III that the singularity at the origin in $C(N)$ makes an actual contribution. We turn now to an analysis of region III, and in particular to the diffracted wave produced.

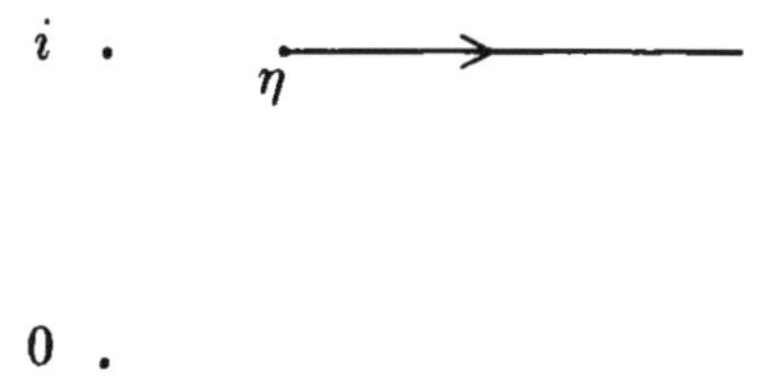

FIGURE 1

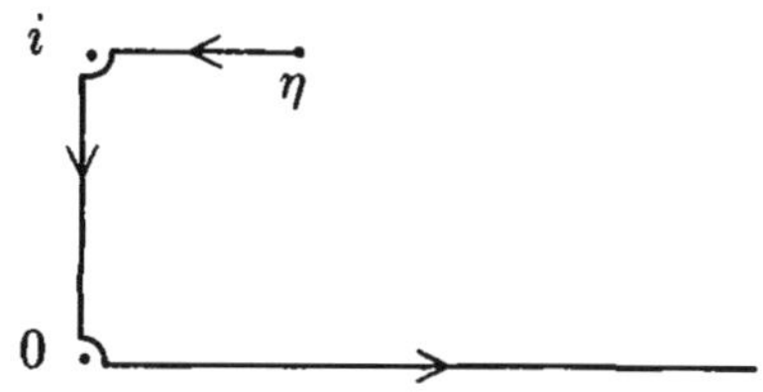

FIGURE 2

In order to make this analysis, it is convenient to make simultaneous use both of (2.16) and of another formula for the wave kernel in this region, obtained by choosing another path from η to ∞ in the integral representation (2.13). The formula (2.16) is obtained by taking a horizontal line segment; see Figure 1. If instead we take the path indicated in Figure 2, we obtain the following formula for $(-\Delta)^{-1/2}\sin t(-\Delta)^{1/2}$ in region III:

(2.21)
$$(1/\pi)(r_1r_2)^{-(n-2)/2}\left\{\int_0^\pi (t^2-r_1^2-r_2^2+2r_1r_2\cos s)^{-1/2}\cos s\nu\, ds \right.$$
$$\left. - \sin\pi\nu\int_0^{\beta_2}(t^2-r_1^2-r_2^2-2r_1r_2\cosh s)^{-1/2}e^{-s\nu}\,ds\right\}.$$

The operator ν on $R/4\pi\mathbf{Z}$ given by (2.18) has spectrum consisting of the nonnegative integers and the positive half-integers, all the eigenvalues except for 0 occurring with multiplicity two. The formula (2.16) shows the contribution coming from the half integers vanishes, since $\cos\frac{1}{2}\pi n = 0$ if n is an odd integer. Thus we can use formula (2.21) and compose with the projection onto the sum of the eigenspaces of ν with integer spectrum. This projection is given by

$$P = \cos^2\pi\nu \tag{2.22}$$

on $R/4\pi\mathbf{Z}$. Since $\sin\pi n = 0$, in the case $N = R/4\pi\mathbf{Z}$ we can rewrite (2.21) as

$$(1/\pi)(r_1r_2)^{-(n-2)/2}\int_0^\pi (t^2-r_1^2-r_2^2+2r_1r_2\cos s)^{-1/2}P\cos s\nu\,ds. \tag{2.23}$$

In view of the formulas (2.19) and (2.22), we have

(2.24)
$$P\cos s\nu\delta_{\theta_1}(\theta_2) = \tfrac{1}{4}[\delta(\theta_1-\theta_2+s)+\delta(\theta_1-\theta_2-s)+\delta(\theta_1-\theta_2+2\pi+s)$$
$$+\delta(\theta_1-\theta_2+2\pi-s)] \quad \mathrm{mod}\, 4\pi.$$

Thus, in region III, we have for the wave kernel $(-\Delta)^{-1/2}\sin t(-\Delta)^{1/2}$ the formula

$$(1/4\pi)(t^2-r_1^2-r_2^2+2r_1r_2\cos(\theta_1-\theta_2))^{-1/2}. \tag{2.25}$$

Thus, in region III, the value of the wave kernel at points $(r_1,\theta_1),(r_2,\theta_2)$ of the double cover of $R^2\setminus 0$ is given by half the value of the wave kernel on R^2 at

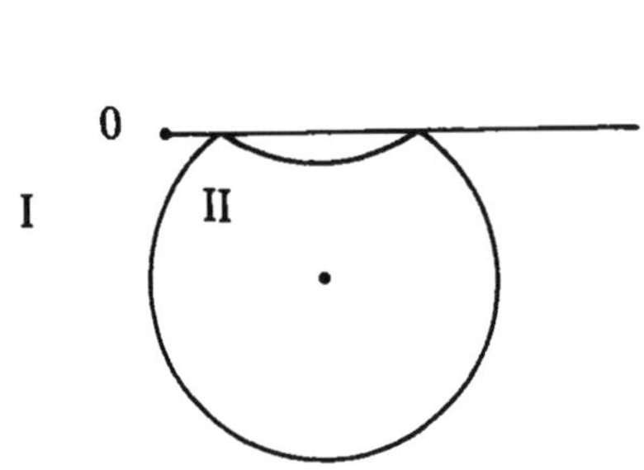

FIGURE 3

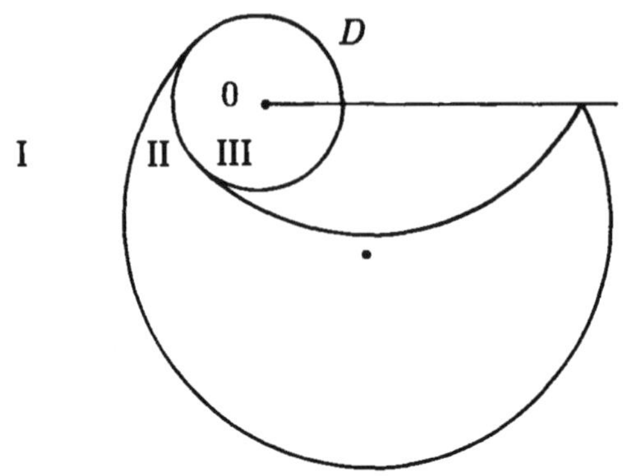

FIGURE 4

values of the image points. The jump in behavior from (2.20) to (2.25) gives rise to the *diffracted wave.*

We have depicted the singularities of the fundamental solution to the wave equation for R^2 minus a slit in Figures 3 and 4. In Figure 3, we have the situation $|t| < r_1$, where no diffraction has occurred, and region III is empty. In Figure 4 we have a typical situation for $|t| > r_1$, with the diffracted wave labeled D.

This diffraction problem was first treated by Sommerfeld [**220**] and was the first diffraction problem to be rigorously analyzed. For a more complete study of the general case, we refer to [**37**]. For other approaches to the diffraction problem presented here, see [**66**, **223**]. We also mention the reference [**237**].

CHAPTER 8

SL(2, R)

The group SL(2, R) is the simplest of the noncompact semisimple Lie groups. We present its representation theory and some harmonic analysis here. A lengthier study is given in Lang [**147**]. SL(2, R) does not provide as effective a tool for the study of general semisimple Lie groups as the mighty Heisenberg group does in the study of nilpotent Lie groups, though it does play important roles. The principal and discrete series, which occur for SL(2, R), have analogues in the general case that provide "most" irreducible unitary representations, it turns out.

1. Introduction to SL(2, R)**.** The group SL(2, R) is the group of all 2×2 real matrices of determinant 1. Thus SL(2, R) is a three-dimensional connected Lie group. Its Lie algebra sl(2, R) consists of all 2×2 real matrices of trace zero. The group SL(2, R) contains the one-dimensional compact subgroup

$$K = \left\{ \begin{pmatrix} \cos\theta & \sin\theta \\ -\sin\theta & \cos\theta \end{pmatrix} : \theta \in R \right\} = \mathrm{SO}(2), \tag{1.1}$$

which is $\exp(\theta Z)$, where $Z \in \mathrm{sl}(2, R)$ is

$$Z = \begin{pmatrix} 0 & 1 \\ -1 & 0 \end{pmatrix}. \tag{1.2}$$

We can form a basis $\{Z, A, B\}$ of sl(2, R) by setting

$$A = \frac{1}{2}\begin{pmatrix} -1 & 0 \\ 0 & 1 \end{pmatrix}, \qquad B = \frac{1}{2}\begin{pmatrix} 0 & 1 \\ 1 & 0 \end{pmatrix}. \tag{1.3}$$

Note that

$$[Z, A] = 2B, \qquad [Z, B] = -2A, \qquad [A, B] = -\tfrac{1}{2}Z. \tag{1.4}$$

There are some other important descriptions of groups naturally isomorphic to SL(2, R), or a quotient group. First, since the volume element on R^2 is also a symplectic form, we have

$$\mathrm{SL}(2, R) = \mathrm{Sp}(1, R). \tag{1.5}$$

Also, the group $\mathrm{SL}(2,R)$ acts naturally on the upper half plane

$$\Omega_+ = \{z \in \mathbf{C} : \operatorname{Im} z > 0\} \tag{1.6}$$

by

$$g \cdot z = (az+b)/(cz+d), \tag{1.7}$$

where

$$g = \begin{pmatrix} a & b \\ c & d \end{pmatrix} \in \mathrm{SL}(2,R). \tag{1.8}$$

These transformations are all holomorphic diffeomorphisms, and they act transitively on Ω_+. There is a metric, called the Poincaré metric,

$$ds^2 = y^{-2}(dx^2 + dy^2), \tag{1.9}$$

which is invariant under all these transformations. Note that $-I$ acts trivially on Ω_+. The component of the identity in the isometry group of Ω_+ is isomorphic to $\mathrm{SL}(2,R)/\{\pm I\} = \mathrm{PSL}(2,R)$.

The Poincaré upper half plane Ω_+ is biholomorphic to the Poincaré disc

$$D = \{z \in \mathbf{C} : |z| < 1\}, \tag{1.10}$$

via the linear fractional transformation $T(z) = (z-i)/(z+i)$. We see that (1.7) also defines a transitive action on D, with $g \in \mathrm{SL}(2,R)$ replaced by $g \in \mathrm{SU}(1,1)$, where

$$\mathrm{SU}(1,1) = \left\{ \begin{pmatrix} \alpha & \beta \\ \overline{\beta} & \overline{\alpha} \end{pmatrix} : \alpha, \beta \in \mathbf{C},\ |\alpha|^2 - |\beta|^2 = 1 \right\}. \tag{1.11}$$

We see that the mapping

$$g \mapsto \begin{pmatrix} 1 & -i \\ 1 & i \end{pmatrix} g \begin{pmatrix} 1 & -i \\ 1 & i \end{pmatrix}^{-1} \tag{1.12}$$

produces an isomorphism of $\mathrm{SL}(2,R)$ with $\mathrm{SU}(1,1)$, so

$$\mathrm{SL}(2,R) \approx \mathrm{SU}(1,1). \tag{1.13}$$

The invariant metric on D is

$$ds^2 = (1-r^2)^{-2}(dx^2 + dy^2). \tag{1.14}$$

Now Ω_+ and D are two models of the two-dimensional simply connected space of constant negative curvature. This space also arises in the following way. Consider R^3 with the Lorentz metric

$$ds^2 = dz^2 - dx^2 - dy^2. \tag{1.15}$$

The group of linear transformations of R^3 preserving this Lorentz metric is denoted $\mathrm{O}(2,1)$, and the connected component of the identity by $\mathrm{SO}_e(2,1)$. The orbits of $\mathrm{O}(2,1)$ consist of hyperboloids

$$\mathcal{O}_\lambda = \{(x,y,z) \in R^3 : z^2 - x^2 - y^2 = \lambda\}. \tag{1.16}$$

If $\lambda < 0$, $\mathcal{O}_\lambda$ is a hyperboloid of one sheet, and the induced metric is a Lorentz metric. If $\lambda > 0$, $\mathcal{O}_\lambda$ is a hyperboloid of two sheets, and the induced metric is Riemannian. Such $\mathcal{O}_\lambda$ is a homogeneous space, and thus has constant curvature, which is negative (since $\mathcal{O}_\lambda$ is not compact). We have

$$\text{SO}_e(2,1) \approx \text{PSL}(2,R). \tag{1.17}$$

We refer to Chapters 9 and 10 for more general studies of $\text{SO}(n,1)$ acting as a group of isometries on hyperbolic space $\mathcal{H}^n$, and also as a group of conformal transformations on S^{n-1}. (In the present case $S^{n-1} = S^1$ has a trivial conformal structure.)

The isomorphism (1.17) can be described in the following explicit fashion. The space $M_2(R)$ of real 2×2 matrices has a natural quadratic form of signature $(2,2)$, namely $Q(g) = \det g$. This, restricted to $\text{sl}(2,R) = \{g \in M_2(R)\colon \text{tr}\, g = 0\}$, is a metric of signature $(2,1)$, i.e., a Lorentz metric, and $\text{SL}(2,R)$ acts by conjugation:

$$g \cdot u = gug^{-1}, \qquad g \in \text{SL}(2,R),\ u \in \text{sl}(2,R). \tag{1.18}$$

Clearly (1.18) preserves the determinant, and hence the Lorentz metric on $\text{sl}(2,R)$. The matrices $\pm I \in \text{SL}(2,R)$ act trivially.

Having taken a few snapshots of $\text{SL}(2,R)$, we now look more closely at its Lie algebra, which has just been assigned a Lorentz metric, which we will denote $\langle\ ,\ \rangle$. The conjugation invariance of this metric is equivalent to its invariance under the adjoint representation of $\text{SL}(2,R) = G$ on $\text{sl}(2,R) = \mathfrak{g}$:

$$\langle \text{Ad}(g)X, \text{Ad}(g)Y\rangle = \langle X,Y\rangle \quad \text{for } g \in \text{SL}(2,R),\ X,Y \in \text{sl}(2,R). \tag{1.19}$$

Since the adjoint representation produces the isomorphism

$$\text{PSL}(2,R) \approx \text{SO}_e(2,1),$$

and it is clear that $\text{SO}_e(2,1)$ has no invariant subspaces in R^3, it follows that $\text{sl}(2,R)$ has no ideals. The Lie algebra $\text{sl}(2,R)$ is *simple*.

Recall from Chapter 0 that the Killing form, defined by

$$B(X,Y) = \text{tr}(\text{ad}\, X\, \text{ad}\, Y), \tag{1.20}$$

is of interest on a simple, or more generally semisimple Lie algebra. We claim that the two forms (1.19) and (1.20) are proportional:

$$B(X,Y) = c\langle X,Y\rangle. \tag{1.21}$$

This can be verified by a routine computation, and we leave the evaluation of c as an exercise. We can justify (1.21) on general principles as follows. Since $\langle\ ,\ \rangle$ is nondegenerate, we have $B(X,Y) = \langle T(X),Y\rangle$ for some $T \in \text{End}(\mathfrak{g})$. The invariance (1.19) together with the same Ad-invariance for the Killing form, imply $T\,\text{Ad}(g) = \text{Ad}(g)T$ for each $g \in \text{SL}(2,R)$. T must have at least one real eigenvalue, since it is given by a real 3×3 matrix; call one such eigenvalue c. Then $\ker(T - cI)$ is invariant under $\text{Ad}(g)$ for all $g \in \text{SL}(2,R)$, hence an ideal of $\text{sl}(2,R)$, so in fact $T = cI$, and (1.21) follows.

The group SL(2, R) acts on sl(2, R) via the adjoint representation, as a group of inner automorphisms. An automorphism of sl(2, R) which is not inner is defined by

$$\alpha(A) = B, \qquad \alpha(B) = A, \qquad \alpha(Z) = -Z. \tag{1.22}$$

This gives rise to an automorphism of SL(2, R), which we also denote α.

In the representation theory of SU(2) and other compact semisimple Lie groups, we made use of the complexified Lie algebra and the universal enveloping algebra, particularly of elements in the center of the universal enveloping algebra. We will also use such objects here. Define elements $X_\pm \in \mathfrak{g}_C$ for $\mathfrak{g} = \mathrm{sl}(2, R)$ by

$$X_+ = A - iB, \qquad X_- = A + iB. \tag{1.23}$$

The commutation relations (1.4) are equivalent to

$$[Z, X_\pm] = \pm 2iX_\pm, \qquad [X_+, X_-] = -iZ. \tag{1.24}$$

The first of these identities suggests that $\pi(X_\pm)$ can play the role of raising and lowering operators for a representation π, and this will be pursued in §2. Of special interest is the *Casimir operator*

$$\Box = Z^2 - 4A^2 - 4B^2. \tag{1.25}$$

It is routine to verify that $\Box$ commutes with Z, A, and B:

$$[Z, \Box] = [A, \Box] = [B, \Box] = 0, \tag{1.26}$$

and hence the operator $\Box$ belongs to the center $\mathfrak{z}$ of the universal enveloping algebra $\mathfrak{U}(\mathrm{sl}(2, R))$. It is also possible to prove that $\Box$ generates $\mathfrak{z}$, when $\mathfrak{g} = \mathrm{sl}(2, R)$, but we will not make use of this. In terms of the elements $X_\pm$, we have

$$\Box = Z^2 - 2(X_+X_- + X_-X_+). \tag{1.27}$$

It is useful that (1.27) together with the last part of (1.24) provide expressions for the sum and difference of X_+X_- and X_-X_+ in $\mathfrak{U}(\mathfrak{g})$. Solving, we have the identities

$$4X_+X_- = Z^2 - 2iZ - \Box, \tag{1.28}$$

$$4X_-X_+ = Z^2 + 2iZ - \Box, \tag{1.29}$$

in the universal enveloping algebra of sl(2, R). These will also be useful in the representation theory developed in the next section.

To end this section, we recall the local isomorphism of SO(4) with SO(3) $\times$ SO(3) and with SU(2) $\times$ SU(2). Similarly, we claim that $\mathrm{SO}_e(2,2)$, the connected component of the identity in the group of linear transformations of R^4 preserving a metric of signature (2, 2), is covered by SU(1, 1) $\times$ SU(1, 1). One way to see this is the following. Via (1.11), we can identify SU(1, 1) with the subset of $\mathbf{C}^2$ given by $\{(w, z) : |w|^2 - |z|^2 - 1\}$, and also think of SU(1, 1) as acting on $\mathbf{C}^2$, preserving the quadratic form $|w|^2 - |z|^2$, which we think of as a metric on R^4 of signature $(2,2)$. Thus we see that $\mathrm{SU}(1,1)$ is an orbit for the action of $\mathrm{SO}_e(2,2)$ on R^4. The Lorentz metric which $\mathrm{SU}(1,1)$ inherits from this imbedding in R^4, a bi-invariant metric agreeing up to a constant factor with the metric induced by the Killing form, is of course preserved by the action of $\mathrm{SO}_e(2,2)$, which hence acts as a group of isometries on $\mathrm{SU}(1,1)$. $\mathrm{SU}(1,1) \times \mathrm{SU}(1,1)$ acts as a group of isometries also, by

$$(1.30) \qquad (g_1, g_2) \cdot x = g_1 x g_2^{-1}, \qquad g_j, x \in \mathrm{SU}(1,1),$$

and we have a group homomorphism

$$(1.31) \qquad \tau: \mathrm{SU}(1,1) \times \mathrm{SU}(1,1) \to \mathrm{SO}_e(2,2).$$

Note that $(g_1, g_2) \in \ker \tau$ implies $g_1 = g_2 = \pm I$. We can verify that τ is onto, e.g., by a dimension count, so

$$(1.32) \qquad \mathrm{SO}_e(2,2) \approx \mathrm{SU}(1,1) \times \mathrm{SU}(1,1)/\{\pm(I,I)\}.$$

2. Classification of irreducible unitary representations. We aim to classify the irreducible unitary representations of $\mathrm{SL}(2,R)$. Let π be such a representation, on a Hilbert space H. Let us denote the action on the basis $\{A, B, Z\}$ of the Lie algebra $\mathrm{sl}(2,R)$, discussed in §1, by

$$(2.1) \qquad E = \pi(Z), \qquad A_1 = \pi(A), \qquad B_1 = \pi(B).$$

In analogy with the representation theory of $\mathrm{SU}(2)$, we find it convenient to consider the image of $X_\pm = A \mp iB$; say

$$(2.2) \qquad R_\pm = \pi(X_\pm).$$

The basic commutator relations for $\mathrm{sl}(2,R)$, given by (1.4), (1.24), yield

$$(2.3) \qquad [E, R_\pm] = \pm 2i R_\pm, \qquad [R_+, R_-] = -iE$$

on $C^\infty(\pi)$, and if A_1 and B_1 are skew symmetric, we have

$$(2.4) \qquad R_+^* = -R_-$$

on $C^\infty(\pi)$.

One fundamental tool in the analysis will be the fact that, since $\Box = Z^2 - 2(X_+X_- + X_-X_+)$ belongs to the center of the enveloping algebra of $\mathrm{sl}(2,R)$, $\pi(\Box)$ acts as a scalar;

$$(2.5) \qquad \pi(\Box) = \lambda I$$

for some $\lambda \in R$ (since $\Box$ is symmetric). The identity (2.5) holds a priori on $C^\infty(\pi)$, but of course this implies $\pi(\Box)$ extends by continuity to the scalar λ on H. We will exploit (2.5) in several ways. In addition to (2.5), the algebraic workhorses in our analysis will be the following identities:

$$(2.6) \qquad 4R_+R_- = E^2 - 2iE - \pi(\Box),$$

$$4R_-R_+ = E^2 + 2iE - \pi(\Box), \tag{2.7}$$

$$ER_\pm = R_\pm E \pm 2iR_\pm, \tag{2.8}$$

on $C^\infty(\pi)$. The identities (2.6) and (2.7) follow from the identities $X_+X_- + X_-X_+ = \frac{1}{2}Z^2 - \frac{1}{2}\Box$ and $X_+X_- - X_-X_+ = -iZ$ in the universal enveloping algebra of $\mathfrak{sl}(2, R)$, derived in §1, the latter being in fact translated into the last identity in (2.3). The identities (2.8) follow from the first set of identities in (2.3).

The element Z generates the compact subgroup $K \approx S^1$ of SL(2, R) described in §1. It is natural to decompose the action of $\pi(K)$ on H. Clearly we have the orthogonal direct sum

$$H = \overline{\bigoplus_{k\in\mathbf{Z}} V_k} \tag{2.9}$$

where

$$\pi(\exp sZ) = e^{iks} \quad \text{on } V_k, \tag{2.10}$$

i.e.,

$$E = ik \quad \text{on } V_k. \tag{2.11}$$

Of course, some of the V_k could be zero. We will demonstrate that each V_k consists of smooth vectors, even of analytic vectors (and furthermore, if not zero, is one-dimensional), but for the moment we do not know that. Set

$$V_k^0 = V_k \cap C^\infty(\pi). \tag{2.12}$$

Note that the orthogonal projection onto V_k is given by

$$P_k = (2\pi)^{-1} \int_0^{2\pi} e^{-iks}\pi(\exp sZ)\, ds. \tag{2.13}$$

It is clear that P_k preserves the Gårding space $\mathcal{G}(\pi)$, introduced in Chapter 0, so V_k^0 is certainly dense in V_k. We have the following description of the action of $R_\pm$ on V_k^0.

LEMMA 2.1. *We have*

$$R_\pm : V_k^0 \to V_{k\pm 2}^0. \tag{2.14}$$

PROOF. If $v \in V_k^0$, then by (2.8), we have

$$E(R_\pm v) = R_\pm(Ev) \pm 2iR_\pm v = (k \pm 2)i(R_\pm v), \tag{2.15}$$

by (2.11), and the proof is complete.

In other words, R_+ and R_- are playing the roles of raising and lowering operators, analogous to such operators in the study of the representations of SU(2). In order to study continuity properties of $R_\pm|_{V_k}$, we note that (2.6)–(2.7), in concert with (2.5), give

$$-4R_+R_- = k^2 - 2k + \lambda = (k-1)^2 + \lambda - 1, \tag{2.16}$$

and

$$(2.17)\qquad -4R_-R_+ = k^2 + 2k + \lambda = (k+1)^2 + \lambda - 1,$$

on V_k^0. Since $R_+^* = -R_-$ on $C^\infty(\pi)$, this implies

$$(2.18)\qquad \|R_-\|_{\mathcal{L}(V_k^0, V_{k-2}^0)} = \tfrac{1}{2}[(k-1)^2 + \lambda - 1]^{1/2}$$

and

$$(2.19)\qquad \|R_+\|_{\mathcal{L}(V_k^0, V_{k+2}^0)} = \tfrac{1}{2}[(k+1)^2 + \lambda - 1]^{1/2}.$$

We see that

$$\pi(A) = R_+ + R_-, \qquad \pi(B) = i(R_+ - R_-): V_k^0 \to V_{k-2}^0 \oplus V_{k+2}^0$$

have unique continuous extensions to maps of V_k into $V_{k+2} \oplus V_{k-2}$. Since these operators are closed on their natural domains, we have

$$(2.20)\qquad V_k \subset \mathcal{D}(\pi(A)) \cap \mathcal{D}(\pi(B)),$$

and

$$(2.21)\qquad \pi(A), \pi(B): V_k \to V_{k-2} \oplus V_{k+2}.$$

This makes it clear that each V_k is contained in $C^\infty(\pi)$; in other words,

$$(2.22)\qquad V_k = V_k^0.$$

We indicate schematically the situation at which we have arrived:

$$(2.23)\qquad \cdots V_{k-2} \underset{R_-}{\overset{R_+}{\rightleftarrows}} V_k \underset{R_-}{\overset{R_+}{\rightleftarrows}} V_{k+2} \cdots.$$

In fact, we have more than just $V_k \subset C^\infty(\pi)$; every element of V_k is an *analytic vector*. In other words, with $X_1, X_2, X_3 = E, R_+, R_-$, $X^\alpha = X_{\alpha(1)} \cdots X_{\alpha(K)}$, $|\alpha| = K$, we have estimates

$$(2.24)\qquad \|X^\alpha u\|_H \le C_1(C_2|\alpha|)^{|\alpha|}, \qquad u \in V_k.$$

These estimates follow from (2.18)–(2.19), together with (2.14) and (2.11). In general, $C_1 = C_1(k)$, but we note that C_2 can be taken independent of k. In other words, the formal power series in (t_1, t_2, t_3) for

$$(2.25)\qquad \exp(t_1 Z + t_2 A + t_3 B)u$$

has a radius of convergence $\ge r_0 > 0$, independent of k. One consequence of this is the following. If the spectrum of $(1/i)E$, which must consist of integers, contained both even and odd integers, if we set $H = H_e \oplus H_o$ with

$$H_e = \overline{\bigoplus_{k \text{ even}} V_k}, \qquad H_o = \overline{\bigoplus_{k \text{ odd}} V_k},$$

then both H_e and H_o would be invariant under the operators (2.25) for (t_1, t_2, t_3) close to 0, and hence invariant under $\pi(g)$ for all $g \in \mathrm{SL}(2, R)$. Since π is assumed to be irreducible, this means that the spectrum of $(1/i)E$ is contained in either the set of even integers or the set of odd integers.

Note by (2.16)–(2.17) that, in (2.23), each $R_+: V_k \to V_{k+2}$ is a scalar multiple of a unitary operator, as is each $R_-: V_k \to V_{k-2}$. Such a multiple could be zero. However, if $V_k \neq 0$ and $R_+ = 0$ on V_k, then, comparing (2.18) and (2.19), we see that also $R_- = 0$ on V_{k+2}. Thus the sequence (2.23) is severed at V_k, and both $\overline{\bigoplus_{l\le k} V_l}$ and $\overline{\bigoplus_{l>k} V_l}$ would be invariant under π. If π is irreducible, this forces $V_l = 0$ for all $l > k$, provided $R_+ = 0$ on V_k, $V_k \neq 0$. By the same reasoning, if $R_- = 0$ on V_k and $V_k \neq 0$, then $V_l = 0$ for all $l < k$. In other words, (2.23) is either two-sided infinite, with all R_+ and R_- isomorphisms, or one-sided infinite, of the form $\overline{\bigoplus_{j\ge 0} V_{k-2j}}$, or of the form $\overline{\bigoplus_{j\ge 0} V_{k+2j}}$, with $R_+|_{V_k} = 0$ or $R_-|_{V_k} = 0$, in these respective cases, and all other $R_\pm$ isomorphisms.

The remaining possibility, that (2.23) is finite, does not ever occur, except when π is the trivial representation. Indeed, from (2.18)–(2.19), if $R_-|_{V_k} = 0$ then $(k-1)^2 = 1-\lambda$, while if $R_+|_{V_{k+l}} = 0$ then $(k+l+1)^2 = 1-\lambda$. That would imply $k+l+1 = \pm(k-1)$, i.e., either $l = -2$ or $l = -2k$. Since $l \ge 0$, $l = -2$ is not possible. If $l = -2k$, then the string (2.23) is $V_{-|k|} \oplus \cdots \oplus V_{|k|}$. We will reach a contradiction after making a general comment.

Since $R_+^* = -R_-$, the scalars in (2.16)–(2.17) must be ≥ 0, for all $k \in \operatorname{spec}(1/i)E$. If $\operatorname{spec}(1/i)E$ contains $k = 1$, or $k = -1$, this forces $\lambda \ge 1$. If $\operatorname{spec}(1/i)E$ contains $k = 0$, it forces $\lambda \ge 0$.

If π is finite-dimensional, the identity $(k-1)^2 = 1-\lambda$, which implies $(|k|+1)^2 = 1-\lambda$, forces $\lambda < 0$ unless $|k| = 0$. Thus π must be representation on which E and $R_\pm$, hence A_1 and B_1, act trivially, and thus a trivial representation.

We remark that it is easy to see directly from simple general principles that SL(2, R) cannot have any nontrivial finite-dimensional irreducible unitary representation. Such a representation would induce a Lie algebra homomorphism

$$\pi: \mathrm{sl}(2, R) \to \mathrm{u}(n). \tag{2.26}$$

Since $\ker \pi$ is an ideal and sl(2, R) is simple, this means π is injective on sl(2, R) if π is not trivial. Now any bi-invariant Riemannian metric on U(n) induces an Ad-invariant positive definite form on u(n), which by (2.26) would pull back to an Ad-invariant positive definite form on sl(2, R). However, as mentioned in §1, any such form must be a multiple of the Killing form, which is certainly not definite, as it has Lorentz signature. This same sort of argument shows that a general noncompact (connected) semisimple Lie group has no nontrivial finite-dimensional unitary representations.

To return to an analysis of the infinite-dimensional irreducible unitary representations of SL(2, R), we have four possibilities for the spectrum of $(1/i)E$:

$$\operatorname{spec}(1/i)E = \{n, n+2, n+4, \dots\}, \tag{2.27}$$

$$\operatorname{spec}(1/i)E = \{\dots, n-4, n-2, n\}, \tag{2.28}$$

$$\operatorname{spec}(1/i)E = \{\dots, -4, -2, 0, 2, 4, \dots\}, \tag{2.29}$$

$$\operatorname{spec}(1/i)E = \{\dots, -5, -3, -1, 1, 3, 5, \dots\}. \tag{2.30}$$

All the maps $R_+ : V_k \to V_{k+2}$, $R_- : V_k \to V_{k-2}$, between nonzero eigenspaces of E, are isomorphisms. We are now ready to prove that

$$\dim V_k = 1 \quad \text{if } k \in \operatorname{spec}(1/i)E. \tag{2.31}$$

In fact, pick any $k_0 \in \operatorname{spec}(1/i)E$, pick a nonzero $v_{k_0} \in V_{k_0}$, consider its iterated images under R_+, belonging to V_k for $k > k_0$, and its iterated images under R_-, belonging to V_k for $k < k_0$. The representation π induces a representation of the Lie algebra sl$(2, R)$ on the finite linear span of these vectors. By the analyticity (2.24) of all the vectors in each V_k, this generates a unitary representation of $\widetilde{\mathrm{SL}(2,R)}$ on the closure of this linear span (see Appendix D). In fact, this representation simply covers π. Since π is supposed to be irreducible, this closed linear span must be all of H, so (2.31) is established.

We are now ready for a case by case analysis of (2.27)–(2.30), which will produce the classification of the irreducible unitary representations of SL$(2, R)$. In each case, pick a unit vector $v_k \in V_k$ for each $k \in \operatorname{spec}(1/i)E$, producing an orthonormal basis of H, uniquely determined up to phase, which can be adjusted arbitrarily. We may as well specify

$$R_+ v_k = \alpha_k v_{k+2} \tag{2.32}$$

where α_k is any complex number whose absolute value is equal to the quantity computed in (2.19), i.e.,

$$|\alpha_k| = \tfrac{1}{2}[(k+1)^2 + \lambda - 1]^{1/2}. \tag{2.33}$$

Then the action of R_- on v_{k+2} is uniquely determined:

$$R_- v_{k+2} = \beta_k v_k, \tag{2.34}$$

where the identity (2.17) forces $-4\alpha_k\beta_k = (k+1)^2 + \lambda - 1 = 4|\alpha_k|^2$, i.e.,

$$\beta_k = -\overline{\alpha}_k. \tag{2.35}$$

Picking β_k so this holds also makes (2.16) satisfied, and of course (2.16)–(2.17) imply $-4[R_+, R_-] = -4k$ on V_k, so we have the commutation relation given by the last part of (2.3). It is routine to verify that (2.32)–(2.35) implies all the commutation relations (2.3) are satisfied on the finite linear span $\bigoplus_k V_k$, and also (2.35) implies $R_+^* = -R_-$, so whenever (2.32)–(2.35) hold we have a Lie algebra representation of sl$(2, R)$ by skew-symmetric operators on $H_0 = \bigoplus_k V_k$, a dense subspace of H, consisting of analytic vectors.

In the case (2.27), we have $R_- = 0$ on V_n, which, by (2.18), implies

$$(n-1)^2 = 1 - \lambda, \tag{2.36}$$

i.e.,

$$\pi(\Box) = 1 - (n-1)^2 = \lambda. \tag{2.37}$$

If n were ≤ 0, then either 0 or -1 would have to belong to $\operatorname{spec}(1/i)E$. If -1 belongs, then (2.33) forces $\lambda > 1$, and if 0 belongs it forces $\lambda > 0$, in both cases contradicting (2.37). Thus, for the case (2.27) to occur, we need

$$n \geq 1. \tag{2.38}$$

Conversely, as long as (2.38) holds, then (2.33) can be implemented for $k = n+2j$, $j = 0, 1, 2, \ldots$, i.e.,

$$|\alpha_k| = \tfrac{1}{2}[(k+1)^2 - (n-1)^2]^{1/2}, \tag{2.39}$$

and a Lie algebra representation of $\mathrm{sl}(2, R)$ by skew symmetric operators is defined. As we will see, this can be exponentiated to a representation of $\mathrm{SL}(2, R)$, for each integer n satisfying (2.38). The representations so obtained are denoted π_n^+ and are called members of the *holomorphic discrete series* of representations of $\mathrm{SL}(2, R)$.

The case (2.28) is handled similarly. We have $R_+ = 0$ on V_n, so by (2.19), $(n+1)^2 = 1 - \lambda$, i.e.,

$$\pi(\Box) = 1 - (n+1)^2 = \lambda. \tag{2.40}$$

The same argument as before forbids 0 or 1 from belonging to $\mathrm{spec}(1/i)E$, so we must have

$$n \leq -1, \tag{2.41}$$

and conversely if (2.41) holds then we can implement (2.33) for $k = n - 2j$, $j = 0, 1, 2, \ldots$, choosing α_k so that

$$\begin{aligned} |\alpha_k| &= \tfrac{1}{2}[(k+1)^2 - (n+1)^2]^{1/2} \\ &= \tfrac{1}{2}[(|k|-1)^2 - (|n|-1)^2]^{1/2} \end{aligned} \tag{2.42}$$

for $k < n$. This can be exponentiated to a representation of $\mathrm{SL}(2, R)$, for each integer n satisfying (2.41), denoted π_n^-, and said to be a member of the *anti-holomorphic discrete series.* It is easy to see that π_n^+ and π_{-n}^- are related via the automorphism α of $\mathrm{SL}(2, R)$ given in (1.22):

$$\pi_{-n}^-(g) = \pi_n^+(\alpha(g)). \tag{2.43}$$

Actually, π_1^+ and π_{-1}^- are called "mock discrete series" representations, rather than discrete series representations, for reasons briefly discussed in §4.

Consider now the case when $\mathrm{spec}(1/i)E$ is given by (2.29). Since 0 belongs to $\mathrm{spec}(1/i)E$, (2.33) implies

$$\pi(\Box) = \lambda > 0. \tag{2.44}$$

Conversely, whenever (2.44) holds, we can implement (2.33). In fact, we could take

$$\alpha_k = \tfrac{1}{2}(k + 1 + is) \tag{2.45}$$

with $s = \pm\sqrt{\lambda - 1}$, provided

$$\lambda \geq 1. \tag{2.46}$$

In such a case, the representation is denoted

$$\pi_{is}^e, \quad \lambda = 1 + s^2, \qquad s \in R, \tag{2.47}$$

and called a member of the *first principal series*; one typically takes both signs of s in (2.47), but notes that

$$\pi^e_{is} \approx \pi^e_{-is}. \tag{2.48}$$

If $0 < \lambda < 1$, then we still get a representation of sl(2, R) by skew symmetric operators, but we cannot use (2.45) in order to achieve the desired identity

$$|\alpha_k| = \tfrac{1}{2}[(k+1)^2 - s^2]^{1/2}, \tag{2.49}$$

where $1 - \lambda = s^2$. In this case, the representation is denoted

$$\pi^e_s, \quad \lambda = 1 - s^2, \qquad s \in (-1,1)\backslash\{0\}, \tag{2.50}$$

and called a member of the *complementary series*. Again by convention we take s of both signs, but note

$$\pi^e_s = \pi^e_{-s}. \tag{2.51}$$

Finally, there is the case when spec$(1/i)E$ is given by (2.30). Since -1 belongs to spec$(1/i)E$, (2.33) implies

$$\pi(\square) = \lambda > 1. \tag{2.52}$$

Conversely, whenever (2.52) holds we can implement (2.33). In fact, we could take

$$\alpha_k = \tfrac{1}{2}(k + 1 + is) \tag{2.53}$$

with $s = \pm\sqrt{\lambda - 1}$. In such a case, the representation is denoted

$$\pi^o_{is}, \quad \lambda = 1 + s^2, \qquad s \in R\backslash\{0\}, \tag{2.54}$$

and called a member of the *second principal series*. Again one takes both signs of s, and

$$\pi^o_{is} \approx \pi^o_{-is}. \tag{2.55}$$

We remark that you can formally pass to the limit in (2.54) as $s \to 0$. In such a case, $R_+ : V_{-1} \to V_1$ vanishes, and the representation splits into a sum of representations considered in cases (2.27) and (2.28), with $n = \pm 1$, i.e.,

$$\pi^o_0 = \pi^+_1 \oplus \pi^-_{-1}. \tag{2.56}$$

As we have stated, each of the Lie algebra representations of sl(2, R) constructed above exponentiates to a representation of SL(2, R). In fact, we will give explicit realizations of these representations in the next three sections, and this will provide one justification of such a statement. We can also use the general results discussed in Appendix D, to exploit the fact that the $v_k \in V_k$ are all analytic vectors. This implies each such representation of sl(2, R) considered above exponentiates to a unitary representation $\tilde{\pi}$ of the universal covering group $\widetilde{\mathrm{SL}(2,R)}$. Note that the kernel of the natural projection $\widetilde{\mathrm{SL}(2,R)} \to \mathrm{SL}(2,R)$ consists precisely of $\widetilde{\exp}2\pi kZ$, $k \in \mathbf{Z}$. By (2.11), $\tilde{\pi}(g) = I$ for g in the kernel, so $\tilde{\pi}$

gives rise to a representation π of SL$(2, R)$, as desired. The dense set of analytic vectors also implies the uniqueness of the exponentiated unitary representation for any such given Lie algebra representation.

We now state the main result on the classification of the irreducible unitary representations of SL$(2, R)$, first derived by Bargmann [**12**].

THEOREM 2.2. *Any nontrivial irreducible unitary representation of* SL$(2, R)$ *is unitarily equivalent to one of the following types:*

$$\pi_n^+, \quad n \in \mathbf{Z}^+, \qquad \pi_{-n}^-, \quad n \in \mathbf{Z}^+,$$
$$\pi_{is}^e, \quad s \in R, \qquad \pi_{is}^o, \quad s \in R\backslash\{0\},$$
$$\pi_s^e, \quad s \in (-1, 1)\backslash\{0\}.$$

Each of these representations is irreducible, and there are no unitary equivalences except for (2.48), (2.51), (2.55). *In particular, an irreducible unitary representation π of* SL$(2, R)$ *is determined uniquely by the spectra of* $\pi(\Box)$ *and* $\pi(Z)$.

The only point remaining to be proved is the actual irreducibility of all the representations on this list, and we turn to that task. From (2.10) and (2.31), we deduce that any closed invariant linear subspace of the representation space H of such a representation must be an orthogonal direct sum of some collection of V_k's. Thus, if any such π were not irreducible, there would exist $v_k \in V_k$, $v_l \in V_l$, $l > k$, such that

$$(\pi(g)v_k, v_l) = 0 \quad \text{for all } g \in \mathrm{SL}(2, R). \tag{2.57}$$

Differentiating repeatedly and evaluating at $g = e$, which is permissible since $v_k \in C^\infty(\pi)$, yields $(\pi(P)v_k, v_l) = 0$ for each P in the universal enveloping algebra of sl$(2, R)$. In particular,

$$(R_+^{(l-k)/2} v_k, v_l) = 0. \tag{2.58}$$

But by the construction (2.32)–(2.35) this is impossible. The irreducibility of each representation listed in Theorem 2.2 is proved.

We close this section with a compact relisting of the action of each of these representations on the Casimir operator, from (2.37), (2.40), (2.47), (2.50), (2.54):

$$\pi_n^+(\Box) = 1 - (n-1)^2, \qquad n \in \mathbf{Z}^+, \tag{2.59}$$
$$\pi_{-n}^-(\Box) = 1 - (n-1)^2, \qquad n \in \mathbf{Z}^+, \tag{2.60}$$
$$\pi_{is}^e(\Box) = 1 + s^2, \qquad s \in R, \tag{2.61}$$
$$\pi_{is}^o(\Box) = 1 + s^2, \qquad s \in R\backslash\{0\}, \tag{2.62}$$
$$\pi_s^e(\Box) = 1 - s^2, \qquad s \in (-1,1)\backslash\{0\}. \tag{2.63}$$

3. The principal series. The simple form

$$\alpha_k = \tfrac{1}{2}(k + 1 + is) \tag{3.1}$$

for the coefficients occurring in the construction (2.32)–(2.35) of the principal series representations π^e_{is} and π^o_{is}, derived in (2.45) and (2.53), makes it tempting and easy to realize these representations of the Lie algebra sl(2, R) as a Lie algebra of vector fields (plus zero order terms) on the circle S^1, if we let $e^{ik\theta} \in L^2(S^1)$ (with square-norm $(2\pi)^{-1}\int_0^{2\pi}|f(\theta)|^2\,d\theta$) correspond to v_k. Consequently, the representation $\pi_{is} = \pi^e_{is} \oplus \pi^o_{is}$ of sl(2, R) on the space of trigonometric polynomials on S^1 is given by

$$\pi_{is}(Z) = \partial/\partial\theta, \tag{3.2}$$

and

$$\pi_{is}(X_+) = (1/2i)e^{2i\theta}\partial/\partial\theta + \tfrac{1}{2}(1+is)e^{2i\theta}, \tag{3.3}$$

$$\pi_{is}(X_-) = -(1/2i)e^{-2i\theta}\partial/\partial\theta + \tfrac{1}{2}(1+is)e^{2i\theta}, \tag{3.4}$$

and hence, since $2A = X_+ + X_-$, $-2iB = X_+ - X_-$,

$$2\pi_{is}(A) = \sin 2\theta\partial/\partial\theta + (1+is)\cos 2\theta, \tag{3.5}$$

and

$$2\pi_{is}(B) = \cos 2\theta\partial/\partial\theta - (1+is)\sin 2\theta. \tag{3.6}$$

These last two operators are formally skew symmetric first order differential operators with analytic coefficients, as is $\pi(Z)$, so it is clear that the analytic vectors for this Lie algebra are precisely the real-analytic functions on S^1, which of course includes the set of trigonometric polynomials. Rather than directly tackle the exponentiation of this Lie algebra representation, which would in principle just amount to integration along orbits of appropriate vector fields, we will take a detour on the way to explicitly realizing the associated representation of SL(2, R).

Namely, the forms of (3.5)–(3.6) make it natural to consider these operators as given by *real vector fields* on R^2, acting on functions of the form

$$r^{-(1+is)}f(\theta), \tag{3.7}$$

especially in view of the role of the Mellin transform, discussed below. In fact, acting on (3.7), the operators (3.2), (3.5), (3.6) coincide respectively with

$$R(Z) = \partial/\partial\theta, \qquad 2R(A) = \sin 2\theta\partial/\partial\theta - (\cos 2\theta)r\partial/\partial r, \tag{3.8}$$
$$2R(B) = \cos 2\theta\partial/\partial\theta + (\sin 2\theta)r\partial/\partial r,$$

or, in Cartesian coordinates on R^2, $x = r\cos\theta$, $y = r\sin\theta$,

$$R(Z) = -y\partial/\partial x + x\partial/\partial y, \qquad 2R(A) = -x\partial/\partial x + y\partial/\partial y, \tag{3.9}$$
$$2R(B) = y\partial/\partial x + x\partial/\partial y.$$

Now this Lie algebra representation of sl(2, R) by vector fields on R^2 with linear coefficients exponentiates explicitly to the *regular representation* of SL(2, R) on $L^2(R^2)$, defined by

$$R(g)f(x) = f(g^t \cdot x), \qquad g \in \mathrm{SL}(2,R),\ x \in R^2. \tag{3.10}$$

We now decompose this representation into the irreducibles π^e_{is}, π^o_{is}. Note that $R(g)$ commutes with the group of dilations

$$D(t)f(x) = e^t f(e^t x), \tag{3.11}$$

and with the inversion map $Jf(x) = f(-x)$. We show that the spectral decomposition of $D(t)$, plus splitting functions into even and odd parts, will effect an irreducible decomposition of $R(g)$.

The spectral decomposition of $D(t)$ can be effected via the Mellin transform:

$$\mathcal{M}\beta(s) = \int_0^\infty \beta(t)t^{is}\,dt. \tag{3.12}$$

Note that a change of variable $t = e^y$ gives

$$\mathcal{M}\beta(s) = \int_{-\infty}^\infty \beta(e^y)e^y e^{isy}\,dy, \tag{3.13}$$

so the Fourier inversion formula and Plancherel theorem give, respectively,

$$\beta(t) = (1/2\pi)\int_{-\infty}^\infty \mathcal{M}\beta(s)t^{-is-1}\,ds, \tag{3.14}$$

and

$$\int_{-\infty}^\infty |\mathcal{M}\beta(s)|^2\,ds = 2\pi\int_0^\infty |\beta(t)|^2 t\,dt. \tag{3.15}$$

If we define

$$P_s f(x) = \int_0^\infty f(tx)t^{is}\,dt, \tag{3.16}$$

we see that, for $f \in C_0^\infty(R^2)$, $P_s f$ belongs to the space

$$\mathcal{H}_s = \left\{g \in L^2_{\text{loc}}(R^2\backslash 0): g(rx) = r^{-is-1}g(x),\ r > 0, \text{ and } \int_{S^1}|g|^2 < \infty\right\}. \tag{3.17}$$

We make $\mathcal{H}_s$ a Hilbert space, with norm $\int_{S^1}|g|^2$; note that the homogeneity condition

$$g(rx) = r^{-is-1}g(x) \tag{3.18}$$

makes $g \in \mathcal{H}_s$ uniquely determined by its restriction to S^1. We see from (3.15) that

$$L^2(R^2) \approx \int_{-\infty}^{\infty}{}_{\oplus} \mathcal{H}_s\,ds. \tag{3.19}$$

There is the "Plancherel theorem"

$$\|f\|^2_{L^2(R^2)} = (1/2\pi)\int_{-\infty}^\infty \|P_s f\|^2_{\mathcal{H}_s}\,ds, \tag{3.20}$$

and the inversion formula

$$f(r\omega) = (1/2\pi)\int_{-\infty}^\infty P_s f(\omega)r^{-is-1}\,ds. \tag{3.21}$$

The representation $R(g)$ decomposes as a direct integral of representations ρ_s on $\mathcal{H}_s$:

$$\rho_s(g) = f(g^t \cdot x), \qquad f \in \mathcal{H}_s. \tag{3.22}$$

We can realize all these representations on $L^2(S^1)$. Indeed, the restriction map $f \mapsto f|_{S^1}$ takes $\mathcal{H}_s$ isometrically onto $L^2(S^1)$. Using (3.9), we see that this unitary map intertwines ρ_s with π_{is}, defined by

$$\pi_{is}(g)f(\omega) = \|g^t \cdot \omega\|^{-is-1} f(A(g^t)\cdot\omega), \qquad f \in L^2(S^1), \tag{3.23}$$

where $A(g)\colon S^1 \to S^1$ for each $g \in \mathrm{SL}(2,R)$ is defined by

$$A(g)\omega = g\omega/\|g\omega\|. \tag{3.24}$$

In (3.23), it is clear that $L^2_{\text{even}}(S^1)$ and $L^2_{\text{odd}}(S^1)$ are invariant. Note that the group SO(2), generated by Z, operates by rotation:

$$\pi_{is}(k)f(\omega) = f(k^{-1}\omega), \qquad k \in \mathrm{SO}(2). \tag{3.25}$$

Thus it is clear that, if we write

$$\pi_{is} = \pi_{is}^e \oplus \pi_{is}^o, \tag{3.26}$$

then π_{is}^e and π_{is}^o are explicit realizations of the principal series representations described in §2.

Note that the image of the Casimir operator $\Box$ under R is given by

$$R(\Box) = -X^2 + 1, \tag{3.27}$$

where X generates $D(t)$, so $X = r\partial/\partial r + 1$.

These representations π_{is}^e and π_{is}^o of $\mathrm{SL}(2,R)$ on spaces of functions on S^1, on which $\mathrm{SL}(2,R)$ acts transitively, can be seen to be examples of *induced representations*. For more on this perspective, see Chapter 9, §3.

If we have a principal series representation of the form π_{is}^e, acting on a Hilbert space $\mathcal{H}_s^e$, pick f_0, a generator of the one-dimensional space on which $K = \mathrm{SO}(2)$ acts trivially, and for each $f \in \mathcal{H}_s^e$ consider the function on $G = \mathrm{SL}(2,R)$ defined by

$$(T_s f)(g) = (f, \pi_{is}^e(g) f_0). \tag{3.28}$$

Then it is clear that $(T_s f)(gk) = (T_s f)(g)$ for all $k \in K$, so we define a function $\tau_s f \in C(G/K) = C(\Omega_+)$ by

$$\tau_s f(x) = (T_s f)(g), \qquad x = gK \in \Omega_+. \tag{3.29}$$

Note that the map $\tau_s\colon \mathcal{H}_s^e \to C(\Omega_+)$ is injective. Indeed, $\ker \tau_s$ must be invariant under π_{is}^e, and cannot be all of $\mathcal{H}_s^e$, since $f_0 \notin \ker \tau_s$, so the irreducibility of π implies $\ker \tau_s = 0$.

Note that

$$\tau_s(\Box f) = -4\Delta(\tau_s f) \tag{3.30}$$

where $\square$ is the Casimir operator on G and Δ is the Laplace operator on Ω_+, with its Poincaré metric (1.9), i.e.,

$$\Delta = y^{-2}(\partial^2/\partial x^2 + \partial^2/\partial y^2). \tag{3.31}$$

In view of the results (2.61) on $\pi^e_{is}(\square)$, we see that

$$\Delta(\tau_s f) = -\tfrac{1}{4}(1+s^2)(\tau_s f), \qquad f \in \mathcal{H}^e_s. \tag{3.32}$$

In other words, the image of $\mathcal{H}^e_s$ under τ_s consists of C^∞ functions u on Ω_+ satisfying the equation

$$\Delta u = -\tfrac{1}{4}(1+s^2)u. \tag{3.33}$$

If we denote the space of C^∞ functions on Ω_+ satisfying (3.33) by $\mathcal{E}_s$, we have

$$\tau_s \colon \mathcal{H}^e_s \to \mathcal{E}_s, \quad \text{injective}. \tag{3.34}$$

Note that

$$\tau_s(\pi^e_{is}(g)f)(x) = \tau_s(f)(g^{-1}\cdot x). \tag{3.35}$$

Let f_n span the one-dimensional subspace of $\mathcal{H}^e_s$ on which $K = \mathrm{SO}(2)$ acts as the character $e^{in\theta}$, and let $F_n = \tau_s(f_n)$. Then, in geodesic polar coordinates centered at the origin $eK \in \Omega_+$,

$$F_n(x) = e^{in\theta}F^b_n(r), \qquad \Delta F_n = -\tfrac{1}{4}(1+s^2)F_n. \tag{3.36}$$

It is not hard to show that the space of smooth functions on Ω_+ satisfying (3.36) is one-dimensional, so these conditions characterize F_n up to some scalar factor, which depends on s and n. If one expresses the Laplace operator in geodesic polar coordinates (it is easier to think of geodesic polar coordinates on the Poincaré *disc*, centered at the origin), one sees that

$$F_n(x) = C_{n,s}e^{in\theta}P^n_{-1/2+is}(\cosh r), \tag{3.37}$$

where P^n_ν is the associated Legendre function.

It will follow from the inversion formula to be proved in §7, that if R_1 denotes the regular representation of SL(2, R) on $L^2(\Omega_+)$,

$$R_1(g)f(x) = f(g^{-1}\cdot x), \qquad g \in \mathrm{SL}(2,R),\ x \in \Omega_+, \tag{3.38}$$

then

$$R_1 \approx \int_{0\,\oplus}^{\infty} \pi^e_{is}\,ds, \tag{3.39}$$

with the corresponding decomposition of Hilbert spaces

$$L^2(\Omega_+) \approx \int_{0\,\oplus}^{\infty} \mathcal{H}^e_s\,ds. \tag{3.40}$$

If we compare (3.39) and (3.19) and recall the equivalence $\pi_{is} \approx \pi_{-is}$, we see that the even part of R, defined by (3.10), acting on $L^2_{\text{even}}(R^2)$, is equivalent to a sum of two copies of the regular representation R_1 defined by (3.38).

An explicit operator intertwining these representations can be constructed by the following device, used by Ehrenpreis [**54**, **55**]. See also Strichartz [**228**, **229**] for related ideas. First, we can identify even functions on R^2 with functions on C, the forward light cone in R^3 with the Minkowski metric. Then the intertwining operator takes a pair of functions on Ω_+, identified with the upper sheet of a two sheeted hyperboloid in R^3, and uses this pair of functions as Cauchy data to solve the wave equation within the forward light cone, then taking the limiting values on the light cone. Inverting this process involves solving a Goursat problem for the wave equation. For details we refer to [**54**].

4. The discrete series. There is a natural representation of SL(2, R) on $L^2(R^2)$ which contains each member of the holomorphic discrete series, suggested by some calculations mentioned at the end of §1 of Chapter 7, on harmonic analysis on cones. If we view R^2 as the cone over S^1, these observations specialize to the following. Let

$$X = r\partial/\partial r + 1, \qquad U = i\Delta, \qquad V = ir^2. \tag{4.1}$$

Then X, U, V are skew adjoint operators, satisfying the commutator relations

$$[X, U] = -2U, \qquad [X, V] = 2V, \qquad [U, V] = 4X. \tag{4.2}$$

Let

$$\begin{aligned} Z &= -\tfrac{1}{2}(U - V) = \tfrac{1}{2}i(-\Delta + |x|^2), \\ A &= -\tfrac{1}{2}X = -\tfrac{1}{2}(r\partial/\partial r + 1), \\ B &= -\tfrac{1}{4}(U + V) = -\tfrac{1}{4}i(\Delta + |x|^2). \end{aligned} \tag{4.3}$$

Then we have

$$[Z, A] = 2B, \qquad [Z, B] = -2A, \qquad [A, B] = -\tfrac{1}{2}Z, \tag{4.4}$$

so these skew adjoint operators on $L^2(R^2)$, span a Lie algebra isomorphic to sl(2, R). Thus they generate a representation of $\tilde{G}$, the universal covering group of SL(2, R). Note that the kernel of the covering map $\tilde{G} \to \mathrm{SL}(2, R)$ consists of $\exp(2\pi k Z)$, $k \in \mathbf{Z}$. The first formula in (4.3) shows that the spectrum of $(1/i)Z$ consists of all positive integers, since, as we saw in Chapter 1, the harmonic oscillator hamiltonian $-\Delta + x^2$ on R^2 has as its spectrum all even positive integers. Thus $\exp(2\pi k Z)$ is represented by the identity operator on $L^2(R^2)$, so the representation of $\tilde{G}$ actually comes from a representation of SL(2, R). Let us denote this representation by γ.

Note that all the operators (4.3) commute with rotations. Thus we can decompose γ on $L^2(R^2)$ by

$$L^2(R^2) = \bigoplus_{k=-\infty}^{\infty} H_k, \tag{4.5}$$

where

$$H_k = \{e^{ik\theta} g(r) \in L^2(R^2)\}. \tag{4.6}$$

As already shown in Chapter 7 on cones, the $ax + b$ group, generated by X and $U = i\Delta$, already acts irreducibly on each H_k, this representation being intertwined with a "standard model" via a Hankel transform. It follows a fortiori that SL(2, R) acts irreducibly on each H_k, via γ. Since spec$(1/i)Z$ consists of positive integers, it is clear from the classification of §2 that each irreducible component γ_k must be a holomorphic discrete series representation.

We claim that, on H_k, the smallest element of spec$(1/i)Z$ is $|k| + 1$. In fact, consider our operators on Bargmann-Fok space $\mathcal{H}$, a Hilbert space of entire functions on $\mathbf{C}^2$ with orthonormal basis $(2/\alpha!)^{1/2}z^\alpha$, $\alpha = (\alpha_1, \alpha_2) \geq 0$, on which $(1/i)Z$ acts as multiplication by $|\alpha| + 1$ (see Chapter 1, §6). Then the rotation group action $f(x) \mapsto f(kx)$, $k \in \mathrm{SO}(2)$, $f \in L^2(R^2)$, is intertwined to a subgroup of the natural SU(2) action on $\mathcal{H}$, $u(z) \mapsto u(\sigma z)$, $\sigma \in \mathrm{SU}(2)$. As we know, each irreducible representation of SU(2) is contained exactly once in $\mathcal{H}$, on one of the spaces E_j = span of $\{z^\alpha : |\alpha| = j\}$. On the space E_j, the operator $(1/i)Z$ is the scalar $j + 1$ and the skew adjoint generator of the SO(2) action on E_j takes the values $-j, -j + 1, \ldots, j - 1, j$, multiplied by i. This proves that the smallest eigenvalue of $(1/i)Z$ on the $\pm ij$ eigenspace of the generator of the SO(2) action is $j + 1$, as asserted.

It follows that

$$\gamma_k \approx \pi^+_{|k|+1}. \tag{4.7}$$

Thus γ contains π^+_n exactly twice, for each $n \geq 2$, and it contains π^+_1 once.

Via the identity $\pi^-_{-n} = \pi^+_n \circ \alpha$, where α is the automorphism of SL(2, R) given in §1, we obtain the antiholomorphic discrete series.

Another realization of the holomorphic discrete series is given on

$$\tilde{\mathcal{H}}_n = \left\{ u(z) \text{ holomorphic in } D\text{: } \int_D |u(z)|^2(1 - r^2)^{n-2} r\, dr\, d\theta < \infty \right\} \tag{4.8}$$

where D is the unit disc

$$D = \{z \in \mathbf{C} : |z| < 1\}. \tag{4.9}$$

We can write

$$\tilde{\pi}^+_n(g)u(z) = (\overline{\beta} z + \overline{\alpha})^{-n} u(g^{-1} \cdot z) \tag{4.10}$$

where

$$g^{-1} = \begin{pmatrix} \alpha & \beta \\ \overline{\beta} & \overline{\alpha} \end{pmatrix} \in \mathrm{SU}(1,1) \approx \mathrm{SL}(2, R), \tag{4.11}$$

with action on D given by the linear fractional transformation, as described in §1. Note that the compact subgroup K is given by

$$K = \left\{ k_\theta = \begin{pmatrix} e^{i\theta} & 0 \\ 0 & e^{-i\theta} \end{pmatrix} : \theta \in R/2\pi\mathbf{Z} \right\}, \tag{4.12}$$

and

$$\pi_n^+(k_\theta)\,\mathrm{u}(z) = e^{in\theta}\,\mathrm{u}(e^{2i\theta}z). \tag{4.13}$$

Thus spec$(1/i)Z$ has n as its smallest element. The Hilbert space $\tilde{\mathcal{H}}_n$ is nontrivial precisely for $n \geq 2$, so we obtain all the holomorphic discrete series representations π_n^+ with $n \geq 2$. We do not get π_1^+ in this fashion.

In fact, one usually says that π_1^+ is not a discrete series representation, but rather a mock discrete series representation. The reason for this is that, for any f, f' in the representation space of π_n^+, the function

$$\varphi_{f,f'}(g) = (\pi_n^+(g)f, f') \tag{4.14}$$

belongs to $L^2(\mathrm{SL}(2,R))$ if $n \geq 2$, but not if $n = 1$. We will skip the details on this, but remark that it implies $\pi_{\pm n}^{\pm}$ will occur in the Plancherel formula on SL(2, R), discretely, for $n \geq 2$, while $\pi_{\pm 1}^{\pm}$ do not make such an appearance.

The Plancherel formula is an identity of the form

$$\|f\|^2_{L^2(G)} = \int_{\hat{G}} \|\pi(f)\|^2_{\mathrm{HS}}\, d\mu(\pi) \tag{4.15}$$

where $\hat{G}$ denotes the set of equivalence classes of irreducible unitary representations of G, and μ is a certain measure on G, called the Plancherel measure. $\|T\|^2_{\mathrm{HS}} = \mathrm{tr}(T^*T)$ is the squared Hilbert-Schmidt norm. A general formula of this type holds for a broad class of groups, the unimodular type I groups, a class which contains SL(2, R) and, indeed, all semisimple Lie groups. This general result was derived by Segal [**212**] and Mautner [**168**]. The explicit Plancherel formula for SL(2, R) is the following:

$$\begin{aligned} \|f\|^2_{L^2(G)} = &\int_0^\infty \|\pi^e_{is}(f)\|^2_{\mathrm{HS}} s\tanh \pi s\, ds \\ &+ \int_0^\infty \|\pi^o_{is}(f)\|^2_{\mathrm{HS}} s\coth \pi s\, ds \\ &+ \sum_{n=2}^\infty \frac{1}{2}(n-1)\left[\|\pi_n^+(f)\|^2_{\mathrm{HS}} + \|\pi^-_{-n}(f)\|^2_{\mathrm{HS}}\right]. \end{aligned} \tag{4.16}$$

We will not give a proof of this Plancherel formula. We refer to Lang [**147**] for a treatment of this formula.

The Plancherel formulas (4.15)–(4.16) might be called Plancherel formulas for the decomposition of the regular representation of G on $L^2(G)$ into irreducibles. In §7 we will establish a Plancherel formula for the "regular" representation of SL(2, R) on $L^2(\Omega_+)$, where Ω_+ is the Poincaré upper half plane. This formula will be similar to (4.16), but simpler, as no discrete series representations occur.

5. The complementary series. The complementary series π^e_s, $s \in (0,1)$, seems a bit more mysterious than the other series. We will confine ourselves to a brief description of one realization of π^e_s. Namely, let

$$H_s = \left\{u \in L^2_{\mathrm{loc}}(R) \colon \iint |x-y|^{-s-1}u(x)\overline{u(y)}\,dx\,dy < \infty\right\}, \tag{5.1}$$

and set

$$\pi_s^e(g)u(x) = |cx+d|^{s-1}u(g^{-1}\cdot x) \tag{5.2}$$

where

$$g^{-1} = \begin{pmatrix} a & b \\ c & d \end{pmatrix} \tag{5.3}$$

and

$$g^{-1}\cdot x = (ax+b)/(cx+d). \tag{5.4}$$

6. The spectrum of $L^2(\Gamma\backslash \mathrm{PSL}(2,R))$, in the compact case. Suppose Γ is a discrete subgroup of $G = \mathrm{PSL}(2,R)$ such that $X = \Gamma\backslash G$ is *compact.* We also assume Γ acts on G/K without fixed points. Then $M = X/K = \Gamma\backslash G/K$ is an arbitrary compact surface of constant negative curvature, i.e., an arbitrary compact Riemann surface whose first Betti number exceeds two, and X is the unit circle bundle over M. X is a homogeneous space for G. It has a natural Lorentz metric, inherited from the Killing form on G, which is preserved by the G action, and hence its natural volume element is preserved by the G action. Thus we have a unitary representation of G on $L^2(X)$:

$$\rho(g)f(x) = f(g^{-1}\cdot x), \qquad g\in G,\ x\in X. \tag{6.1}$$

In this section we want to say something about the decomposition of ρ into irreducibles.

Let us denote the operators $\rho(Z)$ and $\rho(\Box)$ on X by Z and $\Box$. Since $\Box$ belongs to the center of the universal enveloping algebra of G, these operators on X commute. Note that the operaor $2Z^2-\Box$ is *elliptic.* Thus it has a discrete spectrum, with eigenspaces of finite dimension, and both Z and $\Box$ leave each of these eigenspaces invariant. Thus

$$L^2(X) = \bigoplus_{n,\lambda} V_{n,\lambda} \tag{6.2}$$

where

$$V_{n,\lambda} = \{f\in C^\infty(X)\colon Zf = \mathrm{inf},\ \Box f = \lambda f\}, \tag{6.3}$$

and we know the sum is countable and each $V_{n,\lambda}$ is finite-dimensional.

Let us denote by Σ the set of (n,λ) such that $V_{n,\lambda}\neq 0$. For each λ, define

$$E_\lambda = \bigoplus_n \{V_{n,\lambda}\colon (n,\lambda)\in\Sigma\}. \tag{6.4}$$

Then we have

$$L^2(X) = \bigoplus_\lambda E_\lambda \tag{6.5}$$

and $\Box$ acts on E_λ as the scalar λ. Thus each E_λ in (6.5) is invariant under the action of ρ, and if we denote by ρ_λ the restriction of ρ to E_λ, each ρ_λ is a finite sum of irreducible representations of G. From our computation in §2, we see that the representation which occurs in ρ_λ is

$$\begin{aligned} \pi_n^+ \text{ or } \pi_{-n}^- &\quad \text{if } \lambda = 1-(n-1)^2 \leq 0, \\ \pi_s^e &\quad \text{if } \lambda = 1-s^2 \in (0,1), \\ \pi_{is}^e &\quad \text{if } \lambda = 1+s^2 \in [1,\infty). \end{aligned} \tag{6.6}$$

An important question is to determine the multiplicity with which such an irreducible occurs in ρ_λ. We describe here a classical result of Gelfand and Graev relating this multiplicity to some phenomena arising from the compact Riemann surface M.

First, suppose $\lambda > 0$ and π_s^e or π_{is}^e occurs in ρ_λ, where either $\lambda = 1 - s^2$ belongs to $(0,1)$ or $\lambda = 1 + s^2 \in [1,\infty)$. Then we know that each irreducible component has a one-dimensional subspace where $Z = 0$, so

$$\text{multiplicity in } \rho_\lambda = \dim V_{0,\lambda}. \tag{6.7}$$

Now elements of $V_{0,\lambda}$ are invariant under the SO(2) action, so there is an isomorphism

$$V_{0,\lambda} \to \tilde{V}_{0,\lambda} \tag{6.8}$$

to a linear subspace of $C^\infty(M) = C^\infty(X/K)$, which is a restriction of the isomorphism $L^2(X,K) \to L^2(M)$, where $L^2(X,K)$ denotes the space of elements of $L^2(X)$ invariant under K. Note that the Casimir operator $\square$ on $C^\infty(X,K)$ is taken to -4Δ on $C^\infty(M)$, where Δ is the Laplace operator on M, with respect to its natural metric. Thus

$$\tilde{V}_{0,\lambda} = \{f \in C^\infty(M) : \Delta f = -\tfrac{1}{4}\lambda\}, \tag{6.9}$$

or in other words $V_{0,\lambda}$ is an eigenspace of the Laplace operator on M. Those eigenvalues of Δ in $(-\infty, -\frac{1}{4}]$ give rise to principal series representations of G on $L^2(X)$, and those eigenvalues in $(-\frac{1}{4}, 0)$ give rise to complementary series representations. In either case,

$$\text{multiplicity in } \rho_\lambda = \text{multiplicity of the eigenvalue } -\lambda/4, \quad \text{for } \lambda > 0. \tag{6.10}$$

We remark that there is considerable interest in determining for which compact Riemann surfaces M does Δ have spectrum in $(-\frac{1}{4}, 0)$. Of course, the trivial representation of G occurs once in $L^2(X)$, and this corresponds to the 0 eigenvalue of Δ on M.

Next suppose $\lambda \leq 0$. Then the irreducible representation occurring in ρ_λ must be a discrete series representation, π_n^+ or π_{-n}^-, where $n = 1 + \sqrt{1-\lambda}$; $n = 2m$ must be even. Suppose for a moment it is π_n^+; the other case is handled similarly. Then we know that each irreducible component in ρ_λ has a one-dimensional space spanned by a lowest weight vector, so in this case

$$\text{multiplicity in } \rho_\lambda = \dim V_{n,\lambda}. \tag{6.11}$$

Now elements of $V_{n,\lambda}$ are not constant on the fibers of $X \to M$, but they belong to the space $C^\infty(X,n) = \{f \in C^\infty(X): f(k_\theta \cdot x) = e^{in\theta}f(x)\}$. Since SL(2, R) is a double cover of PSL(2, R), this space is naturally isomorphic to the space

$$C^\infty(M, \kappa(m)) \tag{6.12}$$

of smooth sections of the vector bundle $\kappa(m) \to M$ obtained from the principal SO(2) bundle $X \to M$ via the representation $e^{i\theta} \mapsto e^{im\theta}$ of SO(2) on C. $\kappa(m)$ is a holomorphic line bundle on the Riemann surface M, the mth power of the canonical bundle κ.

In fact, the subspace $V_{n,\lambda}$ of $C^\infty(X,n)$ is characterized by the property of being annihilated by the lowering operator $R_- = \rho(A + iB)$. This gives rise to a first order operator on $C^\infty(M, \kappa(m))$, which is verified to coincide with the $\bar{\partial}$ operator. Thus

$$V_{n,\lambda} \approx \mathcal{O}(M, \kappa(m)), \tag{6.13}$$

the space of holomorphic sections of $\kappa(m)$. Hence, when $n = 2m$,

$$\text{multiplicity of } \pi_n^+ = \dim \mathcal{O}(M, \kappa(m)). \tag{6.14}$$

Similarly we see the multiplicity of π_{-n}^- is equal to the dimension of the space of antiholomorphic elements of $C^\infty(M, \overline{\kappa(m)})$. Thus π_n^+ and π_{-n}^- occur with equal multiplicity.

The dimension of $\mathcal{O}(M, \kappa(m))$ can be deduced from the Riemann-Roch theorem. We have

$$\dim \mathcal{O}(M, \kappa) = g, \qquad \dim \mathcal{O}(M, \kappa(m)) = (2m-1)(g-1), \quad m \geq 2, \tag{6.15}$$

where

$$g = \tfrac{1}{2} \dim H^1(M, \mathbf{C}) \tag{6.16}$$

is the genus of M. See Gunning, *Lectures on Riemann Surfaces* [**90**], page 111.

To close this section, we note a group theoretic interpretation of the geodesic flow on the compact Riemann surface $M = X/K$. In fact, the geodesic flow on the Poincaré disc $D = G/K$ is a one parameter family F_t of transformations on the unit tangent bundle $T_1 D \approx G$, which commutes with the left G-action of G. In other words, the infinitesimal generator of F_t is a left invariant vector field on G, so there is a one parameter subgroup $\gamma(t)$ of G such that

$$F_t g = g \cdot \gamma(t), \qquad g \in G. \tag{6.17}$$

If we identify G with $SU(1,1)/\{\pm I\}$, acting on D by (1.7), we see that if we identify the unit tangent vector over $0 \in D$ pointing along the real axis with $e \in G$, then

$$\gamma(t) = F_t e = \begin{pmatrix} \cosh t & \sinh t \\ \sinh t & \cosh t \end{pmatrix} \in SU(1,1), \tag{6.18}$$

with infinitesimal generator

$$A' = \begin{pmatrix} 0 & 1 \\ 1 & 0 \end{pmatrix} \in \mathrm{su}(1,1) \tag{6.19}$$

that is taken by the isomorphism (1.12) to

$$-2A = \begin{pmatrix} 1 & 0 \\ 0 & -1 \end{pmatrix} \in \mathrm{sl}(2,R). \tag{6.20}$$

Using these observations, we can establish that the geodesic flow F_t on T_1M is *ergodic* for any compact surface M of constant negative curvature.

PROPOSITION 6.1. *If $u \in L^2(X)$ is invariant under F_t, $t \in R$, then u is constant.*

PROOF. In view of the discrete decomposition of $L^2(X)$ into irreducibles, it suffices to show that if π is any nontrivial irreducible unitary representation of $G = \mathrm{PSL}(2,R)$,

$$\ker \pi(A) = 0. \tag{6.21}$$

Given the classification of these representations discussed in §§2–5, this is a straightforward task, whose details we omit.

A great deal of work has been done on $L^2(\Gamma\backslash G)$, for quotients $\Gamma\backslash G$ either compact or noncompact, but with finite volume, work centered around the Selberg trace formula. We refer to Hejhal [**99**], Lang [**147**], Lax-Phillips [**151**], and Terras [**238**] for more on this large topic.

7. Harmonic analysis on the Poincaré upper half plane. We wish to produce the spectral decomposition of the Laplace operator Δ on the Poincaré upper half plane Ω_+. From comments in §3, it follows that a decomposition of the regular representation of $\mathrm{SL}(2,R)$ on $L^2(\Omega_+)$ into irreducibles will be achieved.

The spectral decomposition of Δ, or equivalently of the operator

$$\nu = (-\Delta - \tfrac{1}{4})^{1/2}, \tag{7.1}$$

is given formally by

$$P_\sigma = \delta(\nu - \sigma) = \int_{-\infty}^{\infty} e^{i(\nu-\sigma)s}\, ds. \tag{7.2}$$

If we use the fact that $\operatorname{spec} \nu \subset [0,\infty)$, then, for $\sigma \geq 0$, we can write

$$\begin{aligned} P_\sigma &= \int_{-\infty}^{\infty} \cos s\nu e^{-i\sigma s}\, ds = -i\sigma \int_{-\infty}^{\infty} \nu^{-1} \sin s\nu e^{-i\sigma s}\, ds \\ &= 2\sigma \int_0^{\infty} \nu^{-1} \sin s\nu \sin \sigma s\, ds. \end{aligned} \tag{7.3}$$

Now we use the formula for the operator $\nu^{-1}\sin t\nu$, the fundamental solution to the wave equation $\partial^2 u/\partial t^2 - (\Delta + \frac{1}{4})u = 0$, derived in Chapter 4, §2:

$$\nu^{-1}\sin t\nu = \begin{cases} C_2(\operatorname{sgn} t)(2\cosh t - 2\cosh r)^{-1/2} & \text{if } r < |t|, \\ 0 & \text{if } r > |t|, \end{cases} \tag{7.4}$$

where r denotes the geodesic distance between the source point and the observation point. We obtain

$$P_\sigma u(x) = \int_{\Omega_+} u(y) p_\sigma(x, y)\, d\operatorname{vol}(y) \tag{7.5}$$

with

$$p_\sigma(x, y) = p_\sigma(r), \qquad r = \operatorname{dist}(x, y), \tag{7.6}$$

where

$$p_\sigma(r) = 2\sigma C_2 \int_r^\infty (2\cosh s - 2\cosh r)^{-1/2} \sin \sigma s\, ds. \tag{7.7}$$

This is a Legendre function. In fact, there is the integral formula

$$P_\nu(\cosh r) = (2/\pi)\cot(\nu + \tfrac{1}{2})\pi \int_r^\infty (2\cosh s - 2\cosh r)^{-1/2} \sinh(\nu + \tfrac{1}{2}) s\, ds; \tag{7.8}$$

see Lebedev [**154**], page 173, valid for $r \in R$, $-1 < \operatorname{Re} \nu < 0$, and this implies

$$P_{-1/2+i\sigma}(\cosh r) = -(2/\pi)\coth \pi\sigma \int_r^\infty (2\cosh s - 2\cosh r)^{-1/2} \sin \sigma s\, ds, \tag{7.9}$$

so (7.7) is equivalent to

$$p_\sigma(r) = -C_2 \pi\sigma \tanh \pi\sigma P_{-1/2+i\sigma}(\cosh r). \tag{7.10}$$

Let $o \in \Omega_+$ denote the point eK. (One usually takes $o = i \in \Omega_+$, but it is more convenient to visualize o as the origin in the Poincaré disc.) Using geodesic polar coordinates centered at o, let

$$H_n = \{f \in L^2(\Omega_+) : f = e^{in\theta} f^b(r)\}. \tag{7.11}$$

Then if E_n is the orthogonal projection on H_n, we have E_n and $e^{i\sigma\nu}$ commuting, so $E_n P_\sigma E_n = P_\sigma E_n$, whose range is the one-dimensional subspace of the eigenspace $\mathcal{E}_\sigma \in C^\infty(\Omega_+)$ consisting of elements satisfying (3.27), with $s = \sigma$. It follows that

$$p_\sigma(x, y) = \sum_{n=-\infty}^{\infty} l_n(\sigma) e^{in(\theta - \theta')} p_{n\sigma}(r) p_{n\sigma}(r') \tag{7.12}$$

where $x = (r, \theta)$, $y = (r', \theta')$ in geodesic polar coordinates, $l_n(\sigma) \in \mathbf{C}$, and, by (3.28),

$$p_{n\sigma}(r) = P^n_{-1/2+i\sigma}(\cosh r). \tag{7.13}$$

Note that $y = o$ gives

$$P_\sigma \delta_o(x) = p_\sigma(x, o) = l_o(\sigma) P_{-1/2+i\sigma}(\cosh r), \tag{7.14}$$

and comparison with (7.6), (7.10) shows that

$$l_o(\sigma) = \pi\sigma \tanh \pi\sigma \tag{7.15}$$

since $P_\nu(1) = 1$.

We can get an inversion formula from this. Note that

$$I = \int_0^\infty \delta(\nu - \sigma)\, d\sigma = \int_0^\infty P_\sigma\, d\sigma, \tag{7.16}$$

so (7.12) implies

$$u(x) = \int_0^\infty \sum_{n=-\infty}^{\infty} l_n(\sigma) \int_{\Omega_+} P^n_{-1/2+i\sigma}(\cosh r) P^n_{-1/2+i\sigma}(\cosh r') \cdot e^{in(\theta-\theta')} u(y)\, d\,\mathrm{vol}(y)\, d\sigma, \tag{7.17}$$

where, in geodesic polar coordinates, $x = (r, \theta)$, $y = (r', \theta')$. In particular, if $u(x) = u(r)$ is radial, using the fact that the area element on the Poincaré upper half plane Ω_+ is $\sinh r\, dr\, d\theta$, we have

$$u(r) = (1/\pi) \int_0^\infty \int_0^\infty \sigma \tanh \pi\sigma P_{-1/2+i\sigma}(\cosh r) P_{-1/2+i\sigma}(\cosh r') \cdot u(r') \sinh r'\, dr'\, d\sigma. \tag{7.18}$$

This is known as the Mehler-Fok inversion formula (see Lebedev [**154**], page 221).

Note that if we multiply (7.17) by $u(x)$ and integrate, we obtain the Plancherel formula

$$\|u\|^2_{L^2(\Omega_+)} = \int_0^\infty \sum_n l_n(\sigma) |(u, p^\#_{n\sigma})|^2\, d\sigma, \tag{7.19}$$

where

$$p^\#_n(x) = e^{in\theta} p_{n\sigma}(r), \qquad x = (r, \theta). \tag{7.20}$$

It is useful to compare this with the general Plancherel formula for representations of unimodular type I groups, which in this context implies

$$\|u\|^2_{L^2(\Omega_+)} = \int_0^\infty \|P_\sigma u\|^2_{\mathcal{H}^e_\sigma}\, d\mu(\sigma), \tag{7.21}$$

where $\mathcal{H}^e_\sigma$ is identified with a subspace of $\mathcal{E}_\sigma$ via the mapping (3.25). To identify the measure $d\mu(\sigma)$, it suffices to consider radial u, and then (7.15) yields

$$d\mu(\sigma) = C\sigma \tanh \pi\sigma\, d\sigma. \tag{7.22}$$

We now note that, in view of the connection between f_0 and $f_n \in \mathcal{H}^e_\sigma$ given by raising and lowering operators, and the recursion formulas connecting P^n_ν to P_ν, it follows that

$$l_n(\sigma) = l_0(\sigma) = C\sigma \tanh \pi\sigma, \tag{7.23}$$

so the inversion formula (7.17) is written more precisely as

$$(7.24)\quad u(x) = \frac{1}{\pi}\int_0^\infty \sum_{n=-\infty}^{\infty} \sigma \tanh \pi\sigma \int_{\Omega_+} P^n_{-1/2+i\sigma}(\cosh r) P^n_{-1/2+i\sigma}(\cosh r') \cdot e^{in(\theta-\theta')} u(y)\, d\,\mathrm{vol}(y)\, d\sigma.$$

The identity (7.16) generalizes to

$$(7.25)\qquad f(\nu) = \int_0^\infty f(\sigma) P_\sigma \, d\sigma,$$

which implies

(7.26)

$$f(\nu)u(x) = \frac{1}{\pi}\int_0^\infty \sum_{n=-\infty}^{\infty} \sigma \tanh \pi\sigma f(\sigma) \int_{\Omega_+} P^n_{-1/2+i\sigma}(\cosh r) P^n_{-1/2+i\sigma}(\cosh r') \cdot e^{in(\theta-\theta')} u(y)\, d\,\mathrm{vol}(y)\, d\sigma.$$

8. The subelliptic operator $\mathcal{L}_\alpha = A^2 + B^2 + \frac{1}{2}i\alpha Z$ on SL(2, R). Here we make a brief study of the operators

$$(8.1)\qquad \mathcal{L}_\alpha = A^2 + B^2 + \tfrac{1}{2}i\alpha Z$$

on SL(2, R). It is known from the general theory of pseudodifferential operators that $\mathcal{L}_\alpha$ is hypoelliptic provided α avoids a certain discrete set of real values. We will see this condition arising from the structure of the image of $\mathcal{L}_\alpha$ under irreducible representations of SL(2, R).

The existence of a parametrix for $\mathcal{L}_\alpha$ depends entirely on the behavior of the discrete series. The proof of this assertion is an exercise in microlocal analysis which is beyond the scope of these notes, but we can state the reason. The operator $\mathcal{L}_\alpha$ is difficult to invert microlocally only where Z dominates A and B. Note that the Casimir operator on SL(2, R)

$$(8.2)\qquad \Box - 1 = Z^2 - 4A^2 - 4B^2 - 1$$

is represented by real scalars on irreducible representation spaces, which are negative for the discrete series and positive for the principal series (and between 0 and -1 for the supplementary series), so distributions with wave front set near the characteristic set of $\mathcal{L}_\alpha$ are synthesized from discrete series representations.

Let us consider the image of $\mathcal{L}_\alpha$ under a holomorphic discrete series representation. We use the representations γ_k of §4, which sum to $\gamma = \bigoplus \gamma_k$ acting on $L^2(R^2)$. Recall that

$$(8.3)\qquad \gamma(Z) = \tfrac{1}{2}i(-\Delta + |x|^2).$$

Since

$$(8.4)\qquad \mathcal{L}_\alpha = -\tfrac{1}{4}\Box + \tfrac{1}{4}Z^2 + \tfrac{1}{2}i\alpha Z$$

and

$$\gamma_k(\Box) = 1 - (|k| - 1)^2, \tag{8.5}$$

we have

$$\gamma_k(\mathcal{L}_\alpha) = \tfrac{1}{4}[(|k| - 1)^2 - \tfrac{1}{4}(-\Delta + |x|^2)^2 - \alpha(-\Delta + |x|^2) - 1]. \tag{8.6}$$

This operator acts on a subspace of $L^2(R^2)$ which is invariant under the unitary group generated by $\gamma(Z) = \frac{1}{2}i(-\Delta + |x|^2)$, and on which the eigenvalues of $(1/i)\gamma(Z)$ are greater than or equal to $|k| + 1$. In particular, the eigenvalues of $\frac{1}{4}(-\Delta + |x|^2)^2$ are greater than or equal to $(|k| + 1)^2$. It is now straightforward to determine invertibility of the right side of (8.6), as long as α avoids the set of negative odd integers. Similar analysis of the antiholomorphic discrete series produces the condition that α should avoid the set of positive odd integers. We shall not go into details as to how to construct a microlocal parametrix for $\mathcal{L}_\alpha$ when (8.6) is uniformly invertible for large k.

The reader who recalls the treatment of other subelliptic operators in Chapters 1 and 2 will observe the ubiquity of the harmonic oscillator Hamiltonian. All these operators have symplectic double characteristics and general analyses, given, e.g., by Sjöstrand [**218**], Boutet de Monvel [**24**] or Taylor [**235**], apply, so this fact need not be an enormous surprise. Nevertheless, it is remarkable that such diverse treatments of these operators all lead one to the harmonic oscillator Hamiltonian.

CHAPTER 9

SL(2, C) and More General Lorentz Groups

As we saw in the last chapter, there is a double covering $\mathrm{SL}(2,R) \to \mathrm{SO}_e(2,1)$. This happy situation is repeated for the Lorentz group on four-dimensional Minkowski space. We have a double covering $\mathrm{SL}(2,\mathbf{C}) \to \mathrm{SO}_e(3,1)$, as will be shown in §1. Another happy coincidence which will be exploited heavily is that the complexification of the Lie algebra $\mathrm{sl}(2,\mathbf{C})$ is isomorphic to a direct sum of two copies of the complexification of $\mathrm{sl}(2,R)$. This will make it easy to understand the structure of the universal enveloping algebra of $\mathrm{sl}(2,\mathbf{C})$. In §2 we describe the irreducible unitary representations of $\mathrm{SL}(2,\mathbf{C})$. Our treatment is adapted from that of Gelfand, Minlos, and Shapiro [**73**], but emphasizes the role of the Casimir operators in $\mathfrak{U}(\mathrm{sl}(2,\mathbf{C}))$ and identities involving these operators. Further results on harmonic analysis on $\mathrm{SL}(2,\mathbf{C})$ are given in [**73**], and also in the books [**71, 74**] of Gelfand et al. §3 of this chapter discusses the structure and a little of the representation theory of more general Lorentz groups $\mathrm{SO}_e(n,1)$.

1. Introduction to $\mathrm{SL}(2,\mathbf{C})$. The group $\mathrm{SL}(2,\mathbf{C})$ is the group of all 2×2 complex matrices of determinant 1. Thus $\mathrm{SL}(2,\mathbf{C})$ is a six-dimensional connected Lie group. Its Lie algebra $\mathrm{sl}(2,\mathbf{C})$ consists of all 2×2 complex matrices of trace zero. The group $\mathrm{SL}(2,\mathbf{C})$ contains the compact subgroup $\mathrm{SU}(2)$, which plays a role in some ways analogous to that played by $\mathrm{SO}(2)$ in $\mathrm{SL}(2,R)$, since in both cases they are maximal compact subgroups, and in some ways different, since $\mathrm{SU}(2)$ is not commutative (and hence is not a "Cartan subgroup" of $\mathrm{SL}(2,\mathbf{C})$). As a basis for $\mathrm{sl}(2,\mathbf{C})$, we can take

$$
\begin{aligned}
&Z=\begin{pmatrix}0&-1\\1&0\end{pmatrix}, \quad A=\frac{1}{2}\begin{pmatrix}1&0\\0&-1\end{pmatrix}, \quad B=\frac{1}{2}\begin{pmatrix}0&1\\1&0\end{pmatrix},\\
&Z'=i\begin{pmatrix}0&-1\\1&0\end{pmatrix}, \quad A'=\frac{1}{2}i\begin{pmatrix}1&0\\0&-1\end{pmatrix}, \quad B'=\frac{1}{2}i\begin{pmatrix}0&1\\1&0\end{pmatrix},
\end{aligned}
\tag{1.1}
$$

and we have the commutation relations

$$
[Z,A]=2B, \qquad [Z,B]=-2A, \qquad [A,B]=-\tfrac{1}{2}Z, \tag{1.2}
$$

as in Chapter 8, and if one term on the left of these formulas is replaced by its prime, replace the right term by its prime, and if both terms on the left are replaced by their primes, leave the right side unprimed but change its sign. Of course, Z and Z' commute, as do A, A' and also B, B'.

The Lie algebra sl(2, C) possesses the structure of a complex Lie algebra, with $Z' = iZ$, etc. However, we are viewing SL(2, C) as a real Lie group, and sl(2, C) as a real Lie algebra, in which context such an identity is meaningless. Since we will also want to examine the complexification $\mathfrak{g}_{\mathbf{C}}$ of $\mathfrak{g} = \mathrm{sl}(2,\mathbf{C})$ (in $\mathfrak{g}_{\mathbf{C}}$, iZ is defined, but $iZ \neq Z'$), it is useful to represent sl(2, C) as an algebra of 4×4 real matrices. Indeed, replacing i by $\left(\begin{smallmatrix} 0 & -1 \\ 1 & 0 \end{smallmatrix}\right) = J$, we can replace (1.1) by

$$(1.3)\qquad \begin{aligned} Z &= \begin{pmatrix} 0 & -I \\ I & 0 \end{pmatrix}, & A &= \frac{1}{2}\begin{pmatrix} I & 0 \\ 0 & -I \end{pmatrix}, & B &= \frac{1}{2}\begin{pmatrix} 0 & I \\ I & 0 \end{pmatrix}, \\ Z' &= \begin{pmatrix} 0 & -J \\ J & 0 \end{pmatrix}, & A' &= \frac{1}{2}\begin{pmatrix} J & 0 \\ 0 & -J \end{pmatrix}, & B' &= \frac{1}{2}\begin{pmatrix} 0 & J \\ J & 0 \end{pmatrix}. \end{aligned}$$

The group SL(2, C) covers the Lorentz group $\mathrm{SO}_e(3,1)$. This can be seen as follows. Consider the set of selfadjoint matrices

$$(1.4)\qquad X = \begin{pmatrix} x_0 - x_3 & x_2 - ix_1 \\ x_2 + ix_1 & x_0 + x_3 \end{pmatrix}.$$

Then

$$(1.5)\qquad \det X = x_0^2 - x_1^2 - x_2^2 - x_3^2,$$

and if $g \in \mathrm{SL}(2,\mathbf{C})$, then g^*Xg is selfadjoint, provided X is, and $\det(g^*Xg) = \det X$. Since $g^*Xg = X$ for all such X if and only if $g = \pm I$, we have

$$(1.6)\qquad \mathrm{SO}_e(3,1) \approx \mathrm{SL}(2,\mathbf{C})/\{\pm I\}.$$

Note that $\mathrm{SO}_e(3,1)$ acts as a group of isometries of the three-dimensional hyperbolic space, which we can describe as the orbit in R^4 satisfying

$$(1.7)\qquad x_0^2 - x_1^2 - x_2^2 - x_3^2 = 1, \qquad x_0 > 0,$$

with the metric (of constant negative curvature), induced from the Minkowski metric (1.5) on R^4.

We now want to look a little at the structure of the complexified Lie algebra C sl(2, C), and its universal enveloping algebra. Clearly $\{Z, A, B\}$ generates $\mathrm{sl}(2, R)$ as a real Lie subalgebra of sl(2, C). We will see $\mathrm{Csl}(2, R)$ as a subalgebra of Csl(2, C) in various other ways. Let

$$(1.8)\qquad Z_\pm = Z \pm iZ', \qquad A_\pm = A \pm iA', \qquad B_\pm = B \pm iB'.$$

Then $\{Z_+, A_+, B_+\}$ and $\{Z_-, A_-, B_-\}$ span over $\mathbf{C}$ a pair of mutually commuting subalgebras, each isomorphic to $\mathrm{Csl}(2, R)$, in view of the easily verified commutation relations

$$(1.9)\qquad [Z_\pm, A_\pm] = 4B_\pm, \qquad [Z_\pm, B_\pm] = -4A_\pm, \qquad [A_\pm, B_\pm] = -Z_\pm,$$

with coherently chosen signs, and

$$[xZ_+ + yA_+ + zB_+, rZ_- + sA_- + tB_-] = 0. \tag{1.10}$$

Consequently, if we denote $\mathbf{C}\mathfrak{g}_+$ the linear span over $\mathbf{C}$ of $\{Z_+, A_+, B_+\}$, and define $\mathbf{C}\mathfrak{g}_-$ similarly,

$$\mathrm{Csl}(2, \mathbf{C}) = \mathbf{C}\mathfrak{g}_+ + \mathbf{C}\mathfrak{g}_- \approx \mathrm{Csl}(2, R) + \mathrm{Csl}(2, R). \tag{1.11}$$

In particular, the construction in Chapter 8 of the Casimir operator $\Box$ carries over to produce a pair of "Casimir operators,"

$$\Box_\pm = Z_\pm^2 - 4A_\pm^2 - 4B_\pm^2, \tag{1.12}$$

belonging to the center of the universal enveloping algebra $\mathfrak{U} = \mathfrak{U}(\mathrm{sl}(2, \mathbf{C}))$.

We can construct "raising" and "lowering" operators in $\mathbf{C}\mathfrak{g}_+$ and $\mathbf{C}\mathfrak{g}_-$, analogous to the elements X_+ and X_- in the complexified Lie algebra of $\mathrm{sl}(2, R)$ constructed in Chapter 8. Set

$$\begin{aligned} X_+^+ &= A_+ - iB_+, \qquad & X_-^+ &= A_+ + iB_+, \\ X_+^- &= A_- - iB_-, \qquad & X_-^- &= A_- + iB_-. \end{aligned} \tag{1.13}$$

We have

$$\begin{aligned} [Z_+, X_+^+] &= 4iX_+^+, \qquad & [Z_+, X_-^+] &= -4iX_-^+, \\ [Z_-, X_+^-] &= 4iX_+^-, \qquad & [Z_-, X_-^-] &= -4iX_-^-. \end{aligned} \tag{1.14}$$

As for commutation relations among $X_\pm^\pm$, we have

$$[X_+^+, X_-^+] = -2iZ_+, \qquad [X_-^-, X_+^-] = 2iZ_-. \tag{1.15}$$

Other commutation relations follow from the fact that everything in $\mathbf{C}\mathfrak{g}_+$ commutes with everything in $\mathbf{C}\mathfrak{g}_-$. In direct analogy with (1.27) of Chapter 8, we have

$$\Box_+ = Z_+^2 - 2(X_+^+X_-^+ + X_-^+X_+^+), \qquad \Box_- = Z_-^2 - 2(X_-^-X_+^- + X_+^-X_-^-). \tag{1.16}$$

In concert with (1.15), this yields

$$4X_+^+X_-^+ = Z_+^2 - 4iZ_+ - \Box_+, \qquad 4X_-^+X_+^+ = Z_+^2 + 4iZ_+ - \Box_+, \tag{1.17}$$

with analogous identities for $X_+^-X_-^-$ and $X_-^-X_+^-$, respectively, in terms of Z_- and $\Box_-$.

The identities (1.15) and (1.17) do not play the same role in the analysis of irreducible unitary representations of $\mathrm{SL}(2, \mathbf{C})$ as did the identities (2.6)–(2.8) of Chapter 8 in the analysis of $\mathrm{SL}(2, R)$, for the following reason. In (2.6)–(2.8) of Chapter 8, $\pi(Z)$ is skew adjoint and $\pi(X_+)^* = -\pi(X_-)$ on $C^\infty(\pi)$. The behavior of the adjoints of representations of $Z_\pm, X_\pm^\pm$, is quite different, so (1.17) does not lead immediately to identities for the operator norms of X_+^+ or X_-^+ on linear subspaces, for example.

In fact, define a conjugation on $\mathfrak{U}(\mathfrak{g})$ as follows. For $X + iY \in \mathbf{C}\mathfrak{g}$, $X, Y \in \mathfrak{g}$, set $(X + iY)^* = -X + iY$, and extend to $\mathfrak{U}(\mathfrak{g})$ so $(AB)^* = B^*A^*$. Then a unitary representation π of G gives a representation of $\mathfrak{U}(\mathfrak{g})$ such that $\pi(T^*) = \pi(T)^*$ on $C^\infty(\pi)$, for $T \in \mathfrak{U}(\mathfrak{g})$. With this definition of $^*: \mathfrak{U}(\mathfrak{g}) \to \mathfrak{U}(\mathfrak{g})$, note that

$$(1.18) \qquad Z_+^* = -Z_-, \qquad A_+^* = -A_-, \qquad B_+^* = -B_-,$$

so

$$(1.19) \qquad (X_+^+)^* = -X_-^-, \qquad (X_-^+)^* = -X_+^-.$$

Note also that

$$(1.20) \qquad \Box_+^* = \Box_-.$$

Now it is clear that, unlike (2.7) of Chapter 8, which is an expression for $-4R_+^*R_+$, the identities (1.17) do not constitute an expression for $-4(X_+^+)^*X_+^+$.

As in the case of SL(2, R), an important tool in the analysis of an irreducible unitary representation of SL(2, C) is the study of the decomposition of the restriction to a maximal compact subgroup K. For $G =$ SL(2, C), we take K to be SU(2). Note that a basis of the Lie algebra $\mathfrak{k} = \mathrm{su}(2)$ is given by Z, A', B'. Recalling our analysis of the representations of SU(2), we set

$$(1.21) \qquad \begin{aligned} R_+ &= A' - iB' = (X_+^+ - X_+^-)/2i, \\ R_- &= A' + iB' = (X_-^+ - X_-^-)/2i. \end{aligned}$$

Then $\{Z, R_+, R_-\}$ spans over $\mathbf{C}$ the complexification Csu(2), and we have

$$(1.22) \qquad [Z, R_+] = 2iR_+, \qquad [Z, R_-] = -2iR_-, \qquad [R_+, R_-] = iZ.$$

Note that

$$(1.23) \qquad R_+^* = -R_-.$$

Since we have noted the special properties of $X_\pm^\pm$, it is useful to understand their interactions with $R_\pm$. A straightforward calculation gives

$$(1.24) \qquad \begin{aligned} &[R_+, X_+^+] = 0, \quad [R_+, X_-^+] = -Z_+, \quad [R_+, X_+^-] = 0, \quad [R_+, X_-^-] = Z_-, \\ &[R_-, X_+^+] = Z_+, \quad [R_-, X_-^+] = 0, \quad [R_-, X_+^-] = -Z_-, \quad [R_-, X_-^-] = 0. \end{aligned}$$

To capture the essence of these commutator identities, it is convenient to supplement the set $\{Z, R_+, R_-\}$ with three more elements of Csl(2, C), to obtain a basis. This additional set should contain raising operators formed from $X_\pm^\pm$. In view of (1.21), symmetry suggests looking at

$$(1.25) \qquad Y_+ = X_+^+ + X_+^-, \qquad Y_- = X_-^+ + X_-^-,$$

and since we are loath to choose Z_+ or Z_-, we should throw in Z'. So we consider the basis

$$(1.26) \qquad \{Z, R_+, R_-, Z', Y_+, Y_-\}$$

of Csl(2, C). In addition to (1.22), by routine calculation, we have

$$(1.27) \qquad [Z', Y_+] = -2R_+, \qquad [Z', Y_-] = -2R_-, \qquad [Y_+, Y_-] = -4iZ,$$

and as for the interaction of the first three basis elements with the last three, we have

$$(1.28)\quad \begin{aligned} &[Z,Y_+]=2iY_+, && [Z,Y_-]=-2iY_-, && [Z,Z']=0,\\ &[R_+,Y_+]=0, && [R_+,Y_-]=-2iZ', && [R_+,Z']=iY_+,\\ &[R_-,Y_+]=2iZ', && [R_-,Y_-]=0, && [R_-,Z']=-iY_-. \end{aligned}$$

Note that, if $\mathfrak{a}$ denotes the complex linear span of $\{Z, R_+, R_-\}$, and $\mathfrak{b}$ that of $\{Z', Y_+, Y_-\}$, we have

$$(1.29)\quad [\mathfrak{a},\mathfrak{a}]\subset\mathfrak{a},\qquad [\mathfrak{a},\mathfrak{b}]\subset\mathfrak{b},\qquad [\mathfrak{b},\mathfrak{b}]\subset\mathfrak{a}.$$

Note that, as a consequence of (1.19),

$$(1.30)\quad Y_+^*=-Y_-.$$

In terms of the basis (1.26), the Casimir operators (1.12) are given by

(1.31)
$$\begin{aligned}\Box_+ &= (Z+iZ')^2-4i(Z+iZ')-Y_+Y_- -2iR_+Y_- -2iY_+R_- +4R_+R_-\\ &= (Z+iZ')^2+4i(Z+iZ')-Y_-Y_+ -2iR_-Y_+ -2iY_-R_+ +4R_-R_+,\end{aligned}$$

and

(1.32)
$$\begin{aligned}\Box_- &= (Z-iZ')^2-4i(Z-iZ')-Y_+Y_- +2iR_+Y_- +2iY_+R_- +4R_+R_-\\ &= (Z-iZ')^2+4i(Z-iZ')-Y_-Y_+ +2iR_-Y_+ +2iY_-R_+ +4R_-R_+,\end{aligned}$$

in view of (1.17) and the identities

$$(1.33)\quad \begin{aligned} 2X_+^+ &= Y_+ +2iR_+, & 2X_+^- &= Y_+ -2iR_+,\\ 2X_-^+ &= Y_- +2iR_-, & 2X_-^- &= Y_- -2iR_-.\end{aligned}$$

It is useful to separate $\Box_+$ and $\Box_-$ into symmetric and skew symmetric parts. Let

$$(1.34)\quad \Box_1=\Box_+ +\Box_-,\qquad \Box_2=(1/i)(\Box_+ -\Box_-),$$

so

$$(1.35)\quad \Box_1^*=\Box_1,\qquad \Box_2^*=\Box_2.$$

In view of the identities

$$(1.36)\quad Z_+^2+Z_-^2=2(Z^2-Z'^2),\qquad Z_+^2-Z_-^2=4iZZ',$$

we deduce from (1.31)–(1.32) that

$$(1.37)\quad \begin{aligned}\Box_1 &= 2(Z^2-Z'^2)-8iZ-2Y_+Y_- +8R_+R_-\\ &= 2(Z^2-Z'^2)+8iZ-2Y_-Y_+ +8R_-R_+,\end{aligned}$$

and

$$(1.38)\quad \begin{aligned}\Box_2 &= 4ZZ'-8iZ'-4R_+Y_- -4Y_+R_-\\ &= 4ZZ'+8iZ'-4R_-Y_+ -4Y_-R_+.\end{aligned}$$

Since we expect to decompose the restriction of a representation of SL(2, C) to $K = \mathrm{SU}(2)$, we should expect the element

$$K = Z^2 + 4A'^2 + 4B'^2 \tag{1.39}$$

of the center of the universal enveloping algebra of su(2), which is represented as a scalar on each K-irreducible representation space, to play an important role. In terms of the basis (1.26), we have

$$K = Z^2 - 2iZ + 4R_+R_- = Z^2 + 2iZ + 4R_-R_+. \tag{1.40}$$

Commutation relations of Y_+ and Y_- with K are not as simple as with Z, but they will exhibit Y_+ and Y_- as raising operators, with respect to K, at least on the hightest (resp., lowest) weight vectors for the SU(2) action. A short computation using (1.28) gives

$$[Y_+, K] = -4iY_+Z + 8Y_+ - 8iZ'R_+ \tag{1.41}$$

and

$$[Y_-, K] = 4iY_-Z + 8Y_- + 8iZ'R_-. \tag{1.42}$$

These identities are equivalent to

$$KY_+ = Y_+(K + 4iZ - 8) + 8iZ'R_+ \tag{1.43}$$

and

$$KY_- = Y_-(K - 4iZ - 8) - 8iZ'R_-. \tag{1.44}$$

We also note the following relation between K and the Casimir operator $\Box_1$:

$$\begin{aligned} \Box_1 &= 2K - 2Z'^2 - 2Y_-Y_+ + 4iZ \\ &= 2K - (2Z'^2 + Y_+Y_- + Y_-Y_+), \end{aligned} \tag{1.45}$$

which follows directly from the formulas (1.37) and (1.40). This expresses $\Box_1$ as a difference of two negative semidefinite operators. In the next section we will use this observation to obtain operator norm estimates which imply that certain vectors are analytic.

2. Representations of SL(2, C). We want to analyze the structure of a general irreducible unitary representation π of $G = \mathrm{SL}(2, \mathbf{C})$ on a Hilbert space H. We decompose H into spaces on which $K = \mathrm{SU}(2)$ acts as copies of various irreducible unitary representations of K:

$$H = \overline{\bigoplus_{l \geq 0} H_l} \tag{2.1}$$

where, on H_l, K acts as copies of its standard representation ρ_l on $\mathbf{C}^{2l+1} = V_l$ ($l \geq 0$ is an integer or half integer). We have

$$V_l = \bigoplus_{m=-l}^{l} V_{l,m} \tag{2.2}$$

where

$$\rho_l(Z) = 2im \quad \text{on } V_{l,m}. \tag{2.3}$$

For the sake of compactness, we will denote $\pi(Z)$ simply by Z, and similarly $\pi(T)$ by T, for other elements T of $\mathfrak{g} = \mathrm{sl}(2, \mathbf{C})$, or even its universal enveloping algebra. Eventually we will show that each $H_{l,m}$ consists of C^∞, even analytic, vectors, and also has dimension one, if not zero, but provisionally let us set

$$H_{l,m}^0 = H_{l,m} \cap C^\infty(\pi), \qquad H_l^0 = H_l \cap C^\infty(\pi). \tag{2.6}$$

It is clear that $H_{l,m}^0$ is dense in $H_{l,m}$, and that $H_{l,m} \cap H_l^0 = H_{l,m}^0$.

The nature of the action of K, and hence of the operators Z, R_+, R_-, on H_l, has been thoroughly discussed in Chapter 2. In addition to (2.3), which implies $Z = 2im$ on $H_{l,m}$, we have

$$H_{l,-l} \underset{R_-}{\overset{R_+}{\rightleftarrows}} H_{l,-l+1} \underset{R_-}{\overset{R_+}{\rightleftarrows}} \cdots \underset{R_-}{\overset{R_+}{\rightleftarrows}} H_{l,l-1} \underset{R_-}{\overset{R_+}{\rightleftarrows}} H_{l,l}. \tag{2.7}$$

The structure of R_+ and R_- is revealed in the following identity for the action of $\mathcal{K}$, defined by (1.39)–(1.40):

$$\mathcal{K} = Z^2 - 2iZ + 4R_+R_- = Z^2 + 2iZ + 4R_-R_+. \tag{2.8}$$

Applying $\mathcal{K}$ to an element of $H_{l,l}$, knowing that $\mathcal{K}$ acts as a scalar on H_l, we have

$$\mathcal{K}u = -2l(2l+2)u \quad \text{for } u \in H_l^0 \tag{2.9}$$

and hence

$$\begin{aligned} 4R_-R_+u &= [-2l(2l+2) + 2m(2m+2)]u, \\ 4R_+R_-u &= [-2l(2l+2) + 2m(2m-2)]u, \end{aligned} \qquad \text{for } u \in H_{l,m}^0. \tag{2.10}$$

The next task confronting us is to analyze the action of Z', Y_+, and Y_- on these spaces. First we see how Y_+ ties together various highest weight vectors for the K-action. Let

$$W_l^+ = H_{l,l}^0 \tag{2.11}$$

denote the space of (smooth) highest weight vectors for the action of K on H_l^0:

$$H_{l,l}^0 = \{u \in H_l^0 : R_+u = 0\}. \tag{2.12}$$

Similarly, let $W_l^- = H_{l,-l}^0 = \{u \in H_l^0 : R_-u = 0\}$.

LEMMA 2.1. *We have*

$$Y_+ : W_l^+ \to W_{l+1}^+, \tag{2.13}$$

and similarly

$$Y_- : W_l^- \to W_{l+1}^-. \tag{2.14}$$

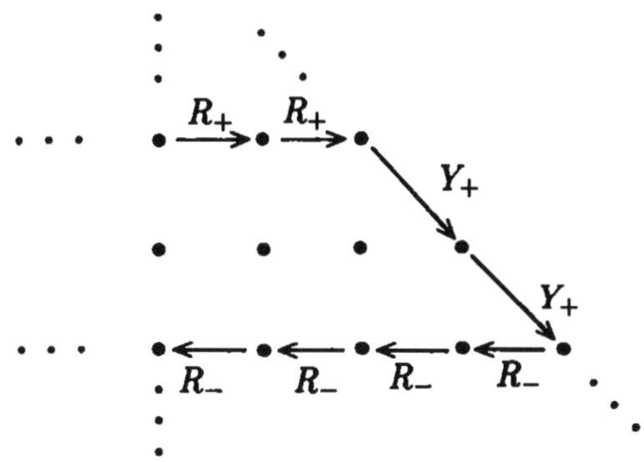

FIGURE 2.1

PROOF. To say $v \in W_l^+$ is to say that $v \in C^\infty(\pi)$ and that

$$R_+ v = 0, \qquad Zv = 2ilv, \qquad \mathcal{K}v = -2l(2l+2)v. \tag{2.15}$$

Since $[Y_+, R_+] = 0$, by (1.28), we see that, given (2.15),

$$R_+(Y_+ v) = 0. \tag{2.16}$$

Using the commutation relation $[Z, Y_+] = 2iY_+$, from (1.28), we have

$$Z(Y_+ v) = Y_+ Zv + 2iY_+ v = 2i(l+1)(Y_+ v). \tag{2.17}$$

To analyze $\mathcal{K}(Y_+ v)$, use (1.43). Since $R_+ v = 0$, we have

$$\begin{aligned} \mathcal{K}(Y_+ v) &= Y_+(\mathcal{K} + 4iZ - 8)v \\ &= Y_+(-4l^2 - 4l - 8l - 8)v \\ &= -(2l+2)(2l+4)Y_+ v. \end{aligned} \tag{2.18}$$

These calculations, in fact any two of the three identities (2.16)–(2.18), prove the lemma for Y_+, and the proof of Y_- is similar.

Let us schematically indicate our situation as follows. Each dot will represent some space $H_{l,m}$; adjacent horizontal dots are connected by R_+ and R_-, and span H_l; l increases as you go down (see Figure 2.1).

We next show that the norm of $Y_+ : W_l^+ \to W_{l+1}^+$ is determined, for each l, by the values of the Casimir operators under the representation. Irreducibility of π implies that, for some real scalars μ_1 and μ_2,

$$\Box_1 u = \mu_1 u, \quad \Box_2 u = \mu_2 u, \qquad u \in C^\infty(\pi). \tag{2.19}$$

The identities (1.37), (1.38) for the Casimir operator lead to the following two identities:

$$\mu_1 u = 2(Z^2 - Z'^2)u + 8iZu - 2Y_- Y_+ u + 8R_- R_+ u, \tag{2.20}$$

$$\mu_2 u = 4ZZ'u + 8iZ'u - 4R_- Y_+ u - 4Y_- R_+ u, \tag{2.21}$$

for $u \in C^{\infty}(\pi)$. For $v \in W_l^+$, these specialize to

$$2Z'^2v = (-\mu_1 + 2Z^2 + 8iZ)v - 2Y_-Y_+v, \qquad v \in W_l^+, \tag{2.22}$$

and

$$4(Z + 2i)Z'v = \mu_2 v + 4R_-Y_+v, \qquad v \in W_l^+. \tag{2.23}$$

Since $Zv = 2ilv$ for $v \in W_l^+$, we can write equivalently

$$2Z'^2v = -(\mu_1 + 8l^2 + 16l)v - 2Y_-Y_+v, \qquad v \in W_l^+, \tag{2.24}$$

$$8i(l + 1)Z'v = \mu_2 v + 4R_-Y_+v, \qquad v \in W_l^+. \tag{2.25}$$

These two identities will simultaneously determine $\|Z'v\|$ and $\|Y_+v\|$. If we take the inner product of both sides of (2.24) with v we obtain

$$2\|Z'v\|^2 = (\mu_1 + 8l^2 + 16l)\|v\|^2 - 2\|Y_+v\|^2, \tag{2.26}$$

while if we note that the right side of (2.25) is an orthogonal decomposition, we have

$$64(l + 1)^2\|Z'v\|^2 = \mu_2^2\|v\|^2 + 32(l + 1)\|Y_+v\|^2 \tag{2.27}$$

in view of the fact that $\|R_-\xi\|^2 = (2l + 2)\|\xi\|^2$ for $\xi \in W_{l+1}^+$, which is a consequence of (2.10). Comparing (2.26) and (2.27), we obtain

$$32[2(l + 1)^2 + l + 1]\|Y_+v\|^2 = [32(l + 1)^2(\mu_1 + 8l^2 + 16l) - \mu_2^2]\|v\|^2, \tag{2.28}$$

for $v \in W_l^+$. Note that the coefficient of $\|Y_+v\|^2$ is > 0, in fact ≥ 96, for all $l \geq 0$. Thus the coefficient of $\|v\|^2$ must be ≥ 0 for all l such that $H_l \neq 0$, i.e.,

$$32(l + 1)^2(\mu_1 + 8l^2 + 16l) - \mu_2^2 \geq 0 \quad \text{for all } l \text{ such that } H_l \neq 0. \tag{2.29}$$

We also have the formula for $\|Z'v\|^2$:

$$16[4(l + 1)^2 + 2(l + 1)]\|Z'v\|^2 = [16(l + 1)(\mu_1 + 8l^2 + 16l) + \mu_2^2]\|v\|^2, \tag{2.30}$$

for $v \in W_l^+$.

Suppose l_0 is the smallest number (integer or half-integer) such that H_{l_0} in (2.1) is nonzero. We say ρ_{l_0} is the *lowest K-type* of $K = \mathrm{SU}(2)$ occurring in the representation π of G. Pick a unit vector $e_{l_0,l_0} \in H_{l_0,l_0}$. Let $e_{l_0,m} \in H_{l_0,m}$ be the image of e_{l_0,l_0} under repeated powers of R_-, normalized by positive scalars so $\|e_{l_0,m}\| = 1$. Let $e_{l_0+j,l_0+j} \in H_{l_0+j,l_0+j}$ denote the image of e_{l_0,l_0} under Y_+^j, normalized, so

$$(\eta_{l_0+j-1} \cdots \eta_{l_0})e_{l_0+j,l_0+j} = Y_+^j e_{l_0,l_0}, \tag{2.32}$$

where $\eta_{l_0+k} \geq 0$ is defined by $\|Y_+v\| = \eta_{l_0+k}\|v\|$, $v \in W_{l_0+k}^+$, which in turn is defined by (2.28). If $\eta_{l_0+k} = 0$, pick $e_{l_0+k,l_0+k} \in H_{l_0+k}^0$ arbitrarily, with unit norm; shortly we will show that this possibility cannot arise. Then let e_{l_0+j-k,l_0+j}, $0 \leq k \leq l_0 + j$, denote the image of e_{l_0+j,l_0+j} under R_-^k, normalized by multiplying by a positive quantity so that $\|e_{l_0+j-k,l_0+j}\| = 1$. Let

$$\mathcal{H} = \bigoplus_{l,m} \mathcal{H}_{l,m}; \qquad \mathcal{H}_{l,m} = \text{linear span of } \{e_{l,m}\}. \tag{2.33}$$

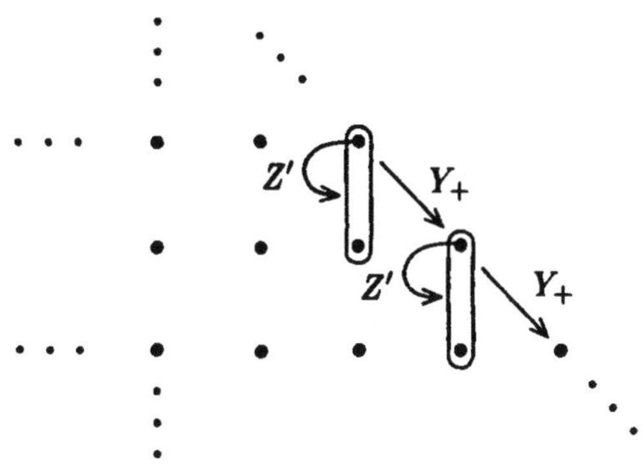

FIGURE 2.2

We will eventually show $\mathcal{H} = \bigoplus_{l,m} H_{l,m} = \bigoplus_l H_l$. We first specify the action of $\mathfrak{g}$ on $\mathcal{H}$; the action of $\{Z, R_+, R_-\}$ on $\mathcal{H}$ already being given as a consequence of (2.5)–(2.10). From now on, the dots in such figures as Figure 2.1 stand for $\mathcal{H}_{l,m}$. So far we have Y_+ on $e_{l,l}$:

$$Y_+e_{ll} = \eta_l e_{l+1,l+1}, \tag{2.34}$$

where $\eta_l \geq 0$, and, by (2.28),

$$\eta_l^2 = [32(l+1)^2(\mu_1 + 8l^2 + 16l) - \mu_2^2]/32[2(l+1)^2 + l + 1]. \tag{2.35}$$

In other words, we have defined Y_+ on $\mathcal{H} \cap \ker R_+$. Furthermore, the identity (2.25) specifies Z' on $\mathcal{H} \cap \ker R_+$. We have

$$\begin{aligned} 8i(l+1)Z'e_{ll} &= \mu_2 e_{ll} + 4R_-Y_+e_{ll} \\ &= \mu_2 e_{ll} + 4\sqrt{2l+2}\eta_l e_{l+1,l}, \end{aligned} \tag{2.36}$$

or

$$Z'e_{ll} = \alpha_l e_{ll} + \beta_l e_{l+1,l}, \quad \alpha_l = \mu_2/8i(l+1), \quad \beta_l = \eta_l/i\sqrt{2l+2}. \tag{2.37}$$

In Figure 2.2, we schematically indicate the operators Y_+ and Z' specified, so far, on $\ker R_+$.

Our next objective will be to specify Z' on all of $\mathcal{H}$. To this end, following [73], we will use the identity

$$[R_+, [R_-, Z']] = -2Z', \tag{2.38}$$

a consequence of (1.28), which is equivalent to

$$R_+Z'R_- + R_-Z'R_+ = R_+R_-Z' + Z'R_-R_+ + 2Z'. \tag{2.39}$$

This will provide a three-term recurrence relation, for $Z'e_{l,m-1}$ in terms of $Z'e_{lm}$ and $Z'e_{l,m+1}$. We will use this in concert with the identity

$$[Z, Z'] = 0, \tag{2.40}$$

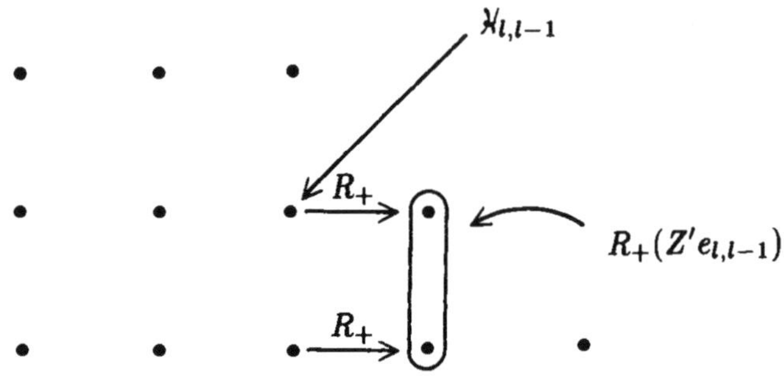

FIGURE 2.3

which implies that

$$Z(Z'e_{lm}) = 2imZ'e_{lm}. \tag{2.41}$$

We first apply (2.39) to analyze $Z'e_{l,l-1}$; applying both sides of (2.39) to e_{ll} and using

$$R_- e_{ll} = \sqrt{2l}e_{l,l-1},$$

we have

$$\sqrt{2l}R_+Z'e_{l,l-1} = R_+R_-Z'e_{ll} + 2Z'e_{ll}, \tag{2.42}$$

into which we substitute (2.36). Using

$$R_+R_-e_{ll} = -2le_{ll},$$
$$R_+R_-e_{l+1,l} = -2(2l+1)e_{l+1,l},$$

from (2.10), we have an explicit formula for $R_+(Z'e_{l,l-1})$:

$$R_+Z'e_{l,l-1} = -(2(l-2)/\sqrt{2l})\alpha_l e_{ll} - 2\beta_l\sqrt{2l}e_{l+1,l}, \tag{2.43}$$

where α_l and β_l are defined in (2.37). We indicate our situation in Figure 2.3.

Note that $R_+(Z'e_{l,l-1}) \in \mathcal{H}_{ll} \oplus \mathcal{H}_{l+1,l}$. Now R_+ is injective on each H^0_{lm} except the highest weight spaces H^0_{ll}. In view of (2.41), we see that (2.43) uniquely determines $Z'e_{l-1,l}$, modulo an element of $H^0_{l-1,l-1}$. We want to identify this element, and show that it actually belongs to $\mathcal{H}_{l-1,l-1}$, i.e., is a multiple of $e_{l-1,l-1}$.

Specifying the component of $Z'e_{l,l-1}$ in $\mathcal{H}_{l-1,l-1}$ is easy; skew symmetry of Z' implies

$$(Z'e_{l,l-1}, e_{l-1,l-1}) = -(e_{l,l-1}, Z'e_{l-1,l-1}), \tag{2.44}$$

and the right side of (2.44) is defined by (2.36), with l replaced by $l-1$:

$$8ilZ'e_{l-1,l-1} = \mu_2 e_{l-1,l-1} + 4\sqrt{2l}\eta_{l-1}e_{l,l-1},$$

so the right side of (2.44) is equal to $-i\eta_{l-1}/\sqrt{2l}$. On the other hand, suppose $f_{l-1,l-1} \in H^0_{l-1,l-1}$ is orthogonal to $e_{l-1,l-1}$. We have

$$(Z'e_{l,l-1}, f_{l-1,l-1}) = -(e_{l,l-1}, Z'f_{l-1,l-1}). \tag{2.45}$$

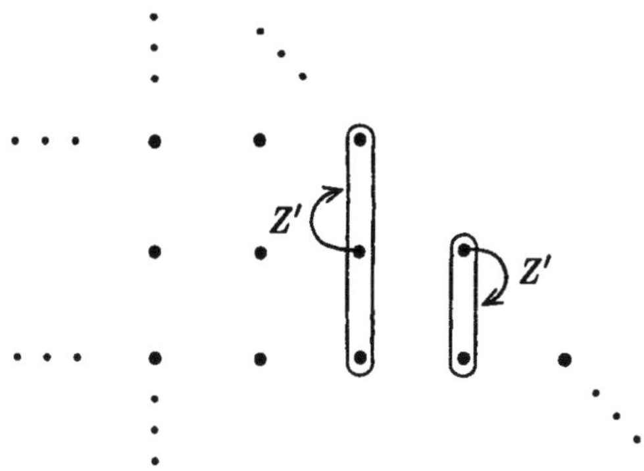

FIGURE 2.4

Now $Z'f_{l-1,l-1}$ is analyzed as in (2.36), and via (2.28) is seen to be orthogonal to $e_{l,l-1}$, so (2.45) vanishes. This proves

$$Z'\colon \mathcal{H}_{l,l-1} \to \mathcal{H}_{l-1,l-1} \oplus \mathcal{H}_{l,l-1} \oplus \mathcal{H}_{l+1,l-1}. \tag{2.46}$$

In Figure 2.4, we record schematically our progress in understanding Z'.

From here, for each $m \leq l-2$, the recurrence (2.39) uniquely determines the component of $Z'e_{lm}$ in $\mathcal{H}_{l-1,m} \oplus \mathcal{H}_{lm} \oplus \mathcal{H}_{l+1,m}$. We claim this is everything, i.e.,

$$Z'\colon \mathcal{H}_{lm} \to \mathcal{H}_{l-1,m} \oplus \mathcal{H}_{lm} \oplus \mathcal{H}_{l+1,m}. \tag{2.47}$$

Indeed, any extra components of $Z'e_{lm}$ must belong to $\bigoplus_{\lambda\leq l-2} H^0_\lambda$. The analysis we have so far shows $Z'(H^0_l) \perp H^0_\mu$ for $\mu \geq l+2$, and so skew-symmetry of Z' implies $Z'(H^0_l) \perp H^0_\lambda$ for $\lambda \leq l-2$. This proves (2.47) and shows that Z' has been uniquely determined on $\mathcal{H}$. Then, Y_+ and Y_- are uniquely specified on $\mathcal{H}$ by the identities

$$Y_+ = -i[R_+, Z'], \qquad Y_- = i[R_-, Z'], \tag{2.48}$$

from (1.28). Note that, by (2.47),

$$Y_+\colon \mathcal{H}_{lm} \to \mathcal{H}_{l-1,m+1} \oplus \mathcal{H}_{l,m+1} \oplus \mathcal{H}_{l+1,m+1} \tag{2.49}$$

and

$$Y_-\colon \mathcal{H}_{lm} \to \mathcal{H}_{l-1,m-1} \oplus \mathcal{H}_{l,m-1} \oplus \mathcal{H}_{l+1,m-1}. \tag{2.50}$$

Thus, the complexified Lie algebra $\mathrm{Csl}(2,\mathbf{C})$ acts on $\mathcal{H}$.

LEMMA 2.2. *Each $e_{lm} \in \mathcal{H}_{lm}$ is an analytic vector for the action of* $\mathrm{Csl}(2,\mathbf{C})$ *induced by π.*

PROOF. This follows from estimates of operator norms of $R_\pm, Z, Y_\pm, Z'$ on the spaces $\mathcal{H}_{lm}$. We know the norms of $R_\pm$ and Z. To estimate those of $Y_\pm$ and Z', we use the following consequence of (1.45):

$$-2Z'^2 - (Y_+Y_- + Y_-Y_+) = \mu_1 - 2\mathcal{K} = \mu_1 + 8l(l+1). \tag{2.51}$$

Applying this to e_{lm} and taking the inner product with e_{lm}, using $Z'^* = -Z'$ and $Y_+^* = -Y_-$, we have

$$2\|Z'e_{lm}\|^2 + \|Y_-e_{lm}\|^2 + \|Y_+e_{lm}\|^2 = \mu_1 + 8l(l+1), \tag{2.52}$$

which provides adequate estimates to prove analyticity. Details are left to the reader.

Since power series expansions for $\exp(s_1Z+s_2A+s_3B+s_4Z'+s_5A'+s_6B')e_{lm}$ are consequently convergent for $\sum |s_j|^2$ small, it follows that $\overline{\mathcal{H}}$, the closure of $\mathcal{H}$ in H, is invariant under π. If π is irreducible, this implies $\overline{\mathcal{H}} = H$, so we have proved

$$\mathcal{H}_{lm} = H^0_{lm} = H_{lm}, \tag{2.53}$$

and in particular

$$\begin{aligned} &\dim H_{lm} = 1 \quad \text{if } H_{lm} \neq 0, \\ &\dim H_l = 2l+1 \quad \text{if } H_l \neq 0. \end{aligned} \tag{2.54}$$

We see that $H_l \neq 0$ only for integers (resp., nonintegral half-integers) $\geq l_0$, if l_0 is an integer (resp., nonintegral half-integer). Furthermore, if π is not the trivial representation, $H_l \neq 0$ for each such l. For if not, say if a possibility l_1 is excluded, then $\bigoplus_{l_0\leq l\leq l_1-1} H_l$ is a (finite-dimensional) proper invariant subspace for π, which would have to be all of H. But an argument as in Chapter 8 shows that SL(2, C) cannot have any nontrivial finite-dimensional irreducible unitary representation, so this possibility is excluded.

In our analysis so far, we have not excluded the possibility that some $\eta_{l_0+k} = 0$, i.e., $Y_+e_{ll} = 0$ for some $l = l_0 + k$. We now take care of this.

LEMMA 2.3. *If l_0 is minimal such that $H_{l_0} \neq 0$, $k \in \mathbf{Z}^+ \cup \{0\} = \{0, 1, 2, \ldots\}$, $l = l_0 + k$, then*

$$Y_+e_{ll} \neq 0, \tag{2.55}$$

i.e., Y_+e_{ll} is a nonzero multiple of $e_{l+1,l+1}$.

PROOF. If $\eta_l = 0$, then, by (2.36), $Z'e_{ll} \perp e_{l+1,l}$, so, by skew-symmetry, $Z'e_{l+1,l} \perp e_{ll}$. Hence Z' maps both $e_{l+1,l+1}$ and $e_{l+1,l}$ to $H_{l+1} \oplus H_{l+2}$ in such a case, and hence, by (2.48), $Y_+, Y_-: H_{l+1,l+1} \to H_{l+1} \oplus H_{l+2}$. Inductively, one has that the Lie algebra action leaves $\bigoplus_{\lambda\geq l+1} H_\lambda$ invariant, so the G-action must leave its closure invariant. If π is irreducible, this is impossible, so the lemma is proved.

Consequently, the inequality (2.29) can be strengthened to a strict inequality:

$$32(l+1)^2(\mu_1 + 8l^2 + 16l) - \mu_2^2 > 0 \quad \text{for } l = l_0 + k,\ k = 0, 1, 2, \ldots,$$

which in turn is equivalent to its special case

$$32(l_0+1)^2(\mu_1 + 8l_0^2 + 16l_0) - \mu_2^2 > 0. \tag{2.56}$$

Thus the action of Csl(2, C) on H defined by a given irreducible unitary representation of SL(2, C) is determined by the action of $\{Z, R_\pm, Z', Y_\pm\}$ on e_{lm}

for $l = l_0 + k$, $k = 0, 1, 2, \ldots$, $m \in \{-l, -l+1, \ldots, l-1, l\}$. This in turn has been specified in terms of $\{l_0, \mu_1, \mu_2\}$ in the argument presented above. There is one further analysis to be made, giving μ_2 in terms of l_0 and μ_1. Before we get to this, we make some comments on specific formulas for the action of these Lie algebra elements on e_{lm}. Of course, we have

$$Ze_{lm} = 2ime_{lm}, \tag{2.57}$$

and, by (2.10), we can write

$$R_+e_{lm} = \rho^+_{lm}e_{l,m+1}, \qquad R_-e_{lm} = \rho^-_{lm}e_{l,m-1} \tag{2.58}$$

with

$$\rho^+_{lm} = -\sqrt{l(l+1) - m(m+1)}, \qquad \rho^-_{lm} = \sqrt{l(l+1) - m(m-1)}. \tag{2.59}$$

To start the recursion for the action of Y_+ and Z', recall that

$$Y_+e_{ll} = \eta_l e_{l+1,l+1}, \tag{2.60}$$

where $\eta_l > 0$ satisfies (2.37), and, by (2.36),

$$Z'e_{ll} = \alpha_l e_{ll} + \beta_l e_{l+1,l} \tag{2.61}$$

where

$$\alpha_l = -i\mu_2/8(l+1), \qquad \beta_l = -i\eta_l/\sqrt{2l+2}. \tag{2.62}$$

The analysis of Z' on $\mathcal{H}_{l,l-1}$ discussed above yields the formula

$$Z'e_{l,l-1} = -\beta_{l-1}e_{l-1,l-1} + \sigma_l e_{l,l-1} + \tau_l e_{l+1,l-1} \tag{2.63}$$

for $l \geq l_0 + 1$, where β_{l-1} is defined by (2.62), and

$$\sigma_l = \alpha_l(l-2)/l, \qquad \tau_l = 2\beta_l\sqrt{2l}/\sqrt{4l+2}. \tag{2.64}$$

We have

$$Z'e_{l_0,l_0-1} = \sigma_{l_0}e_{l_0,l_0-1} + \tau_{l_0}e_{l_0+1,l_0-1}. \tag{2.65}$$

These formulas determine $Y_-e_{ll} = i[R_-, Z']e_{ll}$; we have

$$\begin{aligned} Y_-e_{ll} = & -i\rho^-_{ll}\beta_{l-1}e_{l-1,l-1} + i\rho^-_{ll}(\alpha_l - \sigma_l)e_{l,l-1} \\ & + i(\beta_l\rho^-_{l+1,l} + \tau_l\rho^-_{ll})e_{l+1,l-1}. \end{aligned} \tag{2.66}$$

Note that $-i\rho^-_{ll}\beta_{l-1} = -\eta_{l-1}$. Again, for $l = l_0$, the first term on the right is taken to be 0. If $l = l_0 = 0$, then of course σ_0 and τ_0 are to be replaced by 0, in both (2.65) and (2.66); one would have

$$Y_-e_{00} = i\beta_0\rho^-_{10}e_{1,-1} = \eta_0 e_{1,-1}, \tag{2.67}$$

which parallels the identity $Y_+e_{00} = \eta_0 e_{11}$, from (2.60).

Note that, in the event that $l_0 = 0$, if we apply the identities $-2iZ' = [R_+, Y_-]$ and $2iZ' = [R_-, Y_+]$ to e_{00}, we obtain $2iZ'e_{00} = R_-Y_+e_{00} = \eta_0\rho^-_{11}e_{10}$, and $-2iZ'e_{00} = R_+Y_-e_{00} = \eta_0\rho^+_{1,-1}e_{10}$; in either case, $Z'e_{00} = -i(\eta_0/\sqrt{2})e_{10}$.

Comparing the formula (2.61) for $l = 0$ yields $\alpha_0 = 0$. Since, by (2.62), $\alpha_0 = -i\mu_2/8$, we have established

$$l_0 = 0 \Rightarrow \mu_2 = 0. \tag{2.68}$$

This is the first case of a constraint on μ_2 implied by a specification of l_0; the constraint for general $l_0 > 0$ is given by the following result. Define η_l^2, for any real $l > -1$, by the formula (2.35).

LEMMA 2.4. *If the lowest K-type of π is the representation on H_{l_0}, then*

$$\eta_{l_0-1}^2 = 0. \tag{2.69}$$

PROOF. We will deduce this by examining the identity $[Y_+, Y_-] = -4iZ$, applied to e_{ll}, for $l \geq l_0$:

$$[Y_+, Y_-]e_{ll} = 8le_{ll}. \tag{2.70}$$

For $l > l_0$, the left side, $(Y_+Y_- - Y_-Y_+)e_{ll}$, is a sum of four terms:

(2.71) component of e_{ll} in $Y_-(-\eta_l e_{l+1,l+1})$,

(2.72) component of e_{ll} in $Y_+(e_{l+1,l-1}$ component of $Y_-e_{ll})$,

(2.73) component of e_{ll} in $Y_+(e_{l,l-1}$ component of $Y_-e_{ll})$,

(2.74) component of e_{ll} in $Y_+(e_{l-1,l-1}$ component of $Y_-e_{ll})$.

All these coefficients are algebraic functions of l, and they sum to $8l$ for each $l = l_0 + k$, $k \in \{1, 2, 3, \ldots\}$. The fourth term is equal to $-\eta_{l-1}^2$. For $l = l_0$, the left side of (2.70) is the sum (2.71)–(2.73), with (2.74) omitted. Now, for $l = l_0$, the sum (2.71)–(2.74) must also continue to equal $8l_0$ if (2.74) is replaced by $-\eta_{l_0-1}^2$. The only way this can happen is for the conclusion (2.60) to hold.

The identity (2.69) is equivalent to

$$\mu_2^2 = 32l_0^2(\mu_1 + 8l_0^2 - 8). \tag{2.75}$$

This is also consistent with the result (2.68) derived in case $l_0 = 0$. If $l_0 > 0$, it is convenient to set

$$s^2 = (\mu_1 + 8l_0^2 - 8)/8, \tag{2.76}$$

which is ≥ 0. Then we can write μ_1 and μ_2 as

$$\mu_1 = 8(s^2 + 1 - l_0^2), \tag{2.77}$$

$$\mu_2 = 16l_0 s. \tag{2.78}$$

As we will see shortly, for any choice of $s \in R$ and $l_0 \in \{0, \frac{1}{2}, 1, \frac{3}{2}, \ldots\}$, this produces members of the *principal series* of representations of SL(2, **C**). In case $l_0 = 0$, when μ_1, μ_2 satisfy (2.77)–(2.78), the associated representation is also a member of the principal series. But the inequality (2.56) only requires $\mu_1 > 0$ in this case, whereas (2.77), with s real, requires $\mu_1 \geq 8$. We also have irreducible unitary representations such that (2.77)–(2.78) hold, with $l_0 = \mu_2 = 0$, and $s = it$, $t \in (-1, 1)$, called representations of the *complementary series*.

Whenever this condition on μ_2 holds, the construction above based on $\{l_0, \mu_1, \mu_2\}$ yields an irreducible unitary representation of SL(2, C). The fact that it yields a Lie algebra representation of sl(2, R) by skew-symmetric operators could be checked by explicitly solving the recursion formula (2.29) for $Z'e_{lm}$; such explicit formulas are given in [**73**]. Rather than produce such formulas here, we will be content with the realizations of the principal and supplementary series given below, which establish the existence of all irreducible unitary representations described by the parametrization $\{l_0, \mu_1, \mu_2\}$ above, subject to (2.77)–(2.78), for $s \in R$ or $is \in (-1, 1)$. We state the result on the classification of the irreducible unitary representations of SL(2, C), first derived by Gelfand et al. [**74**].

THEOREM 2.5. *Each nontrivial unitary irreducible representation of the group* SL(2, C) *is equivalent to one of the following sort:*

$$\text{(2.79)} \qquad \textit{Principal series: } \pi_{l_0,s},$$

where $l_0 \in \{0, \frac{1}{2}, 1, \frac{3}{2}, \ldots\}$, $s \in R$. *This is determined by the condition that* (2.77)–(2.78) *hold.*

$$\text{(2.80)} \qquad \textit{Complementary series: } \pi_{0,it} \qquad (t \neq 0),$$

where $t \in (-1, 1)$; $\mu_2 = 0$ *and* (2.77) *holds with* $s = it$. *These representations are mutually inequivalent, except* $\pi_{0,s} \approx \pi_{0,-s}$ *and* $\pi_{0,it} \approx \pi_{0,-it}$, *and all irreducible.*

The mutual inequivalence is clear since for different parameters the triples $\{l_0, \mu_1, \mu_2\}$ differ. Other than the realizations of these representations, which we will undertake shortly, the only point of the theorem which remains to be established is the irreducibility of each of these representations. This can be accomplished in the same fahion as the proof of irreducibility for SL(2, R), in Theorem 2.2 of Chapter 8. Namely, by (2.54), each G-invariant subspace of $\overline{\bigoplus_{l \geq l_0} H_l}$ must be a direct sum of certain H_l's, so if the representation $\pi = \pi_{l_0,s}$ of $G = \mathrm{SL}(2, \mathbf{C})$ were not irreducible, there must be an identity of the form

$$\text{(2.81)} \qquad (\pi(g)e_{lm}, e_{l'm'}) = 0 \quad \text{for all } g \in \mathrm{SL}(2, \mathbf{C}),$$

for some pair $e_{lm}, e_{l'm'}$, which in turn implies

$$\text{(2.82)} \qquad (Te_{lm}, e_{l'm'}) = 0 \quad \text{for all } T \in \mathfrak{A}(\mathrm{sl}(2, \mathbf{C})).$$

Setting T equal to an appropriate product of powers of R_+, Y_+, and R_- produces a contradiction, thus proving irreducibility.

We turn to the construction of realizations of the representations in the principal series. In parallel with the case for SL(2, R), we can construct these by decomposing the regular representation of SL(2, C) on $L^2(\mathbf{C}^2) = L^2(R^4)$:

$$\text{(2.83)} \qquad R(g)f(z) = f(g^t z), \qquad g \in \mathrm{SL}(2, \mathbf{C}),\ z \in \mathbf{C}^2.$$

In this case, $R(g)$ commutes with the group of complex dilations:

$$\text{(2.84)} \qquad D(a)f(z) = |a|^2 f(az), \qquad a \in \mathbf{C}^* = \mathbf{C} \setminus \{0\}.$$

We expect to decompose R into irreducibles via the spectral decomposition of D. Since $\mathbf{C}^* = S^1 \times R^+$, this decomposition is accomplished by combining Fourier series and the Mellin transform. Thus, given $\beta(\theta, t)$ on $S^1 \times R^+$, set

$$\mathfrak{s}\beta(n,s) = \int_0^\infty \int_{S^1} \beta(\theta,t) e^{-in\theta} t^{-is} \, d\theta \, dt, \tag{2.85}$$

for $n \in \mathbf{Z}$, $s \in R$. Then we get the inversion formula

$$\beta(\theta,t) = (2\pi)^{-2} \sum_{n=-\infty}^{\infty} \int_{-\infty}^{\infty} \mathfrak{s}\beta(n,s) e^{in\theta} t^{is-1} \, ds, \tag{2.86}$$

and the Plancherel formula

$$\int_{R^+} \int_{S^1} |\beta(\theta,t)|^2 t \, d\theta \, dt = (2\pi)^{-2} \sum_{n=-\infty}^{\infty} \int_{-\infty}^{\infty} |\mathfrak{s}\beta(n,s)|^2 \, ds. \tag{2.87}$$

Thus, if we define

$$P_{n,s} f(z) = \int_0^\infty \int_{S^1} f(re^{i\theta} z) e^{-in\theta} r^{-is} \, d\theta \, dr, \tag{2.88}$$

we see that, for $f \in C_0^\infty(R^4)$, $P_{n,s} f$ belongs to the space

$$\mathcal{H}_{n,s} = \Big\{ g \in L^2_{\mathrm{loc}}(R^4 \setminus 0) : g(re^{i\theta} z) = r^{is-1} e^{in\theta} g(z), \quad r > 0,\ e^{i\theta} \in S^1, \text{ and } \int_{S^3} |g|^2 < \infty \Big\}. \tag{2.89}$$

We make $\mathcal{H}_{n,s}$ into a Hilbert space, with norm square $\int_{S^3} |g|^2$; note that the homogeneity condition

$$g(re^{i\theta} z) = r^{is-1} e^{in\theta} g(z) \tag{2.90}$$

makes $g \in \mathcal{H}_{n,s}$ uniquely determined by its restriction to S^3. In fact, $\mathcal{H}_{n,s}$ is naturally isomorphic to the space of L^2 sections of the line bundle

$$L_n \to S^2 \tag{2.91}$$

gotten from the principal S^1-bundle $S^3 \to S^2$ via the representation of S^1 on $\mathbf{C}$ given by $e^{i\theta} \mapsto e^{in\theta}$.

One realization of the principal series representation $\pi_{n,s}$ of SL(2, C) described in Theorem 2.5 is as a representation on $\mathcal{H}_{n,s}$ given by

$$\pi_{n,s}(g) f(z) = f(g^t z), \qquad f \in \mathcal{H}_{n,s}. \tag{2.92}$$

The fact that this has the Lie algebra action described above can be deduced from the expression of the action of R, defined by (2.83), on the Lie algebra sl(2, C), in parallel with the analysis for SL(2, R) in Chapter 8. We omit the details.

Another way to describe the homogeneity condition (2.90) is to write

$$g(az) = a^\lambda \bar{a}^\mu g(z), \qquad a \in \mathbf{C}^*, \tag{2.93}$$

with

$$\lambda = \tfrac{1}{2}(is - 1 + n), \qquad \mu = \tfrac{1}{2}(is - 1 - n). \tag{2.94}$$

By restricting the argument of an element of $\mathcal{H}_{n,s}$ to the hyperplane $z_2 = 1$ rather than to S^3, we can make $\mathcal{H}_{n,s}$ unitarily equivalent to a representation, denoted $D_{(\lambda+1,\mu+1)}$ in [**71**], of SL(2, C) on $L^2(\mathbf{C}) = L^2(R^2)$, given by

$$D_{(\lambda+1,\mu+1)}(g)u(z) = (bz + d)^{\lambda}(\bar{b}\bar{z} + \bar{d})^{\mu}u((az + c)/(bz + d)), \tag{2.95}$$

where

$$g = \begin{pmatrix} a & b \\ c & d \end{pmatrix} \in \mathrm{SL}(2, \mathbf{C}). \tag{2.96}$$

As long as (2.94) holds, this is a principal series representation.

The complementary series is realized as follows. It is given by the formula (2.95) with

$$\lambda + 1 = \mu + 1 = t \tag{2.97}$$

and is unitary provided

$$-1 < t < 1, \tag{2.98}$$

if we use the inner product

$$(u, v) = (i/2)^2 \int_{\mathbf{C}^2} |z_1 - z_2|^{-2t-2} u(z_1)\overline{v(z_2)}\, dz_1\, d\bar{z}_1\, dz_2\, d\bar{z}_2. \tag{2.99}$$

For further analysis of the complementary series, see [**71**, **73**].

3. The Lorentz groups SO(n, 1). The group O(n, 1) is the group of linear transformations of R^{n+1} preserving the metric

$$g(x, x) = x_1^2 + \cdots + x_n^2 - x_{n+1}^2. \tag{3.1}$$

The subgroup SO(n, 1) consists of elements with determinant 1, and has two connected components. We denote the component of the identity by $\mathrm{SO}_e(n, 1)$. It is useful to note some special subgroups of $\mathrm{SO}_e(n, 1)$, the study of which is crucial to our understanding of $\mathrm{SO}_e(n, 1)$, which we will also call G in the rest of this section. First, we have the subgroup consisting of rotations in the variables $(x_1, \ldots, x_n)$,

$$K = \mathrm{SO}(n). \tag{3.2}$$

K is a maximal compact subgroup of G. Another important subgroup is the group of Lorentz transformations affecting only the variables (x_n, x_{n+1}),

$$A = \mathrm{SO}_e(1, 1). \tag{3.3}$$

The interaction of K and A is obviously important, and we single out the subgroup of K consisting of elements which *commute* with A, namely the rotations in the variables $(x_1, \ldots, x_{n-1})$ alone,

$$M = \mathrm{SO}(n - 1). \tag{3.4}$$

There are a couple of other special subgroups of G we need to pick out, which are best produced by looking at the action of $\mathfrak{a}$, the Lie algebra of A, on $\mathfrak{g}$, the Lie algebra of G. The algebra $\mathfrak{a}$ is one-dimensional here, with generator

$$a_0 = \begin{pmatrix} 0 & \vdots & 0 \\ \cdots & \cdots & \cdots \\ 0 & \vdots & \begin{matrix} 0 & 1 \\ 1 & 0 \end{matrix} \end{pmatrix} \tag{3.5}$$

if we identify $\mathfrak{g}$ with a subalgebra of the $(n+1)\times(n+1)$ matrices in the standard fashion. If we diagonalize the action of $\operatorname{ad} a_0$ on $\mathfrak{g}$, we find that

$$\mathfrak{g} = \mathfrak{g}_{-1} \oplus \mathfrak{g}_0 \oplus \mathfrak{g}_1 \tag{3.6}$$

where

$$\mathfrak{g}_\beta = \{X \in \mathfrak{g} \colon \operatorname{ad} a_0(X) = \beta X\}. \tag{3.7}$$

It is easy to verify that

$$\mathfrak{g}_0 = \mathfrak{m} \oplus \mathfrak{a} \tag{3.8}$$

where $\mathfrak{m}$ is the Lie algebra of M. A complementary space to $\mathfrak{g}_0$ in $\mathfrak{g}$, invariant under $\operatorname{ad} a_0$, is

$$\mathfrak{c} = \left\{ X = \begin{pmatrix} 0 & \vdots & V_1 & V_2 \\ \cdots & \cdots & \cdots & \cdots \\ -V_1^t & \vdots & 0 & 0 \\ V_2^t & \vdots & 0 & 0 \end{pmatrix} : V_j \text{ column vectors in } R^{n-1} \right\}. \tag{3.9}$$

The condition $[a_0, X] = X$ is equivalent to $V_1 = -V_2$ and the condition $[a_0, X] = -X$ is equivalent to $V_1 = V_2$, so we have

$$\mathfrak{n} = \mathfrak{g}_1 = \{X \in \mathfrak{c} \colon V_1 = -V_2\} \tag{3.10}$$

and

$$\bar{\mathfrak{n}} = \mathfrak{g}_{-1} = \{X \in \mathfrak{c} \colon V_1 = V_2\}. \tag{3.11}$$

We have introduced the standard notations $\mathfrak{n}$ and $\bar{\mathfrak{n}}$ for these (abelian) subalgebras. Note that

$$[\mathfrak{g}_\alpha, \mathfrak{g}_\beta] \subset \mathfrak{g}_{\alpha+\beta}. \tag{3.12}$$

As we will see in Chapter 13, these constructions apply to other semisimple groups, though in general $\mathfrak{a}$ has dimension greater than one and $\mathfrak{n}$ need not be abelian, but it will always be nilpotent.

Here, it is easy to verify that

$$\mathfrak{g} = \mathfrak{k} \oplus \mathfrak{a} \oplus \mathfrak{n}, \tag{3.13}$$

where $\mathfrak{k}$ is the Lie algebra of $K = \mathrm{SO}(n)$. With some effort one can show that

$$G = KAN, \tag{3.14}$$

where N is the connected Lie group in G generated by the Lie algebra $\mathfrak{n}$. This is a special case of the "Iwasawa decomposition." For details, see Helgason [**100**]. The Iwasawa decomposition is discussed further in Chapter 13.

There is another decomposition of G, known as the Cartan decomposition, which we need to describe. It can be given a very geometrical interpretation. Consider the action of $G = \mathrm{SO}_e(n,1)$ on the upper sheet $x_{n+1} > 0$ of the 2-sheeted hyperboloid

$$x_1^2 + \cdots + x_n^2 = x_{n+1}^2 - 1. \tag{3.15}$$

This is hyperbolic space $\mathcal{H}^n$, with the induced metric, and G acts transitively on $\mathcal{H}^n$ by a group of isometries. $K = \mathrm{SO}(n)$ is the subgroup fixing the pole $p = (0, \ldots, 0, 1) \in \mathcal{H}^n$, so

$$\mathcal{H}^n = G/K. \tag{3.16}$$

The tangent space to $\mathcal{H}^n$ at p is naturally isomorphic to $\mathfrak{g}/\mathfrak{k}$. In fact, there is a natural subspace of $\mathfrak{g}$, complementary to $\mathfrak{k}$, which we denote $\mathfrak{p}$. It is the orthogonal complement to $\mathfrak{k}$ in $\mathfrak{g}$, with respect to the following nondegenerate bilinear form,

$$B(X,Y) = -\mathrm{tr}(XY), \tag{3.17}$$

where we view $X \in \mathfrak{g}$ as an $(n+1) \times (n+1)$ matrix. This is proportional to the Killing form, introduced in Chapter 0. We have

$$\mathfrak{k} = \{X \in \mathfrak{g} : X = -X^t\}, \qquad \mathfrak{p} = \{X \in \mathfrak{g} : X = X^t\}. \tag{3.18}$$

Thus

$$\mathfrak{g} = \mathfrak{k} \oplus \mathfrak{p}. \tag{3.19}$$

It is easy to verify the relations

$$[\mathfrak{k}, \mathfrak{k}] \subset \mathfrak{k}, \qquad [\mathfrak{k}, \mathfrak{p}] \subset \mathfrak{p}, \qquad [\mathfrak{p}, \mathfrak{p}] \subset \mathfrak{k}. \tag{3.20}$$

Furthermore, there is an automorphism θ of $\mathfrak{g}$, known as the Cartan involution, defined for $\mathfrak{g} = \mathrm{so}(n,1)$ by

$$\theta(X) = -X^t, \tag{3.21}$$

such that

$$\mathfrak{k} = \{X \in \mathfrak{g} : \theta(X) = X\}, \tag{3.22}$$

and

$$\mathfrak{p} = \{X \in \mathfrak{g} : \theta(X) = -X\}. \tag{3.23}$$

This involution defines an involutory automorphism $\tilde{\theta}$ of $G = \mathrm{SO}_e(n,1)$, which is the identity on K, and hence there is defined an involution on $G/K = \mathcal{H}^n$. In fact, the involution of $\mathcal{H}^n$ is inversion through the point $p = (0, \ldots, 0, 1) \in \mathcal{H}^n$, sending $(x_1, \ldots, x_n, x_{n+1}) \in \mathcal{H}^n$ to $(-x_1, \ldots, -x_n, x_{n+1})$. We remark that $\theta(\mathfrak{n}) = \overline{\mathfrak{n}}$.

The decomposition (3.19) is the Cartan decomposition of $\mathfrak{g}$. On the Lie group level, one has the decomposition

$$G = K(\exp \mathfrak{p}), \tag{3.24}$$

mapping $K \times \mathfrak{p}$ diffeomorphically onto G. This is different from the Iwasawa decomposition (3.13)–(3.14). $\mathfrak{p}$ is not a subalgebra of $\mathfrak{g}$, as is $\mathfrak{a} \oplus \mathfrak{n}$, in view of the inclusion

$$[\mathfrak{a}, \mathfrak{n}] \subset \mathfrak{n}, \tag{3.25}$$

which follows from (3.12), or even (3.7). AN is a solvable subgroup of G, while $\exp \mathfrak{p}$ is not a subgroup. Let us note that, in light of the fact that $\mathfrak{m}$ centralizes $\mathfrak{a}$, we have

$$\operatorname{ad} \mathfrak{m} : \mathfrak{g}_\alpha \to \mathfrak{g}_\alpha \tag{3.26}$$

for each $\mathfrak{g}_\alpha$, and hence

$$[\mathfrak{m}, \mathfrak{n}] \subset \mathfrak{n}. \tag{3.27}$$

Thus $\mathfrak{m} \oplus \mathfrak{a} \oplus \mathfrak{n}$ is a subalgebra of $\mathfrak{g}$, and in fact

$$B = MAN \tag{3.28}$$

is a subgroup of G. Note that

$$G/B = K/M = S^{n-1}. \tag{3.29}$$

It turns out that, if S^{n-1} is given its unique (up to a constant) K-invariant metric, i.e., the standard metric, then $G = \mathrm{SO}_e(n,1)$ acts as a group of conformal automorphisms of S^{n-1}. For more on this, see Chapter 10.

We give a brief description of some of the irreducible unitary representations of $\mathrm{SO}_e(n,1)$. First, there is the principal series. An element of this series is constructed by taking a certain finite dimensional unitary representation of the group $B = MAN$, and constructing the induced representation on G. More precisely, let $\lambda \in \hat{M}$ be an irreducible unitary representation of M and let $\nu \in \mathfrak{a}^*$. Define a unitary representation (λ, ν) of MAN by

$$(\lambda, \nu)(man) = \lambda(m) e^{i\nu(\log a)}. \tag{3.30}$$

Here log is the inverse of the exponential map of $\mathfrak{a}$ onto A. From (3.25)–(3.27) and the fact that m and a commute, we see that this is a representation of B. So define

$$\pi_{(\lambda,\nu)} = \operatorname{Ind}_B^G(\lambda, \nu). \tag{3.31}$$

Note that the representation space for $\pi_{(\lambda,\nu)}$ is the space of L^2 sections of a vector bundle over S^{n-1}, with fiber isomorphic to the representation space $V_\lambda \otimes \mathbf{C}$ of $(\lambda, \nu) \in \hat{B}$.

We have seen examples of this before. When the principal series of $\mathrm{SL}(2, R)$, covering $\mathrm{SO}_e(2,1)$, was obtained in Chapter 8, §3, by decomposing the regular action of $\mathrm{SL}(2, R)$ on $L^2(R^2)$ by (purely imaginary) degree of homogeneity, we obtained representations of $\mathrm{SL}(2, R)$ on $L^2(S^1)$, given by formula (3.23) of Chapter 8. In this case, M is trivial for $\mathrm{SO}_e(2,1) = \mathrm{PSL}(2, R)$, and $K \to S^1$ is a trivial fibration. In §2 of this chapter, we made an analogous construction of the principal series for $\mathrm{SL}(2, \mathbf{C})$, covering $\mathrm{SO}_e(3, 1)$, and obtained representations of $\mathrm{PSL}(2, \mathbf{C})$ on sections of line bundles over S^2, corresponding to various representations of the group $M = \mathrm{SO}(2) = S^1$. In this case $K \to S^2$ is the Hopf fibration. One can verify that the definition of the principal series representations of $\mathrm{SL}(2, \mathbf{C})$ given by (2.89)–(2.92) in this chapter make them of the form (3.31).

In the previous discussion of the principal series for $\mathrm{SL}(2, R)$ and $\mathrm{SL}(2, \mathbf{C})$, we saw that, by restricting homogeneous functions to a hyperplane rather than to a sphere, we could represent these groups on $L^2(R^1)$ and $L^2(R^2)$, respectively, rather than on $L^2(S^1)$ and L^2 sections of line bundles over S^2. The geometrical correspondences between R^n and S^n in these two cases are stereographic projections, discussed in more detail in Chapter 10. This alternate description of the principal series can be generalized. It turns out that

$$G = B\overline{N}, \quad \text{modulo a set of measure zero,} \tag{3.32}$$

where $\overline{N} = \exp \overline{\mathfrak{n}}$, and we can define a representation equivalent to $\pi_{(\lambda,\nu)}$ on $L^2(\overline{N}, V_\lambda)$, with Haar measure on $\overline{N}$, by picturing a section of the appropriate vector bundle over S^{n-1} as a function on G with values in V_λ, satisfying appropriate compatibility conditions on the right B-cosets, and restricting such functions to $\overline{N}$. We omit the details; see [**48**] for a discussion of $\mathrm{SO}_e(n, 1)$ in particular, and [**27**, **256**], for a general discussion.

As we have seen, $\mathrm{SL}(2, R)$, which covers $\mathrm{SO}_e(2, 1)$, has discrete series representations. On the other hand, $\mathrm{SL}(2, \mathbf{C})$, which covers $\mathrm{SO}_e(3, 1)$, does not. It turns out that generally $\mathrm{SO}_e(2n, 1)$ has discrete series representations, while $\mathrm{SO}_e(2n + 1, 1)$ does not. We refer to Chapter 13 for a further discussion of the discrete series. For $G = \mathrm{SO}_e(n, 1)$, the principal series and discrete series make up "almost all" the irreducible unitary representations, in the sense that there is a Plancherel measure μ supported on this set Σ of representations such that, for $f \in C_0^\infty(G)$,

$$\|f\|_{L^2}^2 = \int_\Sigma \operatorname{tr}(\pi(f)^* \pi(f))\, d\mu(\pi). \tag{3.33}$$

The measure μ is atomic on discrete series representations. A detailed analysis of this for $\mathrm{SO}_e(2n + 1, 1)$ is given in the last chapter of Wallach [**253**]. There are also certain "supplementary" series of representations, of total Plancherel measure zero. A study of the irreducible unitary representations of $\mathrm{SO}_e(n, 1)$ was made in Hirai [**113**, **114**, **115**].

CHAPTER 10

Groups of Conformal Transformations

Some Riemannian manifolds, particularly the spheres S^n, have groups of conformal automorphisms much larger than their isometry groups. This can be very helpful in solving certain PDEs. These phenomena are explored in this chapter. We do not actually consider unitary representations of conformal groups in this chapter. See [**128**, **189**] for material on this topic.

1. Laplace operators and conformal changes of metric. Let g_{jk} be the metric tensor of a Riemannian manifold M_1, or more generally of a semi-Riemannian manifold, where (g_{jk}) is nondegenerate but perhaps not positive definite, e.g., a Lorentz manifold. Then the Laplace operator Δ_1 on functions is defined by

$$\Delta_1 u = g^{-1/2}\partial_j g^{jk} g^{1/2}\partial_k u, \tag{1.1}$$

where $\partial_j = \partial/\partial x_j, (g^{jk})$ is the inverse matrix to (g_{jk}), $g = \det(g_{jk})$, and in (1.1) we use the summation convention. In this section we examine the relation between the Laplacians associated to conformally related metrics.

A second metric tensor $\tilde{g}_{jk}$ is said to be conformally equivalent to g_{jk} if, for some real-valued $\gamma(x) = e^{2\kappa(x)}$,

$$\tilde{g}_{jk} = \gamma(x) g_{jk}. \tag{1.2}$$

The Laplace operator associated to the new metric is defined by

$$\tilde{\Delta} u = \tilde{g}^{-1/2}\partial_j \tilde{g}^{jk}\tilde{g}^{1/2}\partial_k u. \tag{1.3}$$

The metric tensor $\tilde{g}_{jk}$ may be the pull-back of a metric tensor g_2 on M_2 under a diffeomorphism from M_1 to M_2, so $\tilde{\Delta}$ is the pull-back to M_1 of the Laplace operator Δ_2 on M_2.

In this section, following Ørsted [**189**] and Helgason [**106**], we relate the kernels of L_1 and L_2, where

$$L_j = \Delta_j + (n-2)K_j/4(n-1). \tag{1.4}$$

Here, $n = \dim M_j$, and K_j is the *scalar curvature* of M_j. The principal result is the following. Let $\tau\colon M_1 \to M_2$ be a conformal diffeomorphism, with $\tau^* g_2 = \gamma(x) g_1$. Let

$$U_\tau f(x) = \gamma(x)^{(n-2)/4} f(\tau(x)). \tag{1.5}$$

THEOREM 1.1. *If τ is conformal, then*

$$L_1 U_\tau f = \gamma(x) U_\tau L_2 f. \tag{1.6}$$

In particular,

$$f \in \ker L_2 \Rightarrow U_\tau f \in \ker L_1. \tag{1.7}$$

The proof of Theorem 1.1 divides naturally into two parts, one a straightforward calculation, and the second an application of a differential geometric identity, involving scalar curvatures and the factor in (1.5). Let us isolate the first part.

PROPOSITION 1.2. *With conventions as above, let*

$$\beta(x) = \gamma(x)^{(n-2)/4}. \tag{1.8}$$

Then

$$\Delta_1(\beta f) - \beta\gamma\tilde{\Delta} f = (\Delta_1 \beta) f. \tag{1.9}$$

PROOF. For general $\varphi \in C^\infty$ we have

$$\Delta_1(\varphi f) = \varphi \Delta_1 f + 2(\partial_j \varphi) g^{jk} \partial_k f + (\Delta_1 \varphi) f. \tag{1.10}$$

Hence, if $\gamma = e^{2\kappa}$, so $\beta = e^{(n-2)\kappa/2}$, we have

$$\Delta_1(\beta f) = \beta \Delta_1 f + (n-2)\beta(\partial_j \kappa) g^{jk} \partial_k f + (\Delta_1 \beta) f. \tag{1.11}$$

Now the formula (1.3) together with (1.2) implies

$$\tilde{\Delta} f = \gamma^{-1} \Delta_1 f + (n-2)\gamma^{-1}(\partial_j \kappa) g^{jk} \partial_k f. \tag{1.12}$$

Hence

$$\beta\gamma\tilde{\Delta} f = \beta \Delta_1 f + (n-2)\beta(\partial_j \kappa) g^{jk} \partial_k f. \tag{1.13}$$

Comparing (1.11) and (1.13) yields (1.9).

In order to prove Theorem 1.1, we use the identity (1.9) together with the following identity, due to Yamabe:

$$\begin{aligned}\Delta_1 \beta &= c_n (K_2 \gamma^{(n+2)/4} - K_1 \gamma^{(n-2)/4}) \\ &= c_n (K_2 \gamma\beta - K_1 \beta), \qquad c_n = (n-2)/4(n-1).\end{aligned} \tag{1.14}$$

For a proof we refer to [**266**]. Putting together (1.9) and (1.14) gives

$$\Delta_1(\beta f) - \beta\gamma\tilde{\Delta} f = c_n (K_2 \gamma\beta f - K_1 \beta f), \tag{1.15}$$

which is equivalent to (1.6). This proves Theorem 1.1.

2. Conformal transformations on R^n, S^n, and balls. Recall from complex analysis that an orientation preserving diffeomorphism between two regions of the complex plane **C** is a conformal mapping if and only if it is holomorphic.

Important examples include linear fractional transformations

$$\varphi(z) = (az+b)/(cz+d), \tag{2.1}$$

which extend to give holomorphic maps of the Riemann sphere S^2 to itself. A particular example is the inversion

$$v(z) = 1/z, \tag{2.2}$$

interchanging 0 and ∞. One might recall that such linear fractional transformations map circles to circles or, occasionally, lines. Consider for example the image of the unit circle, centered at the point 1. We claim that its image under the map (2.2) is a line parallel to the vertical axis. This follows from the formula

$$\begin{aligned} v(e^{i\theta}+1) - \tfrac{1}{2} &= (e^{i\theta}+1)^{-1} - \tfrac{1}{2} = \tfrac{1}{2}(e^{i\theta}-1)/(e^{i\theta}+1) \\ &= \tfrac{1}{2} i \tan\theta/2. \end{aligned} \tag{2.3}$$

The interior of the unit disc centered at 1 is mapped by (2.2) onto the half plane $\operatorname{Re} z > \frac{1}{2}$.

While there are not so many conformal transformations on regions of R^n as on $R^2 = \mathbf{C}$, there is a rich enough set to generate interesting results. In addition to the rigid motions of R^n, we have the inversion $V \colon R^n \setminus 0 \to R^n \setminus 0$, defined by

$$V(x) = |x|^{-2}x. \tag{2.4}$$

Note that, for $R^n = R^2 = \mathbf{C}$, $V(z) = 1/\bar{z} = v(\bar{z})$. If we set $y = V(x)$, we see from (2.4) that

$$V^* dy_j = |x|^{-2} dx_j - |x|^{-4} 2x_j r\, dr \tag{2.5}$$

where

$$r\, dr = \sum_k x_k\, dx_k, \tag{2.6}$$

and hence

$$\begin{aligned} V^* \sum dy_j^2 &= |x|^{-8} \sum_j (|x|^2\, dx_j - 2x_j r\, dr)^2 \\ &= |x|^{-4} \sum dx_j^2. \end{aligned} \tag{2.7}$$

Thus V is conformal. Now let B be any ball in R^n whose boundary S contains the origin. We claim the image of B under V is a half space in R^n, the image of $S \setminus 0$ under V being a hyperplane. In fact, if we consider arbitrary two-dimensional subspaces L of R^n, with one basis vector being the center of B, we see that V restricted to L is essentially the map v of (2.2), composed with a reflection across the line through the center of B, so this observation follows from the analysis of (2.2). In particular the map

$$\tau(x) = |x+e_n|^{-2}(x+e_n) - \tfrac{1}{2}e_n \tag{2.8}$$

is a conformal map of $R^n \setminus \{-e_n\}$ into R^n which maps the unit ball B_1, centered at the origin, conformally onto the upper half space $x_n > 0$. e_n is the nth standard unit basis vector of R^n.

So the unit ball and the upper half space in R^n are conformally equivalent. One implication of this is that the group of conformal automorphisms of the ball is transitive, since it is clear that translations and other rigid motions parallel to the boundary together with dilations are all conformal automorphisms of the upper half space, which generate a transitive group there. We will return to this point shortly, but first we want to draw a corollary about the conformal structure of spheres.

It is clear that any conformal diffeomorphism $\varphi\colon M_1 \to M_2$ between (semi) Riemannian manifolds will restrict to a conformal diffeomorphism of any (noncharacteristic) submanifold of M_1 onto its range. In particular, the map (2.8) must map $S^{n-1} \setminus \{-e_n\}$, the punctured sphere, conformally onto R^{n-1}. Note that, if $x = (x_1, \ldots, x_n) = (x', x_n) \in S^{n-1}$, then $|x + e_n|^2 = |x|^2 + 2x_n + 1 = 2(x_n + 1)$, so

$$\begin{aligned}(2.9)\qquad \tau(x) &= \tfrac{1}{2}[(x_n + 1)^{-1}(x + e_n) - e_n] \\ &= \tfrac{1}{2}(x_n + 1)^{-1}(x', 0), \quad x \in S^{n-1}.\end{aligned}$$

It is convenient to change notation a little to present the familiar formula for the *stereographic projection* of $S^n \setminus p$ onto R^n. We will move the pole to e_{n+1}, and dilate R^n, to write, for $(x_1, \ldots, x_n, x_{n+1}) = (x, x_{n+1}) \in S^n$,

$$(2.10)\qquad S(x, x_{n+1}) = (1 - x_{n+1})^{-1} x.$$

If we set $y = S(x, x_{n+1})$, then

$$(2.11)\qquad S^* dy_j = (1 - x_{n+1})^{-1}\, dx_j - (1 - x_{n+1})^{-2} x_j\, dx_{n+1}$$

and since $x_{n+1} = \pm\sqrt{1 - |x|^2} = \pm\sqrt{1 - r^2}$, we have $dx_{n+1} = \pm(1 - r^2)^{-1/2} r\, dr$, so one obtains a direct verification that

$$(2.12)\qquad S^* \sum_1^n dy_j^2 = (1 - x_{n+1})^{-2} \sum_1^{n+1} dx_j^2$$

on $S^n \setminus \{e_{n+1}\}$.

Note that such stereographic projection makes the lower hemisphere of S^n conformally diffeomorphic to the unit ball in R^n. On the other hand, the hemisphere $x_n > 0$ on S^n is mapped conformally onto the upper half space $x_n > 0$ in R^n. Since one hemisphere can be transformed to any other by a rotation, we are led back to a conformal equivalence between the unit ball and a half space in R^n. The sphere is a *conformal compactification* of R^n. Note that the inversion (2.4) on R^n, conjugated by the stereographic projection, is simply reflection across the equator in S^n.

A point we have brought out before is that constructions involving spheres can often be modified (analytically continued) to constructions involving hyperbolic

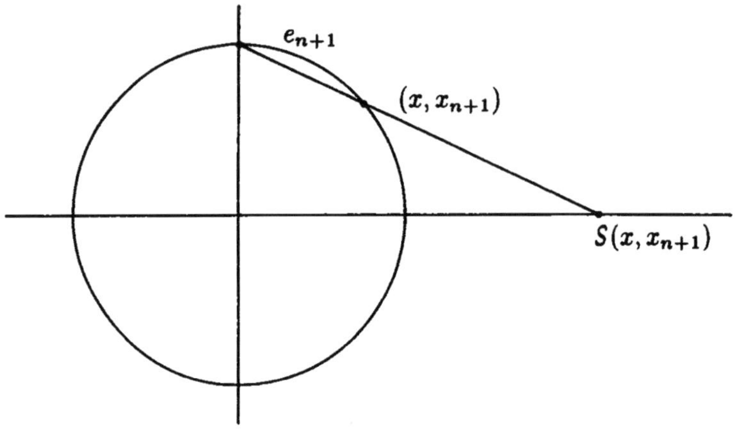

FIGURE 1

space. Here we shall take hyperbolic space $\mathcal{H}^n$ in its purest form, as the sheet $x_{n+1} > 0$ of the hyperboloid $x_{n+1}^2 - x_1^2 - \cdots - x_n^2 = 1$ in Minkowski space $R^{n,1}$, endowed with the Lorentz metric $dx_1^2 + \cdots + dx_n^2 - dx_{n+1}^2$. Since $\mathcal{H}^n$ is an orbit for the proper Lorentz group, homogeneity is manifest. A modification of stereographic projection is produced by the following obvious variant of Figure 1. We set $x = (x_1, \ldots, x_n) \in R^n$.

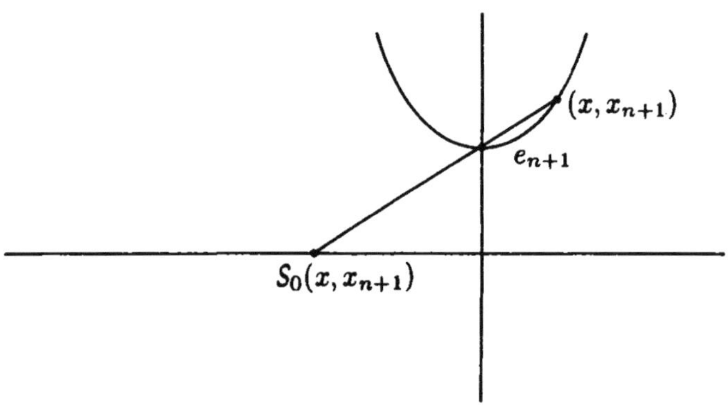

FIGURE 2

We have

$$S_0(x, x_{n+1}) = (1 - x_{n+1})^{-1}x.$$

The image of $\mathcal{H}^n \backslash \{e_{n+1}\}$ is the complement of the unit ball in R^n. We compose with the inversion (2.4) to produce a map of $\mathcal{H}^n$ onto the unit ball:

$$\mathfrak{s}(x, x_{n+1}) = (1 - x_{n+1})|x|^{-2}x. \tag{2.13}$$

In this case,

$$x_{n+1} = \sqrt{1 + |x|^2}, \tag{2.14}$$

and we can rewrite (2.13) as

$$\mathfrak{s}(x, x_{n+1}) = (1 + x_{n+1})^{-1}x. \tag{2.15}$$

If we set $y = \mathfrak{s}(x, x_{n+1})$, we see that

$$\mathfrak{s}^* dy_j = (1 + x_{n+1})^{-1}\, dx_j - (1 + x_{n+1})^{-2} x_j\, dx_{n+1}, \tag{2.16}$$

and a straightforward calculation gives

$$\mathfrak{s}^* \sum dy_j^2 = (1 + x_{n+1})^{-2} \left(\sum_1^n dx_j^2 - dx_{n+1}^2 \right). \tag{2.17}$$

Note that from $x_{n+1} = \sqrt{1 + |x|^2}$ and $1 + x_{n+1} = |x|/|y|$ follows the identity $|x| = 2|y|/(1 - |y|^2)$, which in turn gives

$$(1 + x_{n+1})^2 = (1 + \sqrt{1 + |x|^2})^2 = 4(1 - |y|^2)^{-2}. \tag{2.18}$$

Thus we can rewrite (2.17) as

$$\mathfrak{s}^* 4(1 - |y|^2)^{-2} \sum dy_j^2 = \sum_1^n dx_j^2 - dx_{n+1}^2. \tag{2.19}$$

Thus we see that hyperbolic space $\mathcal{H}^n$ is conformally equivalent to the unit ball in R^n, and more, the left side of (2.19) gives an explicit hyperbolic metric on the unit ball. Since $\mathrm{SO}_e(n, 1)$ is a transitive group of isometries of $\mathcal{H}^n$, we see that $\mathrm{SO}_e(n, 1)$ acts as a transitive group of conformal automorphisms of the unit ball in R^n.

Let us relate the ball to the upper half space again. We have the conformal map τ, defined by (2.8), from the unit ball to the upper half space in R^n. We see what happens to the hyperbolic metric on the ball when it is pulled back to the half space. If $y = \tau(x)$, then (2.8) gives

$$\tau^* \sum dy_j^2 = |x + e_n|^{-4} \sum dx_j^2, \tag{2.20}$$

in analogy with (2.7). Now the hyperbolic metric on the ball is

$$4(1 - |x|^2)^{-2} \sum dx_j^2,$$

and, using $x + e_n = |y + \frac{1}{2}e_n|^{-2}(y + \frac{1}{2}e_n)$, we see after a short calculation that $(1 - |x|^2)^{-2}|x + e_n|^4 = (1/2y_n)^2$, and hence (2.20) gives

$$\tau^* \left(y_n^{-2} \sum dy_j^2 \right) = 4(1 - |x|^2)^{-2} \sum dx_j^2. \tag{2.21}$$

Thus we obtain $y_n^{-2} \sum_1^n dy_j^2$ as the hyperbolic metric on the upper half space in R^n. Note that this metric is invariant under dilations, and also under rigid motions parallel to the boundary. It is easy to see that any metric on the upper half space with this property would have to be of the form

$$C y_n^{-2} \left(\sum_1^{n-1} dy_j^2 + \alpha^2\, dy_n^2 \right),$$

and changing the y_n variable we can make $\alpha = 1$, so the hyperbolic metric on the upper half space could also be derived that way. What is not so clear when one looks at this metric alone is the appearance of a group of metric preserving rotations about any fixed point of the upper half space, which is manifest for the center of the ball.

On the other hand, we can gain insight by pulling the dilations of the upper half space back to the ball. So we want to conjugate the dilation action

$$D_\sigma x = \sigma x, \qquad \sigma > 0, \tag{2.22}$$

on the upper half space by the map τ, given by (2.8). Note that

$$\tau^{-1}(y) = |y + \tfrac{1}{2}e_n|^{-2}(y + \tfrac{1}{2}e_n) - e_n. \tag{2.23}$$

One obtains for $\tilde{D}_\sigma = \tau^{-1}D_\sigma\tau$, acting on the unit ball in R^n, the formula

$$\tilde{D}_\sigma(x) = \frac{\sigma|x+e_n|^{-2}(x+e_n) + \frac{1}{2}(1-\sigma)e_n}{|\sigma|x+e_n|^{-2}(x+e_n) + \frac{1}{2}(1-\sigma)e_n|^2} - e_n. \tag{2.24}$$

Note that

$$\begin{aligned} \tilde{D}_\sigma(0) &= [(\tfrac{1}{2}\sigma + \tfrac{1}{2})^{-1} - 1]e_n \\ &= [(1-\sigma)/(1+\sigma)]e_n, \end{aligned} \tag{2.25}$$

so as $\sigma \downarrow 0$, $\tilde{D}_\sigma(0)$ approaches e_n on the boundary of the ball; as $\sigma \uparrow \infty$, $\tilde{D}_\sigma(0)$ approaches $-e_n$.

We can use the results of §1 to solve the Dirichlet problem for functions harmonic on the unit ball $B_1 \subset R^n$:

$$u = 0 \quad \text{on } B_1, \qquad u|_{S^{n-1}} = f. \tag{2.26}$$

First recall that harmonic functions satisfy the mean value principle, so

$$u(0) = \operatorname*{Avg}_{S^{n-1}} f(x). \tag{2.27}$$

Now, as a special case of the results of §1, we see that, if $u(x)$ is harmonic in B_1, so is

$$U_\sigma u(x) = \gamma_\sigma(x)^{(n-2)/4}u(\tilde{D}_\sigma(x)). \tag{2.28}$$

In this case, since $\tilde{D}_\sigma$ preserves the hyperbolic metric

$$4(1-|x|^2)^{-2}\sum dx_j^2 = h(x)\sum dx_j^2, \tag{2.29}$$

we have

$$\begin{aligned} \gamma_\sigma(x) &= h(x)/h(\tilde{D}_\sigma(x)) \\ &= (1-|\tilde{D}_\sigma(x)|^2)^2/(1-|x|^2)^2. \end{aligned} \tag{2.30}$$

We will apply the mean value property to (2.28), in order to evaluate u at $\tilde{D}_\sigma(0)$. Let $r = (1-\sigma)/(1+\sigma)$, so $\tilde{D}_\sigma(0) = re_n$. We obtain the formula

$$\begin{aligned} u(re_n) &= \gamma_\sigma(0)^{-(n-2)/4} \operatorname*{Avg}_{S^{n-1}} \gamma_\sigma(x)^{(n-2)/4} f(\tilde{D}_\sigma(x)) \\ &= (1-r^2)^{1-n/2} \operatorname*{Avg}_{S^{n-1}} f(x)\gamma_\sigma(\tilde{D}_\sigma^{-1}(x))^{(n-2)/4}\gamma_{1/\sigma}(x)^{(n-1)/2}. \end{aligned} \tag{2.31}$$

Note that

$$\gamma_\sigma(\tilde{D}_\sigma^{-1}(x)) = h(\tilde{D}_{1/\sigma}(x))/h(x) = \gamma_{1/\sigma}(x)^{-1}, \tag{2.32}$$

so

$$u(re_n) = (1-r^2)^{1-n/2} \operatorname*{Avg}_{S^{n-1}} f(x)\gamma_{1/\sigma}(x)^{n/4}. \tag{2.33}$$

In order to utilize the formula (2.30) for $\gamma_\sigma(x)$, we have to pass to the limit $x \in S^{n-1}$ from $|x| < 1$. A fairly straightforward calculation from the formula (2.24) gives

$$(1-|\tilde{D}_\sigma(x)|^2)/(1-|x|^2) = \sigma[-\tfrac{1}{2}\sigma^2 x_n + \tfrac{1}{2}x_n + \sigma + \tfrac{1}{4}(1-\sigma)^2(|x|^2+1)]^{-1}, \tag{2.34}$$

and in the limit $|x| = 1$ we get

$$2\sigma[\sigma^2(1-x_n) + (1+x_n)]^{-1}, \qquad x \in S^{n-1}. \tag{2.35}$$

Since $\sigma = (1-r)/(1+r)$, we obtain

$$\gamma_\sigma(x) = (1-r^2)^2[1+r^2+2rx\cdot e_n]^{-2}, \qquad x \in S^{n-1}, \tag{2.36}$$

and hence

$$\gamma_{1/\sigma}(x) = (1-r^2)^2[1+r^2-2rx\cdot e_n]^{-2}, \qquad x \in S^{n-1}. \tag{2.37}$$

Thus (2.33) gives

$$u(re_n) = (1-r^2) \operatorname*{Avg}_{S^{n-1}} f(x)(1+r^2-2rx\cdot e_n)^{-n/2}. \tag{2.38}$$

In light of rotational invariance, we immediately obtain the following formula for the solution to (2.26):

$$u(r\omega) = (1-r^2)V_{n-1}^{-1} \int_{S^{n-1}} f(x)(1+r^2-2rx\cdot\omega)^{-n/2}\,dx, \tag{2.39}$$

where V_{n-1} is the volume of S^{n-1}, $\omega \in S^{n-1}$, $r < 1$. This is Poisson's formula, which we used in Chapter 4, §1. Note that, for our argument to be complete, we need Yamabe's identity, which in this case is

$$\Delta\beta = 0,$$

where β is obtained by raising (2.34) to the power $(n-2)/2$. Any derivation of the Poisson integral formula seems to have a grubby calculation somewhere, and this is it. We leave it as a challenge to the reader to verify this identity.

We have seen the action of $\mathrm{SO}_e(n,1)$ on the unit ball in R^n, by conformal automorphisms. This action induces a group of conformal diffeomorphisms of the boundary S^{n-1}. The ball can be pictured as the lower hemisphere in S^n. Here we will sketch an action of $\mathrm{SO}_e(n+1,1)$, as a group of conformal automorphisms of S^n, such that the natural subgroup $\mathrm{SO}_e(n,1)$ leaves the equator, and hence each hemisphere, of S^n, invariant.

In fact, consider $R^{n+2} = R^{n+1,1}$, with a Lorentz metric. Let Γ^+ be the forward light cone, $x_{n+2} > 0$, $x_1^2 + \cdots + x_{n+1}^2 - x_{n+2}^2 = 0$. There is a natural R^+ action, by dilations, on Γ^+, and

$$\Gamma^+/R^+ = S^n. \tag{2.40}$$

The group $\mathrm{SO}_e(n+1,1)$, acting on $R^{n+1,1}$, preserving the Lorentz metric, leaves Γ^+ invariant, and acts as a group of "isometries" of Γ^+. Now the Minkowski metric on $R^{n+1,1}$, restricted to Γ^+, is degenerate in the radial direction. However it induces a conformal structure on S^n, via (2.40), coinciding with the usual conformal structure given by the standard metric on S^n, and $\mathrm{SO}_e(n+1,1)$ is seen to act as a group of conformal transformations on $\Gamma^+/R^+ = S^n$.

Similarly $\mathrm{SO}_e(n,1)$, acting on $R^{n,1}$, acts on the forward light cone Γ^+ and hence on S^{n-1}. Let us consider the subgroup of $\mathrm{SO}_e(n,1)$ fixing some point $p \in S^{n-1}$, which we will identify with the ray in Γ^+ through the point $(0,\ldots,0,1,1)$. Clearly this is fixed by the subgroup $M = \mathrm{SO}(n-1)$, acting only on the first $n-1$ variables, and by the subgroup $A = \mathrm{SO}_e(1,1)$, acting on the last two variables. Consider now N, the connected subgroup of $\mathrm{SO}_e(n,1)$ with Lie algebra $\mathfrak{n}$, given by (3.9)–(3.10) of Chapter 9. If such an element X of $\mathfrak{n}$ as described by these formula acts on $(0,\ldots,0,1,1) \in R^{n+1}$, one obtains 0, so $\exp tX$ fixes this point. Hence from the material developed in Chapter 9, §3, we see that MAN is precisely the subgroup of $\mathrm{SO}_e(n,1)$ fixing $p \in S^{n-1}$. If we use a stereographic projection with p as the north pole, we see that MAN acts as a group of conformal automorphisms of Euclidean space R^{n-1}. In this representation MN is the Euclidean group and A is the group of dilations.

This last description of the action of $\mathrm{SO}_e(n+1,1)$ on $\Gamma^+/R^+ = S^n$, which is the conformal compactification of R^n, can be generalized to the action of $\mathrm{SO}_e(p+1,q+1)$ on the conformal compactification of $R^{p,q}$, with the metric $\|x\|^2 = x_1^2 + \cdots + x_p^2 - x_{p+1}^2 - \cdots - x_{p+q}^2$, as follows. If $\sigma\colon R^{n+2}\setminus 0 \to P^{n+1}(R)$ is the canonical map onto projective space, define

$$J(x) = \sigma(1 - \tfrac{1}{4}\|x\|^2, x, 1 + \tfrac{1}{4}\|x\|^2), \qquad x \in R^{p,q},\ p+q = n. \tag{2.41}$$

Then J maps $R^{p,q}$ into the set

$$\overline{M} = \{\sigma(\xi)\colon \xi \in R^{p+1,q+1}\setminus 0\colon \|\xi\|^2 = 0\}. \tag{2.42}$$

The set $\overline{M}$ gets a natural conformal structure, from the degenerate metric on the set $\{\xi \in R^{p+1,q+1}\colon \|\xi\|^2 = 0\}$, making it a conformal compactification of $R^{p,q}$. In particular, $\mathrm{SO}_e(4,2)$ acts as a group of conformal transformations on the conformal compactification of Minkowski space $R^{3,1}$. For more on this, see **[189]**.

CHAPTER 11

The Symplectic Group and the Metaplectic Group

The symplectic group $\mathrm{Sp}(n, R)$ arose in Chapter 1 as a group of automorphisms of $\mathbf{H}^n$. Here we make a more detailed study of $\mathrm{Sp}(n, R)$. In Chapter 1 we produced a unitary representation of the universal covering group $\widetilde{\mathrm{Sp}(n, R)}$, related to this group of automorphisms. It is an important and beautiful fact that one actually obtains a representation of the double cover, denoted $\mathrm{Mp}(n, R)$, and we present the details on this here.

1. Symplectic vector spaces and the symplectic group. A symplectic structure on a real vector space is a bilinear form

$$\sigma: V \times V \to R \tag{1.1}$$

which is skew symmetric ($\sigma(u, v) = -\sigma(v, u)$) and nondegenerate, i.e., if $\sigma(u, v) = 0$ for all $v \in V$, then $u = 0$. If α_1 is a nonzero element of V, there is a $\beta_1 \in V$ such that $\sigma(\alpha_1, \beta_1) = 1$. If W_1 denotes the linear span of α_1 and β_1 and if V_1 is the subspace of V consisting of vectors orthogonal to W_1 with respect to σ (i.e., $v \in V_1$ if and only if $\sigma(v, w) = 0$ for all $w \in W_1$) then V_1 is also a symplectic vector space with respect to $\sigma|_{V_1 \times V_1}$. We can repeat this argument on V_1. Thus, inductively, we can choose a basis $\alpha_1, \ldots, \alpha_n, \beta_1, \ldots, \beta_n$ of V such that

$$\sigma(\alpha_j, \alpha_k) = \sigma(\beta_j, \beta_k) = 0, \qquad \sigma(\alpha_j, \beta_k) = \delta_{jk}. \tag{1.2}$$

Such a basis of V is called a symplectic basis. It follows that the dimension of V is even, and any two symplectic vector spaces of the same dimension are symplectically isomorphic. Thus we can consider the standard model, $V = R^{2n} = R^n \oplus R^n$ with symplectic form

$$\sigma(u, v) = u_1 \cdot v_2 - u_2 \cdot v_1 \tag{1.3}$$

where $u = (u_1, u_2)$, $v = (v_1, v_2)$, $u_j, v_j \in R^n$, and $u_j \cdot v_k$ is the standard inner product on R^n. Alternatively, with $u \cdot v$ representing the standard inner product on R^{2n}, we can write

$$\sigma(u, v) = u \cdot Jv \tag{1.4}$$

where

$$J = \begin{pmatrix} 0 & I \\ -I & 0 \end{pmatrix}, \tag{1.5}$$

I denoting the $n \times n$ identity matrix.

The group of linear transformations on R^{2n} preserving the symplectic form is called the symplectic group, and is denoted $\mathrm{Sp}(n, R)$:

$$\mathrm{Sp}(n, R) = \{A \in \mathrm{End}(R^{2n}) : \sigma(Au, Av) = \sigma(u, v)\}. \tag{1.6}$$

If we denote "standard" coordinates on R^{2n} by $(x_1, \ldots, x_n, \xi_1, \ldots, \xi_n)$, then, as an element of $\bigwedge^2(R^{2n})'$, the symplectic form can be written

$$\sigma = \sum d\xi_j \wedge dx_j. \tag{1.7}$$

Then σ^n is a volume element on R^{2n}, preserved by each $A \in \mathrm{Sp}(n, R)$, so

$$\mathrm{Sp}(n, R) \subset \mathrm{SL}(2n, R). \tag{1.8}$$

Note that $\mathrm{Sp}(1, R) = \mathrm{SL}(2, R)$.

The Lie algebra $\mathrm{sp}(n, R)$ of $\mathrm{Sp}(n, R)$ is identified with the set of linear transformations T on R^{2n} which are skew symmetric with respect to σ, i.e., such that

$$\sigma(Tu, v) = -\sigma(u, Tv), \tag{1.9}$$

or, equivalently, such that

$$T^t J = -JT, \tag{1.10}$$

where J is given by (1.5) and T^t denotes the transpose of T:

$$Tu \cdot v = u \cdot T^t v. \tag{1.11}$$

We can give the following alternative characterization of $\mathrm{sp}(n, R)$. For any smooth function f on R^{2n}, there is associated the Hamiltonian vector field

$$H_f = \sum_{j=1}^{n} [(\partial f/\partial \xi_j)(\partial/\partial x_j) - (\partial f/\partial x_j)(\partial/\partial \xi_j)].$$

The flow generated by H_f preserves the form σ. Now this flow consists of linear transformations on R^{2n} precisely when $f(x, \xi)$ is a second order homogeneous polynomial in x and ξ. Thus $\mathrm{sp}(n, R)$ is isomorphic to hp_2, defined by

$$hp_2 = \{Q(x, \xi) : Q \text{ homogeneous polynomial of degree } 2\}. \tag{1.12}$$

The Lie bracket on $\mathrm{sp}(n, R)$ corresponds to the Poisson bracket on hp_2. The Poisson bracket of two smooth functions is given by

$$\{f, g\} = H_f g = \sum [(\partial f/\partial \xi_j)(\partial g/\partial x_j) - (\partial f/\partial x_j)(\partial g/\partial \xi_j)]. \tag{1.13}$$

An explicit Lie algebra isomorphism from hp_2 to the set of linear transformations on R^{2n} satisfying (1.9) is given by

$$Q \mapsto F_Q \tag{1.14}$$

where F_Q is defined by

$$Q(u,v) = \sigma(F_Q u, v) \tag{1.15}$$

where $Q(u,v)$ denotes the symmetric bilinear form on R^{2n} polarizing Q, i.e., such that $Q(u,u) = Q(u)$. The map F_Q is the Hamilton map associated with Q, already considered in Chapter 1, §6. One way to look at (1.13) is the following. Suppose

$$Q(x,\xi) = Ax\cdot x + 2Bx\cdot\xi + C\xi\cdot\xi, \tag{1.16}$$

with A and C symmetric. If we identify H_Q, a vector field on R^{2n} with linear coefficients, in a natural fashion with a linear transformation $\mathfrak{X}_Q$ on R^{2n}, we have

$$\mathfrak{X}_Q = \begin{pmatrix} B & -A \\ C & -B^t \end{pmatrix} \tag{1.17}$$

since $H_Q = (Q_\xi, -Q_x)$. If Q denotes the bilinear form associated with $Q(u)$,

$$Q = \begin{pmatrix} A & B \\ B^t & C \end{pmatrix}, \tag{1.18}$$

and these formulas clearly give

$$\mathfrak{X}_Q = Q\circ J. \tag{1.19}$$

We leave it to the reader to verify that the Lie algebra structures are preserved.

To grasp the structure of the Lie algebra $\mathrm{sp}(n,R)$, it is useful to consider the following subalgebras:

$$\mathfrak{m} = \left\{ \begin{pmatrix} 0 & 0 \\ X & 0 \end{pmatrix} : X = X^t,\ n\times n \text{ real matrix} \right\}, \tag{1.20}$$

$$\overline{\mathfrak{m}} = \left\{ \begin{pmatrix} 0 & X \\ 0 & 0 \end{pmatrix} : X = X^t,\ n\times n \text{ real matrix} \right\}, \tag{1.21}$$

$$\mathfrak{n} = \left\{ \begin{pmatrix} Y & 0 \\ 0 & -Y^t \end{pmatrix} : Y \text{ is } n\times n \text{ real matrix} \right\}. \tag{1.22}$$

If we identify $\mathrm{sp}(n,R)$ with hp_2, we can write

$$\mathfrak{m} = \text{linear span of } x_j x_k,\qquad 1\le j,\ k\le n = (x_j x_k : 1\le j,\ k\le n), \tag{1.23}$$

$$\overline{\mathfrak{m}} = (\xi_j\xi_k : 1\le j,\ k\le n), \tag{1.24}$$

$$\mathfrak{n} = (x_j\xi_k : 1\le j,\ k\le n). \tag{1.25}$$

Then $\mathrm{sp}(n,R)$ is the linear direct sum of $\mathfrak{m}$, $\overline{\mathfrak{m}}$, and $\mathfrak{n}$;

$$\mathrm{sp}(n,R) = \mathfrak{m} + \overline{\mathfrak{m}} + \mathfrak{n}. \tag{1.26}$$

We see that $\mathfrak{m}$ and $\overline{\mathfrak{m}}$ are commutative subalgebras of $\mathrm{sp}(n, R)$, and $\mathfrak{n}$ is a subalgebra of $\mathrm{sp}(n, R)$ isomorphic to $\mathrm{gl}(n, R)$. Furthermore, simple calculations give

$$(1.27)\qquad [\mathfrak{m}, \overline{\mathfrak{m}}] \subset \mathfrak{n}, \qquad [\mathfrak{m}, \mathfrak{n}] \subset \mathfrak{m}, \qquad [\overline{\mathfrak{m}}, \mathfrak{n}] \subset \overline{\mathfrak{m}}.$$

For example,

$$(1.28)\qquad \left[\begin{pmatrix} 0 & X \\ 0 & 0 \end{pmatrix}, \begin{pmatrix} 0 & 0 \\ Y & 0 \end{pmatrix}\right] = \begin{pmatrix} XY & 0 \\ 0 & -YX \end{pmatrix}$$

and

$$(1.29)\qquad \left[\begin{pmatrix} 0 & 0 \\ X & 0 \end{pmatrix}, \begin{pmatrix} Y & 0 \\ 0 & -Y^t \end{pmatrix}\right] = \begin{pmatrix} 0 & 0 \\ XY + Y^t X & 0 \end{pmatrix}.$$

The action of $\mathrm{gl}(n, R)$ on $\mathfrak{m}$, identified with the linear space of symmetric real $n \times n$ matrices, devolving from (1.27), is the differential of the action of $\mathrm{Gl}(n, R)$ on this linear space given by $A \cdot X = AXA^t$, as is easy to verify.

Another subalgebra of $\mathrm{sp}(n, R)$ is spanned by

$$(1.30)\qquad \lambda_{jk}(x, \xi) = x_j x_k + \xi_j \xi_k, \qquad \mu_{jk}(x, \xi) = x_j \xi_k - x_k \xi_j.$$

In the picture of $\mathrm{sp}(n, R)$ given by (1.10), this subalgebra is of the form

$$(1.31)\qquad \left\{\begin{pmatrix} A & B \\ B & -A \end{pmatrix} : A, B \text{ real symmetric } n \times n \text{ matrices}\right\}.$$

This generates a subgroup of $\mathrm{Sp}(n, R)$ isomorphic to the unitary group $\mathrm{U}(n)$. Indeed, let $(x, \xi) \mapsto x + i\xi$ identify R^{2n} with $\mathbf{C}^n$. Then the symplectic form (1.3) becomes

$$(1.32)\qquad \sigma(z_1, z_2) = \mathrm{Im}\, z_1 \cdot \overline{z}_2.$$

The group generated by (1.31) leaves invariant the Hermitian form $(z_1, z_2) = z_1 \cdot \overline{z}_2$, and thus $\mathrm{U}(n)$ is exhibited as a subgroup of $\mathrm{Sp}(n, R)$. It is a maximal compact subgroup of $\mathrm{Sp}(n, R)$, as we will see in the next section.

The map $(x, \xi) \mapsto (\xi, -x)$, which preserves the symplectic form on R^{2n}, induces an involutive automorphism θ of $hp_2 \approx \mathrm{sp}(n, R)$:

$$(1.33)\qquad (\theta Q)(x, \xi) = Q(\xi, -x).$$

Note that θ is the identity on the Lie algebra of $\mathrm{U}(n)$, spanned by (1.30), which we will denote $\mathfrak{k}$. Note also that $\theta = -I$ on the linear space $\mathfrak{p}$, spanned by

$$(1.34)\qquad x_j x_k - \xi_j \xi_k \quad \text{and} \quad x_j \xi_k + x_k \xi_j, \qquad 1 \le j, k \le n.$$

Note that

$$(1.35)\qquad hp_2 = \mathfrak{k} + \mathfrak{p}$$

and that

$$(1.36)\qquad [\mathfrak{k}, \mathfrak{k}] \subset \mathfrak{k}, \qquad [\mathfrak{k}, \mathfrak{p}] \subset \mathfrak{p}, \qquad [\mathfrak{p}, \mathfrak{p}] \subset \mathfrak{k}.$$

There is a further condition, which will be discussed in Chapter 13, which makes θ a special case of a *Cartan involution*. As a consequence, we have the following result, analogous to polar decomposition. A proof, in a more general setting, will be given in Chapter 13.

THEOREM 1.1. *Let G be any connected Lie group with Lie algebra* $\mathrm{sp}(n, R)$. *Denote the exponential map by* $\exp\colon \mathrm{sp}(n, R) \to G$. *Let K be the connected Lie group in G with Lie algebra $\mathfrak{k}$. Then K is closed and the map $\mathfrak{p} \times K$ to G given by*

$$(X, k) \mapsto (\exp X)k \tag{1.37}$$

is a surjective diffeomorphism.

This result applies to the group $\mathrm{Sp}(n, R)$ in view of

PROPOSITION 1.2. $\mathrm{Sp}(n, R)$ *is connected.*

We leave the proof of this to the reader. Note that it gives us topological information on $\mathrm{Sp}(n, R)$, as it follows that $\mathrm{Sp}(n, R)$ is homeomorphic to $R^N \times \mathrm{U}(n)$, $N = n^2 + n$. Thus $\mathrm{Sp}(n, R)$ is homotopically equivalent to $\mathrm{U}(n)$. In particular, its fundamental group is isomorphic to $\mathbf{Z}$. Thus, for each k, there is a unique connected Lie group which is a k-fold cover of $\mathrm{Sp}(n, R)$. The 2-fold cover is called the metaplectic group and denoted $\mathrm{Mp}(n, R)$.

2. Symplectic inner product spaces and compact subgroups of the symplectic group. If a real vector space V is endowed with a symplectic form $\sigma(u, v)$ and a positive definite inner product $Q(u, v)$, we will call V a symplectic inner product space. Note that any compact subgroup K of $\mathrm{Sp}(n, R)$ must leave invariant some such inner product, and hence K is contained in the compact group K_Q defined by

$$K_Q = \{g \in \mathrm{Sp}(n, R)\colon Q(gu, gv) = Q(u, v) \text{ for all } u, v \in V\}. \tag{2.1}$$

As before, we have the Hamilton map F_Q on V, defined by

$$Q(u, v) = \sigma(F_Q u, v). \tag{2.2}$$

We also introduce the linear map

$$A_Q = (-F_Q^2)^{1/2}, \tag{2.3}$$

the unique square root of $-F_Q^2$ with positive spectrum. If we choose an orthonormal basis E_j, F_j, $1 \le j \le n$, of V such that

$$F_Q E_j = \mu_j F_j, \quad F_Q F_j = -\mu_j E_j, \qquad \mu_j > 0, \tag{2.4}$$

whose existence is proved in Lemma 6.2 of Chapter 1, then

$$A_Q E_j = \mu_j E_j, \qquad A_Q F_j = \mu_j F_j. \tag{2.5}$$

Now consider the map $J\colon V \to V$ defined by

$$J = A_Q^{-1} F_Q. \tag{2.6}$$

Note that

$$J^2 = -I. \tag{2.7}$$

In fact, in the basis E_j, F_j above,

$$JE_j = F_j, \qquad JF_j = -E_j. \tag{2.8}$$

Thus J defines a complex structure on V. Note that J and F_Q commute and

$$Q(Ju, v) = -Q(u, Jv). \tag{2.9}$$

Define a new quadratic form $\tilde{Q}$ by

$$\tilde{Q}(u, v) = \sigma(Ju, v). \tag{2.10}$$

Note that

$$\tilde{Q}(u, v) = Q(A_Q^{-1}u, v), \tag{2.11}$$

so $\tilde{Q}$ also defines a positive inner product on V. It is easy to see that $F_{\tilde{Q}} = J$, and hence, if $\tilde{J}$ is associated with $\tilde{Q}$ by the process above, then $\tilde{J} = J$.

LEMMA 2.1. *We have*

$$K_Q \subset K_{\tilde{Q}}. \tag{2.12}$$

PROOF. Clearly, from (2.1), we have

$$K_Q = \{g \in \mathrm{Sp}(n, R) : g \text{ commutes with } F_Q\}. \tag{2.13}$$

Also

$$g \text{ commutes with } F_Q \Rightarrow g \text{ commutes with } J. \tag{2.14}$$

This makes (2.12) apparent.

Now define an R-bilinear form $(\ ,\ ): V \times V \to \mathbf{C}$ by

$$\begin{aligned}(u, v) &= \sigma(Ju, v) + i\sigma(u, v) = \tilde{Q}(u, v) - i\tilde{Q}(Ju, v) \\ &= \tilde{Q}(u, v) + i\tilde{Q}(u, Jv).\end{aligned} \tag{2.15}$$

In the last identity, we have used (2.9) with Q replaced by $\tilde{Q}$. We see that

$$(Ju, v) = \tilde{Q}(Ju, v) + i\tilde{Q}(u, v) = i(u, v) \tag{2.16}$$

and

$$(u, Jv) = \tilde{Q}(u, Jv) - i\tilde{Q}(u, v) = -i(u, v). \tag{2.17}$$

Also, since $\sigma(Ju, v) = -\sigma(u, Jv)$,

$$(v, u) = \sigma(Jv, u) - i\sigma(u, v) = \sigma(Ju, v) - i\sigma(u, v) = \overline{(u, v)}, \tag{2.18}$$

and furthermore

$$(u, u) = \tilde{Q}(u, u) \geq 0. \tag{2.19}$$

Thus (,) is a Hermitian inner product on the complex vector space (V, J). Note that

$$K_{\tilde{Q}} = \{g \in \mathrm{Sp}(n, R): g \text{ commutes with } J\} \\ = \{g \in \mathrm{Gl}(V): g \text{ preserves } \tilde{Q} \text{ and commutes with } J\}, \tag{2.20}$$

so (2.15)–(2.17) imply

$$K_{\tilde{Q}} = \{g \in \mathrm{Gl}(V): g \text{ commutes with } J \text{ and } g \text{ preserves } (\ ,\ )\} \\ = \mathrm{U}(n). \tag{2.21}$$

In light of the observation of §1 that $\mathrm{U}(n) \subset \mathrm{Sp}(n, R)$, we have

THEOREM 2.2. *$\mathrm{U}(n)$ is a maximal compact subgroup of* $\mathrm{Sp}(n, R)$.

Thus we see that any symplectic inner product space (V, σ, Q) has a complex structure such that (by (2.9)–(2.11))

$$\sigma(Ju, Jv) = \sigma(u, v). \tag{2.22}$$

There is furthermore associated a new symplectic inner product space $(V, \sigma, \tilde{Q})$, whose symmetry group is enlarged from K_Q to $K_{\tilde{Q}} = \mathrm{U}(n)$. In fact, (2.10) produces a one-to-one correspondence between symplectic inner product spaces $(V, \sigma, \tilde{Q})$ with maximal symmetry group $K_{\tilde{Q}} = \mathrm{U}(n)$ and complex symplectic spaces (V, J, σ), satisfying (2.22), such that

$$\sigma(Ju, u) \geq 0. \tag{2.23}$$

Then (2.15) produces a one-to-one correspondence between each of these classes and complex Hermitian spaces $(V, J, (\ ,\ ))$.

We could generalize, discarding the positivity hypothesis (2.23). If J is any complex structure on V satisfying (2.22), then the quadratic form (2.10) and the Hermitian form (2.15) are still nondegenerate. It is clear that the Hermitian form (,) could have any signature (p, q) such that $p + q = n$. Such a form gives an inclusion

$$\mathrm{U}(p, q) \subset \mathrm{Sp}(n, R), \qquad p + q = n. \tag{2.24}$$

3. The metaplectic representation. The metaplectic representation on the Lie algebra level is quite easily described. If $Q \in hp_2 \approx \mathrm{sp}(n, R)$, we set

$$\omega(Q) = iQ(X, D), \tag{3.1}$$

where $Q(X, D)$ is the second order differential operator associated to $Q(x, \xi)$ via the Weyl calculus. Explicitly, we have

$$\begin{aligned} \omega(x_j x_k)u(x) &= i x_j x_k u(x), \\ \omega(\xi_j \xi_k)u(x) &= i\partial^2 u/\partial x_k \partial x_j, \\ \omega(x_j \xi_k)u(x) &= \tfrac{1}{2}(x_j \partial u/\partial x_k + \partial(x_j u)/\partial x_k). \end{aligned} \tag{3.2}$$

It is easy to verify that $\omega(\{Q_1, Q_2\}) = [\omega(Q_1), \omega(Q_2)]$. Now general considerations, discussed in Appendix D, imply that ω exponentiates to a unitary representation $\tilde{\omega}$ of the universal covering group $\widetilde{\mathrm{Sp}(n,R)}$. This representation and some of its important consequences were discussed in Chapter 1, on the Heisenberg group. A more subtle result is that ω actually gives a representation of the metaplectic group $\mathrm{Mp}(n,R)$, the 2-fold cover of $\mathrm{Sp}(n,R)$. We will prove this result here.

First, we see that ω cannot give rise to a representation of $\mathrm{Sp}(n,R)$ itself. Indeed, consider $Q_j \in hp_2 \approx \mathrm{sp}(n,R)$ given by $Q_j(x,\xi) = x_j^2 + \xi_j^2$. Then

$$H_{Q_j} = 2(\xi_j \partial/\partial x_j - x_j \partial/\partial \xi_j) = -2\partial/\partial\theta_j \tag{3.3}$$

where $\partial/\partial\theta_j$ generates rotation in the $x_j - \xi_j$ plane. It is clear that the group of transformations on R^{2n} generated by the vector field H_{Q_j} coincides with the group of symplectic transformations generated by $Q_j \in \mathrm{sp}(n,R)$ under the exponential map. In view of (3.3),

$$\exp(\pi Q_j) = e, \quad \text{the identity in } \mathrm{Sp}(n,R). \tag{3.4}$$

However, as we have seen in Chapter 1, §6, the spectrum of $Q_j(X,D) = -d^2/dx_j^2 + x_j^2$ consists of $\{1,3,5,7,\ldots\}$. Hence

$$\omega(\pi Q_j) = \pi i(-\partial^2/\partial x_j^2 + x_j^2) \tag{3.5}$$

has the property

$$e^{\omega(\pi Q_j)} = -I. \tag{3.6}$$

This of course leaves open the possibility of ω existing on the double cover of $\mathrm{Sp}(n,R)$. We will see that this happens by looking at the Bargmann-Fok representation, introduced in §5 of Chapter 1. Here, $\mathrm{Sp}(n,R)$ is represented on the following Hilbert space of entire functions on $\mathbf{C}^n$:

$$\mathcal{H} = \left\{ u(z) \text{ holomorphic on } \mathbf{C}^n \colon \iint |u(z)|^2 e^{-|z|^2}\, dz\, d\bar{z} < \infty \right\}. \tag{3.7}$$

Let us denote by $\omega^\#$ the representation $K\omega K^{-1}$ where $K\colon L^2(R^n) \to \mathcal{H}$ is the unitary map given in (5.6) of Chapter 1. We have the following key result.

PROPOSITION 3.1. *The representation $\omega^\#$ restricted to $\mathfrak{k}$, the Lie algebra of* $\mathrm{U}(n)$, *exponentiates to a unitary representation of* $\mathrm{MU}(n)$, *the double cover of* $\mathrm{U}(n)$, *to give*

$$\omega^\#(g)u(z) = (\det g)^{-1/2} u(g^{-1}\cdot z), \qquad g \in \mathrm{MU}(n). \tag{3.8}$$

Here $g\cdot z$ is the action of $\mathrm{MU}(n)$ *on* $\mathbf{C}^n$ *defined by* $\mathrm{MU}(n) \to \mathrm{U}(n)$ *and* $(\det g)^{1/2}$ *is the unique smooth square root equal to* 1 *at $g = e$, the identity element.*

PROOF. It is clear that the right side of (3.8) defines a representation of $\mathrm{MU}(n)$. That its derived representation of $\mathfrak{k}$ is given by (3.1), conjugated by K, follows from the proof of Lemma 7.8 of Chapter 1.

We are now ready to define the metaplectic representation. First, if $k \in \mathrm{MU}(n)$, set

$$\omega(k) = K^{-1}\omega^{\#}(k)K \tag{3.9}$$

where $\omega^{\#}(k)$ is defined by (3.8). Then, if $g \in \mathrm{Mp}(n, R)$ is given by

$$g = (\exp Q)k, \qquad Q \in \mathfrak{p},\ k \in \mathrm{MU}(n), \tag{3.10}$$

set

$$\omega(g) = e^{iQ(X,D)}\omega(k). \tag{3.11}$$

By Theorem 1.1, $(Q, k) \mapsto (\exp Q)k$ defines a surjective (analytic) diffeomorphism of $\mathfrak{p} \times \mathrm{MU}(n)$ onto $\mathrm{Mp}(n, R)$, so ω is a well-defined strongly continuous function on $\mathrm{Mp}(n, R)$, taking values in the set of unitary operators on $L^2(R^n)$.

THEOREM 3.2. *The formula* (3.11) *defines a representation of* $\mathrm{Mp}(n, R)$.

PROOF. What we need to show is that

$$\omega(g_1)\omega(g_2)f = \omega(g_1g_2)f, \qquad f \in L^2(R^n),\ g_j \in \mathrm{Mp}(n, R). \tag{3.12}$$

We will use the fact that (3.1) does exponentiate to a representation of $\widetilde{\mathrm{Sp}(n, R)}$; call it $\tilde{\omega}$ (see Appendix D). We can identify a small neighborhood U of the identity element in $\mathrm{Sp}(n, R)$ with a neighborhood U of the identity element in $\mathrm{Mp}(n, R)$, and view $\tilde{\omega}$ as defining a *local representation* of U. Thus, if V is a neighborhood of $e \in \mathrm{Mp}(n, R)$ so small that $V \cdot V \subset U$, we have

$$\tilde{\omega}(g_1)\tilde{\omega}(g_2) = \tilde{\omega}(g_1g_2), \qquad g_j \in V. \tag{3.13}$$

Now clearly if $g \in U$ is of the form $g = (\exp Q)k$, then $\tilde{\omega}(g)$ must be given by the right side of (3.11), so $\tilde{\omega} = \omega$ on U. Thus, in view of (3.13), we see that (3.12) holds for $g_j \in V$. Now if f is an analytic vector (with respect to $\omega(hp_2)$), both sides of (3.12) are real analytic in (g_1, g_2), so the identity (3.12) must hold for all $g_j \in \mathrm{Mp}(n, R)$. Since the set of analytic vectors is dense in $L^2(R^n)$ (see Appendix D), it follows that (3.12) holds in general. This completes the proof.

The representation ω of $\mathrm{Mp}(n, R)$ is called the metaplectic representation. As was remarked in Chapter 1, it is not irreducible. The subspaces of $L^2(R^n)$ consisting of functions which are, respectively, even or odd under the involution $f(x) \mapsto f(-x)$, are invariant under ω; ω acts irreducibly on each of them. We leave these assertions as an exercise.

For g belonging to a large subset of $\mathrm{Mp}(n, R)$, $\omega(g)$ has an explicit integral representation, which we will now derive. Let g^0 denote the image of g in $\mathrm{Sp}(n, R)$. Suppose $g^0 \in \mathrm{Sp}(n, R)$ has a generating function $A(x, x')$ such that

$$g^0(x', \xi') = (x, \xi) \tag{3.14}$$

is equivalent to

$$\xi = (\partial A/\partial x)(x, x'), \qquad \xi' = -(\partial A/\partial x')(x, x'). \tag{3.15}$$

$A(x, x')$ is a homogeneous second order polynomial in (x, x'). We can hence write

$$A(x,x') = \tfrac{1}{2}Px\cdot x - Lx\cdot x' + \tfrac{1}{2}Qx'\cdot x', \tag{3.16}$$

where P and Q are symmetric linear maps on R^n, $L \in \mathrm{End}(R^n)$. We will say $g^0 \in \mathrm{Sp}_{\mathrm{reg}}(n, R)$ if this holds with $\det L \neq 0$. Such g^0 has two preimages in $\mathrm{Mp}(n, R)$, and a choice of preimage coincides with a choice of $(\det L)^{1/2}$. We denote by $\mathrm{Mp}_{\mathrm{reg}}(n, R)$ the preimage of $\mathrm{Sp}_{\mathrm{reg}}(n, R)$. Clearly $\mathrm{Sp}_{\mathrm{reg}}(n, R)$ is open and dense in $\mathrm{Sp}(n, R)$.

PROPOSITION 3.3. *Let $g \in \mathrm{Mp}_{\mathrm{reg}}(n, R)$ determine A and $(\det L)^{1/2}$, via (3.14)–(3.16). Then*

$$\omega(g)u(x) = (2\pi)^{-n/2}(\det L)^{1/2}\int e^{iA(x,y)}u(y)\,dy. \tag{3.17}$$

PROOF. We analyze the right side of (3.17) as a composition of several unitary operators in the metaplectic representation, applied to u. Write

$$e^{iA(x,y)}u(y) = e^{(1/2)iPx\cdot x}e^{-iLx\cdot y}e^{(1/2)iQy\cdot y}u(y). \tag{3.18}$$

Then the right side of (3.17) is equal to

$$\mathcal{P}\mathcal{L}\mathcal{Q}u(x) \tag{3.19}$$

where

$$\mathcal{Q}u(x) = e^{(1/2)iQx\cdot x}u(x), \tag{3.20}$$

$$\mathcal{L}u(x) = (2\pi)^{-n/2}(\det L)^{1/2}\int e^{iLx\cdot y}u(y)\,dy, \tag{3.21}$$

and

$$\mathcal{P}u(x) = e^{(1/2)iPx\cdot x}u(x). \tag{3.22}$$

Note that if $Q(x,\xi) = Qx\cdot x$, then $Q(X, D)u(x) = Qx\cdot xu(x)$ and $e^{iQ(X,D)}u(x) = e^{iQx\cdot x}u(x)$, so the unitary operators (3.20) and (3.22) are in the metaplectic representation. We can write (3.21) as

$$\mathcal{L}u = L^*\mathcal{F}u \tag{3.23}$$

where $\mathcal{F}$ is the Fourier transform:

$$\mathcal{F}u(x) = (2\pi)^{-n/2}\int e^{ix\cdot y}u(y)\,dy, \tag{3.24}$$

and

$$L^*v(x) = (\det L)^{1/2}v(Lx). \tag{3.25}$$

The operator L^* is clearly unitary. It is also in the metaplectic representation. In fact, the last line of (3.2) defines a representation of the Lie algebra of

$$\mathrm{Gl}(n, R) \subset \mathrm{Sp}(n, R) \quad \left(\text{via } A \mapsto \begin{pmatrix} A & 0 \\ 0 & -A^t \end{pmatrix}\right)$$

which gives rise to the representation of $\mathrm{Ml}(n,R)$, the 2-fold cover of $\mathrm{Gl}(n,R)$ lying in $\mathrm{Mp}(n,R)$, defined by (3.25). Also the Fourier transform belongs to the metaplectic representation. In fact, the map $J \in \mathrm{Sp}(n,R)$ given by $J(x,\xi) = (\xi,-x)$ defines an automorphism α of the Heisenberg group such that the representations π_1 and $\pi_1 \circ \alpha$ of $\mathbf{H}^n$ are conjugate by $\omega(g)$ if g is a preimage of J (i.e., $\pi(g) = J$ where $\pi\colon \mathrm{Mp}(n,R) \to \mathrm{Sp}(n,R)$ is the natural projection). The Fourier transform accomplishes this and is so characterized to within a scalar. Indeed, a simple calculation, using $J = \exp(\frac{1}{2}\pi Q)$, $Q(x,\xi) = |x|^2 + |\xi|^2$, gives

$$\omega(g) = \pm\mathcal{F}, \quad \text{if } \pi(g) = J. \tag{3.26}$$

Now the right side of equation (3.17) is consequently unitary, and is equal to $\omega(g_1g_2g_3g_4)$ where $g_j \in \mathrm{Mp}(n,R)$ have just been described. A straightforward calculation shows that $g_1g_2g_3g_4 \in \mathrm{Mp}(n,R)$ is equal to the $g \in \mathrm{Mp}(n,R)$ corresponding to $(A,(\det L)^{1/2})$, and the proposition is proved.

We make note of

PROPOSITION 3.4. *Any $g \in \mathrm{Mp}(n,R)$ can be written as a product of two terms*

$$g = g_1g_2, \qquad g_j = \mathrm{Mp}_{\mathrm{reg}}(n,R). \tag{3.27}$$

We leave the proof to the reader.

As an application of Proposition 3.3, we will derive formulas for e^{isH}, where $H = -\Delta + |x|^2 = Q(X,D)$ is the harmonic oscillator Hamiltonian. Formulas for e^{isH} and e^{-sH} have already been derived in Chapter 1, by other methods. In view of (3.3), we have

$$H_Q = -2\sum \partial/\partial\theta_j \tag{3.28}$$

in this case, so the group of symplectic linear transformations generated by Q is

$$\begin{aligned}(\exp sH_Q)(x,\xi) &= ((\cos 2s)x + (\sin 2s)\xi, -(\sin 2s)x + (\cos 2s)\xi) \\ &= g_0(s)(x,\xi).\end{aligned} \tag{3.29}$$

We see that generating functions $A_s(x,x')$ satisfying (3.14), (3.15) are given by

$$A_s(x,y) = (\sin 2s)^{-1}\left[-\tfrac{1}{2}(\cos 2s)x\cdot x + x\cdot y - \tfrac{1}{2}(\cos 2s)y\cdot y\right]. \tag{3.30}$$

Thus, according to (3.17),

$$\begin{aligned}e^{isH}u(x) &= (2\pi i)^{-n/2}(\sin 2s)^{-n/2} \\ &\quad\cdot \int dy u(y)\exp\left\{-i\left[\tfrac{1}{2}(\cos 2s)(|x|^2+|y|^2) + x\cdot y\right]/\sin 2s\right\}.\end{aligned} \tag{3.31}$$

This is valid whenever s is not an integer multiple of $\pi/2$, where the correct square root in $(\sin 2s)^{-n/2}$ must be taken for n odd. Note that (3.31) analytically continues to give

$$\begin{aligned}e^{-sH}u(x) &= (2\pi)^{-n/2}(\sinh 2s)^{-n/2} \\ &\quad\cdot \int dy u(y)\exp\left\{-\left[\tfrac{1}{2}(\coth 2s)(|x|^2+|y|^2) - x\cdot y\right]/\sinh 2s\right\}.\end{aligned} \tag{3.32}$$

Compare this formula with formulas (7.14)–(7.15) from Chapter 1.

CHAPTER 12

Spinors

In previous chapters, we have seen the double coverings

$$\mathrm{SU}(2) \to \mathrm{SO}(3), \qquad \mathrm{SU}(2) \times \mathrm{SU}(2) \to \mathrm{SO}(4),$$

and

$$\mathrm{SL}(2, R) \to \mathrm{SO}_e(2,1), \qquad \mathrm{SL}(2, \mathbf{C}) \to \mathrm{SO}_e(3,1).$$

Generally, the orthogonal groups $\mathrm{SO}(n)$ and their noncompact analogues $\mathrm{SO}(p, q)$ have double coverings, denoted $\mathrm{Spin}(n)$, and more generally $\mathrm{Spin}(p, q)$. We study these spin groups here. We also consider spinor bundles and Dirac operators on manifolds with a spin structure.

1. Clifford algebras and spinors. To a real vector space V ($\dim V = n$) equipped with a quadratic form $Q(v)$, with associated bilinear form $Q(u, v)$, we associate a Clifford algebra $\mathrm{Cl}(V, Q)$, which is an algebra with unit, containing V, with the property that, for $v \in V$,

$$v \cdot v = -Q(v) \cdot 1. \tag{1.1}$$

We can define $\mathrm{Cl}(V, Q)$ as the quotient of the tensor algebra $\bigotimes V$ by the two-sided ideal J generated by $v \otimes v + Q(v) \cdot 1$, $v \in V$:

$$\mathrm{Cl}(V, Q) = \bigotimes V / J. \tag{1.2}$$

Note that, for $u, v \in V$,

$$u \cdot v + v \cdot u = -2Q(u, v) \cdot 1. \tag{1.3}$$

By construction, $\mathrm{Cl}(V, Q)$ has the following universal property. Let A_0 be any associative algebra over R, with unit, containing V as a linear subset, generated by V, and such that (1.1) holds in A_0, for all $v \in V$. Then there is a natural surjective homomorphism

$$\alpha: \mathrm{Cl}(V, Q) \to A_0. \tag{1.4}$$

If $\{e_1, \dots, e_n\}$ is a basis of V, any element of $\mathrm{Cl}(V, Q)$ can be written as a polynomial in the e_j. Since $e_j e_k = -e_k e_j - 2Q(e_j, e_k) \cdot 1$ and in particular

$e_j^2 = -Q(e_j) \cdot 1$, we can, starting with terms of highest order, rearrange each monomial in such a polynomial so the e_j appear with j in ascending order, and no exponent greater than one occurs on any e_j. In other words, each element $w \in \mathrm{Cl}(V, Q)$ can be written in the form

$$w = \sum_{i_j = 0 \text{ or } 1} a_{i_1 \cdots i_n} e_1^{i_1} \cdots e_n^{i_n}. \tag{1.5}$$

We claim this representation is unique, i.e., if (1.5) is equal to 0 in $\mathrm{Cl}(V, Q)$, then all coefficients $a_{i_1 \cdots i_n}$ vanish. This can be seen as follows. Denote by A the linear space of expressions of the form (1.5), so $\dim A = 2^n$. We have a natural linear map

$$\beta_Q: A \to \mathrm{Cl}(V, Q) \tag{1.6}$$

and by the discussion above β_Q is surjective. We want to show β_Q is injective.

First consider the (most degenerate) case $Q = 0$. In such a case, $\mathrm{Cl}(V, 0)$ is the *exterior algebra* $\bigwedge^* V$; (1.5) is customarily written

$$w = \sum_{i_j = 0 \text{ or } 1} a_{i_1 \cdots i_n} e_1^{i_1} \wedge \cdots \wedge e_n^{i_n}. \tag{1.7}$$

The rule $e_j \wedge e_k = -e_k \wedge e_j$ defines an algebraic structure on A, in this case. By the universal property mentioned above, we have a surjective homomorphism of algebras $\alpha: \mathrm{Cl}(V, 0) \to A$. To compose this with the linear map $\beta_0: A \to \mathrm{Cl}(V, 0)$, which we also see to be a homomorphism of algebras, we note that $\alpha\beta_0$ and $\beta_0\alpha$ are clearly the identity on V, which generates each algebra, so α and β_0 are inverses of each other. Hence (1.6) is seen to be an isomorphism of algebras for $Q = 0$:

$$\beta_0: A \xrightarrow{\approx} \mathrm{Cl}(V, 0) = \textstyle\bigwedge^* V. \tag{1.8}$$

Using this, we will identify A and $\bigwedge^* V$ as linear spaces. If we write

$$\textstyle\bigwedge^* V = \displaystyle\bigoplus_{k=0}^{n} \textstyle\bigwedge^k V, \tag{1.9}$$

where $\bigwedge^k V$ is the linear span of (1.7) for $i_1 + \cdots + i_n = k$, we have a natural identification of $\bigwedge^k V$ with the space of antisymmetric k-linear forms $V' \times \cdots \times V' \to R$ (V' = dual space to V), via

$$(v_1 \wedge \cdots \wedge v_k)(X_1, \ldots, X_k) = (1/k!) \sum_{\sigma \in S_k} (\operatorname{sgn} \sigma) \langle v_1, X_{\sigma(1)} \rangle \cdots \langle v_k, X_{\sigma(k)} \rangle, \tag{1.10}$$

for $v_j \in V$, $X_l \in V'$.

In order to show (1.6) is an isomorphism for general Q, we produce an algebraic structure on $\bigwedge^* V = A$, for each quadratic form Q on V. Note that such Q defines a linear map

$$\mathcal{Q}: V \to V' \tag{1.11}$$

by

$$\langle w, \mathcal{Q}(v)\rangle = Q(v,w), \qquad v, w \in V. \tag{1.12}$$

We will use both the *exterior product* on $\bigwedge^* V$, i.e., the product $v \wedge w$ in $\mathrm{Cl}(V,0)$, and the *interior product*, defined as follows. First, for $X \in V'$, we define

$$\iota_X : \bigwedge^k V \to \bigwedge^{k-1} V \tag{1.13}$$

by

$$(\iota_X \omega)(X_1, \dots, X_{k-1}) = \omega(X, X_1, \dots, X_{k-1}), \tag{1.14}$$

where we have identified $\omega \in \bigwedge^k V$ with an antisymmetric k-linear form, as described above. Then, for $v \in V$, the interior product

$$\jmath_v : \bigwedge^k V \to \bigwedge^{k-1} V \tag{1.15}$$

is defined by

$$\jmath_v = \iota_{\mathcal{Q}(v)}, \tag{1.16}$$

with $\mathcal{Q}$ given by (1.11)–(1.12). Let us consider the algebraic relations among these operators and the operators

$$\wedge_v : \bigwedge^k V \to \bigwedge^{k+1} V, \qquad v \in V, \tag{1.17}$$

defined by

$$\wedge_v (v_1 \wedge \dots \wedge v_k) = v \wedge v_1 \wedge \dots \wedge v_k. \tag{1.18}$$

The algebraic properties of these operators are the following. Clearly

$$\wedge_v \wedge_w = - \wedge_w \wedge_v \tag{1.19}$$

for all $v, w \in V$, and also a simple calculation shows that $\iota_X \iota_X = 0$ for all $X \in V'$, hence $\jmath_v \jmath_v = 0$ for all $v \in V$, and thus

$$\jmath_v \jmath_w = -\jmath_w \jmath_v \tag{1.20}$$

for all $v, w \in V$. Finally, a calculation yields

$$(\iota_X \wedge_w + \wedge_w \iota_X)\omega = \langle X, w\rangle \omega \tag{1.21}$$

and hence

$$\jmath_v \wedge_w + \wedge_w \jmath_v = Q(v,w) I. \tag{1.22}$$

In particular, if we define linear operators M_v on $\bigwedge^* V$ by

$$M_v = \wedge_v - \jmath_v, \qquad v \in V, \tag{1.23}$$

we have the anticommutation relations

$$M_v M_w + M_w M_v = -2Q(v,w) I, \tag{1.24}$$

for $v, w \in V$, as a consequence of (1.19)–(1.22). Note also that

$$M_v v = -Q(v) \cdot 1 \tag{1.25}$$

and

$$M_v 1 = v. \tag{1.26}$$

Thus the algebra $\mathcal{M}$ in $\operatorname{End}(\bigwedge^* V)$ generated by $\{M_v : v \in V\}$ naturally contains V, and by the anticommutation relations (1.24) and the universal property of $\mathrm{Cl}(V, Q)$, we conclude that there is a natural surjective homomorphism of algebras

$$\mu: \mathrm{Cl}(V, Q) \to \mathcal{M}, \tag{1.27}$$

extending $v \mapsto M_v$. Define a linear map

$$\gamma: \mathrm{Cl}(V, Q) \to \textstyle\bigwedge^* V \tag{1.28}$$

by

$$\gamma(x) = \mu(x)(1).$$

If we use (1.8) to identify A and $\bigwedge^* V$, we have a commutative diagram

$$\begin{array}{ccc} \mathrm{Cl}(V,Q) & \xrightarrow{\gamma} & \bigwedge^* V \\ {\scriptstyle\beta_Q}\uparrow & & \uparrow{\scriptstyle\eta} \\ A & \underset{\beta_0}{\xrightarrow{\approx}} & \bigwedge^* V \end{array}$$

defining a map $\eta: \bigwedge^* V \to \bigwedge^* V$. We see that

$$\gamma\beta_Q(x) - \beta_0(x) \in \bigoplus_{l<k} \textstyle\bigwedge^l V \quad \text{if } \beta_0(x) \in \bigwedge^k V.$$

Hence $\eta = I + \sigma$ with σ nilpotent on $\bigwedge^* V$. This implies η is a linear isomorphism. The commutativity of the diagram above then implies that β_Q is injective, hence an isomorphism, so γ is an isomorphism. This implies μ is injective, hence an isomorphism. We have proved

PROPOSITION 1.1. *The maps* (1.6), (1.27), *and* (1.28) *are all isomorphisms.*

Thus we make the identification

$$\mathrm{Cl}(V, Q) = \textstyle\bigwedge^* V, \tag{1.29}$$

via (1.28). In particular

$$\dim \mathrm{Cl}(V, Q) = 2^n \quad \text{if } \dim V = n. \tag{1.30}$$

From now on, we will suppose Q is *nondegenerate*, though not necessarily positive definite.

The Clifford algebra has a natural $\mathbf{Z}_2$ grading. Set

$$\begin{aligned} \mathrm{Cl}^0(V, Q) &= \text{linear span of } \{v_1 \cdots v_k : v_j \in V,\ k \text{ even}\}, \\ \mathrm{Cl}^1(V, Q) &= \text{linear span of } \{v_1 \cdots v_k : v_j \in V,\ k \text{ odd}\}. \end{aligned} \tag{1.31}$$

It follows that

$$\begin{aligned} \mathrm{Cl}^0 \cdot \mathrm{Cl}^0 \subset \mathrm{Cl}^0, \qquad \mathrm{Cl}^1 \cdot \mathrm{Cl}^0 \subset \mathrm{Cl}^1, \\ \mathrm{Cl}^0 \cdot \mathrm{Cl}^1 \subset \mathrm{Cl}^1, \qquad \mathrm{Cl}^1 \cdot \mathrm{Cl}^1 \subset \mathrm{Cl}^0. \end{aligned} \tag{1.32}$$

We are now going to define the spinor groups $\mathrm{Pin}(V,Q)$ and $\mathrm{Spin}(V,Q)$. We set

$$\mathrm{Pin}(V,Q) = \{v_1 \cdots v_k \in \mathrm{Cl}(V,Q) : v_j \in V,\ Q(v_j) = \pm 1\}, \tag{1.33}$$

with the induced multiplication. Since $(v_1 \cdots v_k)(v_k \cdots v_1) = \pm 1$, it follows that $\mathrm{Pin}(V,Q)$ is a group. We can define an action of $\mathrm{Pin}(V,Q)$ on V as follows. If $u \in V$ and $x \in V$, then $ux + xu = -2Q(x,u) \cdot 1$ implies

$$uxu = -xuu - 2Q(x,u)u = Q(u)x - 2Q(x,u)u. \tag{1.34}$$

Note that, if also $y \in V$,

$$\begin{aligned} Q(uxu, uyu) &= Q(u)^2 Q(x,y) \\ &= Q(x,y) \quad \text{if } Q(u) = \pm 1. \end{aligned} \tag{1.35}$$

Thus, if $u = v_1 \cdots v_k \in \mathrm{Pin}(V,Q)$, and if we define a conjugation on $\mathrm{Cl}(V,Q)$ by

$$u^* = v_k \cdots v_1, \qquad v_j \in V, \tag{1.36}$$

it follows that

$$x \mapsto uxu^*, \qquad x \in V, \tag{1.37}$$

is a Q-isometry on V for each $u \in \mathrm{Pin}(V,Q)$. It will be more convenient to use

$$u^{\#} = u^*(Q(v_1) \cdots Q(v_k))^{-1} \tag{1.38}$$

for $u = v_1 \cdots v_k \in \mathrm{Pin}(V,Q)$. Then we have a group homomorphism

$$\tau : \mathrm{Pin}(V,Q) \to \mathrm{O}(V,Q), \tag{1.39}$$

where $\mathrm{O}(V,Q)$ denotes the orthogonal group, defined by

$$\tau(u)x = uxu^{\#}, \qquad x \in V,\ u \in \mathrm{Pin}(V,Q). \tag{1.40}$$

Note that if $u \in V$, then, by (1.34),

$$\tau(u)x = x - Q(u,u)^{-1}Q(x,u)u, \tag{1.41}$$

which is reflection across the hyperplane in V orthogonal to u. We will next show that any element of $\mathrm{O}(V,Q)$ can be written as a product of isometries of the type (1.41), so τ is onto.

LEMMA 1.2. *Any $g \in \mathrm{O}(V,Q)$ can be written as a product of the reflections* (1.41).

PROOF. If we pick a coordinate system in which Q is diagonal, we see that any $g \in \mathrm{O}(V,Q)$ is a product of reflections across coordinate hyperplanes and an element of the connected component of the identity $\mathrm{O}_e(V,Q) \approx \mathrm{SO}_e(p,q)$, so we can suppose $g \in \mathrm{SO}_e(p,q)$. We first consider the case of Q definite, so $g \in \mathrm{O}_e(V,Q) = \mathrm{SO}(n)$. In this case, we can write $g = \exp X$ for some $X \in \mathrm{so}(n)$, a skew symmetric real $n \times n$ matrix. Choosing an orthonormal basis of R^n with respect to which X is an orthogonal sum of 2×2 matrices, plus perhaps a zero matrix, we can reduce our considerations to a product of planar rotations, hence to a single planar rotation, which is trivially representable as a product of two reflections.

Now we consider the case of a general nondegenerate Q, i.e., $g \in \mathrm{SO}_e(p,q)$, $p+q=n$. As described in Chaper 9, we have a maximal compact subgroup $K \approx \mathrm{SO}(p) \times \mathrm{SO}(q)$. For any $g \in K$, the fact that g is a product of reflections follows from the argument in the previous paragraph. Now, for any $h \in G = \mathrm{SO}_e(p,q)$, the conjugated operator $h^{-1}gh$ can be regarded as the operator g viewed in another coordinate system. Since the description (1.41) is coordinate invariant, this implies any conjugate g' in

$$\mathfrak{A} = \{h^{-1}gh : g \in K,\ h \in G = \mathrm{SO}_e(p,q)\}$$

is a product of reflections (1.41). However, it is easy to see that this set contains an *open* set $\mathcal{O}$ in G. Since, for general $X \in \mathfrak{k}$, $Y \in \mathfrak{g}$, we have

$$\begin{aligned}\exp(sY)\exp(tX)\exp(-sY) &= \exp(te^{s\,\mathrm{ad}\,Y}X)\\ &= \exp(tX + st[Y,X] + O(s^2)),\end{aligned}$$

this openness can be deduced from the implicit function theorem, together with the observation that $\{X + [X,Y] : X \in \mathfrak{k},\ Y \in \mathfrak{g}\} = \mathfrak{g}$ for $\mathfrak{g} = \mathrm{so}(p,q)$. Thus any element of $\mathcal{O}$, and hence also any element of $\mathcal{O}^{-1}$, is a product of reflections (1.41). But any element of G is a finite product of elements of $\mathcal{O}$ and $\mathcal{O}^{-1}$, for any open set $\mathcal{O}$, so the proof is complete.

Note that each isometry (1.41) is orientation reversing. Thus, if we define

$$\begin{aligned}\mathrm{Spin}(V,Q) &= \{v_1 \cdots v_k : v_j \in V,\ Q(v_j) = \pm 1,\ k \text{ even}\}\\ &= \mathrm{Pin}(V,Q) \cap \mathrm{Cl}^0(V,Q)\end{aligned} \tag{1.42}$$

then

$$\tau : \mathrm{Spin}(V,Q) \to \mathrm{SO}(V,Q), \tag{1.43}$$

and in fact $\mathrm{Spin}(V,Q)$ is the inverse image of $\mathrm{SO}(V,Q)$ under τ in (1.39). We now show that τ is a two-fold covering map.

PROPOSITION 1.3. *τ is a two-fold covering map. In fact,* $\ker\tau = \{\pm 1\}$.

PROOF. Note that $\pm 1 \in \mathrm{Spin}(Q,V) \subset \mathrm{Cl}(V,Q)$ and ± 1 acts trivially on V, via (1.40). Now, if $u = v_1 \cdots v_k \in \ker\tau$, we know k is even (since orientation is preserved), so $uu^\# = 1$. Also, since $uxu^\# = x$ for all $x \in V$, we have $ux = xu$, so $xux = -Q(x)u$, $x \in V$. Now pick an orthonormal basis $e_1, \ldots, e_n$ of V, so $Q(e_1) = \cdots = Q(e_p) = 1$, and $Q(e_{p+1}) = \cdots = Q(e_{p+q}) = -1$, $n = p+q$. We have $Q(e_j)u = -e_jue_j$, if $u \in \ker\tau$. If we write $u = \sum a_{i_1\cdots i_n} e_1^{i_1} \cdots e_n^{i_n}$ where each i_j is equal to 0 or 1, $i_1 + \cdots + i_n$ even, we have, for all j,

$$u = \sum (-1)^{i_j} a_{i_1\cdots i_n} e_1^{i_1} \cdots e_n^{i_n} \quad \text{if } u \in \ker\tau.$$

Hence $i_j = 0$ for all j, so u is a scalar. Hence $u = \pm 1$.

We next consider the connectivity properties of $\mathrm{Spin}(V,Q)$.

PROPOSITION 1.4. *$\operatorname{Spin}(V,Q)$ is the connected 2-fold cover of $\operatorname{SO}(V,Q)$ if Q is positive definite or negative definite.*

PROOF. It suffices to connect $-1 \in \operatorname{Spin}(V,Q)$ to the identity element $1 \in \operatorname{Spin}(V,Q)$ via a continuous curve in $\operatorname{Spin}(V,Q)$. In fact, pick orthogonal unit vectors e_1, e_2, with $Q(e_1) = Q(e_2) = \pm 1$, and set

$$\gamma(t) = e_1 \cdot (-\cos te_1 + \sin te_2), \qquad 0 \le t \le \pi. \tag{1.44}$$

If Q is nondegenerate but not definite, then it is easy to see that $\operatorname{SO}(V,Q)$ has two connected components, rather than one. In this case, let $\operatorname{SO}_e(V,Q)$ denote the connected component of the identity in $\operatorname{SO}(V,Q)$, and let $\operatorname{Spin}_e(V,Q)$ denote the connected component of the identity in $\operatorname{Spin}(V,Q)$.

PROPOSITION 1.5. *Except in the case where Q has signature $(1,1)$,*

$$\tau \colon \operatorname{Spin}_e(V,Q) \to \operatorname{SO}_e(V,Q)$$

is a 2-fold covering map.

PROOF. You can still pick orthogonal vectors e_1, e_2 with $Q(e_1) = Q(e_2) = \pm 1$, and then (1.44) still gives a curve connecting 1 to -1 in $\operatorname{Spin}_e(V,Q)$.

Note that if Q has signature $(1,1)$, then $\operatorname{SO}_e(V,Q) \approx R$ is simply connected, so a 2-fold cover cannot be connected.

If Q is definite then, as is well known, $\operatorname{SO}(V,Q)$ has fundamental group $\mathbf{Z}_2$, and $\operatorname{Spin}(V,Q)$ is hence simply connected. It does not follow that $\operatorname{Spin}_e(V,Q)$ is simply connected if Q is indefinite. For example, as we saw in Chapter 8, $\operatorname{SL}(2,R)/\mathbf{Z}_2 \approx \operatorname{SO}_e(2,1)$, so $\operatorname{Spin}_e(2,1) \approx \operatorname{SL}(2,R)$. However, $\operatorname{SL}(2,R)$ is not simply connected; it is homeomorphic to $S^1 \times R^2$.

Before defining the spaces of spinors, we want to take a brief look at the structure of the complexified Clifford algebra

$$\mathbb{C}l(V,Q) = \mathbf{C} \otimes \operatorname{Cl}(V,Q). \tag{1.45}$$

If $V = R^n$, this is just the Clifford algebra over $\mathbf{C}$ of $\mathbf{C}^n$ with a nondegenerate bilinear form. Since on $\mathbf{C}^n$ all nondegenerate bilinear forms are equivalent, the algebraic structures of $\mathbb{C}l(V,Q)$ and $\mathbb{C}l(V,Q')$ are isomorphic, for any two nondegenerate forms Q, Q'. We can write

$$\mathbb{C}l(V,Q) \approx \mathbb{C}l(n), \quad n = \dim V. \tag{1.46}$$

PROPOSITION 1.6. *We have the isomorphisms*

$$\mathbb{C}l(1) \approx \mathbf{C} + \mathbf{C}, \tag{1.47}$$

$$\mathbb{C}l(2) \approx \operatorname{End}_{\mathbf{C}}(\mathbf{C}^2), \tag{1.48}$$

$$\mathbb{C}l(n+2) \approx \mathbb{C}l(n) \otimes \mathbb{C}l(2). \tag{1.49}$$

PROOF. We leave (1.47) and (1.48) as exercises. As for (1.49), if you embed R^{n+2} into $\mathbb{C}l(n) \otimes \mathbb{C}l(2)$ by picking an orthonormal basis $e_1, \dots, e_{n+2}$ and taking

$$\begin{aligned} e_j &\mapsto e_j \otimes e_{n+1}e_{n+2} && \text{for } 1 \le j \le n, \\ e_j &\mapsto 1 \otimes e_j && \text{for } j = n+1, n+2, \end{aligned}$$

the universal property of $\mathbf{Cl}(n+2)$ gives the isomorphism (1.49).

By induction, we deduce the following.

COROLLARY 1.7. *We have the isomorphisms of algebras*

$$\mathrm{Cl}(2k) \approx \mathrm{End}(\mathbf{C}^{2^k}), \tag{1.50}$$

$$\mathrm{Cl}(2k+1) \approx \mathrm{End}(\mathbf{C}^{2^k}) \oplus \mathrm{End}(\mathbf{C}^{2^k}). \tag{1.51}$$

There is also an analysis of the algebraic structure of $\mathrm{Cl}(V, Q)$, which we shall not give in detail. If $V = R^n$, $n = p + q$, and Q is defined by

$$Q(a_1e_1 + \cdots + a_ne_n) = a_1^2 + \cdots + a_p^2 - a_{p+1}^2 - \cdots - a_{p+q}^2,$$

let us adopt the notation

(1.52)

$$\mathrm{Cl}(p, q) = \mathrm{Cl}(R^n, Q), \quad \mathrm{Spin}(p, q) = \mathrm{Spin}(R^n, Q), \quad \mathrm{Spin}(n) = \mathrm{Spin}(n, 0).$$

The proof of (1.49) shows that

$$\mathrm{Cl}(p, 0) \otimes \mathrm{Cl}(0, 2) \approx \mathrm{Cl}(p+2, 0). \tag{1.53}$$

From this it is possible to deduce

$$\mathrm{Cl}(n+8, 0) \approx \mathrm{Cl}(n, 0) \otimes \mathrm{Cl}(8, 0). \tag{1.54}$$

In low dimensions, one has

$$\begin{gathered} \mathrm{Cl}(1,0) \approx \mathbf{C}, \qquad \mathrm{Cl}(0,1) \approx R \oplus R, \\ \mathrm{Cl}(2,0) \approx \mathbf{H}, \qquad \mathrm{Cl}(0,2) \approx \mathrm{End}(R^2), \\ \mathrm{Cl}(4,0) \approx \mathbf{H}(2), \qquad \mathrm{Cl}(3,1) \approx \mathbf{H}(2), \qquad \mathrm{Cl}(2,2) \approx \mathrm{End}(R^4). \end{gathered} \tag{1.55}$$

For a more complete description, see Lawson and Michelson [**149**]. Here, $\mathbf{H}$ is the quarternionic field, and $\mathbf{H}(k)$ is the ring of $k \times k$ quarternionic matrices.

We want to associate a space of spinors with (V, Q). In fact, one needs to impose more structure to obtain a spinor space; there is no canonically defined $S(V, Q)$. This fact will have a profound influence on the structure of spinor bundles, as we will see in the following sections.

As one example, suppose $\dim V = 2k$ is even, and that V has a complex structure J (i.e., a linear map $J: V \to V$ with $J^2 = -I$) which is an isometry relative to Q, i.e., $Q(u, v) = Q(Ju, Jv)$. This implies $Q(u, Jv) = -Q(Ju, v)$. Denote the complex vector space (V, J) by $\mathcal{V}$. On $\mathcal{V}$ we have the Hermitian form (nondegenerate, but perhaps indefinite if Q is)

$$(u, v) = Q(u, v) + iQ(u, Jv). \tag{1.56}$$

Form the complex exterior algebra

$$\Lambda_{\mathbf{C}}^* \mathcal{V} = \bigoplus_{j=0}^{k} \Lambda_{\mathbf{C}}^j \mathcal{V} \tag{1.57}$$

(note that $\bigwedge^* V = \bigoplus_{j=0}^{2k} \bigwedge^j V$) with its natural Hermitian form. For $v \in \mathcal{V}$, one has the exterior product $v\wedge: \bigwedge_{\mathbf{C}}^j \to \bigwedge_{\mathbf{C}}^{j+1} \mathcal{V}$; denote its conjugate by the interior product $j_v: \bigwedge_{\mathbf{C}}^{j+1} \mathcal{V} \to \bigwedge_{\mathbf{C}}^j \mathcal{V}$. Set

$$M_v\varphi = v \wedge \varphi - j_v\varphi, \qquad v \in \mathcal{V},\ \varphi \in \textstyle\bigwedge_{\mathbf{C}}^* \mathcal{V}. \tag{1.58}$$

Note that $v \wedge \varphi$ is $\mathbf{C}$-linear in v, $j_v\varphi$ conjugate linear in v; so M_v is R-linear in v. As in the analysis of (1.24) we obtain $M_vM_v\varphi = -Q(v)\varphi$, so M_v extends to a homomorphism of $\mathbf{C}$-algebras:

$$\mu: \mathrm{Cl}(V,Q) \to \mathrm{End}_{\mathbf{C}}\left(\textstyle\bigwedge_{\mathbf{C}}^* \mathcal{V}\right). \tag{1.59}$$

Now $\ker \mu$ would have to be a two-sided ideal, and since clearly $v \neq 0$, $v \in V \Rightarrow v \notin \ker \mu$, and since $\mathbf{Cl}(V,Q)$ is isomorphic to $\mathrm{End}(\mathbf{C}^{2^k})$, which has no proper two-sided ideals, it follows that μ is injective. Since both sides of (1.59) have the same dimension, μ is an isomorphism.

Using the inclusions

$$\mathrm{Pin}(V,Q) \subset \mathrm{Cl}(V,Q) \subset \mathbf{Cl}(V,Q), \tag{1.60}$$

we have the representation

$$\rho: \mathrm{Pin}(V,Q) \to \mathrm{Aut}\left(\textstyle\bigwedge_{\mathbf{C}}^* \mathcal{V}\right). \tag{1.61}$$

Thus, when V has a complex hermitian structure, we can associate the spinor space

$$S(V,Q,J) = \textstyle\bigwedge_{\mathbf{C}}^* \mathcal{V}, \tag{1.62}$$

and we have canonically defined a representation of $\mathrm{Pin}(V,Q)$ on $S(V,Q,J)$.

PROPOSITION 1.8. *The representation ρ of* $\mathrm{Pin}(V,Q)$ *is irreducible.*

PROOF. Since the $\mathbf{C}$-subalgebra of $\mathbf{Cl}(V,Q)$ generated by $\mathrm{Pin}(V,Q)$ is all of $\mathbf{Cl}(V,Q)$, the isomorphism (1.59) makes this clear.

The restriction of ρ to $\mathrm{Spin}(V,Q)$ is not irreducible. In fact, set

$$\begin{aligned} S_+(V,Q,J) &= \bigoplus_{j\,\text{even}} \textstyle\bigwedge_{\mathbf{C}}^j \mathcal{V}, \\ S_-(V,Q,J) &= \bigoplus_{j\,\text{odd}} \textstyle\bigwedge_{\mathbf{C}}^j \mathcal{V}. \end{aligned} \tag{1.63}$$

It is clear that under ρ, the action of $\mathrm{Spin}(V,Q)$ preserves both S_+ and S_-. In fact, we have (1.59) restricting to

$$\mu: \mathrm{Cl}^0(V,Q) \to \mathrm{End}_{\mathbf{C}}(S_+(V,Q,J)) \oplus \mathrm{End}_{\mathbf{C}}(S_-(V,Q,J)), \tag{1.64}$$

this map being an isomorphism. On the other hand, if $z \in \mathrm{Cl}^1(V,Q)$, then

$$\mu(z): S_+ \to S_- \quad \text{and} \quad \mu(z): S_- \to S_+. \tag{1.65}$$

From (1.64), we get representations

$$D_{1/2}^{\pm}: \mathrm{Spin}(V,Q) \to \mathrm{Aut}(S_\pm(V,Q,J)) \tag{1.66}$$

which are irreducible.

If $V = R^{2k}$ with its standard (positive) Euclidean metric, standard orthonormal basis $e_1, \ldots, e_{2k}$, we impose the complex structure

$$Je_i = e_{i+k}, \quad Je_{i+k} = -e_i, \qquad 1 \leq i \leq k, \tag{1.67}$$

and we set

$$\begin{aligned} S(2k) &= S(R^{2k}, \| \|^2, J), \\ S_\pm(2k) &= S_\pm(R^{2k}, \| \|^2, J), \end{aligned} \tag{1.68}$$

and (1.66) defines representations

$$D^\pm_{1/2} \text{ of } \mathrm{Spin}(2k) \text{ on } S_\pm(2k). \tag{1.69}$$

If $V = R^{2k-1}$, we use the isomorphism

$$\mathrm{Cl}(2k-1) \to \mathrm{Cl}^0(2k) \tag{1.70}$$

produced by the map

$$v \mapsto ve_{2k}, \qquad v \in R^{2k-1}. \tag{1.71}$$

Then the inclusion

$$\mathrm{Spin}(2k-1) \subset \mathrm{Cl}(2k-1) \approx \mathrm{Cl}^0(2k) \tag{1.72}$$

gives an inclusion

$$\mathrm{Spin}(2k-1) \subset \mathrm{Spin}(2k), \tag{1.73}$$

so we have a representation

$$D^+_{1/2}: \mathrm{Spin}(2k-1) \to \mathrm{Aut}\, S_+(2k). \tag{1.74}$$

We also have a representation $D^-_{1/2}$ of $\mathrm{Spin}(2k-1)$ on $S_-(2k)$, but these two representations are equivalent. They are intertwined by the map $\mu(e_{2k}): S_+(2k) \to S_-(2k)$.

We produce the spinor representation of $\mathrm{Spin}(p,q)$ via the inclusions

$$\mathrm{Spin}(p,q) \subset \mathrm{Cl}^0(p,q) \subset \mathbf{C} \otimes \mathrm{Cl}^0(p,q) \approx \mathbf{Cl}(n); \qquad n = p+q, \tag{1.75}$$

and the action of $\mathbf{Cl}(n)$ on $S_\pm(2k)$, $n = 2k$ or $2k-1$.

Let us remark that the definition (1.58) of M_v on $S(V,Q,J)$ shows that M_v is skew adjoint with respect to the natural Hermitian metric on $S(V,Q,J) = \bigwedge^*_{\mathbf{C}} \mathcal{V}$, for any $v \in V$. It follows that, if $\varphi, \psi \in \bigwedge^*_{\mathbf{C}} \mathcal{V}$, $(v\varphi, v\psi) = -(\varphi, vv\psi) = Q(v)(\varphi,\psi)$. In particular this implies $\rho: \mathrm{Spin}_e(V,Q) \to \mathrm{U}\left(\bigwedge^*_{\mathbf{C}} \mathcal{V}, (\ ,\ )\right)$, the space of unitary maps. If Q is positive definite, so is the Hermitian metric on $\bigwedge^*_{\mathbf{C}} \mathcal{V}$, and hence the representations $D^\pm_{1/2}$ of $\mathrm{Spin}(2k)$ are unitary. Of course, since $\mathrm{Spin}(2k)$ and $\mathrm{Spin}(2k-1)$ are compact, their representations would necessarily be unitarizable. Since the representations $D^\pm_{1/2}$ are irreducible, the invariant Hermitian metric on the representation space is unique, up to a scalar multiple.

2. Spinor bundles and the Dirac operator. Let M be a smooth manifold with a nondegenerate metric tensor g. Say $n = \dim M$ and g has signature (p,q), $p+q=n$. Associated with the tangent bundle TM is the bundle of orthonormal frames, which is a principal $\mathrm{O}(p,q)$ bundle. Suppose we can reduce the group to $\mathrm{SO}_e(p,q)$. In case the metric is positive definite, this amounts to putting an orientation on M; if M is Lorentzian, it amounts to also imposing a causal structure on M. So we have the principal $\mathrm{SO}_e(p,q)$ bundle

$$P \to M. \tag{2.1}$$

Now recall the double covering

$$\mathrm{Spin}_e(p,q) \to \mathrm{SO}_e(p,q). \tag{2.2}$$

Certainly, locally, the principal $\mathrm{SO}_e(p,q)$ bundle can be lifted to a principal $\mathrm{Spin}_e(p,q)$ bundle, doubly covering P. There are topological obstructions to implementing this globally, which we shall not discuss here (see [**196**]), beyond mentioning that the obstruction is an element of $H^2(M,\mathbf{Z}_2)$, whose vanishing guarantees the existence of a lift, and in general there are inequivalent bundles lifting $P \to M$, parametrized by elements of $H^1(M,\mathbf{Z}_2)$. We will suppose that

$$\tilde{P} \to M \tag{2.3}$$

is a given lift of P to a $\mathrm{Spin}_e(p,q)$ bundle.

The Levi-Civita connection determined by the metric tensor g gives a connection on $P \to M$, which in turn determines a unique connection on $\tilde{P} \to M$.

Using the representation

$$\rho\colon \mathrm{Spin}_e(p,q) \to \mathrm{Aut}(S(2k)) \tag{2.4}$$

from §1, where $p+q=n=2k$ or $2k-1$, we form the bundle of spinors associated with P:

$$S(\tilde{P}) = \tilde{P} \times_\rho S(2k). \tag{2.5}$$

Thus $S(\tilde{P})$ is a vector bundle over M. The sections of $S(\tilde{P})$ are in natural one-to-one correspondence with the functions on $\tilde{P}$, taking values in the vector space $S(2k)$, which satisfy the compatibility conditions

$$f(gp_0) = \rho(g)f(p_0), \qquad p_0 \in \tilde{P},\ g \in \mathrm{Spin}_e(p,q). \tag{2.6}$$

We also have the Clifford bundle

$$\mathrm{Cl}(TM,g) = \tilde{P} \times_\tau \mathrm{Cl}(p,q) \tag{2.7}$$

where the representation τ of $\mathrm{Spin}_e(p,q)$ on $\mathrm{Cl}(p,q)$ is given by

$$\tau(u)x = uxu^{\#}, \qquad u \in \mathrm{Spin}_e(p,q),\ x \in \mathrm{Cl}(p,q). \tag{2.8}$$

Compare with (1.16).

Recall from §1 that $S(2k)$ is a $\mathrm{Cl}(p,q)$ module. It is an extremely important fact that this extends to the bundle level.

PROPOSITION 2.1. *The spinor bundle* $S(\tilde{P})$ *is a natural* $\mathrm{Cl}(TM, g)$ *module.*

PROOF. Given a section u of $\mathrm{Cl}(TM, g)$ and a section φ of $S(\tilde{P})$, we need to define $u \cdot \varphi$ as a section of $S(\tilde{P})$. We regard u as a function on $\tilde{P}$ with values in $\mathrm{Cl}(p, q)$ and φ as a function on $\tilde{P}$ with values in $S(2k)$. Then $u \cdot \varphi$ is a function on $\tilde{P}$ with values in $S(2k)$; we need to verify the compatibility condition (2.6). Indeed, for $p_0 \in \tilde{P}$, $g \in \mathrm{Spin}_e(p, q)$,

$$\begin{aligned} u \cdot \varphi(gp_0) &= \tau(g)u(p_0) \cdot \rho(g)\varphi(p_0) \\ &= gu(p_0)g^{\#}g\varphi(p_0) = gu(p_0) \cdot \varphi(p_0) \end{aligned} \tag{2.9}$$

since $gg^{\#} = 1$ for $g \in \mathrm{Spin}_e(p, q)$. This completes the proof.

The natural connection on $\tilde{P}$ gives a covariant derivative on any vector bundle $E = \tilde{P} \times_\gamma V$, where γ is a representation of $\mathrm{Spin}_e(p, q)$ on V. In fact, if X is a vector field on M, X^H its horizontal lift to $\tilde{P}$, and if a section ω of E is identified with a function $\omega^{\#}$ on $\tilde{P}$ with values in V, satisfying the compatibility condition analogous to (2.6), then $\nabla_X \omega$, as a section of E, is identified with $X^H \omega^{\#}$. In particular, we have covariant derivatives on the bundles $\mathrm{Cl}(TM, g)$ and $S(\tilde{P})$, and it is clear that the appropriate derivation identities hold. For example, if u is a section of $\mathrm{Cl}(TM, g)$, φ a section of $S(\tilde{P})$,

$$\nabla_X(u \cdot \varphi) = (\nabla_X u) \cdot \varphi + u \cdot (\nabla_X \varphi). \tag{2.10}$$

We are now in a position to define the Dirac operator on $\Gamma(M, S(\tilde{P}))$, the space of smooth sections of $S(\tilde{P})$. The covariant derivative defines

$$\nabla: \Gamma(M, S(\tilde{P})) \to \Gamma(M, T^*M \otimes S(\tilde{P})). \tag{2.11}$$

Using the metric g we identify T^*M with TM, and write

$$\nabla: \Gamma(M, S(\tilde{P})) \to \Gamma(M, TM \otimes S(\tilde{P})). \tag{2.12}$$

Meanwhile, we have the natural inclusion

$$TM \hookrightarrow \mathrm{Cl}(TM, g), \tag{2.13}$$

and hence Clifford multiplication induces a map

$$m: TM \otimes S(\tilde{P}) \to S(\tilde{P}). \tag{2.14}$$

DEFINITION. The Dirac operator

$$\not{D}: \Gamma(M, S(\tilde{P})) \to \Gamma(M, S(\tilde{P})) \tag{2.15}$$

is given by

$$\not{D}\varphi = m(\nabla\varphi). \tag{2.16}$$

If we pick a local orthonormal frame field $e_1, \ldots, e_n$ on TM, so $g(e_j, e_j)$ is 1 for $1 \le j \le p$, -1 for $p + 1 \le j \le p + q$, then, locally,

$$\not{D}\varphi = \sum_j e_j \cdot \nabla_{e_j}\varphi. \tag{2.17}$$

Note that

$$\mathcal{D}\colon \Gamma(M, S_\pm(\tilde{P})) \to \Gamma(M, S_\mp(\tilde{P})) \tag{2.18}$$

where $S_\pm(\tilde{P}) = \tilde{P} \times_\rho S_\pm(2k)$.

The operator $\mathcal{D}$ is selfadjoint with respect to the natural inner product on the space of sections of $S(\tilde{P})$, as we will now show. If $\langle\ ,\ \rangle$ denotes the natural inner product on $S(\tilde{P})$, or on $S(2k)$, we define

$$(\varphi, \psi) = \int_M \langle \varphi, \psi\rangle(x)\, d\,\mathrm{vol}(x) = \int_{\tilde{P}} \langle \varphi(p), \psi(p)\rangle\, d\,\mathrm{vol}(p). \tag{2.19}$$

PROPOSITION 2.2. *We have, for φ or ψ with compact support,*

$$(\mathcal{D}\varphi, \psi) = (\varphi, \mathcal{D}\psi). \tag{2.20}$$

PROOF. Write

$$(\mathcal{D}\varphi, \psi) - (\varphi, \mathcal{D}\psi) = \int_M [\langle \mathcal{D}\varphi, \psi\rangle - \langle \varphi, \mathcal{D}\psi\rangle]\, d\,\mathrm{vol}(x). \tag{2.21}$$

The integrand is equal to

$$\sum_j \langle e_j \nabla_{e_j} \varphi, \psi\rangle - \langle \varphi, e_j \nabla_{e_j} \psi\rangle, \tag{2.22}$$

for any local orthonormal frame $e_1, \dots, e_n$. For a given $x_0 \in M$, pick the orthonormal frame field $e_1, \dots, e_n$ on a neighborhood of x_0 so that, at x_0, $\nabla_{e_j} e_k = 0$. Thus, at x_0, the expression (2.22) is equal to

$$-\sum_j \langle \nabla_{e_j} \varphi, e_j \psi\rangle + \langle \varphi, \nabla_{e_j}(e_j \psi)\rangle = -\sum_j \nabla_{e_j} \langle \varphi, e_j \psi\rangle. \tag{2.23}$$

Now the codifferential $\delta = *d*$, acting on 1-forms, is given by

$$\delta\mu = -\sum_j (\nabla_{e_j} \mu)(e_j). \tag{2.24}$$

Thus, if $\mu_{\varphi,\psi}$ is the 1-form on M defined by

$$\mu_{\varphi,\psi}(X) = \langle \varphi, X\psi\rangle, \qquad X \in \Gamma(M, TM), \tag{2.25}$$

we see that

$$(\mathcal{D}\varphi, \psi) - (\varphi, \mathcal{D}\psi) = \int_M *\delta\mu_{\varphi,\psi} = \pm \int_M d(*\mu_{\varphi,\psi}) = 0 \tag{2.26}$$

by Stokes's theorem. This completes the proof.

If M is a Riemannian manifold (g positive definite) then the inner product $(\ ,\ )$ is a positive definite (pre-Hilbert) inner product, and in this case (2.20) says $\mathcal{D}$ is a formally selfadjoint operator. If M is complete, $\mathcal{D}$ is selfadjoint. For M compact, this is elementary; for noncompact complete M, see Chernoff [**38**]. We remark that (2.16) also serves to define the Dirac operator $\mathcal{D}$ on sections of $S(\tilde{P})$ tensored with any vector bundle with connection.

There are useful notions of Dirac operators even on a manifold M which does not have a spin structure. Suppose for simplicity that M is an even dimensional

Riemannian manifold. We assume M is oriented. We can associate to $TM \to M$ the bundle of Clifford algebras

$$\text{Cliff}\, M \to M. \tag{2.27}$$

If a connection ∇ is chosen for this vector bundle, we can define a "Dirac operator"

$$D^{\#}: \Gamma(M, \text{Cliff}\, M) \to \Gamma(M, \text{Cliff}\, M) \tag{2.28}$$

by

$$D^{\#}u = m(\nabla u) \tag{2.29}$$

in analogy with (2.16), where here

$$m: TM \otimes \text{Cliff}\, M \to \text{Cliff}\, M \tag{2.30}$$

is given by Clifford multiplication, with $TM \hookrightarrow \text{Cliff}\, M$. More generally, if $E \to M$ is a vector bundle with connection ∇, which is a Cliff M *module*, we can define

$$D^{\#}_E: \Gamma(M, E) \to \Gamma(M, E) \tag{2.31}$$

by the identity (2.29), where

$$m: TM \otimes E \to E \tag{2.32}$$

is given by the action of $TM \subset \text{Cliff}\, M$ on E.

Under a topological restriction on M, substantially weaker than that required for M to give a spin structure, there may exist a vector bundle $S \to M$ such that, for all $x \in M$, S_x is an *irreducible* Cliff M_x module. In such a case, we say an oriented manifold M is a *spin*^{c} *manifold*; $S \to M$ is a spin^c structure. Clearly a spin structure, yielding the bundle $S(\tilde{P})$, gives a spin^c structure. Given a connection on S, we can define a Dirac operator

$$D: \Gamma(M, S) \to \Gamma(M, S) \tag{2.33}$$

as before, and also, given a vector bundle $F \to M$, we can regard $E = F \otimes S$ as a Cliff M module and, bringing in a connection on this vector bundle, we have an associated Dirac operator, as above.

Dirac operators on spin^c manifolds retain a lot of important properties of the special case of spin manifolds. This is important in index theory, particularly since spin^c manifolds exist in fair profusion. For example, as the construction (1.57)–(1.62) implies, any almost complex manifold M, i.e., any M with a complex structure on the tangent bundle, not necessarily integrable, has a natural spin^c structure. Symplectic manifolds are examples of almost complex manifolds, since as shown in Chapter 11 a symplectic vector space with an inner product imposed has a complex structure. Thus a Riemannian metric on a symplectic manifold determines an almost complex structure, and hence a spin^c structure.

There are similar constructions for $\dim M$ odd, replacing Cliff M by its even part.

3. Spinors on four-dimensional Riemannian manifolds. Let M be a four-dimensional oriented Riemannian manifold, with metric tensor g, whose principal SO(4) frame bundle lifts to a Spin(4) bundle,

$$\tilde{P} \to M. \tag{3.1}$$

Special properties of the spinor bundle $S(\tilde{P}) = \tilde{P} \times_{\mathrm{Spin}(4)} S(4)$ arise from special properties of a four-dimensional oriented vector space V with a positive definite quadratic form Q, which we will investigate in this section.

Recall from §1 that, in general, for $\dim V = 2k$, if a complex structure J is imposed such that $Q(Ju) = Q(u)$, $u \in V$, and if the associated k-dimensional complex vector space is denoted by $\mathcal{V}$, then $\mathrm{Cl}(V, Q)$ acts on $\bigwedge_{\mathbf{C}}^{*} \mathcal{V} = S$. If we set

$$S_{+} = \bigoplus_{j\,\text{even}} \textstyle\bigwedge_{\mathbf{C}}^{j} \mathcal{V}, \qquad S_{-} = \bigoplus_{j\,\text{odd}} \textstyle\bigwedge_{\mathbf{C}}^{j} \mathcal{V}, \tag{3.2}$$

we have maps

$$P_1 \colon \mathbf{C} \otimes V \to \mathrm{Hom}(S_{+}, S_{-}), \tag{3.3}$$

$$P_2 \colon \mathbf{C} \otimes V \to \mathrm{Hom}(S_{-}, S_{+}). \tag{3.4}$$

Recall that $\dim_{\mathbf{C}} S_{+} = \dim_{\mathbf{C}} S_{-} = 2^{k-1}$, and

$$\dim \mathrm{Hom}(S_{+}, S_{-}) = \dim \mathrm{Hom}(S_{-}, S_{+}) = 2^{2k-2}.$$

In particular, if $\dim V = 4 = 2k$, then both sides of (3.3) and (3.4) have the same dimension. Thus the following should be no surprise.

PROPOSITION 3.1. *If* $\dim V = 4$, P_1 *and* P_2 *are isomorphisms.*

PROOF. It is easy to see that in general P_1 and P_2 are injective, so this follows from the remarks above on dimensions.

In addition to the maps (3.1), (3.2), in general, for $\dim V = 2k$, we have the maps

$$L_{\pm} \colon \textstyle\bigwedge^2 V \otimes \mathbf{C} \to \mathrm{End}(S_{\pm}). \tag{3.5}$$

Here the left side has complex dimension $\binom{2k}{2} = k(2k-1)$ and the right side has dimension 2^{2k-2}. If $2k = 4$, these dimensions are, respectively, six and four. We will show that L_{+} and L_{-} are isomorphisms from appropriate linear subspaces of $\bigwedge^2 V \otimes \mathbf{C}$ to certain subspaces of $\mathrm{End}(S_{+})$ and $\mathrm{End}(S_{-})$, respectively. The Hodge star operator plays a role in this analysis.

Generally, for $n = \dim V$, the Hodge star operator

$$* \colon \textstyle\bigwedge^k V \to \bigwedge^{n-k} V, \tag{3.6}$$

is defined by

$$*\alpha \wedge \beta = \langle \alpha, \beta \rangle \omega \tag{3.7}$$

where $\langle\ ,\ \rangle$ is the natural inner product on $\bigwedge^* V$ extending the inner product $Q(\ ,\ )$ on V, and ω is the positively oriented unit element of $\bigwedge^n V$. If Q is positive definite, we have

$$**=(-1)^{k(n-k)} \quad \text{on } \textstyle\bigwedge^k V. \tag{3.8}$$

In particular, if $\dim V = 4$, we have

$$*\colon \textstyle\bigwedge^2 V \to \bigwedge^2 V \quad \text{and} \quad **=1 \text{ on } \bigwedge^2 V. \tag{3.9}$$

Since $*$ is readily verified to be an isometry with respect to the natural inner product on $\bigwedge^* V$, we see that, if $\dim V = 4$,

$$\textstyle\bigwedge^2 V = \bigwedge^2_+ V \oplus \bigwedge^2_- V \tag{3.10}$$

with

$$* = \pm 1 \quad \text{on } \textstyle\bigwedge^2_\pm V. \tag{3.11}$$

Calculations which we shall perform shortly will show that, for $\dim V = 4$,

$$L_+(*\alpha) = L_+(\alpha), \quad L_-(*\alpha) = -L_-(\alpha), \qquad \alpha \in \textstyle\bigwedge^2 V. \tag{3.12}$$

Thus, $L_\pm$ annihilates $\bigwedge^2_\mp V$. What we will then show is that if, for a vector space E,

$$\operatorname{End}(E)^0 = \{T \in \operatorname{End}(E)\colon \operatorname{tr} T = 0\}, \tag{3.13}$$

then we have

PROPOSITION 3.2. *If* $\dim V = 4$, *we have isomorphisms*

$$L_\pm\colon \textstyle\bigwedge^2_\pm V \otimes \mathbf{C} \to \operatorname{End}(S_\pm)^0. \tag{3.14}$$

Let us now make some explicit computations of the action of V and $\bigwedge^2 V$ on $S = \bigwedge^*_{\mathbf{C}} \mathcal{V}$. Let e_1, e_2, e_3, e_4 be an orthonormal basis of V, positively oriented. A complex structure is defined on V by

$$Je_1 = e_3, \qquad Je_2 = e_4. \tag{3.15}$$

Thus $\mathcal{V}$ has basis e_1, e_2, and $e_3 = ie_1$, $e_4 = ie_2$. We have

$$S_+ = \mathbf{C} \oplus \textstyle\bigwedge^2_{\mathbf{C}} \mathcal{V}, \qquad S_- = \mathcal{V}. \tag{3.16}$$

A basis for S_+ is

$$1,\ e_1 \wedge e_2, \tag{3.17}$$

and a basis for S_- is

$$e_1,\ e_2. \tag{3.18}$$

Recall the action of V on $\bigwedge^*_{\mathbf{C}} \mathcal{V}$ is given by

$$v \cdot \varphi = v \wedge \varphi - j_v \varphi. \tag{3.19}$$

In particular we have, for $v \in V$,

$$v \cdot 1 = v,$$
$$(3.20)\qquad v \cdot w = v \wedge w - (w, v) \cdot 1, \qquad w \in \mathcal{V},$$
$$v \cdot (w \wedge x) = (x, v)w - (w, v)x, \qquad w, x \in \mathcal{V}.$$

Here $(\ ,\ )$ denotes the natural Hermitian inner product on $\Lambda^*_{\mathbf{C}} \mathcal{V}$.

With respect to the bases (3.17) and (3.18), we can write $P_1(v)$ and $P_2(v)$ as 2×2 complex matrices, for each $v \in V$. If

$$(3.21)\qquad v = a_1 e_1 + a_2 e_2 + a_3 e_3 + a_4 e_4,$$

a straightforward computation gives

$$(3.22)\qquad P_1(v) = \begin{pmatrix} a_1 + ia_3 & a_2 + ia_4 \\ a_2 - ia_4 & -a_1 + ia_3 \end{pmatrix}$$

and

$$(3.23)\qquad P_2(v) = \begin{pmatrix} -a_1 + ia_3 & -a_2 - ia_4 \\ -a_2 + ia_4 & a_1 + ia_3 \end{pmatrix}.$$

To record this differently, we have

$$(3.24)\qquad \begin{aligned} P_1(e_1) &= \begin{pmatrix} 1 & 0 \\ 0 & -1 \end{pmatrix}, & P_2(e_1) &= \begin{pmatrix} -1 & 0 \\ 0 & 1 \end{pmatrix}, \\ P_1(e_2) &= \begin{pmatrix} 0 & 1 \\ 1 & 0 \end{pmatrix}, & P_2(e_2) &= \begin{pmatrix} 0 & -1 \\ -1 & 0 \end{pmatrix}, \\ P_1(e_3) &= \begin{pmatrix} i & 0 \\ 0 & i \end{pmatrix}, & P_2(e_3) &= \begin{pmatrix} i & 0 \\ 0 & i \end{pmatrix}, \\ P_1(e_4) &= \begin{pmatrix} 0 & i \\ -i & 0 \end{pmatrix}, & P_2(e_4) &= \begin{pmatrix} 0 & -i \\ i & 0 \end{pmatrix}. \end{aligned}$$

Note that $P_1(e_j) = -P_2(e_j)$ (in this 2×2 matrix representation) if $j \neq 3$, $P_1(e_3) = P_2(e_3)$. Simple calculations show $P_1(e_j)P_2(e_j) = -I$, consistent with the fact that $P_1(v)P_2(v)$ is equal to the action of $v \cdot v$ on S_-, while $v \cdot v = -\|v\|^2 \cdot 1$.

These calculations clearly reprove Proposition 3.1. Let us turn to calculations of the action of $\Lambda^2 V$ on $S_\pm$. In $\mathrm{Cl}(V, Q)$, we have $v \cdot w = v \wedge w - Q(v, w) \cdot 1$, so

$$(3.25)\qquad (v \wedge w) \cdot \varphi = v \cdot (w \cdot \varphi) + Q(v, w)\varphi$$

for $v, w \in V$, $\varphi \in \Lambda^*_{\mathbf{C}} \mathcal{V}$. In particular

$$(3.26)\qquad (e_j \wedge e_k) \cdot \varphi = e_j \cdot (e_k \cdot \varphi), \qquad \varphi \in S_\pm.$$

Thus

$$(3.27)\qquad \begin{aligned} (e_j \wedge e_k) \cdot \varphi &= P_2(e_j)P_1(e_k)\varphi, & \varphi &\in S_+, \\ (e_j \wedge e_k) \cdot \varphi &= P_1(e_j)P_2(e_k)\varphi, & \varphi &\in S_-. \end{aligned}$$

Using (3.24), we easily find that the map $L_+ \colon \bigwedge^2 V \otimes \mathbf{C} \to \mathrm{End}(S_+)$ has matrix representation

$$L_+(e_1 \wedge e_2) = \begin{pmatrix} 0 & -1 \\ 1 & 0 \end{pmatrix} = L_+(e_3 \wedge e_4),$$

$$(3.28) \qquad L_+(e_1 \wedge e_3) = \begin{pmatrix} -i & 0 \\ 0 & i \end{pmatrix} = L_+(e_4 \wedge e_2),$$

$$L_+(e_1 \wedge e_4) = \begin{pmatrix} 0 & -i \\ -i & 0 \end{pmatrix} = L_+(e_2 \wedge e_3).$$

This calculation directly verifies the first part of (3.12), and a similar calculation verifies the analogous assertion for L_-, since we have

$$(3.29) \quad *(e_1 \wedge e_2) = e_3 \wedge e_4, \quad *(e_1 \wedge e_3) = e_4 \wedge e_2, \quad *(e_1 \wedge e_4) = e_2 \wedge e_3.$$

We can also read off a proof of Proposition 3.2 directly from these calculations. Note that all the matrices in (3.28) are skew adjoint. We thus have

PROPOSITION 3.3. *If* $\dim V = 4$ *(with positive definite inner product) we have the R-isomorphisms*

$$(3.30) \qquad L_\pm \colon \textstyle\bigwedge^2_\pm V \to \mathrm{Skew}(S_\pm)^0$$

where $\mathrm{Skew}(S_\pm)^0$ *denotes the space of traceless endomorphisms of* $S_\pm$*, skew adjoint with respect to the natural inner product on* $S_\pm$.

We next want to see explicitly the action of Spin(4) on S_+ and S_-. It is convenient to derive first the action of the Lie algebra of $\mathrm{Spin}(V,Q)$ on S_+ and S_-. We need therefore to identify this Lie algebra. Note that $\mathrm{Spin}(V,Q)$ is a Lie subgroup of $\mathrm{Cl}^\times(V,Q)$, the group of invertible elements of $\mathrm{Cl}(V,Q)$. Since $\mathrm{Cl}(V,Q)$ is an associative algebra with unit, $\mathrm{Cl}^\times(V,Q)$ is open in $\mathrm{Cl}(V,Q)$, and its Lie algebra is naturally identified with $\mathrm{Cl}(V,Q)$, with $[\varphi,\psi] = \varphi\cdot\psi - \psi\cdot\varphi$. We want to identify the Lie algebra of $\mathrm{Spin}(V,Q)$ as a subspace of $\mathrm{Cl}(V,Q)$. Using the natural identification $\mathrm{Cl}(V,Q) \approx \bigwedge^* V$, we have

PROPOSITION 3.4. *The Lie algebra of* $\mathrm{Spin}(V,Q)$ *is equal to*

$$\textstyle\bigwedge^2 V \subset \mathrm{Cl}(V,Q).$$

This holds for a form Q of signature (p,q), $p+q=n=\dim V$.

PROOF. It is easy to check that $\bigwedge^2 V$ is a Lie subalgebra of $\mathrm{Cl}(V,Q)$. Denote by G_0 the connected Lie group it generates. One can verify that $y \in \bigwedge^2 V$ implies $[y,x] \in V$ for all $x \in V$, and $y + y^* = 0$. If $u = \exp(y) \in G_0$, then $x \mapsto uxu^*$ gives a homomorphism from $\mathrm{Cl}(V,Q)$ to itself, since $uu^* = 1$ in this case, and V is preserved. Furthermore, the restriction to V preserves Q. The universal property of $\mathrm{Cl}(V,Q)$ implies such a homomorphism is uniquely determined by its action on V. But there is a $v \in \mathrm{Spin}_e(V,Q)$ giving such an action on V, since

$\mathrm{Spin}_e(V,Q)$ covers $\mathrm{SO}_e(V,Q)$, so $vxv^* = uxu^*$ for all $x \in \mathrm{Cl}(V,Q)$. This implies $v = \pm u$. It follows that $G_0 = \mathrm{Spin}_e(V,Q)$, and the proof is complete.

From Proposition 3.4 we can read off the action of spin(4) on S_+ from (3.28), and similarly analyze the action on S_-. Using Proposition 3.3, we have

PROPOSITION 3.5. *The representation*

$$D^+_{1/2} \oplus D^-_{1/2}\colon \mathrm{Spin}(4) \to \mathrm{Aut}(S_+) \times \mathrm{Aut}(S_-) \tag{3.31}$$

gives an isomorphism

$$\mathrm{Spin}(4) \approx \mathrm{SU}(2) \times \mathrm{SU}(2). \tag{3.32}$$

PROOF. Since a traceless skew adjoint operator on $S_\pm$ generates an element of SU(2), the map (3.31) defines a homomorphism of Spin(4) onto SU(2) × SU(2) which is a local isomorphism. Since all these groups are simply connected, this homomorphism is an isomorphism.

In addition to invariant inner products on $S_\pm$, there are invariant complex volume elements, i.e., invariant elements $\omega_\pm \in \bigwedge^2_{\mathbf{C}} S_\pm$, because the group actions on $S_\pm$ are given by SU(2). The elements $\omega_\pm$ give rise to isomorphisms

$$\tilde{\omega}_\pm\colon S_\pm \to S'_\pm, \tag{3.33}$$

where $S'_\pm$ is the dual of $S_\pm$, the space of $\mathbf{C}$-linear maps from $S_\pm$ to $\mathbf{C}$. Note that these maps intertwine each of the representations $D^\pm_{1/2}$ with their adjoints, and are uniquely determined by this property, up to a (complex) scalar. These isomorphisms in turn give rise to isomorphisms

$$\Omega_\pm\colon \mathrm{End}(S_\pm)^0 \to \mathrm{Symm}(S_\pm), \tag{3.34}$$

the space of *symmetric* $\mathbf{C}$-bilinear forms on $S_\pm$. Thus Proposition 3.2 yields isomorphisms

$$\textstyle\bigwedge^2_\pm V \otimes \mathbf{C} \approx \mathrm{Symm}(S_\pm). \tag{3.35}$$

4. Spinors on four-dimensional Lorentz manifolds. Let M be a four-dimensional Lorentz manifold, with orientation and causal structure, and with metric tensor g, whose principal $\mathrm{SO}_e(3,1)$ bundle lifts to a $\mathrm{Spin}_e(3,1)$ bundle

$$\tilde{P} \to M. \tag{4.1}$$

As in §3, special properties of the spin bundle $S(\tilde{P}) = \tilde{P} \times_{\mathrm{Spin}_e(3,1)} S(4)$ arise from special properties of a four-dimensional vector space V with a quadratic form Q of Lorentz signature, which we will investigate in this section.

It is convenient to think of $R^{3,1}$ as a real linear subspace of $R^4 \otimes \mathbf{C}$, which is equipped with a complex bilinear form induced from the Euclidean inner product on R^4. For example, we could take the real linear span of e_1, e_2, e_3, and ie_4. However, in order for our computations analogous to (3.24) to take a form which is "standard" in the literature, we will let $R^{3,1}$ be the real linear span of ε_0, ε_1, ε_2, ε_3, where

$$(4.2) \qquad \varepsilon_0 = -ie_3, \qquad \varepsilon_1 = e_2, \qquad \varepsilon_2 = e_4, \qquad \varepsilon_3 = e_1.$$

Thus $\|\varepsilon_0\|^2 = -1$, and $\|\varepsilon_1\|^2 = \|\varepsilon_2\|^2 = \|\varepsilon_3\|^2 = 1$. We can thus identify the complexified Clifford algebra $\mathbf{C} \otimes \mathrm{Cl}(3,1)$ with $\mathbf{C} \otimes \mathrm{Cl}(4) = \mathrm{Cl}(4)$, and hence $\mathrm{Cl}(3,1)$ acts by Clifford multiplication on $S = \bigwedge_{\mathbf{C}}^{*} \mathcal{V} = \bigwedge_{\mathbf{C}}^{*} \mathbf{C}^2$.

As in §3, we set $S_+ = \bigoplus_{j\,\mathrm{even}} \bigwedge_{\mathbf{C}}^{j} \mathcal{V}$, $S_- = \bigoplus_{j\,\mathrm{odd}} \bigwedge_{\mathbf{C}}^{j} \mathcal{V}$, with the bases (3.17) and (3.18). The inclusion $V = R^{3,1} \hookrightarrow \mathrm{Cl}(3,1)$ induces maps analogous to (3.3), (3.4):

$$(4.3) \qquad \begin{aligned} P_1'&: \mathbf{C} \otimes R^{3,1} \to \mathrm{Hom}(S_+, S_-), \\ P_2'&: \mathbf{C} \otimes R^{3,1} \to \mathrm{Hom}(S_-, S_+). \end{aligned}$$

With respect to the bases we have chosen for S_+ and S_-, $P_j'(v)$ are represented by 2×2 matrices. In fact, we see that

$$(4.4) \qquad \begin{aligned} P_j'(\varepsilon_0) &= -iP_j(e_3), & P_j'(\varepsilon_1) &= P_j(e_2), \\ P_j'(\varepsilon_2) &= P_j(e_4), & P_j'(\varepsilon_3) &= P_j(e_1). \end{aligned}$$

Hence the calculations (3.24) yield

$$(4.5) \qquad \begin{aligned} P_1'(\varepsilon_0) &= \begin{pmatrix} 1 & 0 \\ 0 & 1 \end{pmatrix} = \sigma^0, & P_1'(\varepsilon_1) &= \begin{pmatrix} 0 & 1 \\ 1 & 0 \end{pmatrix} = \sigma^1, \\ P_1'(\varepsilon_2) &= \begin{pmatrix} 0 & i \\ -i & 0 \end{pmatrix} = \sigma^2, & P_1'(\varepsilon_3) &= \begin{pmatrix} 1 & 0 \\ 0 & -1 \end{pmatrix} = \sigma^3. \end{aligned}$$

The matrices $\sigma^0, \sigma^1, \sigma^2, \sigma^3$ are called *Pauli matrices*. As in §3, we have

$$(4.6) \qquad P_2'(\varepsilon_0) = P_1'(\varepsilon_0); \qquad P_2'(\varepsilon_j) = -P_j'(\varepsilon_j) \quad \text{for } 1 \le j \le 3.$$

These computations immediately prove

PROPOSITION 4.1. *The maps P_1' and P_2' are isomorphisms.*

We next investigate the action of $\bigwedge^2 R^{3,1}$ on $S_\pm$. We have

$$(4.7) \qquad L_\pm' : \textstyle\bigwedge^2 R^{3,1} \otimes \mathbf{C} \to \mathrm{End}(S_\pm),$$

and with respect to the bases we have chosen for these spaces they are given explicitly by

$$(4.8) \qquad \begin{aligned} L_+'(\varepsilon_j \wedge \varepsilon_k) &= P_2'(\varepsilon_j) P_1'(\varepsilon_k), \\ L_-'(\varepsilon_j \wedge \varepsilon_k) &= P_1'(\varepsilon_j) P_2'(\varepsilon_k), \end{aligned}$$

in analogy with (3.27). Using (4.5) and (4.6), we obtain

$$
\begin{aligned}
L'_+(\varepsilon_0 \wedge \varepsilon_1) &= \begin{pmatrix} 0 & 1 \\ 1 & 0 \end{pmatrix} = -iL'_+(\varepsilon_2 \wedge \varepsilon_3), \\
L'_+(\varepsilon_0 \wedge \varepsilon_2) &= \begin{pmatrix} 0 & i \\ -i & 0 \end{pmatrix} = iL'_+(\varepsilon_1 \wedge \varepsilon_3), \\
L'_+(\varepsilon_0 \wedge \varepsilon_3) &= \begin{pmatrix} 1 & 0 \\ 0 & -1 \end{pmatrix} = -iL'_+(\varepsilon_1 \wedge \varepsilon_2),
\end{aligned} \tag{4.9}
$$

and analogous results for L'_-. We deduce the following result, which is in interesting contrast with Proposition 3.2.

PROPOSITION 4.2. *We have R-linear isomorphisms*

$$L'_\pm: \bigwedge^2 R^{3,1} \to \mathrm{End}_{\mathbf{C}}(S_\pm)^0. \tag{4.10}$$

Of course, the kernel of $L'_\pm$ on $\bigwedge^2 R^{3,1} \otimes \mathbf{C}$ is a complex three-dimensional linear space, but its intersection with the real vector space $\bigwedge^2 R^{3,1}$ is zero. The difference is partly due to the different behavior of the Hodge star operator on $\bigwedge^* R^{3,1}$. Instead of having (3.9), as is readily verified, one has

$$*: \bigwedge^2 R^{3,1} \to \bigwedge^2 R^{3,1} \quad \text{and} \quad ** = -1 \text{ on } \bigwedge^2 R^{3,1}. \tag{4.11}$$

In other words, $*$ imposes a complex structure on $\bigwedge^2 R^{3,1}$. The computation (4.9) and its analogue for L'_- show that, instead of (3.12), one has

$$L'_+(*\alpha) = iL'_+(\alpha), \quad L'_-(*\alpha) = -iL'_-(\alpha), \qquad \alpha \in \bigwedge^2 R^{3,1}. \tag{4.12}$$

The result (4.10) on the action of $\bigwedge^2 R^{3,1}$ on $\mathrm{End}(S_\pm)$, together with the identification of $\bigwedge^2 R^{3,1}$ with the Lie algebra of $\mathrm{Spin}_e(3,1)$ given by Proposition 3.4, yields

PROPOSITION 4.3. *The representations*

$$D^\pm_{1/2}: \mathrm{Spin}_e(3,1) \to \mathrm{Aut}(S_\pm) \tag{4.13}$$

both give isomorphisms

$$\mathrm{Spin}_e(3,1) \approx \mathrm{SL}(2,\mathbf{C}). \tag{4.14}$$

PROOF. In view of the fact that, for a matrix A, $\det(e^{sA}) = e^{s\,\mathrm{tr}\,A}$, we see that $\mathrm{Spin}_e(3,1)$ is represented by elements of $\mathrm{SL}(2,\mathbf{C})$ in both cases (4.13). The isomorphism (4.10) shows that the maps $D^\pm_{1/2}: \mathrm{Spin}(3,1) \to \mathrm{SL}(2,\mathbf{C})$ are local isomorphisms. Since both $\mathrm{Spin}_e(3,1)$ and $\mathrm{SL}(2,\mathbf{C})$ are connected double covers of $\mathrm{SO}_e(3,1)$, it follows that these maps are isomorphisms.

It can be verified that the two representations $D^\pm_{1/2}$ of $\mathrm{Spin}_e(3,1)$ on $\mathbf{C}^2$ are not equivalent. Their characters are complex conjugates of each other. Thus $D^-_{1/2}$ is the adjoint of $D^+_{1/2}$.

We see that S_+ and S_- do not possess a $\mathrm{Spin}_e(3,1)$ invariant inner product. However, by Proposition 4.3, they do possess $\mathrm{Spin}_e(3,1)$ invariant complex *volume elements*, i.e., invariant elements $\omega_\pm$ of $\bigwedge^2_{\mathbf{C}} S_\pm$. These induce isomorphisms

$$\tilde{\omega}_\pm : S_\pm \to S'_\pm \tag{4.15}$$

where $S'_\pm$ denotes the dual of $S_\pm$, the space of complex linear maps from $S_\pm$ to $\mathbf{C}$. In other words, an antisymmetric second order "tensor" on $S_\pm$ is used to "lower (or raise) indices" of spinors. The maps (4.15) in turn give isomorphisms

$$\Omega_\pm : \mathrm{End}(S_\pm)^0 \to \mathrm{Symm}(S_\pm), \tag{4.16}$$

the space of *symmetric* $\mathbf{C}$-bilinear forms on $S_\pm$. If we let

$$\Gamma_\pm = \Omega_\pm \circ L'_\pm \tag{4.17}$$

we deduce, from Proposition 4.2,

COROLLARY 4.4. *We have the R-linear isomorphisms*

$$\Gamma_\pm : \textstyle\bigwedge^2 R^{3,1} \to \mathrm{Symm}(S_\pm). \tag{4.18}$$

Thus 2-forms on $R^{3,1}$ are represented by symmetric 2-component spinors.

From the analysis of the action of $R^{3,1}$ on $\mathrm{Hom}(S_+, S_-)$ and on $\mathrm{Hom}(S_-, S_+)$ derived in this section, we can derive the following explicit representation of the Dirac operator on flat Minkowski space $R^{3,1}$. With $S = S_+ \oplus S_- \approx \mathbf{C}^4$, we have

$$D : \Gamma(R^{3,1}, S) \to \Gamma(R^{3,1}, S) \tag{4.19}$$

given by

$$D = \sum_{\mu=0}^{3} \gamma^\mu (\partial/\partial x_\mu) \tag{4.20}$$

where $(x_0, \dots, x_3)$ denote coordinates on $R^{3,1}$ with respect to the basis $\varepsilon_0, \dots, \varepsilon_3$, and γ^μ are given by

$$\gamma^0 = \begin{pmatrix} 0 & \sigma^0 \\ \sigma^0 & 0 \end{pmatrix}, \qquad \gamma^j = \begin{pmatrix} 0 & -\sigma^j \\ \sigma^j & 0 \end{pmatrix} \quad \text{for } 1 \le j \le 3. \tag{4.21}$$

The 4×4 matrices $\gamma^0, \dots, \gamma^3$ are known as the *Dirac matrices*. Recall that $\sigma^0, \dots, \sigma^3$ are the Pauli matrices, given by (4.5).

CHAPTER 13

Semisimple Lie Groups

This final chapter is devoted to a short introduction to the huge and growing theory of noncompact semisimple Lie groups, of which all the groups discussed in Chapter 8 on, except for Spin(n), are examples. We show how the constructions of principal and discrete series, derived in Chapter 8 for SL($2, R$), can be extended to more general semisimple Lie groups. Doing this requires some general results on the structure of semisimple Lie groups, such as the Cartan decomposition and the Iwasawa decomposition, sketched in §1; our exposition here owes a good deal to that of Duistermaat [**49**]. The reader who has come this far will be prepared, I hope, to tackle the more extensive treatments of analysis on semisimple Lie groups, given in [**253, 256, 100, 136**], and their generalizations, reductive Lie groups, given in [**248, 251**], as well as survey articles referenced in the text.

1. Introduction to semisimple Lie groups. Recall from Chapter 0 that a connected Lie group G, and its Lie algebra $\mathfrak{g}$, are said to be simple if $\mathfrak{g}$ has no proper ideals, and semisimple if $\mathfrak{g}$ has no proper abelian ideals. We list some basic cases of semisimple Lie groups, many of which have been encountered in previous chapters:

$$\mathrm{SL}(n, R), \tag{1.1}$$

$$\mathrm{SO}_e(p, q), \tag{1.2}$$

$$\mathrm{Sp}(n, R), \tag{1.3}$$

$$\mathrm{SU}(p, q). \tag{1.4}$$

Of these, SL(n, R) is the group of real $n \times n$ matrices of determinant one. The example $n = 2$ was studied in Chapter 8. The group $\mathrm{SO}_e(p, q)$ is the identity component of the group of linear transformations on R^{p+q} preserving the metric $Q(x, x) = x_1^2 + \cdots + x_p^2 - \cdots - x_{p+q}^2$. If $q = 0$, then $\mathrm{SO}_e(p, 0) = \mathrm{SO}(p)$ is compact. The groups $\mathrm{SO}_e(n, 1)$ were studied in Chapter 9. The group SU(p, q) is the group of linear transformations of $\mathbf{C}^{p+q}$, with unit determinant, preserving the Hermitian form $|z_1|^2 + \cdots + |z_p|^2 - \cdots - |z_{p+q}|^2$. Actually, of these, all the Lie algebras are simple, except $\mathrm{so}(4) \approx \mathrm{so}(3) \oplus \mathrm{so}(3)$, and $\mathrm{so}(2, 2) \approx \mathrm{so}(2, 1) \oplus \mathrm{so}(2, 1)$.

We have seen from time to time in the previous chapters that it can be useful to consider the complexification of a real Lie algebra. The complexifications of the Lie algebras of (1.1)–(1.4) are, respectively,

$$\text{(1.5)} \qquad \mathrm{sl}(n,\mathbf{C}), \quad \mathrm{so}(p+q,\mathbf{C}), \quad \mathrm{sp}(n,\mathbf{C}), \quad \text{and} \quad \mathrm{sl}(p+q,\mathbf{C}).$$

All of these are simple complex Lie algebras, except $\mathrm{so}(4,\mathbf{C})$, and they form an exhaustive list of the complex simple Lie algebras, together with the exceptional simple Lie algebras, denoted

$$\text{(1.6)} \qquad \mathfrak{e}_6, \quad \mathfrak{e}_7, \quad \mathfrak{e}_8, \quad \mathfrak{f}_4, \quad \mathfrak{g}_2.$$

This remarkable classification result was established by E. Cartan. For a thorough discussion of this classification, and the exceptional Lie algebras (1.6), see [**100, 159**].

Of course, the Lie algebras (1.5) can be viewed as real Lie algebras. All complex semisimple Lie algebras can be shown to be semisimple when viewed as real Lie algebras. We studied $\mathrm{sl}(2,\mathbf{C})$ as a real Lie algebra in Chapter 9. Note that $\mathrm{sl}(2,\mathbf{C})$ is simple, both as a complex Lie algebra and as a real Lie algebra, but its complexification $\mathrm{Csl}(2,\mathbf{C})$ seen in Chapter 9 to be isomorphic to $\mathrm{Csl}(2,R)\oplus\mathrm{Csl}(2,R)$, is not simple, only semisimple. Generally, $\mathrm{Csl}(n,\mathbf{C})\approx\mathrm{Csl}(n,R)\oplus\mathrm{Csl}(n,R)\approx\mathrm{sl}(n,\mathbf{C})\oplus\mathrm{sl}(n,\mathbf{C})$.

A Lie algebra $\mathfrak{g}$ whose complexification is isomorphic to a certain complex Lie algebra $\mathfrak{g}'$ is said to be a *real form* of $\mathfrak{g}'$. We have seen examples, particularly $\mathrm{so}(3)$ and $\mathrm{sl}(2,R)$, of nonisomorphic real semisimple Lie algebras which have isomorphic complexifications. The following is a list of the real Lie algebras whose complexifications are the complex Lie algebras (1.5).

Complexification	Real forms
$\mathrm{sl}(n,\mathbf{C})$	$\mathrm{sl}(n,R)$; $\mathrm{su}(p,q)$, $p+q=n$; $\mathrm{sl}(k,H)$, $n=2k$.
$\mathrm{so}(n,\mathbf{C})$	$\mathrm{so}(p,q)$, $p+q=n$; $\mathrm{so}^*(2k)$, $n=2k$.
$\mathrm{sp}(n,\mathbf{C})$	$\mathrm{sp}(n,R)$; $\mathrm{sp}(p,q)$, $p+q=n$.

The additions to our list, beyond (1.1)–(1.4), are three types. $\mathrm{sl}(k,H)$ denotes the Lie algebra of $k\times k$ matrices of *quarternions*, of trace zero.

$$\mathrm{so}^*(2k)=\left\{\begin{pmatrix} Z_1 & Z_2 \\ -\overline{Z}_2 & \overline{Z}_1\end{pmatrix} : Z_1 \text{ and } Z_2 \text{ are } k\times k \text{ complex matrices,} \quad Z_1 \text{ skew, } Z_2 \text{ hermitian}\right\}.$$

Finally, $\mathrm{sp}(p,q)$ is the Lie algebra of $\mathrm{Sp}(p,q)$, the subgroup of $\mathrm{Sp}(p+q,\mathbf{C})\subset\mathrm{GL}(2p+2q,\mathbf{C})$ leaving invariant the Hermitian form

$$\sum_{j=1}^{p}[|z_j|^2+|z_{p+q+j}|^2]-\sum_{k=1}^{q}[|z_{p+k}|^2+|z_{2p+q+k}|^2]$$

of signature $(2p, 2q)$ on $\mathbf{C}^{2p+2q}$. In particular, $\mathrm{Sp}(n) = \mathrm{Sp}(n, 0) = \mathrm{Sp}(n, \mathbf{C}) \cap \mathrm{U}(2n)$ is compact. For more details on these additional Lie algebras, and on the list of real forms given above, see [**100**, **159**].

Before discussing general properties of semisimple Lie groups, we will take a look at some elementary properties that the groups (1.1)–(1.4) can be seen to have in common. Each of these groups is a group of matrices. The first three are groups of real matrices and the last a group of $n \times n$ complex matrices, with $n = p + q$, which can be considered a group of $2n \times 2n$ real matrices, identifying a complex number $a + ib$ with the matrix $\left(\begin{smallmatrix} a & b \\ -b & a \end{smallmatrix}\right)$. Note that complex conjugate goes to matrix transpose in this case. The groups (1.2) and (1.3) preserve bilinear forms on R^n $(n = p + q)$, respectively symmetric and skew symmetric. Each of these is of the form

$$\mathfrak{r}(x, y) = x \cdot Ky \tag{1.7}$$

where $x \cdot y$ is the usual dot product on R^n, and K is an $n \times n$ matrix, respectively symmetric or skew symmetric. In fact, in cases (1.2), (1.3), K is also an orthogonal $n \times n$ matrix; we have

$$\text{either} \quad K^t = K^{-1} = K \quad \text{or} \quad K^t = K^{-1} = -K. \tag{1.8}$$

Note that for a linear map $A\colon R^n \to R^n$ to preserve (1.7), it is necessary and sufficient that

$$A^t K A = K. \tag{1.9}$$

Granted (1.8), we see that if (1.9) holds for A, it also holds for A^t. Consequently the groups (1.2) and (1.3) have the property that

$$G \text{ is a Lie group of real matrices such that } A \in G \Rightarrow A^t \in G, \tag{1.10}$$

a property which also holds trivially for the groups (1.1). Such groups that are closed subgroups of $\mathrm{GL}(n, R)$ are called real reductive algebraic groups. Their Lie algebras $\mathfrak{g}$ satisfy the property that

$$\mathfrak{g} \text{ is a Lie algebra of real matrices such that } X \in \mathfrak{g} \Rightarrow X^t \in \mathfrak{g}. \tag{1.11}$$

The groups (1.4) are also easily verified to satisfy (1.10), when viewed as a group of $2n \times 2n$ real matrices $(n = p + q)$. In case (1.4) elements of $\mathrm{SU}(p, q)$ preserve a Hermitian form on $\mathbf{C}^n = R^{2n}$, and hence its real and imaginary parts, which are, respectively, a symmetric bilinear form (an indefinite metric on R^{2n}), and a skew-symmetric bilinear form, a symplectic form on R^{2n}.

For a general real reductive algebraic group G, satisfying (1.10), we single out two subsets. Take a basis giving the matrix representation of G to be orthonormal, and use it to define the usual positive definite inner product on Euclidean space R^n. Then let

$$K = \{g \in G\colon g \text{ is an orthogonal matrix}\}, \tag{1.12}$$

and

$$P = \{g \in G\colon g \text{ is a positive definite matrix}\}. \tag{1.13}$$

Actually, K is a subgroup of G. Its Lie algebra is

$$\mathfrak{k} = \{X \in \mathfrak{g} \colon X = -X^t\}, \tag{1.14}$$

and the tangent space to P at the identity matrix $e = I$ is

$$\mathfrak{p} = \{X \in \mathfrak{g} \colon X = X^t\}. \tag{1.15}$$

Note that $\mathfrak{g} = \mathfrak{k} \oplus \mathfrak{p}$ and that

$$[\mathfrak{k}, \mathfrak{k}] \subset \mathfrak{k}, \qquad [\mathfrak{k}, \mathfrak{p}] \subset \mathfrak{p}, \qquad [\mathfrak{p}, \mathfrak{p}] \subset \mathfrak{k}. \tag{1.16}$$

The set P has the following important property, which will lead to the Cartan decomposition for a group satisfying (1.10). Take $A \in P$; A is a positive definite matrix, so there is a one parameter group of positive definite matrices A^s.

LEMMA 1.1 (CHEVALLEY). *If $A \in P$, then $A^s \in P$ for all $s \in R$.*

We will not give the proof for general G satisfying (1.10); see **[100]**. However, we will sketch a proof of the lemma for the groups (1.1)–(1.4). In fact, given A positive definite, it is clear that

$$\det A = 1 \Rightarrow \det A^s = 1 \tag{1.17}$$

where det denotes the real determinant for (1.1)–(1.3) and the complex determinant for (1.4). Furthermore, if R^n has some nondegenerate bilinear form $\mathfrak{r}(x, y)$, symmetric or antisymmetric, of the form (1.7)–(1.8), then, for A positive definite, by (1.9), we see that A preserves $\mathfrak{r}$ if and only if

$$K^{-1}AK = A^{-1}. \tag{1.18}$$

If $p_k(\lambda)$ is a sequence of polynomials converging to λ^s on $\operatorname{spec} A \cup \operatorname{spec} A^{-1}$ (or indeed a single polynomial equal to λ^s on this finite point set), we have $p_k(A) \to A^s$ and $p_k(A^{-1}) \to A^{-s}$, and since (1.18) implies

$$p_k(A^{-1}) = p_k(K^{-1}AK) = K^{-1}p_k(A)K, \tag{1.19}$$

we deduce that (1.18) implies

$$K^{-1}A^sK = A^{-s}, \tag{1.20}$$

so A^s also preserves $\mathfrak{r}$ if A does. This is enough to verify Lemma 1.1 for all the groups (1.1)–(1.4).

The following corollary is immediate.

COROLLARY 1.2. *The exponential map takes $\mathfrak{p}$ diffeomorphically onto P.*

Indeed, Lemma 1.1 implies that the range of $\mathfrak{p}$ under exp is all of P. That this mapping is a diffeomorphism is clear.

Using this, we obtain the celebrated *Cartan decomposition*.

THEOREM 1.3. *The map* $(k, X) \mapsto k \exp X$ *is a diffeomorphism onto* G:

$$K \times \mathfrak{p} \to G. \tag{1.21}$$

PROOF. Given $g \in G$, write $A = (g^t g)^{1/2}$ and $k = A^{-1}g$, so $kA = g$. Hypothesis (1.10) implies $g^t \in G$, so $g^t g \in P$, and by Lemma 1.1, $A \in P \subset G$, so $k \in G$. It is clear that k is an orthogonal transformation, so by definition (1.12), $k \in K$. This proves that the map (1.21) is onto. That it is one-to-one with surjective differential is elementary.

In Chapter 11, we needed the Cartan decomposition not only for the matrix group $\mathrm{Sp}(n, R)$, but also for its double cover $\mathrm{Mp}(n, R)$, so we need to consider this decomposition for a general covering group $\tilde{G}$ of a semisimple Lie group G satisfying (1.10). Since G is homeomorphic to $K \times \mathfrak{p}$ and $\mathfrak{p}$ is a vector space, G is homotopically equivalent to K. Thus $\tilde{G}$ is homeomorphic to $\tilde{K} \times \mathfrak{p}$, where $\tilde{K} = j^{-1}(K)$, $j: \tilde{G} \to G$ denoting the covering homomorphism. In fact, the group structure on $K \times \mathfrak{p}$ induced by (1.21) lifts to a group structure on $\tilde{K} \times \mathfrak{p}$, making it isomorphic to $\tilde{G}$ and giving the Cartan decomposition for $\tilde{G}$. If $\tilde{G}$ covers a semisimple matrix group, $\tilde{K}$ need not be compact; it will be compact if and only if the center of $\tilde{G}$ is finite.

For a Lie algebra $\mathfrak{g}$ of real matrices satisfying (1.11), the symmetric bilinear form

$$\beta(X, Y) = \operatorname{tr} XY, \qquad X, Y \in \mathfrak{g}, \tag{1.22}$$

is nondegenerate, i.e., if $X \in \mathfrak{g}$ and $\beta(X, Y) = 0$ for all $Y \in \mathfrak{g}$, then $X = 0$. Note that since the trace is conjugation invariant, β satisfies the condition

$$\beta(\mathrm{Ad}(g)X, \mathrm{Ad}(g)Y) = \beta(X, Y) \quad \text{for all } g \in G. \tag{1.23}$$

Differentiating,

$$\beta(\operatorname{ad} ZX, Y) = -\beta(X, \operatorname{ad} ZY), \qquad X, Y, Z \in \mathfrak{g}. \tag{1.24}$$

This invites comparison with another Ad-invariant bilinear form, the Killing form, defined in Chapter 0, and we will return to this shortly. Note that β is negative definite on $\mathfrak{k}$ and positive definite on $\mathfrak{p}$. If we use the map

$$\theta(X) = -X^t, \tag{1.25}$$

which by hypothesis (1.11) preserves $\mathfrak{g}$ and hence defines an automorphism of $\mathfrak{g}$, then the form

$$\beta_\theta(X, Y) = -\beta(X, \theta Y) = \operatorname{tr}(XY^t) \tag{1.26}$$

is positive definite on $\mathfrak{g}$. The involution θ is I on $\mathfrak{k}$ and $-I$ on $\mathfrak{p}$.

Let us return to closed subgroups G of $\mathrm{GL}(n, R)$ satisfying (1.10). It is clear in this case from (1.12) that K is compact. G/K is a homogeneous space, diffeomorphic to P. The decomposition

$$\mathfrak{g} = \mathfrak{k} \oplus \mathfrak{p} \tag{1.27}$$

produces an isomorphism of T_oG/K with $\mathfrak{p}$, where $o = eK \in G/K \approx P$, and since the nondegenerate form β on $\mathfrak{g}$ restricts to a positive definite form on $\mathfrak{p}$, which is $\operatorname{Ad} K$ invariant, we get a natural Riemannian metric on $G/K \approx P$, and G acts on this space by isometries.

Granted hypothesis (1.10), we have an automorphism $\tilde{\theta}$ of G, defined by

$$\tilde{\theta}(g) = (g^t)^{-1}, \tag{1.28}$$

exponentiating the automorphism (1.25). Note that K is fixed by $\tilde{\theta}$ and P is invariant under $\tilde{\theta}$. Thus the map $g \mapsto g^{-1}$ on P is an isometry with respect to the natural metric on G/K. This isometry fixes $e = I$, and its derivative acts as $-I$ on $T_eP = \mathfrak{p}$. A homogeneous space with this property is called a symmetric space.

There is a fixed point theorem which implies that, if M is a symmetric space (or any complete Riemannian manifold) of negative curvature, diffeomorphic to R^n, and if K_1 is a compact group of isometries of M, then there is a point $p \in M$ fixed by all elements of K_1. For a proof, see [**100**] or [**65**]. This fixed point theorem applies to the action of K_1 on G/K in our situation, where K_1 is a compact subgroup of G. It follows that, if K_1 is a compact subgroup of G, then there is an element $g_o \in G$ such that $g_o^{-1}K_1g_0 \subset K$. This implies that K is a *maximal compact* subgroup of G, and also that any two maximal compact subgroups of G are conjugate.

In the cases (1.1)–(1.4), the maximal compact subgroups are, respectively,

$$\mathrm{SO}(n) \subset \mathrm{SL}(n, R), \tag{1.29}$$

$$\left\{\begin{pmatrix} A & 0 \\ 0 & B \end{pmatrix} : A \in \mathrm{SO}(p), B \in \mathrm{SO}(q)\right\} \subset \mathrm{SO}_e(p, q), \tag{1.30}$$

$$\mathrm{U}(n) \subset \mathrm{Sp}(n, R), \tag{1.31}$$

and

$$\left\{\begin{pmatrix} A & 0 \\ 0 & B \end{pmatrix} : A \in \mathrm{U}(p), B \in \mathrm{U}(q), (\det A)(\det B) = 1\right\} \subset \mathrm{SU}(p, q). \tag{1.32}$$

Of these, (1.29) is obvious, and the fact that (1.30)–(1.32) coincide with the general form (1.12) of K is straightforward to check. The fact that (1.31) provides a maximal compact subgroup of $\mathrm{Sp}(n, R)$ was given an independent discussion in Chapter 11. We also note that the maximal compact subgroup of $\mathrm{SL}(n, \mathbf{C})$ is

$$\mathrm{SU}(n) \subset \mathrm{SL}(n, \mathbf{C}). \tag{1.33}$$

There is another sort of intimate relation between noncompact semisimple Lie groups and compact Lie groups. Namely, for each semisimple Lie algebra $\mathfrak{g}$, there is a compact Lie group whose Lie algebra has the same complexification, called a compact real form of $\mathfrak{g}$. Lie algebras of compact groups are found in each row of the table of complexifications and real forms listed above. Explicitly, the compact real forms corresponding to the groups (1.1)–(1.4) are, respectively,

$$\mathrm{SU}(n), \tag{1.34}$$

$$\mathrm{SO}(p+q), \tag{1.35}$$

$$\mathrm{Sp}(n), \tag{1.36}$$

$$\mathrm{SU}(p+q). \tag{1.37}$$

Also, the compact real form of $\mathrm{SL}(n,\mathbf{C})$ is $\mathrm{SU}(n) \times \mathrm{SU}(n)$. There is a simple uniform construction of the compact real form of a semisimple matrix algebra $\mathfrak{g}$ satisfying (1.11), using the Cartan decomposition (1.27). We can consider $\mathfrak{g}$, an algebra of real $n \times n$ matrices, as a (real) Lie algebra in $\mathrm{End}(\mathbf{C}^n)$, by complexifying R^n. In $\mathrm{End}(\mathbf{C}^n)$,

$$\mathfrak{g}_u = \mathfrak{k} + i\mathfrak{p} \tag{1.38}$$

is also a (real) Lie algebra. If $\mathfrak{g}$ is semisimple, so is $\mathfrak{g}_u$. If $\mathbf{C}^n$ gets its standard Hermitian metric, then, according to (1.14)–(1.15), $\mathfrak{g}_u$ consists of skew adjoint transformations. Hence the Lie group U generated by $\mathfrak{g}_u$ is a subgroup of $\mathrm{U}(n)$. It can be shown to be closed, hence compact.

The fact that, for such a semisimple Lie algebra $\mathfrak{g}$, $\mathbf{C}\mathfrak{g}$ is also the complexification of the Lie algebra $\mathfrak{u} = \mathfrak{g}_u$ of a compact Lie group U leads to a relation between the finite-dimensional representations of G, with Lie algebra $\mathfrak{g}$, and of U, known as Weyl's *unitary trick*. Namely, a finite-dimensional representation π_G of G on a complex vector space V defines a Lie algebra representation $\mathfrak{g} \to \mathrm{End}(V)$, hence $\mathbf{C}\mathfrak{g} \to \mathrm{End}(V)$, which restricts to a representation $\mathfrak{g}_u \to \mathrm{End}(V)$. If U is simply connected (a theorem of Weyl assures the universal covering group of U is compact if U is compact and semisimple), this exponentiates to a representation π_U of U on V. Conversely (if G is simply connected), a representation π_U of U on V gives rise to a representation π_G of G on V. This correspondence can be used to show that any finite-dimensional representation of such G can be decomposed into irreducible representations; one merely has to decompose the associated representation of U into irreducibles. Furthermore, the correspondence $\pi_G \leftrightarrow \pi_U$ gives a classification of all the finite-dimensional irreducible representations of G in terms of the irreducible (necessarily unitarizable) representations of U, studied in Chapter 3.

Our discussion so far has applied to matrix Lie algebras satisfying (1.11), particularly the semisimple ones, such as (1.1)–(1.5). It is an important result of E. Cartan that every semisimple Lie algebra can be arranged to be of the form (1.11). This result is part of the foundation of Cartan's structure theory for semisimple Lie algebras. We give a sketch of it, referring to [**100**] for details. The basic ingredient is the construction of a *Cartan involution* on a general semisimple Lie algebra $\mathfrak{g}$, with properties analogous to (1.25)–(1.26). The form $\beta(X,Y)$ defined by (1.22) is replaced by the Killing form

$$B(X,Y) = \operatorname{tr} \operatorname{ad} X \operatorname{ad} Y, \tag{1.39}$$

which is the special case of (1.22) when $\mathfrak{g}$ is represented on $V = \mathfrak{g}$ by the adjoint representation. As we mentioned in Chapter 0, $\mathfrak{g}$ is semisimple if and only if the Killing form is nondegenerate, a result known as Cartan's criterion. A Cartan involution is defined as follows.

DEFINITION. A Cartan involution is a Lie algebra automorphism $\theta: \mathfrak{g} \to \mathfrak{g}$ such that $\theta \circ \theta = \mathrm{id}$ and such that the form

$$B_\theta(X, Y) = -B(X, \theta Y) \tag{1.40}$$

is positive definite.

One of Cartan's fundamental results is that every semisimple Lie algebra $\mathfrak{g}$ has a Cartan involution. For a proof, see [**100**].

If θ is a Cartan involution of $\mathfrak{g}$, it is easy to see that

$$B_\theta(\mathrm{ad}\, ZX, Y) = -B_\theta(X, (\mathrm{ad}\, \theta Z)Y), \qquad X, Y, Z \in \mathfrak{g}, \tag{1.41}$$

as a simple consequence of the identity $B(\mathrm{ad}\, ZX, Y) = -B(X, \mathrm{ad}\, ZY)$. The identity (1.41) implies that, if $\mathfrak{g}$ is given the inner product B_θ, positive definite by hypothesis, and the transpose is with respect to this inner product, then

$$-(\mathrm{ad}\, Z)^t = \mathrm{ad}\, \theta Z. \tag{1.42}$$

Hence the adjoint representation of $\mathfrak{g}$ in $\mathrm{End}(\mathfrak{g})$ satisfies hypothesis (1.11). Thus Cartan's theorem on the existence of a Cartan involution implies that any semisimple Lie algebra is isomorphic to a Lie algebra of matrices satisfying (1.12). The adjoint representation of $\mathfrak{g}$ exponentiates to the adjoint representation of G, so any semisimple Lie group covers a matrix group satisfying (1.10). Furthermore, it can be shown that such a matrix group is closed in $\mathrm{GL}(\mathfrak{g})$. For details, see [**100**].

For a general Lie algebra satisfying (1.11), the nondegenerate form β given by (1.22) need not be proportional to the Killing form (1.39). For example, (1.11) holds for $\mathfrak{g} = \mathrm{gl}(n, R)$, the Lie algebra of all $n \times n$ matrices, which is not semisimple, as the set of multiples of the identity matrix belongs to its center. In this case, β is nondegenerate, but B is not. The two forms cannot be proportional unless $\mathfrak{g}$ is semisimple. Even if $\mathfrak{g}$ is semisimple, if $\mathfrak{g} = \mathfrak{g}_1 \oplus \cdots \oplus \mathfrak{g}_k$, $\mathfrak{g}_j$ simple, $\mathfrak{g}$ could easily be represented as a matrix Lie algebra in such a fashion that β, restricted to $\mathfrak{g}_j$, is a different multiple of the Killing form for each j. The question of whether a Lie algebra $\mathfrak{g}$ has a *unique* Ad-invariant symmetric bilinear form has a simple answer: this happens precisely when the complexification of $\mathfrak{g}$ is simple. In general, the dimension of the space of such bilinear forms on a semisimple Lie algebra $\mathfrak{g}$ is equal to the number of simple complex Lie algebras into which $\mathbf{C}\mathfrak{g}$ decomposes.

Having discussed the Cartan decomposition of a semisimple Lie group and some related matters, we want to describe two other constructions which provide useful tools for the representation theory of such groups, Cartan subalgebras and the Iwasawa decomposition.

A *Cartan subalgebra* of $\mathfrak{g}$ is a maximal element of the class of abelian subalgebras of $\mathfrak{g}$ consisting of semisimple elements, where $X \in \mathfrak{g}$ is said to be semisimple provided the linear transformation $\operatorname{ad} X$ is diagonalizable. A *Cartan subgroup* of G is the centralizer B of a Cartan subalgebra $\mathfrak{b}$. In the case

$$G = \mathrm{SL}(2, R) \tag{1.43}$$

all Cartan subgroups are conjugate to one of the following:

$$B = \mathrm{SO}(2) \quad \text{or} \quad B = \left\{ \begin{pmatrix} a & 0 \\ 0 & a^{-1} \end{pmatrix} : a \in R \setminus \{0\} \right\}. \tag{1.44}$$

More generally, the group

$$G = \mathrm{SL}(n, R) \tag{1.45}$$

has $[\frac{1}{2}n] + 1$ conjugacy classes of Cartan subgroups. For $0 \le j \le [\frac{1}{2}n]$, we can take

$$\begin{aligned} &B_j = \text{set of diagonal sums of } j\ 2 \times 2 \text{ matrices } \begin{pmatrix} a_i & b_i \\ -b_i & a_i \end{pmatrix} \\ &\qquad \text{and a diagonal matrix of size } n - 2j, \text{ belonging to } \mathrm{SL}(n, R). \end{aligned} \tag{1.46}$$

Note that, if $n > 2$, none of the B_j is compact. Only $B = \mathrm{SO}(2)$ in (1.44) is compact. For the group

$$G = \mathrm{SL}(n, \mathbf{C}), \tag{1.47}$$

any Cartan subgroup is conjugate to the group of diagonal matrices in $\mathrm{SL}(n, \mathbf{C})$.

Generally, it can be shown that, for almost all $Y \in \mathfrak{g}$, semisimple Lie algebra, the set of X commuting with Y is a Cartan subalgebra; for details see [**100, 159**]. The identity component of a Cartan subgroup B has the Cartan subalgebra $\mathfrak{b}$ as its Lie algebra, but as the examples (1.44), (1.46) show, B itself need not be connected. If G is a matrix group, B can be shown to be abelian, though more generally it need not be; of course the connected component B_e is abelian.

By inspection, one sees that the Cartan subalgebras of $\mathrm{SL}(n, R)$ and $\mathrm{SL}(n, \mathbf{C})$ described above are invariant under the Cartan involution. In general, as proved by Harish-Chandra (see [**100, 159**]), any Cartan subalgebra is conjugate to one which is invariant under the Cartan involution.

The fact that each $\mathrm{SL}(n, \mathbf{C})$ has only one conjugacy class of Cartan subalgebras is a special case of the general phenomenon that, if $\mathfrak{g}$ is a complex semisimple Lie algebra, then all Cartan subalgebras are conjugate. Generally, if $\mathfrak{b}$ is a Cartan subalgebra of $\mathfrak{g}$, $\mathbf{C}\mathfrak{b}$ can be shown to be a Cartan subalgebra of $\mathbf{C}\mathfrak{g}$; it follows that any two Cartan subalgebras of $\mathfrak{g}$ at least have the same dimension.

Finally, we mention the *Iwasawa decomposition*. This is a unique product representation

$$G = KAN, \tag{1.48}$$

where K is the maximal compact subgroup of G mentioned before, and $S = AN$ is a certain solvable subgroup of G, a semidirect product of an abelian subgroup A and a nilpotent subgroup N. This decomposition is easily described for $G = \mathrm{SL}(n, R)$ and for $G = \mathrm{SL}(n, \mathbf{C})$.

In the case $G = \mathrm{SL}(n, R)$, with $K = \mathrm{SO}(n)$, as usual, we can take A to be the group of diagonal matrices, with positive entries and determinant 1, and N to be the group of lower triangular matrices with ones on the diagonal. Then $AN = S$ is the group of lower triangular matrices of determinant 1, with positive elements on the diagonal. For $\mathrm{sl}(2, R)$, with the basis $\{Z, A, B\}$ given by (1.2)–(1.3) in Chapter 8, K is the group generated by Z, A the group generated by the element denoted A in (1.3), and N the group generated by $B - \frac{1}{2}Z$. The Iwasawa decomposition for $\mathrm{SL}(2, R)$ was not explicitly used in Chapter 8, but a good deal of use of it is made in the book **[147]** of Lang.

In the case $G = \mathrm{SL}(n, \mathbf{C})$, where $K = \mathrm{SU}(n)$, take A to be the group of positive diagonal matrices, of determinant 1, and let N be the group of lower triangular complex matrices, with ones down the diagonal.

There is a uniform construction of the Iwasawa decomposition, which is not hard to describe on the Lie algebra level, for a semisimple Lie algebra $\mathfrak{g}$. If a Cartan involution is chosen, so $\mathfrak{g}$ is represented via its adjoint representation as a matrix algebra satisfying (1.11), and we set $\mathfrak{g} = \mathfrak{k} \oplus \mathfrak{p}$ as before, with $\theta(X) = X$ on $\mathfrak{k}$, $\theta(X) = -X$ on $\mathfrak{p}$, let $\mathfrak{a}$ be a maximal element among the linear subspaces of $\mathfrak{p}$ that are actually abelian subalgebras of $\mathfrak{g}$. It follows from Corollary 1.2 that exp maps $\mathfrak{a}$ diffeomorphically onto a subgroup A of G. By (1.41) it follows that, for $Z \in \mathfrak{a}$, $\mathrm{ad}\, Z$ is selfadjoint with respect to the positive definite inner product B_θ. Since $\mathfrak{a}$ is abelian, we can simultaneously diagonalize $\mathrm{ad}\, \mathfrak{a}$ on $\mathfrak{g}$, so for $\lambda \in \mathfrak{g}^*$, set

$$\mathfrak{g}_\lambda = \{X \in \mathfrak{g} : \mathrm{ad}\, ZX = \lambda(Z)X \quad \text{for all } Z \in \mathfrak{a}\}. \tag{1.49}$$

We have $\mathfrak{g} = \bigoplus_\lambda \mathfrak{g}_\lambda$. Note that

$$[\mathfrak{g}_{\lambda_1}, \mathfrak{g}_{\lambda_2}] \subset \mathfrak{g}_{\lambda_1 + \lambda_2}. \tag{1.50}$$

One says λ is a "restricted root" if $\lambda \neq 0$ and $\mathfrak{g}_\lambda \neq 0$. Choose a hyperplane through the origin in $\mathfrak{g}^*$, not containing any restricted roots, pick one side of it, and say a restricted root λ is positive if it lies on this side. Let

$$\mathfrak{n} = \bigoplus_{\lambda > 0} \mathfrak{g}_\lambda. \tag{1.51}$$

By (1.50), we see that $\mathfrak{n}$ is nilpotent. Also, since $\mathrm{ad}\, \mathfrak{a} : \mathfrak{g}_\lambda \to \mathfrak{g}_\lambda$, we have

$$[\mathfrak{a}, \mathfrak{n}] \subset \mathfrak{n}, \tag{1.52}$$

so $\mathfrak{a} \oplus \mathfrak{n}$ is a solvable subalgebra of $\mathfrak{g}$. We also set

$$\overline{\mathfrak{n}} = \bigoplus_{\lambda < 0} \mathfrak{g}_\lambda. \tag{1.53}$$

Note that, by (1.49), $\theta\mathfrak{g}_\lambda = \mathfrak{g}_{-\lambda}$, where θ is the Cartan involution, since $\theta = -\mathrm{id}$ on $\mathfrak{a} \subset \mathfrak{p}$. It follows that $\bar{\mathfrak{n}} = \theta\mathfrak{n}$. In light of (1.16), and the maximality of $\mathfrak{a}$ with respect to the class of abelian subalgebras of $\mathfrak{g}$ which are contained in $\mathfrak{p}$, it follows that, if we set

$$\mathfrak{m} = \{X \in \mathfrak{k} \colon [\mathfrak{a}, X] = 0\}, \tag{1.54}$$

then

$$\mathfrak{g}_0 = \mathfrak{m} \oplus \mathfrak{a}. \tag{1.55}$$

Thus, by (1.51) and (1.53), we have

$$\mathfrak{g} = \bar{\mathfrak{n}} \oplus \mathfrak{m} \oplus \mathfrak{a} \oplus \mathfrak{n}. \tag{1.56}$$

The following is the Iwasawa decomposition, on the Lie algebra level.

PROPOSITION 1.4. *We have the direct sum decomposition*

$$\mathfrak{g} = \mathfrak{k} \oplus \mathfrak{a} \oplus \mathfrak{n}. \tag{1.57}$$

PROOF. Given $W \in \mathfrak{g}_\lambda$, if we write $W = W_1 + W_2$ with $W_1 \in \mathfrak{k}$, $W_2 \in \mathfrak{p}$, then, by (1.16), $[\mathfrak{a}, W_1] \subset \mathfrak{p}$ and $[\mathfrak{a}, W_2] \in \mathfrak{k}$, so we must have

$$Z \in \mathfrak{a} \Rightarrow \operatorname{ad} Z W_1 = \lambda(Z) W_2, \qquad \operatorname{ad} Z W_2 = \lambda(Z) W_1. \tag{1.58}$$

This implies that the mapping $\mathfrak{n} \to \mathfrak{p}$ assigning to each $X \in \mathfrak{n}$ its projection in $\mathfrak{p}$ is injective, and more generally the natural projection onto $\mathfrak{p}$ annihilating $\mathfrak{k}$ gives an injective map of $\mathfrak{a} + \mathfrak{n}$ into $\mathfrak{p}$. Note that, for $W = W_1 + W_2$ as above, $W_1 + W_2 \in \mathfrak{g}_\lambda$ if and only if $W_1 - W_2 \in \mathfrak{g}_{-\lambda}$. It follows that the natural projection of $\mathfrak{a}+\mathfrak{n}$ into $\mathfrak{p}$ has range coinciding with that of the natural projection of $\bar{\mathfrak{n}}+\mathfrak{a}+\mathfrak{n}$ into $\mathfrak{p}$, which is all of $\mathfrak{p}$, by (1.56). Thus $\dim(\mathfrak{a} + \mathfrak{n}) = \dim \mathfrak{p}$. Since $\mathfrak{k}$ and $\mathfrak{a} + \mathfrak{n}$ have no nonzero elements in common, (1.57) now follows from (1.27).

The factors in the decomposition (1.57) were described explicitly for $\mathfrak{g} = so(n, 1)$ in Chapter 9, §3.

The decomposition (1.57) exponentiates, in a nontrivial fashion, to the following global Iwasawa decomposition. See [**100**] for a proof.

THEOREM 1.5. *The natural map $K \times A \times N \to G$ is a diffeomorphism onto G, where N is the connected Lie group with Lie algebra $\mathfrak{n}$. Thus*

$$G = KAN. \tag{1.59}$$

Furthermore, the exponential map gives a diffeomorphism of $\mathfrak{a} + \mathfrak{n}$ onto the solvable group $S = AN$.

Another important subgroup of G, connected with the Iwasawa decomposition, arises as follows. Let M be the centralizer in K of $\mathfrak{a}$. M may not be connected, but the connected component of the identity has Lie algebra $\mathfrak{m}$, given by (1.54). In the case $G = \mathrm{SL}(n, R)$, $K = \mathrm{SO}(n)$, $\mathfrak{m}$ is trivial, and M is the group with 2^{n-1} elements consisting of diagonal matrices with ± 1 on the diagonal (an even number of -1's). If $G = \mathrm{SL}(n, \mathbf{C})$, $K = \mathrm{SU}(n)$, we see that M consists of the diagonal matrices in $\mathrm{SU}(n)$, so is isomorphic to T^{n-1}. Both for $\mathrm{SL}(n,R)$ and for $\mathrm{SL}(n,\mathbf{C})$, we see that $MA = B$ is a Cartan subgroup. This will not be true in general; M need not be commutative. As we saw in Chapter 9, if $G = \mathrm{SO}_e(n,1)$, we take $K = \mathrm{SO}(n)$ and $M = \mathrm{SO}(n-1)$. As another example, if $G = \mathrm{Sp}(n,R)$, with $K = \mathrm{SU}(n)$, one has $M = \mathrm{SO}(n)$.

Since the action of M on $\mathfrak{g}$ commutes with that of A, N is normalized by MA. Hence

$$P = MAN \tag{1.60}$$

is easily seen to be a subgroup. It is called the "minimal parabolic" subgroup. We have seen one example of this in the study of $\mathrm{SO}_e(n,1)$ in Chapter 9, §3. In the following section we will describe how it is used in the construction of the "principal series" of representations of a semisimple Lie group G.

Chapter 0 ended with a list of some low-dimensional Lie groups. To resume that theme, we end this section with a list of isomorphisms among some low-dimensional semisimple Lie algebras, some of which have been pointed out already in this monograph. For more details, see [**100**].

$$\begin{aligned}
\mathrm{so}(2,1) &= \mathrm{sl}(2,R) = \mathrm{su}(1,1) = \mathrm{sp}(1,R),\\
\mathrm{so}(3,1) &= \mathrm{sl}(2,\mathbf{C}),\\
\mathrm{so}(2,2) &= \mathrm{sl}(2,R) \oplus \mathrm{sl}(2,R),\\
\mathrm{so}(4,1) &= \mathrm{sp}(1,1),\\
\mathrm{so}(3,2) &= \mathrm{sp}(2,R),\\
\mathrm{so}(5,1) &= \mathrm{sl}(2,H),\\
\mathrm{so}(4,2) &= \mathrm{su}(2,2),\\
\mathrm{so}(3,3) &= \mathrm{sl}(4,R),\\
\mathrm{so}(6,2) &= \mathrm{so}^*(8),\\
\mathrm{su}(3,1) &= \mathrm{so}^*(6),\\
\mathrm{so}^*(4) &= \mathrm{su}(2) \oplus \mathrm{sl}(2,R).
\end{aligned}$$

2. Some representations of semisimple Lie groups. We give brief descriptions of some methods of constructing irreducible unitary representations of a semisimple Lie group G. The sorts of representations described here suffice to comprise "almost all" the representations of G, in the sense that the Plancherel measure is supported on them, though we do not give a proof of this fact. We discuss constructions of the principal series first. Then we mention a few results on the construction of the discrete series, and on the construction of various other series associated with each conjugacy class of Cartan subgroups of G. We prove very few results here, and we refer generally to other sources, including [**253, 256, 265**], and other references cited in this section, for proofs of results stated here.

The *principal series* have been described in detail for SL(2, R) and for SL(2, C) in Chapters 8–9. As shown in §3 of Chapter 9, they could be viewed as special cases of a construction of the principal series for $SO_e(n,1)$, which were given as induced representations. Such a construction, based on the Iwasawa decomposition

$$G = KAN, \tag{2.1}$$

can be generalized to any semisimple G, with finite center (so K is compact). Take the minimal parabolic subgroup

$$P = MAN, \tag{2.2}$$

described in (1.60), where M is the centralizer of A in K, so M is a compact subgroup of K. The principal series is indexed by $\hat{M} \times \hat{A}$, where $\hat{M}$ is the set of (equivalence classes of) irreducible unitary representations of M and $\hat{A}$ is the set of characters on A. If $\sigma \in \hat{M}$ and $\tau \in \hat{A}$, we define a representation $\sigma \times \tau$ of P by

$$(\sigma \times \tau)(man) = \sigma(m)\tau(a). \tag{2.3}$$

It is easy to check that this does give a unitary representation of P. The corresponding representation $\pi_{(\sigma,\tau)}$ in the principal series is defined to be the induced representation

$$\pi_{(\sigma,\tau)} = \mathrm{Ind}_P^G(\sigma \times \tau). \tag{2.4}$$

In our analysis of principal series representations π^e_{is} and π^o_{is} of SL(2, R), we saw all were irreducible, except for

$$\pi^o_0 = \pi^+_1 \oplus \pi^-_{-1}. \tag{2.5}$$

In the analysis of principal series representations $\pi_{k,s}$ of SL(2, C), they all turned out to be irreducible. This last phenomenon is a special case of the following, proved by N. Wallach:

PROPOSITION 2.1. *If G has only one conjugacy class of Cartan subgroup, then $\pi_{(\sigma,\tau)}$ is irreducible for all (σ,τ).*

The condition of having only one conjugacy class of Cartan subgroup is satisfied by all complex semisimple Lie groups, and also by $SO_e(2n+1,1)$.

In general, "almost all" the principal series representations are irreducible. There is the following result, due to Bruhat. Let $M^\#$ denote the normalizer of A in K. Then $W = M^\#/M$ is a finite group. It has a natural action on $\hat{M}$, by conjugation. W also has a natural action on $\hat{A}$, from its action by conjugation on A. Bruhat proved the following result.

PROPOSITION 2.2. *The principal series representation $\pi_{(\sigma,\tau)}$ is irreducible provided that, for all $s \in W$, $s \neq \mathrm{id}$, the representation $s\sigma \times s\tau$ is not equivalent to $\sigma \times \tau$.*

It is clear that, for any $\sigma \in \hat{M}$, the set of $\tau \in \hat{A}$ such that $s\sigma \times s\tau \approx \sigma \times \tau$ must have Lebesgue measure 0 in $\hat{A}$. Other results on the possibility of reducing principal series representations include

PROPOSITION 2.3. *For any $\sigma \in \hat{M}$, $\tau \in \hat{A}$, $\pi_{(\sigma,\tau)}$ is a finite sum of irreducible representations. All the irreducible factors in the decomposition of a fixed such $\pi_{(\sigma,\tau)}$ are distinct. Furthermore, $\pi_{(1_M,\tau)}$ is irreducible for all $\tau \in \hat{A}$.*

For proofs of the last three propositions, see [**27, 143, 255**], and references given there.

We look at two other families of examples of principal series representations, those for

$$G = \mathrm{SL}(n, R) \quad \text{and} \quad G = \mathrm{SL}(n, \mathbf{C}). \tag{2.6}$$

For $G = \mathrm{SL}(n, R)$, the minimal parabolic subgroup $P = MAN$ consists of all lower triangular matrices in $\mathrm{SL}(n, R)$. M is a discrete group with 2^{n-1} elements, consisting of diagonal matrices with ± 1 on the diagonal, an even number of -1's. Thus

$$G/P = K/M \tag{2.7}$$

is a quotient of $\mathrm{SO}(n)$ by the discrete group M, and an element of the principal series is defined over L^2 sections of various vector bundles over this quotient. We get 2^{n-1} smooth families of representations, each indexed by $\mathfrak{a}^* \approx R^{n-1}$.

For $G = \mathrm{SL}(n, \mathbf{C})$, $P = MAN$ consists of all lower triangular matrices in $\mathrm{SL}(n, \mathbf{C})$. $M \approx T^{n-1}$ consists of the diagonal matrices in $\mathrm{SU}(n)$, which forms a maximal torus in $\mathrm{SU}(n)$. Thus $\mathrm{SL}(n, \mathbf{C})/P = \mathrm{SU}(n)/T^{n-1}$ has the structure of a complex manifold, as we have seen in Chapter 3. The principal series representations are parametrized by $\mathbf{Z}^{n-1} \times \mathfrak{a}^*$, $\mathfrak{a}^* \approx R^{n-1}$.

For both $\mathrm{SL}(n, R)$ and $\mathrm{SL}(n, \mathbf{C})$, the normalizer $M^\#$ of A is a semidirect product of the group of $n \times n$ permutation matrices and M, so $W = M^\#/M \approx S_n$, the symmetric group. Its natural action on R^n restricts to the hyperplane of vectors the sum of whose components vanishes, and defines the action of W on $\mathfrak{a}^*$ in each case. The action of W on $\hat{M}$ for $G = \mathrm{SL}(n, R)$ coincides with the natural permutation action of S_n, acting on $(\mathbf{Z}_2)^n$, restricted to the subgroup of elements with an even number of factors not the identity, which is isomorphic to $(\mathbf{Z}_2)^{n-1}$. The action of W on $\hat{M}$ for $G = \mathrm{SL}(n, \mathbf{C})$ coincides with the natural action of S_n on the hyperplane $R^{n-1} \subset R^n$ specified above, restricted to the integer lattice. In the special case of $G = \mathrm{SL}(2, R)$, we see that only the representation $\pi_{(\sigma,\tau)}$ with σ the nontrivial representation of $M = \mathbf{Z}_2$ and τ the trivial representation of $A \approx R^+$, i.e., only the representation π_0^o, fails to satisfy the hypotheses of either Proposition 2.2 or the latter part of Proposition 2.3, whose union is consequently sharp in this case. As we have mentioned, all the principal series representations of $\mathrm{SL}(n, \mathbf{C})$ are irreducible, as a special case of Proposition 2.1. For a direct proof in the case of $\mathrm{SL}(n, \mathbf{C})$, see Mackey [**160**].

An irreducible unitary representation π of G is said to belong to the *discrete series* if, for each u, v in the representation space of π,

$$\int_G |(\pi(g)u, v)|^2 \, dg < \infty. \tag{2.8}$$

The representations $\pi^{\pm}_{\pm n}$ of $\mathrm{SL}(2, R)$ constructed in Chapter 8 are discrete series representations. On the other hand, none of the representations of $\mathrm{SL}(2, \mathbf{C})$ constructed in Chapter 9 are discrete series representations. The following result of Harish-Chandra [**96**] characterizes which semisimple Lie groups have discrete series representations.

PROPOSITION 2.4. *A semisimple Lie group G has discrete series representations if and only if G has a compact Cartan subgroup.*

In case G has finite center, so K is compact, an equivalent condition is

$$\operatorname{rank} K = \operatorname{rank} M + \dim A, \tag{2.9}$$

where $\operatorname{rank} K$ is the dimension of a maximal torus of K. Examples of groups which have discrete series representations include

$$\mathrm{SL}(2, R); \quad \mathrm{Sp}(n, R); \quad \mathrm{SU}(m, n); \quad \mathrm{SO}_e(p, q), \ pq \text{ even}.$$

Groups which do not have them include

$$\mathrm{SL}(n, R), n > 2; \quad \mathrm{SO}_e(p, q), \ pq \text{ odd},$$

and complex semisimple Lie groups, such as $\mathrm{SL}(n, \mathbf{C})$.

If a semisimple Lie group has a compact Cartan subgroup, then the Lie algebra $\mathfrak{t}$ of a maximal torus T in K is a Cartan subalgebra of $\mathfrak{g}$. As shown by Harish-Chandra [**96**], there is a natural one-to-one correspondence between the equivalence classes of discrete series representations of G and a certain lattice in $\mathfrak{t}^*$, with some hyperplanes removed, and quotiented out by the natural action of $N(T)/T$. The work in [**96**] does not explicitly construct the discrete series representations, though it computes their characters. Subsequent constructions were given, by Parthasarathy [**193**] and Schmid [**206, 207**], and particularly by Atiyah-Schmid [**7**]. We give a brief discussion of their construction, following [**207**].

Let $\varphi\colon K \to \mathrm{SO}(\mathfrak{p})$ be the map induced from the restriction of the adjoint action of G to K. Assume that φ lifts to $\mathrm{Spin}(\mathfrak{p})$; if not, replace G (hence K) by its double cover. If a maximal torus T in K is also a Cartan subgroup, then the codimension of T in K and also in G is even, so since $\mathfrak{g} = \mathfrak{k} \oplus \mathfrak{p}$, we deduce that $\dim \mathfrak{p}$ is even. Thus $\mathrm{Spin}(\mathfrak{p})$ has two basic representations, on S_+ and S_-, constructed in Chapter 12. The lift of φ gives these a K-module structure.

For any unitary representation of K on a vector space V, there is a homogeneous vector bundle $\mathcal{V}$ over G/K, with principal K-bundle given by $G \to G/K$. There is a natural connection on the principal bundle $G \to G/K$, which induces a connection on the vector bundle $\mathcal{V}$. We have Dirac operators

$$(2.10)\qquad \begin{aligned} D_V^+ &: C^\infty(\mathcal{V}\otimes S_+) \to C^\infty(\mathcal{V}\otimes S_-),\\ D_V^- &: C^\infty(\mathcal{V}\otimes S_-) \to C^\infty(\mathcal{V}\otimes S_+),\end{aligned}$$

where $S_\pm$ are the homogeneous spinor bundles on G/K arising from $S_\pm$. Set

$$(2.11)\qquad H_V^\pm = \ker D_V^\pm \cap L^2(\mathcal{V}\otimes S_\pm).$$

G has a natural unitary action on $H_V^\pm$. The basic result on the construction of the discrete series is

PROPOSITION 2.5. *For an irreducible representation of K on V, we have either $H_V^+ = H_V^- = 0$ or that one of these spaces is irreducible and the other is zero. If H_V^+ or $H_V^- \neq 0$, it gives a discrete series representation. Finally, all discrete series representations of G are realized in this way.*

For a proof, see [7]. Further discussion is given in [**207**]. The parametrization of Harish-Chandra specifies for which irreducible representations of K one obtains discrete series representations of G.

There are other natural ways in which some discrete series representations arise, one of which we mention here. As shown in Chapter 11, if $p+q=n$, then $\mathrm{SU}(p,q)\subset \mathrm{Sp}(n,R)$. Since $\mathrm{SU}(p,q)$ is simply connected, we can identify it with a subgroup of $\mathrm{Mp}(n,R)$. Then the metaplectic representation, when restricted to $\mathrm{SU}(p,q)$, breaks up into a direct sum of discrete series irreducible representations of $\mathrm{SU}(p,q)$. This is discussed in the book [**23**] of Borel and Wallach. The special case $\mathrm{SU}(1,1)\approx \mathrm{SL}(2,R)$ arose in Chapter 8, §4, of this monograph. Also, the special case $\mathrm{SU}(n)\subset \mathrm{Sp}(n,R)$ arose in Chapter 1.

The principal series and the discrete series are the two extremes of certain series of representations associated to each conjugacy class of Cartan subgroup. In general, let B be a Cartan subgroup of G. As mentioned in §1, we can assume the Lie algebra $\mathfrak{b}$ is invariant under the Cartan involution, so

$$(2.12)\qquad \mathfrak{b} = \mathfrak{t}_o + \mathfrak{a}_o, \qquad \mathfrak{t}_o \subset \mathfrak{k}, \qquad \mathfrak{a}_o \subset \mathfrak{a}.$$

In other words, the connected component B_e of the identity in B is of the form

$$(2.13)\qquad B_e = T_o \cdot A_o,$$

where T_o is a torus in K and A_o a subgroup of the group A in the Iwasawa decomposition. The centralizer G_o of A_o in G is of the form

$$(2.14)\qquad G_o = M_o \cdot A_o$$

where M_o can be shown to be a Lie group with a reductive Lie algebra. M_o need not be connected (we will see a family of examples shortly), but it turns out that some quotient M_o' of M_o by a finite subgroup is a connected reductive Lie group with a Cartan subgroup isomorphic to T_o. Furthermore, there is a nilpotent subgroup N_o of N, normalized by G_o, and the product

$$(2.15)\qquad P_o = M_o A_o N_o$$

is a subgroup of G, called a "cuspidal parabolic" subgroup. Even though M_o is not a connected semisimple Lie group, there is also a theory of its discrete series. Suppose σ is an irreducible unitary representation of M_o in the discrete series; we write $\sigma \in \hat{M}_{od}$. If also $\tau \in \hat{A}_o$, we have the unitary representation $\sigma \times \tau$ of P_o, defined by

$$(\sigma \times \tau)(man) = \sigma(m)\tau(a), \tag{2.16}$$

as in (2.3), and we set

$$\pi_{(\sigma,\tau)} = \mathrm{Ind}_{P_o}^{G}(\sigma \times \tau), \tag{2.17}$$

as in (2.4). This is the "principal P_o-series."

In the case where B is a compact Cartan subgroup of G, then (2.15) is $B_e = T_o$, with $A_o = \{e\}$. Thus the centralizer G_o of A_o is G itself, so $P_o = G$ in this case, and the representations described are precisely those in the discrete series. At the other extreme, it turns out that there is up to conjugacy always a unique Cartan subgroup B of G for which $\mathfrak{b} \cap \mathfrak{p} = \mathfrak{a}$, and hence for which A_o is all of A. In this case, $M_o = M$ is compact and the group (2.15) coincides with the minimal parabolic subgroup (2.2), so the representations that arise are the principal series representations described above. It has been shown that no representation in such a principal P_o-series coming from a Cartan subgroup B is equivalent to a representation coming from a Cartan subgroup B', unless B and B' are conjugate. In general, "most" of these representations are irreducible.

To illustrate the construction of these other series, we consider the special case

$$G = \mathrm{SL}(n, R). \tag{2.18}$$

As mentioned in §1, for each $j \in \{0, 1, \ldots, [\frac{1}{2}n]\}$, we have a Cartan subgroup of the form

$$B_j = \text{set of diagonal sums of } j\ 2 \times 2 \text{ matrices } \begin{pmatrix} a_i & b_i \\ -b_i & a_i \end{pmatrix} \tag{2.19}$$
$$\text{and a diagonal matrix of size } n - 2j, \text{ belonging to } \mathrm{SL}(n, R).$$

In this case, T_o is the set of all matrices which are diagonal sums of j 2×2 matrices

$$\begin{pmatrix} \cos\theta_i & \sin\theta_i \\ -\sin\theta_i & \cos\theta_i \end{pmatrix},$$

and the $(n - 2j) \times (n - 2j)$ identity matrix, and

(2.20)

$$A_o = \left\{ \begin{pmatrix} d_{11} & & 0 \\ & \ddots & \\ 0 & & d_{nn} \end{pmatrix} : d_{11} = d_{22},\ d_{33} = d_{44}, \ldots, d_{2j-1,2j-1} = d_{2j,2j} \right\}$$
$$\cap\, \mathrm{SL}(n, R).$$

It follows that the centralizer G_o of A_o has the form

$$G_o = \text{set of diagonal sums of } j \ 2\times 2 \text{ invertible matrices and a diagonal matrix of size } n-2j, \text{ belonging to } \mathrm{SL}(n,R), \tag{2.21}$$

and consequently the factor M_o in (2.14) is of the form

$$M_o = \text{set of diagonal sums of } j \ 2\times 2 \text{ matrices, each in } \mathrm{SL}'(2,R), \text{ and a diagonal matrix of size } n-2j, \text{ consisting of } \pm 1\text{'s}, \text{ belonging to } \mathrm{SL}(n,R), \tag{2.22}$$

where

$$\mathrm{SL}'(2,R) = \{T \in \mathrm{GL}(2,R)\colon \ \det T = \pm 1\}. \tag{2.23}$$

In this case, the subgroup N_o of N, normalized by G_o, is the group of matrices of the form

$$\begin{pmatrix} I & & & & & \\ & \ddots & & & 0 & \\ & & I & & & \\ & & & 1 & & \\ & * & & & \ddots & \\ & & & & & 1 \end{pmatrix}, \qquad I = \begin{pmatrix} 1 & 0 \\ 0 & 1 \end{pmatrix}, \tag{2.24}$$

where there are j copies of the 2×2 identity matrix I, followed by $n-2j$ ones on the diagonal.

In this case, one can find the discrete series of M_o, closely related to a product of j copies of $\mathrm{SL}(2,R)$, by an extension of Bargmann's analysis of $\mathrm{SL}(2,R)$. The representations of $\mathrm{SL}(n,R)$ constructed in this way, via (2.16)–(2.17) were found by Gelfand and Graev. Gelfand and Naimark [**74**] studied representations of the classical groups. The extension to general semisimple Lie groups was analyzed by Harish-Chandra [**92–97**].

The set Σ of irreducible representations so constructed for various conjugacy classes of Cartan subgroups of a semisimple Lie group G with finite center was shown by Gelfand et al. in the case of classical groups, and by Harish-Chandra in general, to be almost all the irreducible unitary representations of G, in the following sense. There is a measure μ on Σ such that, for $f \in C_0^\infty(G)$, $\pi(f)$ is of trace class for all $\pi \in \Sigma$, and

$$\|f\|^2_{L^2(G)} = \int_\Sigma \|\pi(f)\|^2_{\mathrm{HS}}\, d\mu(\pi). \tag{2.25}$$

In other words, the Plancherel measure whose existence was established by Segal [**212**] and Mautner [**168**] is supported in Σ. Explicit formulas for $d\mu(\pi)$ have been constructed, by Gelfand et al. for the classical groups, and by Harish-Chandra for general semisimple G. Computing the trace

$$\operatorname{tr}\pi(f), \qquad f \in C_0^\infty(G) \tag{2.26}$$

for various π is an integral part of producing the Plancherel formula (2.25). If π is a principal series representation, which is given rather simply in terms of a G-action on a homogeneous vector bundle over a compact space, computing (2.26) is a doable exercise in distribution theory; see [**87**] for a nice discussion of particular cases. It is much tougher to compute (2.26) when π is a discrete series representation, though Harish-Chandra [**96**] succeeded in doing this, and subsequently in computing (2.26) for all the P_o-series representations, and in obtaining the Plancherel formula for general semisimple G. We refer to [**93–98**]. See also [**78**]. A treatment of the Plancherel formula for groups with only one conjugacy class of Cartan subgroup, with particular attention to $\mathrm{SO}_e(2n+1,1)$, can be found in the last chapter of Wallach [**253**].

As the cases $\mathrm{SL}(2,R)$ and $\mathrm{SL}(2,\mathbf{C})$ illustrate, such representations do not exhaust all the irreducible unitary representations of a semisimple Lie group. These groups have "complementary series" representations, and so do more general G. More general semisimple Lie groups have further representations, with names like "degenerate series," connoting that they are rather mysterious. Though such representations do not appear explicitly in the Plancherel formula (for the regular representation of G on $L^2(G)$), such representations arise in other situations. Recall from Chapter 8 the possibility of complementary series representations of $G = \mathrm{SL}(2,R)$ occurring in the regular representation of G on $L^2(\Gamma \setminus G)$ in the case of a compact quotient. Determining all the irreducible unitary representations of a semisimple Lie group G remains a major outstanding problem. A great deal of progress has been made in the past decade, particularly as a result of Langland's work on "admissible" irreducible representations of G, and of work by Knapp and Zuckerman [**140**] and Vogan [**250**] building on this. This has recently brought about a complete classification of irreducible unitary representations for a growing list of groups, including $\mathrm{SU}(2,2)$, $\mathrm{SO}_e(3,2)$, $\mathrm{Sp}(n,1)$ and, very recently, Vogan's treatment of $\mathrm{SL}(n,F)$, for $F = \mathbf{R}, \mathbf{C}$, or $\mathbf{H}$. We refer to [**136, 137, 251**] for surveys of this important work, and on this note bring our own introduction to this area to an end.

APPENDIX A

The Fourier Transform and Tempered Distributions

As mentioned in the introduction, the Fourier transform is defined by

$$\hat{f}(\xi) = (2\pi)^{-n/2} \int_{R^n} f(x) e^{-ix\cdot\xi}\, dx. \tag{A.1}$$

We also denote $\hat{f}(\xi)$ by $\mathcal{F}f(\xi)$. If $f \in L^1(R^n)$, it is clear that $\hat{f} \in L^\infty(R^n)$. We want to introduce some other spaces of functions and generalized functions, discuss the behavior of the Fourier transform theorem, and prove the Fourier inversion formula. For a more complete treatment, see Chapter VI of Yosida [**267**], or Chapter I of Taylor [**234**].

One important space is $S(R^n)$, the Schwartz space of rapidly decreasing functions. One says $f \in S(R^n)$ when f is C^∞ and

$$x^\beta D^\alpha f(x) \in L^\infty(R^n) \tag{A.2}$$

for all multi-indices α and β. Here, if $\alpha = (\alpha_1, \dots, \alpha_n)$, we set

$$x^\alpha = x_1^{\alpha_1} \cdots x_n^{\alpha_n}; \qquad D^\alpha = D_1^{\alpha_1} \cdots D_n^{\alpha_n} \tag{A.3}$$

where $D_j = (1/i)\partial/\partial x_j$. It is easy to see that if $f \in S(R^n)$, so if $\hat{f}$, and

$$\xi^\alpha D^\beta \hat{f}(\xi) = \mathcal{F}(D^\alpha x^\beta f)(\xi). \tag{A.4}$$

Thus

$$\mathcal{F}\colon S(R^n) \to S(R^n). \tag{A.5}$$

We denote the space of continuous linear functionals on $S(R^n)$ by $S'(R^n)$. This is called the Schwartz space of tempered distributions.

If we set

$$\mathcal{F}^* f(\xi) = (2\pi)^{-n/2} \int f(x) e^{ix\cdot\xi}\, dx \tag{A.6}$$

we see that $\mathcal{F}^*\colon S(R^n) \to S(R^n)$ and, with respect to the natural inner product $(u, v) = \int u(x)\overline{v(x)}\, dx$, we have

$$(\mathcal{F}u, v) = (u, \mathcal{F}^* v), \qquad u, v \in S(R^n). \tag{A.7}$$

Using this identity, we can extend $\mathcal{F}$ and $\mathcal{F}^*$ to $S'(R^n)$ by duality:

$$\mathcal{F}, \mathcal{F}^* : S'(R^n) \to S'(R^n). \tag{A.8}$$

The following is the Fourier inversion formula.

PROPOSITION A.1. *If $u \in S$, then*

$$\mathcal{F}\mathcal{F}^* u = \mathcal{F}^*\mathcal{F} u = u. \tag{A.9}$$

Hence

$$\mathcal{F}\mathcal{F}^* = \mathcal{F}^*\mathcal{F} = I \quad \text{on } S'(R^n). \tag{A.10}$$

We sketch a proof of this. It suffices to show $\mathcal{F}^*\mathcal{F}u = u$ for $u \in S$. Now

$$\begin{aligned} \mathcal{F}^*\mathcal{F}u(x) &= (2\pi)^{-n} \iint u(y) e^{i(x-y)\cdot\xi}\, dy\, d\xi \\ &= \lim_{\varepsilon\downarrow 0} (2\pi)^{-n} \iint u(y) e^{i(x-y)\cdot\xi - \varepsilon|\xi|^2}\, dy\, d\xi \\ &= \lim_{\varepsilon\downarrow 0} \int u(y) p(\varepsilon, x-y)\, dy \end{aligned} \tag{A.11}$$

where

$$p(\varepsilon, x) = (2\pi)^{-n} \int e^{ix\cdot\xi - \varepsilon|\xi|^2}\, d\xi. \tag{A.12}$$

Note that

$$p(\varepsilon, x) = \varepsilon^{-n/2} q(x/\sqrt{\varepsilon}) \tag{A.13}$$

where $q(x) = p(1, x)$. In a moment we will show that

$$p(\varepsilon, x) = (4\pi\varepsilon)^{-n/2} e^{-|x|^2/4\varepsilon}. \tag{A.14}$$

The derivation of this identity will also show that

$$\int_{R^n} q(x)\, dx = 1. \tag{A.15}$$

From this it is a simple exercise to deduce that

$$\lim_{\varepsilon\downarrow 0} \int u(y) p(\varepsilon, x-y)\, dy = u(x) \tag{A.16}$$

for any $u \in S(R^n)$, and the Fourier inversion formula is proved.

To prove (A.14), we observe that $p(\varepsilon, x)$, defined by (A.12), is an entire analytic function of $x \in \mathbf{C}^n$; it is convenient to verify that

$$p(\varepsilon, ix) = (4\pi\varepsilon)^{-n/2} e^{|x|^2/4\varepsilon}, \qquad x \in R^n, \tag{A.17}$$

from which (A.14) follows by analytic continuation. Now

$$\begin{aligned} p(\varepsilon, ix) &= (2\pi)^{-n} \int e^{-x\cdot\xi - \varepsilon|\xi|^2}\, d\xi \\ &= (2\pi)^{-n} e^{|x|^2/4\varepsilon} \int e^{-|x/2\sqrt{\varepsilon} + \sqrt{\varepsilon}\xi|^2}\, d\xi \\ &= (2\pi)^{-n} \varepsilon^{-n/2} e^{|x|^2/4\varepsilon} \int_{R^n} e^{-|\xi|^2}\, d\xi. \end{aligned} \tag{A.18}$$

So to prove (A.17), it remains to show that

$$\int_{R^n} e^{-|\xi|^2}\, d\xi = \pi^{n/2}. \tag{A.19}$$

Now if $A = \int_{-\infty}^{\infty} e^{-\xi^2}\, d\xi$, then the left side of (A.19) is equal to A^n. But then for $n = 2$ we can use polar coordinates:

$$A^2 = \int_{R^2} e^{-|\xi|^2}\, d\xi = \int_0^{2\pi}\int_0^{\infty} e^{-r^2} r\, dr\, d\theta = \pi.$$

This completes the proof of the identity (A.17), hence of (A.14).

In light of (A.7) and the Fourier inversion formula, we see that, for $u, v \in \mathcal{S}$,

$$(\mathcal{F}u, \mathcal{F}v) = (u, v) = (\mathcal{F}^*u, \mathcal{F}^*v).$$

Thus $\mathcal{F}$ and $\mathcal{F}^*$ extend uniquely from $\mathcal{S}(R^n)$ to $L^2(R^n)$, and are inverse to each other. Thus we have the Plancherel theorem,

PROPOSITION A.2. *The Fourier transform*

$$\mathcal{F}\colon L^2(R^n) \to L^2(R^n) \tag{A.20}$$

is unitary, with inverse $\mathcal{F}^*$.

As mentioned in the introduction, the calculation of the Gaussian integral (A.12) gives the explicit formula for the fundamental solution to the heat equation:

$$e^{t\Delta}\delta(x) = (4\pi t)^{-n/2} e^{-|x|^2/4t}, \qquad t > 0,\ x \in R^n. \tag{A.21}$$

It was also shown in the introduction how to find the Poisson kernel

$$e^{-y\sqrt{-\Delta}}\delta(x) = c_n(y^2 + |x|^2)^{-(n+1)/2}$$

and also the fundamental solution to the wave equation $(\partial^2/\partial t^2 - \Delta)u = 0$, on $R \times R^n$. See (I.12)–(I.17) of the introduction. Here we give some complementary results on solutions to some basic linear PDE on R^n. Namely, we analyze the resolvent of the Laplace operator:

$$(\lambda^2 - \Delta)^{-1}\delta(x) = (2\pi)^{-n}\int_{R^n} (\lambda^2 + |\xi|^2)^{-1} e^{ix\cdot\xi}\, d\xi. \tag{A.22}$$

This Fourier integral is not absolutely convergent, if $n \geq 2$, but can be interpreted in the sense of tempered distributions. It is convenient to use the identity

$$(\lambda^2 - \Delta)^{-1}\delta(x) = \int_0^{\infty} e^{t(\Delta - \lambda^2)}\delta(x)\, dt, \tag{A.23}$$

when $\lambda > 0$, and plug in (A.21), to get

$$(\lambda^2 - \Delta)^{-1}\delta(x) = (4\pi)^{-n/2}\int_0^{\infty} e^{-|x|^2/4t - t\lambda^2} t^{-n/2}\, dt. \tag{A.24}$$

Note that this integral was evaluated in the introduction for $n = 1$ and 3. The identity (I.11) (complemented by the identity (7.89) of Chapter 1) gives, for

$$K(r,\lambda,\nu) = \int_0^\infty e^{-r^2/4t} e^{-t\lambda^2} t^{-1-\nu}\,dt, \tag{A.25}$$

the formulas

$$\begin{aligned} K(r,\lambda,-\tfrac{1}{2}) &= \pi^{-1/2}\lambda^{-1}e^{-\lambda r}, \\ K(r,\lambda,\tfrac{1}{2}) &= 2\pi^{-1/2}r^{-1}e^{-\lambda r}. \end{aligned} \tag{A.26}$$

It is clear that for any *odd* integer n, we can evaluate the integral in (A.24) by differentiating repeatedly (A.26) with respect to r; so on R^{2k+1},

(A.27)

$$\begin{aligned} (\lambda^2 - \Delta)^{-1}\delta(x) &= (4\pi)^{-k-1/2}K(r,\lambda,-\tfrac{1}{2}+k) \\ &= (4\pi)^{-k-1/2}(-2r^{-1}\partial/\partial r)^k(\pi^{-1/2}\lambda^{-1}e^{-\lambda r}) \qquad (r = |x|) \end{aligned}$$

in view of the general identity

$$(\partial/\partial r)K(r,\lambda,\nu) = -\tfrac{1}{2}rK(r,\lambda,\nu+1). \tag{A.28}$$

On R^{2k}, one has

$$(\lambda^2 - \Delta)^{-1}\delta(x) = (4\pi)^{-k}K(r,\lambda,k-1). \tag{A.29}$$

This is not an elementary function. In fact, use of polar coordinates for the Laplace operator as in Chapter 4, §1, shows (A.29) satisfies a modified Bessel equation, as a function of r. We can also see this from (A.25), if we note that, in addition to (A.28), we have

$$(\partial/\partial\lambda)K(r,\lambda,\nu) = -2\lambda K(r,\lambda,\nu-1). \tag{A.30}$$

Also, a change of variable in the integral (A.25) gives

$$K(r,\lambda,\nu) = \lambda^{2\nu}K(\lambda r,1,\nu), \tag{A.31}$$

and if we combine (A.31) and (A.30) we get

$$(\partial/\partial r)K(r,\lambda,\nu) + (2\nu/\lambda)K(r,\lambda,\nu) = -2\lambda K(r,\lambda,\nu-1). \tag{A.32}$$

If we apply $\partial/\partial r$ to both sides of (A.32) and use (A.28), we get

$$(\partial^2/\partial r^2)K + 2\nu\lambda^{-1}(\partial/\partial r)K = \lambda r K(r,\lambda,\nu), \tag{A.33}$$

wich can be converted to the modified Bessel equation. One has the identity

$$\frac{1}{2}(\lambda r/2)^\nu \int_0^\infty e^{-r^2/4t}e^{-t\lambda^2}t^{-1-\nu}\,dt = \lambda^{2\nu}K_\nu(\lambda r); \tag{A.34}$$

see Lebedev [**148**], page 119. Hence, on R^n,

$$(\lambda^2 - \Delta)^{-1}\delta(x) = 2(4\pi)^{-n/2}(2\lambda/r)^{n/2-1}K_{n/2-1}(\lambda r). \tag{A.35}$$

The identity (A.34) gives this for $\lambda > 0$. Then we get it for $\lambda^2 \in \mathbf{C} \setminus R^-$, by analytic continuation.

One could also tackle the resolvent by looking at the Fourier transform on the right side of (A.22). If we use (I.21) to express this Fourier transform as a Hankel transform, the formula (A.35) is seen to be equivalent to the special case when $\nu = \frac{1}{2}n - 1$ of the classical identity

$$\int_0^\infty (\lambda^2 + s^2)^{-1} J_\nu(sr) s^{\nu+1}\, ds = \lambda^\nu K_\nu(\lambda r).$$

See Lebedev [**154**], page 133, for a proof of this identity (in the case of absolute convergence) and further generalizations.

APPENDIX B

The Spectral Theorem

Our purpose here is to give a brief discussion of the spectral theorem. We consider several forms. The best known form is

THE SPECTRAL THEOREM. *If A is a selfadjoint operator on a Hilbert space H, then there is a strongly countably additive projection-valued measure dE such that*

$$Au = \int_{-\infty}^{\infty} \lambda \, dE_\lambda u, \qquad u \in \mathcal{D}(A). \tag{B.1}$$

This result is a consequence of the

STRONG SPECTRAL THEOREM. *If A is a selfadjoint operator on a separable Hilbert space H, there is a σ-compact space Ω, a Borel measure μ on Ω, a unitary map*

$$W: L^2(\Omega, d\mu) \to H, \tag{B.2}$$

and a real-valued measurable function $a(x)$ on Ω, such that

$$W^{-1}AWf(x) = a(x)f(x), \qquad Wf \in \mathcal{D}(A). \tag{B.3}$$

In order to establish this result, we will work with the unitary group

$$U(t) = e^{itA}. \tag{B.4}$$

For a given $\xi \in H$, let H_ξ denote the closed linear span of $U(t)\xi$ for $t \in R$. If $H_\xi = H$, we say ξ is a *cyclic vector.* If H_ξ is not all of H, note that $H_\xi^\perp$ is invariant under $U(t)$. Using this, it is not difficult to decompose H into a (possibly countably infinite) direct sum of spaces H_{ξ_j}. It will suffice to establish the Strong Spectral Theorem on each factor, so we will establish the following result.

PROPOSITION B.1. *If $H = H_\xi$ (so there is a cyclic vector ξ), then we can take $\Omega = R$, and there exists a positive Borel measure μ on R and a unitary map $W: L^2(R, d\mu) \to H$ so that*

$$W^{-1}AWf(x) = xf(x), \qquad Wf \in \mathcal{D}(A), \tag{B.5}$$

or equivalently,

$$W^{-1}U(t)Wf = e^{itx}f(x), \qquad f \in L^2(R, d\mu). \tag{B.6}$$

The measure μ on R will be the Fourier transform

$$\mu = \hat{\varsigma}(\tau) \tag{B.7}$$

where

$$\varsigma(t) = (2\pi)^{-1}(e^{itA}\xi, \xi). \tag{B.8}$$

A priori, it is not at all clear that (B.7) defines a measure, though, since clearly $\varsigma \in L^\infty(R)$, we see that μ is a *tempered distribution.* We will show that μ is indeed a positive measure during the course of our argument. We will in the meantime use notation anticipating that μ is a measure:

$$(e^{itA}\xi, \xi) = \int_{-\infty}^{\infty} e^{it\tau}\, d\mu(\tau), \tag{B.9}$$

keeping in mind that the Fourier transform in (B.9) is to be interpreted in the sense of tempered distributions. As for the map W, we first define

$$W: S(R) \to H, \tag{B.10}$$

where $S(R)$ is the Schwartz space of rapidly decreasing functions, by

$$W(f) = f(A)\xi, \tag{B.11}$$

where we define the operator $f(A)$ by the formula

$$f(A) = (2\pi)^{-1/2}\int_{-\infty}^{\infty} \hat{f}(t)e^{itA}\, dt. \tag{B.12}$$

The reason for this notation will become apparent shortly; see (B.21). Note that, using (B.9), we have

$$\begin{aligned}(f(A)\xi, g(A)\xi) &= (2\pi)^{-1}\left(\int \hat{f}(s)e^{isA}\xi\, ds, \int \hat{g}(t)e^{itA}\xi\, dt\right)\\ &= (2\pi)^{-1}\iint \hat{f}(s)\overline{\hat{g}(t)}(e^{i(s-t)A}\xi, \xi)\, ds\, dt\\ &= (2\pi)^{-1}\iiint \hat{f}(s)\overline{\hat{g}(t)}e^{i(s-t)\tau}\, d\mu(\tau)\, ds\, dt\\ &= \int f(\tau)\overline{g(\tau)}\, d\mu(\tau).\end{aligned} \tag{B.13}$$

For $f, g \in S(R)$, the last "integral" is to be interpreted as $\langle f\overline{g}, \mu\rangle$.

Now, if $g = f$, the left side of (B.13) is $\|f(A)\xi\|^2$, which is ≥ 0. Hence,

$$\langle |f|^2, \mu\rangle \geq 0 \quad \text{for all } f \in S(R). \tag{B.14}$$

This is enough to imply that the tempered distribution μ is actually a positive measure! Knowing this, we can now interpret (B.13) as implying that W has a unique continuous extension

$$W: L^2(R, d\mu) \to H, \tag{B.15}$$

and W is an isometry. Furthermore, since ξ is assumed to be cyclic, it is easy to see the range of W must be dense in H, so W must be unitary.

Now from (B.12) it follows that, if $f \in S(R)$,

$$e^{isA}f(A) = \varphi_s(A) \tag{B.16}$$

where

$$\varphi_s(\tau) = e^{is\tau}f(\tau). \tag{B.17}$$

Hence

$$W^{-1}e^{isA}f(A)\xi = W^{-1}\varphi_s(A)\xi = e^{is\tau}f(\tau). \tag{B.18}$$

Since $\{f(A)\xi \colon f \in S\}$ is dense in H, we obtain

$$W^{-1}e^{isA} = e^{is\tau}W^{-1}, \tag{B.19}$$

which proves (B.6), and hence Proposition B.1.

Note that since $W^{-1}e^{itA}W = e^{it\tau}$, the formula (B.12) implies

$$W^{-1}f(A)W = f(\tau), \tag{B.20}$$

which justifies the notation $f(A)$ in (B.12). In the language of (B.1), we have

$$f(A)u = \int f(\lambda)\, dE_\lambda u. \tag{B.21}$$

If a selfadjoint operator A has the representation (B.5), one says A has *simple spectrum*. It follows from Proposition B.1 that a selfadjoint operator has simple spectrum if and only if it has a cyclic vector.

We remark that a more typical method of showing there is a positive measure μ such that (B.9) holds is to invoke Bochner's theorem, characterizing the Fourier transform of positive measures as positive definite functions. It is well known that, using the theory of tempered distributions, one can give a simple proof of Bochner's theorem. Our proof of the spectral theorem has avoided this detour entirely, working directly with tempered distributions.

In general, the spectral theorem is considered to be nonconstructive in nature. However, especially if A has simple spectrum, or spectrum of low multiplicity, the method of proof of the spectral theorem can sometimes be implemented to give an explicit spectral representation of A. Doing this involves knowing explicitly the unitary group $U(t) = e^{itA}$, and particularly knowing explicitly the function $\varsigma(t) = (e^{itA}\xi, \xi)$ for some cyclic vector ξ, and then finding explicitly the Fourier transform of this function, which is known generally to be some positive measure. We can illustrate this program in the case

$$A = D = id/dx, \tag{B.22}$$

so

$$e^{itD}f(x) = f(x - t). \tag{B.23}$$

If we pick $\xi \in \mathcal{S}(R)$ such that $\hat{\xi}(\tau)$ never vanishes, then

$$\varsigma(t) = (e^{itD}\xi, \xi) = \xi * \check{\xi}(t) \tag{B.24}$$

where $\check{\xi}(t) = \overline{\xi}(-t)$, so we have $\hat{\varsigma}(\tau) = |\hat{\xi}(\tau)|^2$, i.e.,

$$d\mu = |\hat{\xi}(\tau)|^2 \, d\tau \quad \text{on } R. \tag{B.25}$$

In this case the unitary map

$$W\colon L^2(R, |\hat{\xi}(\tau)|^2 \, d\tau) \to L^2(R, dx) \tag{B.26}$$

is given by

$$Wf = f(D)\xi = \hat{f} * \xi, \tag{B.27}$$

and we have $W^{-1}e^{itD}Wf(\tau) = e^{it\tau}f(\tau)$. If we take ξ to be an approximate delta function, then in the limit W becomes the Fourier transform.

Of course, as has been shown several times in this monograph, an explicit knowledge of the unitary group e^{itA} is a very convenient tool for studying the spectral resolution of A. In fact, the spectral measure is given by

$$dE_\lambda = (2\pi)^{-1} \int_{-\infty}^{\infty} e^{itA - i\lambda t} \, dt, \tag{B.28}$$

where the right side is interpreted a priori as an operator valued tempered distribution. Compare the spectral analysis of the Laplace operator on spheres and hyperbolic space, deduced from a knowledge of the fundamental solution of the wave equation, in Chapters 4 and 8.

One can generalize the notion of a cyclic vector to the case of a k-tuple of commuting selfadjoint operators $(A_1, \ldots, A_k)$. More precisely, suppose the unitary groups $U_j(t) = e^{itA_j}$ all commute. We say $\xi \in H$ is cyclic if the closed linear span of $U_1(t_1) \cdots U_k(t_k)\xi$ is all of H. In that case, the proof of Proposition B.1 generalizes to show that there is a positive measure μ on R^k and a unitary map $W\colon L^2(R^k, d\mu) \to H$ such that $W^{-1}A_jWf(x) = x_jf(x)$, $Wf \in \mathcal{D}(A_j)$. We would say $(A_1, \ldots, A_k)$ has simple spectrum. If a single operator $A = A_1$ does not have simple spectrum, it would elucidate its spectral behavior to find a k-tuple of commuting operators such that $(A_1, \ldots, A_k)$ has simple spectrum. More generally, given a selfadjoint operator B, without simple spectrum, we might try to find $(A_1, \ldots, A_k)$, commuting and with simple spectrum, such that $W^{-1}BWf(x) = \beta(x)f(x)$, $Wf \in \mathcal{D}(B)$. For example, for the Laplace operator Δ on R^k, one takes $(A_1, \ldots, A_k) = (D_1, \ldots, D_k)$, $D_j = i\partial/\partial x_j$. Of course, given such B, one would like such $A_1, \ldots, A_k$ to arise in some "natural" fashion. In many cases, such as $B = \Delta$ on the sphere S^n, $n \geq 2$, it is natural to write B as a function of some selfadjoint operators which commute with B, but not with each other, and which generate a noncommutative group of transformations acting irreducibly on each eigenspace of B. Thus noncommutative harmonic analysis arises naturally in the search for a "simple spectrum."

As an application of the spectral theorem, we prove the following result, a version of Schur's lemma, characterizing irreducible group actions, which is used from time to time in this monograph (see Chapter 0, Theorem 1.6).

PROPOSITION B.2. *A group G of unitary operators on a Hilbert space H is irreducible if and only if, for any bounded linear operator A on H,*

$$UA = AU \quad \textit{for all } U \in G \Rightarrow A = \lambda I. \tag{B.29}$$

PROOF. First, if $H_0 \subset H$ is a closed invariant subspace and P is the orthogonal projection on H_0, then (B.29) implies H_0 is 0 or H, so clearly (B.29) implies irreducibility. Conversely, suppose G is irreducible, and let $A \in \mathcal{L}(H)$ be such that

$$UA = AU \quad \text{for all } U \in G. \tag{B.30}$$

The unitarity of U implies the same identity also holds for A^*, so considering $A + A^*$ and $A - A^*$, we can assume without loss of generality that A is selfadjoint in (B.30). Now (B.30) continues to hold with A replaced by any polynomial in A, and then, by the spectral theorem, we can deduce that, if $A = \int \lambda \, dE_\lambda$, then, for any Borel set $\mathcal{B}$, $E(\mathcal{B})U = UE(\mathcal{B})$ for all $U \in G$. But then the irreducibility implies $E(\mathcal{B}) = 0$ or I, since $E(\mathcal{B})$ is a projection on an invariant subspace. Since this holds for any Borel set $\mathcal{B}$, we have $A = \lambda I$ for some λ. This completes the proof.

We say a few words about the general justification of the assertion that for any selfadjoint operator A, the operator iA generates a unitary group (B.4), which is part of Stone's theorem. First, if A is bounded, the operator e^{itA} can be defined by a convergent power series expansion, and power series manipulation gives $e^{isA}e^{itA} = e^{i(s+t)A}$ and $(e^{itA})^* = e^{-itA}$. For unbounded A, there are direct demonstrations that iA generates a unitary group, but it is perhaps simpler to use von Neumann's trick, and consider the unitary operator $V = (A+i)(A-i)^{-1}$. Then $B_1 = (V + V^*)/2$, $B_2 = (V - V^*)/2i$ form a pair of commuting bounded selfadjoint operators, to which the spectral theorem proved above applies. Then the spectral theorem for A is a corollary of that for (B_1, B_2). Of course, granted the spectral theorem for A, the fact that iA generates a unitary group is a simple consequence.

We conclude this appendix by mentioning one important connection between the unitary group e^{itA} and the notion of selfadjointness. A symmetric operator A_0 with domain $\mathcal{D}$ is said to be essentially selfadjoint if there exists a unique selfadjoint operator A such that $\mathcal{D} \subset \mathcal{D}(A)$, the domain of A, and $A_0 u = Au$ for $u \in \mathcal{D}$. For symmetric operators, this extends Proposition 2.2 of Chapter 0.

PROPOSITION B.3. *Let A_0 be an operator on a Hilbert space H, with domain $\mathcal{D}$. Assume $\mathcal{D}$ is dense in H. Let $U(t)$ be a unitary group, with infinitesimal generator iA, so A is selfadjoint, and*

$$U(t) = e^{itA}. \tag{B.31}$$

Suppose $\mathcal{D} \subset \mathcal{D}(A)$ and $A_0 u = Au$ for $u \in \mathcal{D}$, or equivalently,

$$\lim_{t \to 0} t^{-1}(U(t)u - u) = A_0 u \quad \textit{for all } u \in \mathcal{D}. \tag{B.32}$$

Also suppose $\mathcal{D}$ is invariant under $U(t)$:

$$U(t)\mathcal{D} \subset \mathcal{D}. \tag{B.33}$$

Then A_0 is essentially selfadjoint; A is its unique selfadjoint extension. Suppose furthermore that

$$A_0\colon \mathcal{D} \to \mathcal{D}. \tag{B.34}$$

Then A_0^k, with domain $\mathcal{D}$, is essentially selfadjoint, for each positive integer k.

PROOF. We use the following well known criterion for essential selfadjointness (see, e.g., [**202**]). A_0 is essentially selfadjoint if and only if the range of $i \pm A_0$ is dense in H. So suppose $v \in H$ and

$$((i \pm A_0)u, v) = 0 \quad \text{for all } u \in \mathcal{D}. \tag{B.35}$$

Using (B.33) together with the fact that $A_0 = A$ on $\mathcal{D}$, we have

$$((i \pm A_0)u, U(t)v) = 0 \quad \text{for all } t \in R,\ u \in \mathcal{D}. \tag{B.36}$$

Consequently $\int \rho(t)U(t)v\,dt$ is orthogonal to the range of $i \pm A_0$, for any $\rho \in L^1(R)$. Choosing $\rho(t) \in C_0^\infty(R)$ an approximate identity, we approximate v by elements of $\mathcal{D}(A)$, indeed of $\mathcal{D}(A^k)$ for all k. Thus we can suppose in (B.35) that $v \in \mathcal{D}(A)$. Hence, taking adjoints, we have

$$(u, (i \pm A)v) = 0 \quad \text{for all } u \in \mathcal{D}. \tag{B.37}$$

Since $(i \pm A)v \in H$, and $\mathcal{D}$ is dense in H, it follows that (B.37) holds for all $u \in H$, hence for $u = v$. But then the imaginary part is $\|v\|^2$, so (B.35) implies $v = 0$. This proves the first part of the proposition. Granted (B.34), the same proof works with A_0 replaced by A_0^k (but $U(t)$ unaltered), so the proposition is proved.

An example of an application of Proposition B.3 is Chernoff's proof [**38**] that on a complete Riemannian manifold M, with $\mathcal{D} = C_0^\infty(M)$, all powers of the Laplace operator are essentially selfadjoint on $\mathcal{D}$. This is because, with

$$B_0 = \begin{pmatrix} 0 & I \\ \Delta & 0 \end{pmatrix}, \qquad \mathcal{D}(B_0) = C_0^\infty \oplus C_0^\infty, \tag{B.38}$$

the group

$$U(t) = e^{tB_0} \tag{B.39}$$

is the fundamental solution to the wave equation

$$\partial^2 U/\partial t^2 - \Delta U = 0, \tag{B.40}$$

and its unitarity is equivalent to the standard energy conservation law for the wave equation, while *finite propagation speed* for solutions to the wave equation implies $\mathcal{D}(B_0)$ is preserved by $U(t)$. Note that

$$B_0^2 = \begin{pmatrix} \Delta & 0 \\ 0 & \Delta \end{pmatrix}, \tag{B.41}$$

so essential selfadjointness of Δ and its powers follows.

APPENDIX C

The Radon Transform on Euclidean Space

In this appendix we give a brief discussion of the Radon transform on Euclidean space R^n, defined as follows. If H is any $(n-1)$-dimensional linear subspace of R^n, $w \in R^n/H$, set

$$\mathcal{R}f(w,H) = \int_H f(y+w)\,dy, \tag{C.1}$$

where H gets induced Lebesgue measure. This Radon transform is useful in commutative harmonic analysis, and also played a role in Chapter 6, in deriving the Plancherel formula on a nilpotent Lie group. Other variants of the Radon transform include transforms on spheres (see Chapter 4) and also on hyperbolic space and more general symmetric spaces (see Helgason [**102, 103**]). Our goal will be to obtain a Plancherel formula for (C.1), since that was used in Chapter 6. This will be a simple consequence of the Plancherel formula for the Fourier transform.

In fact, if ω is a unit vector orthogonal to H, note that

$$\hat{f}(s\omega) = (2\pi)^{-n/2} \int e^{-isw} \mathcal{R}f(w,H)\,dw, \tag{C.2}$$

provided R^n/H is made isomorphic to R via $\omega + H \mapsto 1$. Now the Plancherel formula $\|f\|_{L^2} = \|\hat{f}\|_{L^2}$ on R^n gives

$$\|f\|_{L^2}^2 = \frac{1}{2}\int_{-\infty}^{\infty}\int_{S^{n-1}} |\hat{f}(s\omega)|^2 |s^{n-1}|\,d\omega\,ds. \tag{C.3}$$

Meanwhile, the Plancherel formula for the Fourier transform on R gives, in view of (C.2),

$$\int_{-\infty}^{\infty} |\hat{f}(s\omega)|^2\,ds = (2\pi)^{-(n-1)/2} \int |\mathcal{R}f(w,H)|^2\,dw. \tag{C.4}$$

In order to combine (C.3) and (C.4), note that

$$\int_{-\infty}^{\infty} |\hat{g}(s)|^2 |s^{\alpha}|\,ds = \int_{-\infty}^{\infty} |\widehat{D^{\alpha/2}g}(s)|^2\,ds, \tag{C.5}$$

so we get

$$\|f\|_{L^2}^2 = \frac{1}{2}(2\pi)^{-(n-1)/2} \int_{\mathcal{G}} \int_{-\infty}^{\infty} |D_w^{(n-1)/2} \mathcal{R}f(w,H)|^2 \, dw \, d\,\mathrm{vol}(H), \tag{C.6}$$

as our Plancherel formula for the Radon transform. Here $d\,\mathrm{vol}(H)$ is the invariant volume element on the Grassmannian manifold of hyperplanes in R^n, normalized so that $\mathrm{vol}(\mathcal{G}) = \mathrm{vol}(S^{n-1})$.

We can polarize (C.6) and get, for $f, g \in L^2(R^n)$,

$$\begin{aligned} (f,g)_{L^2} &= c_n \int_{\mathcal{G}} \int_{\mathbf{R}} |D_w^{(n-1)/2}| \mathcal{R}f(w,H) \overline{|D_w^{(n-1)/2}| \mathcal{R}g(w,H)} \, dw \, d\,\mathrm{vol}(H) \\ &= c_n \int_{\mathcal{G}} \int_{\mathbf{R}} |D_w^{(n-1)}| \mathcal{R}f(w,H) \overline{\mathcal{R}g(w,H)} \, dw \, d\,\mathrm{vol}(H). \end{aligned} \tag{C.7}$$

Note that

$$n \,\mathrm{odd} \Rightarrow |D_w^{(n-1)}| = \pm D_w^{n-1}, \tag{C.8}$$

but if n is even, $|D_w^{n-1}|$ is not a local operator. Taking g to be highly concentrated near a point $p \in R^n$, we see that

$$\mathcal{R}\delta_p(w,H) = \delta(\omega_H \cdot p - w) \tag{C.9}$$

where ω_H is a unit normal to H, determining an isomorphism $R^n/H \approx R$. Thus we get the inversion formula

$$f(p) = c_n \int_{\mathcal{G}} |D_w^{n-1}| \mathcal{R}f(\omega_H \cdot p, H) \, d\,\mathrm{vol}(H). \tag{C.10}$$

The Radon transform provides a nice tool for proving the Huygens principle for constant coefficient hyperbolic systems when n is odd; this arises from the fact that (C.8) is a local operator in that case. It also is a tool for scattering theory. See Lax and Phillips **[150]** for more about these uses of the Radon transform.

APPENDIX D

Analytic Vectors, and Exponentiation of Lie Algebra Representations

We discuss briefly the role of analytic vectors in passing from representations of a Lie algebra to representations of a Lie group.

Let $X_1, \ldots, X_k$ be a set of (unbounded) skew adjoint opertors on a Hilbert space H. We say $u \in H$ is an analytic vector for $(X_1, \ldots, X_k)$ if $u \in \mathcal{D}(X^\alpha)$ for all α, where $X^\alpha = X_{\alpha(1)} X_{\alpha(2)} \cdots X_{\alpha(N)}$, and, for some $C < \infty$, $K < \infty$,

$$\|X^\alpha u\| \le C(K|\alpha|)^{|\alpha|}. \tag{D.1}$$

We will be interested in the case when ω is a representation of a Lie algebra $\mathfrak{g}$ by skew adjoint operators, $X_j = \omega(L_j)$, $L_1, \ldots, L_k$ forming a basis of $\mathfrak{g}$. We denote by $\mathfrak{A}_K$ the set of analytic vectors satisfying (D.1), for some $C < \infty$. We make the following hypothesis:

$$\text{For some } K, \ \mathfrak{A}_K \text{ is dense in } H. \tag{D.2}$$

Note that, if $u \in \mathfrak{A}_K$, then, for $X = \sum a_j X_j$, $\sum |a_j| \le 1$,

$$e^{tX} u = \sum_{j=0}^{\infty} (t^j/j!) X^j u \tag{D.3}$$

is a power series converging, in the norm topology on H, for $|t| < (eK)^{-1}$, even complex t. If X_j are skew adjoint, then e^{tX} is unitary for t real.

We suppose $X_j = \omega(L_j)$ where ω is a representation of the Lie algebra $\mathfrak{g}$ by skew adjoint operators such that

$$\omega([L, M])u = \omega(L)\omega(M)u - \omega(M)\omega(L)u \tag{D.4}$$

for $u \in \mathfrak{A}_K$. Note that $\mathfrak{A}_K$ is invariant under $\omega(L)$, $L \in \mathfrak{g}$. We want to obtain a "local representation" ω_b of a neighborhood U of e in G, a Lie group with Lie algebra, by unitary operators on H. Indeed, if $L = \sum a_j L_j$, $\sum |a_j| < R = (Ke)^{-1}$, set $X = \omega(L)$ and

$$\omega_b(\exp L)u = e^{\omega(L)} u = \sum_{j=0}^{\infty} (1/j!) X^j u, \qquad u \in \mathfrak{A}_K. \tag{D.5}$$

We suppose L belongs to a small enough neighborhood U_0 of $0 \in \mathfrak{g}$ that $\exp : \mathfrak{g} \to G$ maps U_0 diffeomorphically onto a neighborhood U of $e \in G$, and that $L = \sum a_j L_j$, $\sum |a_j| < R$ on U_0.

In such a case, (D.5) defines a function on U taking values in the set of unitary operators on H. We need to show it is a local representation. If $V \subset U$ is a sufficiently small neighborhood of e (in particular, we need $V \cdot V \subset U$), we claim that, under the definition (D.5) of ω_b,

$$\omega_b(g_1 g_2)u = \omega_b(g_1)\omega_b(g_2)u, \qquad g_j \in V, u \in H. \tag{D.6}$$

It suffices to establish this identity for $u \in \mathfrak{A}_K$. This is essentially a consequence of the Campbell-Hausdorff formula, which says

$$\exp X \exp Y = \exp \Phi(X, Y) \tag{D.7}$$

where $\Phi(X, Y)$ is given by a convergent power series for X and Y small. The power series is of a "universal" sort:

$$\Phi(X, Y) = X + Y + \tfrac{1}{2}[X, Y] + \cdots. \tag{D.8}$$

See, e.g., Varadarajan [**246**]. The fact that (D.6) holds when $g_j = \exp X_j$, X_j small, $u \in \mathfrak{A}_K$, is just manipulation of power series, using the fact that the form of (D.8) is preserved on applying the representation ω of $\mathfrak{g}$ to both sides.

Thus we have produced a "local representation" of a neighborhood U of $e \in G$. We want to extend this to a global representation, provided G is simply connected. Any $g \in G$ can be written in the form

$$g = g_1 \cdots g_M, \qquad g_j \in V \ (V \subset U \text{ as above}), \tag{D.9}$$

so we would like to set

$$\omega(g)u = \omega_b(g_1) \cdots \omega_b(g_M)u. \tag{D.10}$$

We need to show that the representation (D.10) is independent of the representation (D.9). What is the same, if

$$e = g_1 \cdots g_N, \qquad g_j \in V, \tag{D.11}$$

we need to show that

$$\omega_b(g_1) \cdots \omega_b(g_N)u = u, \qquad \text{all } u \in H. \tag{D.12}$$

In (D.11) we could express each g_j as a product of terms in a very much smaller neighborhood $V^{\#}$ of e, so in proving (D.12) we can assume that each g_j belongs to as small a neighborhood of e as desired (of course, N gets quite large). The idea is to connect the points $p_0 = e$, $p_1 = g_1, \ldots, p_j = g_1 \cdots g_j, \ldots, p_N = e$ by a closed analytic loop. If G is simply connected, we can analytically contract the loop, leaving it fixed at e. Thus the points p_j trace out analytic curves $p_j(t)$, with $p_j(0) = p_j$, $p_j(1) = e$. So we define $g_j(t)$ by $p_j(t) = p_{j-1}(t) g_j(t)$. We can suppose that each $g_j(t)$ belongs to V for $0 \le t \le 1$. In fact, we can suppose $\operatorname{dist}(g_j(t), e) \le C/N$. Note that $g_1(t) \cdots g_N(t) = e$ for all t. Consider

$$E(t)u = \omega_b(g_1(t)) \cdots \omega_b(g_N(t))u. \tag{D.13}$$

Note that, if t is close to 1, all partial products $g_1(t)\cdots g_J(t)$ belong to V, so, by (D.6), for some $\varepsilon > 0$, we have

$$E(t)u = u \quad \text{for } 1-\varepsilon \le t \le 1. \tag{D.14}$$

Now if $u \in \mathfrak{A}_K$ we claim (D.13) is analytic in t. This gives $E(t)u = u$ for all $t \in [0,1]$, if $u \in \mathfrak{A}_K$, and hence for all u, since $\mathfrak{A}_K$ is dense in H. Thus ω is well defined by (D.10), on all of G, if G is simply connected.

To see that (D.13) is analytic in t for $u \in \mathfrak{A}_K$, note that if $\mathcal{B}_W$ denotes the set of $u \in H$ such that the function $\omega_b(g)u$ (which has been defined for $g \in U$) is analytic for $g \in W$, a neighborhood of $e \in G$, then of course $\mathfrak{A}_K \subset \mathcal{B}_U \subset \mathcal{B}_W$ if $W \subset U$, and if $g \in V$, then

$$\omega_b(g)u \in \mathcal{B}_{\mathrm{Ad}\, gW} \quad \text{if } u \in \mathcal{B}_W, \tag{D.15}$$

as follows from

$$\omega_b(g')\omega_b(g)u = \omega_b(g)\omega_b(g^{-1}g'g)u. \tag{D.16}$$

Now if $\text{dist}(g_j(t), e) \le C/N$, then $\prod_J^N \mathrm{Ad}\, g_j(t)$ has a limited effect, so we can suppose that $u \in \mathcal{B}_W$ and $\omega_b(g_J(t))\cdots\omega_b(g_N(t))u \in \mathcal{B}_W$ for each J. Then the analyticity of (D.13) in t for $u \in \mathfrak{A}_K$ is clear.

Now that $\omega(g)$ has been defined, as a unitary operator, for each $g \in G$, we can show it is a representation of G, i.e.,

$$\omega(g_1 g_2)u = \omega(g_1)\omega(g_2)u, \quad \text{if } g_j \in G. \tag{D.17}$$

Indeed, we know (D.17) holds for $g_j \in V$. But if $u \in \mathfrak{A}_K$ then both sides are analytic in (g_1, g_2), so (D.17) holds for $u \in \mathfrak{A}_K$. Since $\mathfrak{A}_K$ is assumed dense in H, (D.17) holds in general.

Such conclusions as were reached here were applied in Chapters 1 and 11 to the metaplectic representation, where the representation ω of $\mathrm{sp}(n, R)$ was defined by

$$\omega(Q) = iQ(X, D), \tag{D.18}$$

with $Q(x, \xi)$ a second order homogeneous polynomial. We need to know that hypothesis (D.2) is satisfied in this case. Indeed, we have, for some K,

$$\mathfrak{A}_K \supset \{p(x)e^{-|x|^2/2} : p(x) \text{ polynomial in } x\}, \tag{D.19}$$

the linear span of the Hermite functions, discussed in §6 of Chapter 1. We leave (D.19) as an exercise with the hint that it follows from (D.49). Clearly the right side of (D.19) is dense in $L^2(R^n)$.

This method of exponentiating a representation of a Lie algebra $\mathfrak{g}$ to the Lie group G depends on prescribing a dense space $\mathfrak{A}_K$ of analytic vectors. Such a method is never doomed to failure; in fact, we have the following important result, due to Harish-Chandra [**92**], Nelson [**183**], and Gårding [**69**].

THEOREM D.1. *If ω is a strongly continuous unitary representation of G on H, then for some $K < \infty$,*

$$\text{(D.20)} \qquad \mathfrak{A}_K \text{ is dense in } H.$$

We will indicate a proof of this, along the lines of the proof given by Nelson [**183**] and by Gårding [**69**]. We include some details, particularly since this result is often stated in the weaker form

$$\text{(D.21)} \qquad \bigcup_K \mathfrak{A}_K \text{ is dense in } H.$$

A different approach to the stronger result (D.20) is given in [**84**]. The proof involves approximating $u \in H$ by

$$\text{(D.22)} \qquad \rho(t)u = \int_G h(t,x)\omega(x)u\,dx,$$

where $h(t,x)$ is the heat kernel

$$\text{(D.23)} \qquad h(t,x) = e^{t\Delta}\delta(x).$$

Δ is the Laplace operator on G, which we assume endowed with a left invariant metric. It is easy to show that, for any neighborhood U of $e \in G$, $\int_U h(t,x)\,dx \to 1$ as $t \to 0$, while $\int_{G\setminus U} h(t,x)\,dx \to 0$; also $h(t,x) \geq 0$. Thus $\rho(t)u \to u$ in H as $t \to 0$. It remains to show that there exists $K < \infty$ such that, for $0 < t \leq 1$, $\rho(t)u \in \mathfrak{A}_K$ for all $u \in H$. Estimates on $h(t,x)$ were proved in [**69**]. The argument we give here is a special case of an argument from Cheeger, Gromov, and Taylor [**36**]. In this approach, one gets estimates on the Schwartz kernel of $f(\sqrt{-\Delta})$ for a general class of *even* functions $f(\lambda)$, including

$$\text{(D.24)} \qquad f_t(\lambda) = e^{-t\lambda^2},$$

by writing

$$\text{(D.25)} \qquad \begin{aligned} f(\sqrt{-\Delta}) &= (2\pi)^{-1/2}\int_{-\infty}^{\infty} \hat{f}(s)e^{is\sqrt{-\Delta}}\,ds \\ &= (2\pi)^{-1/2}\int_{-\infty}^{\infty} \hat{f}(s)\cos s\sqrt{-\Delta}\,ds, \end{aligned}$$

and exploiting *finite propagation speed* for solutions to the wave equation

$$(\partial^2/\partial s^2 - \Delta)u = 0,$$

which implies

$$\text{(D.26)} \qquad \cos s\sqrt{-\Delta}\delta_p(x) = 0 \quad \text{if } \operatorname{dist}(p,x) > |s|.$$

We get

(D.27)

$$f(\sqrt{-\Delta})\delta_p(x) = (2/\pi)^{1/2}\int_r^{\infty} \hat{f}(s)\cos s\sqrt{-\Delta}\delta_p(x)\,ds, \quad \text{if } \operatorname{dist}(p,x) > r.$$

Now, suppose $f(\lambda)$ satisfies the following estimates:

$$|\hat{f}^{(k)}(s)| \le C_1 (k/A)^k \varphi(s), \tag{D.28}$$

for some $\varphi \in L^1$. We say $f \in \mathcal{F}^{\varphi,A}$. If we set

$$\psi(r) = \int_r^\infty \varphi(s)\, ds, \tag{D.29}$$

then

$$\int_r^\infty |f^{(k)}(s)|\, ds \le C_1 (k/A)^k \psi(r). \tag{D.30}$$

Since we have the elementary L^2 operator norm bound $\| \cos s\sqrt{-\Delta} \| \le 1$, we easily deduce the following. Let $\operatorname{dist}(p,q) = r + 2a$. Then

$$\|\Delta^k f(\sqrt{-\Delta}) \Delta^l u\|_{L^2(B_a(x))} \le C((2k+2l)/A)^{2k+2l} \psi(r) \|u\|_{L^2(B_a(p))} \tag{D.31}$$

provided $u \in L^2$ has support in the ball $B_a(p)$ of radius a, centered at p. Thus, for such u, and for $|y| \le A/2$, the power series

$$\begin{aligned} v(x,y) &= \sum_{k=0}^\infty ((2k)!)^{-1} y^{2k} (-\Delta)^k f(\sqrt{-\Delta}) u \\ &= (\cosh y\sqrt{-\Delta}) f(\sqrt{-\Delta}) u(x) \quad \text{(formally)} \end{aligned} \tag{D.32}$$

converges to an element of $L^2(B_a(x) \times [-A/2, A/2])$. Moreover,

$$(\Delta_G + \partial^2/\partial y^2) v = 0, \tag{D.33}$$

and, by (D.31),

$$\|v\|_{L^2(B_a(x)) \times [-A/2,A/2]} \le C\psi(r) \|u\|_{L^2(B_a(p))}. \tag{D.34}$$

Since G is a homogeneous Riemannian manifold, we can apply regularity estimates for solutions to the elliptic equation (D.33), uniformly, to get

$$\left[\int_{B_a(p)} |f(\sqrt{-\Delta}) \delta_{p'}(x)|^2 \, dp' \right]^{1/2} \le C\psi(r). \tag{D.35}$$

(Most of the effort in [**36**] was devoted to the study of nonhomogeneous manifolds, with semibounded Ricci tensor, where this last piece of reasoning must be replaced by very much more elaborate arguments.) Similarly we obtain

$$\|\Delta_{p'}^l f(\sqrt{-\Delta}) \delta_{p'}(x)\|_{L^2(B_a(p))} \le C(2l/A)^{2l} \psi(r), \tag{D.36}$$

so for each x we can set

$$w(x,p',y) = \sum_{l=0}^\infty ((2l)!)^{-1} y^{2l} \Delta_{p'}^l f(\sqrt{-\Delta}) \delta_{p'}(x) \tag{D.37}$$

and get a harmonic function of (p',y) on $B_a(p) \times [-A/2, A/2]$ whose L^2 norm satisfies

$$\|w(x,\cdot,\cdot)\|_{L^2(B_a(p) \times [-A/2,A/2])} \le C\psi(r), \tag{D.38}$$

and which is equal to $f(\sqrt{-\Delta})\delta_{p'}(x)$ at $y = 0$. Since

$$(\Delta_{p'} + \partial^2/\partial y^2)w(x,p',y) = 0, \tag{D.39}$$

we can again apply the elliptic regularity estimates, uniformly, as before, and deduce

$$|w(x,p,y)| \le C\psi(r). \tag{D.40}$$

Furthermore, since in exponential coordinates (D.39) has analytic coefficients, we can appeal to analytic regularity of solutions to an elliptic equation (see, e.g., [**121**, **179**]), and in light of the homogeneity of G we deduce, for $x' \in B_a(x)$,

$$|D^\alpha_{x'} f(\sqrt{-\Delta})\delta_p(x')| \le C(C|\alpha|)^{|\alpha|}\psi(r). \tag{D.41}$$

Here, $D^\alpha_{x'}$ is computed in exponential coordinates centered at $x \in G$; $\operatorname{dist}(p,x) = r + 2a$.

In order to finish the proof of Theorem D.1, we need to check the functions $f_t(\lambda) = e^{-t\lambda^2}$, to see to what classes $\mathcal{F}^{\varphi,A}$ they belong. We have

LEMMA D.2. *For all $t > 0$, and for all $A > 0$, we have*

$$f_t(\lambda) = e^{-t\lambda^2} \in \mathcal{F}^{\varphi,A}, \tag{D.42}$$

with

$$\varphi(s) = e^{-(1-\varepsilon)s^2/4t}. \tag{D.43}$$

Of course, $\varphi(s)$ depends on t; our notation suppresses this dependence. The important fact is that A is independent of t. This is why we have (D.20) rather than merely the weaker result (D.21). It remains only to establish this lemma. Note that

$$\hat{f}_t(s) = t^{-1/2}\hat{f}_1(t^{-1/2}s), \tag{D.44}$$

so

$$\hat{f}_t^{(k)}(s) = t^{-(k+1)/2}\hat{f}_1^{(k)}(t^{-1/2}s). \tag{D.45}$$

We will estimate $\hat{f}_{1/4}^{(k)}(s)$, since

$$\hat{f}_{1/4}(s) = \pi^{-1/2}e^{-s^2}. \tag{D.46}$$

Then

$$\hat{f}_{1/4}^{(k)}(s) = (-1)^k H_k(s)e^{-s^2}, \tag{D.47}$$

where $H_k(s)$ is the kth Hermite polynomial. In order to estimate (D.47), we will make use of the following generating function identity,

$$\sum_{n=0}^{\infty} H_n(s)^2\eta^n/2^n n! = (1-\eta^2)^{-1/2}\exp(2s^2\eta/(1+\eta)), \qquad |\eta| < 1, \tag{D.48}$$

which is a special case of the identity (7.16) proved in Chapter 1. Pick η positive but fairly small. We obtain that, for any given $\varepsilon > 0$,

$$|H_n(s)|^2/2^n n! \le Ce^{\varepsilon s^2}|\eta|^{-n}. \tag{D.49}$$

It follows that

$$|\hat{f}_{1/4}^{(k)}(s)| \le C[(2/|\eta|)^k k!]^{1/2} e^{-(1-\varepsilon)s^2}. \tag{D.50}$$

Consequently,

$$\begin{aligned} |\hat{f}_t^{(k)}(s)| &\le C[(2/|\eta|)^k k!]^{1/2} t^{-(k+1)/2} e^{-(1-\varepsilon)s^2/4t} \\ &= C[(2/|\eta|t)^k k!]^{1/2} t^{-1/2} e^{-(1-\varepsilon)s^2/4t}. \end{aligned} \tag{D.51}$$

The square root of the term in square brackets justifies the assertion that (D.42) holds for all $A > 0$, and completes the proof of Lemma D.2.

In view of the easily established estimate that the volume of the set of points in G within a distance r of e is bounded by $C_1 e^{C_2 r}$ as $r \to \infty$, the proof of Theorem D.1 now follows.

We include here a brief discussion of the connection between analytic vectors and essential selfadjointness. Let A_0 be a symmetric operator with domain $\mathcal{D}$ in a Hilbert space H. Suppose $\mathcal{D}$ is dense in H. A theorem of Nelson [**183**] states that if $\mathcal{D}$ consists of analytic vectors for A_0, then A_0 is essentially selfadjoint on $\mathcal{D}$. We will give a short proof of the following weaker result, which is nevertheless of use.

PROPOSITION D.3. *Let $\mathfrak{A}_K$ be as defined above with $X_1 = iA_0$. Suppose that, for some K, $\mathcal{D} \subset \mathfrak{A}_K$. Then A_0 is essentially selfadjoint.*

PROOF. Starting from the power series (D.3) for $X = iA_0$, $u \in \mathcal{D}$, we get a vector-valued holomorphic function $U(t)u = e^{itA_0}u$ for $|t| < (eK)^{-1}$, and the argument sketched in the first part of this appendix extends $U(t)$ to $t \in R$ as a unitary group. Thus $U(t) = e^{itA}$ where A is a selfadjoint extension of A_0. If A_1 is any other selfadjoint extension, then it generates a unitary group $U_1(t) = e^{itA_1}$. But $U_1(t)u = U(t)u$ for $u \in \mathcal{D}$, and if $\mathcal{D}$ is dense this implies $U_1(t) = U(t)$, so the selfadjoint extension of A_0 is unique.

We will discuss one final method for exponentiating a Lie algebra representation. This method does not involve analytic vectors, and in fact is closely related to Proposition B.3.

Let $x_1, \dots, x_l$ be a basis for $\mathfrak{g}$ as a linear space; suppose $x_1, \dots, x_{l_0}$ actually *generate* $\mathfrak{g}$ as a Lie algebra. Let $\mathcal{D}$ be a dense linear subspace of H and suppose $\rho\colon \mathfrak{g} \to \mathrm{End}(\mathcal{D})$ is a representation of $\mathfrak{g}$. Thus we suppose

$$\rho(x_j) = iX_j^0\colon \mathcal{D} \to \mathcal{D}, \tag{D.52}$$

and $\rho([x_j, x_k]) = [iX_j^0, iX_k^0]$. Suppose each X_j^0 is a symmetric (possibly unbounded) operator on $\mathcal{D}$.

Now suppose that, for $1 \le j \le l_0$, each X_j^0 has a selfadjoint extension X_j, and

$$e^{itX_j}\colon \mathcal{D} \to \mathcal{D}, \qquad 1 \le j \le l_0. \tag{D.53}$$

We will not a priori make such an assumption on $\rho(x)$ for other elements $x \in \mathfrak{g}$. Note that, by Proposition B.3, (D.53) implies that X_j^0 is essentially selfadjoint on $\mathcal{D}$, for $1 \le j \le l_0$. The hypothesis (D.52) implies $\mathcal{D}$ is a space of smooth vectors for each such X_j.

We will make some further hypotheses on $\mathcal{D}$. We suppose $\mathcal{D}$ has the structure of a Hausdorff locally convex topological vector space such that

$$\mathcal{D} \hookrightarrow H \text{ is continuous.} \tag{D.54}$$

Furthermore, we suppose

$$\text{the maps (D.52) and (D.53) are continuous on } \mathcal{D}, \tag{D.55}$$

and

(D.56)

$$\text{for } v \in \mathcal{D},\ V_j(t) = e^{itX_j}v \text{ is continuous in } t \text{ with values in } \mathcal{D}, 1 \le j \le l_0.$$

From the formula

$$t^{-1}(e^{itX_j}v - v) = it^{-1}\int_0^t e^{isX_j}X_j^0 v\, ds, \qquad v \in \mathcal{D}, \tag{D.57}$$

it follows that, if $\mathcal{D}$ is sequentially complete,

$$v \in \mathcal{D} \Rightarrow t^{-1}(e^{itX_j}v - v) \to iX_j^0 v \quad \text{in } \mathcal{D}, \qquad 1 \le j \le l_0. \tag{D.58}$$

The result we aim to prove is

THEOREM D.4. *Let G be the simply connected Lie group with Lie algebra $\mathfrak{g}$, and let $\rho\colon \mathfrak{g} \to \operatorname{End}(\mathcal{D})$ be a representation of $\mathfrak{g}$ by skew-symmetric operators on a dense subspace $\mathcal{D} \subset H$. Under hypotheses* (D.52)–(D.56), *there is a stongly continuous unitary representation π of G on H such that*

$$x \in \mathfrak{g},\ v \in \mathcal{D} \Rightarrow t^{-1}(\pi(\exp tx)v - v) \to \rho(x)v \quad \textit{as } t \to 0. \tag{D.59}$$

This result is related to a theorem of R. Moore (Bull. Amer. Math. Soc. **71** (1965), 903–908), but has the advantage that (D.53) is assumed only for a set of generators of the Lie algebra, not for all elements of the Lie aglebra. As simple examples show, the condition (D.53) is often much easier to check in interesting cases, than such a stronger condition.

As a tool in proving Theorem D.4, we will need

LEMMA D.5. *For $t \in R$, $1 \le j \le l_0$, $1 \le k \le l$, we have*

$$ie^{itX_j}X_k^0 = \rho(\operatorname{Ad}(\exp tx_j)x_k)e^{itX_j} \quad \textit{on } \mathcal{D}. \tag{D.60}$$

PROOF. Fix $v \in \mathcal{D}$, and let

$$Y_{jk}^{\#}(t) = ie^{itX_j}X_k^0 v, \tag{D.61}$$

and

$$Z^{\#}_{jk}(t) = \rho(\mathrm{Ad}(\exp tx_j)x_k)e^{itX_j}v. \tag{D.62}$$

Then

$$Y^{\#}_{jk}(0) = Z^{\#}_{jk}(0) = X^0_k v. \tag{D.63}$$

Also

$$(d/dt)Y^{\#}_{jk}(t) = iX^0_j Y^{\#}_{jk}(t). \tag{D.64}$$

Write

$$\rho(\mathrm{Ad}(\exp tx_j)x_k) = i\sum_{\nu} \varsigma^{\nu}_{jk}(t)X^0_{\nu} \quad \text{on } \mathcal{D}, \tag{D.65}$$

where $\varsigma^{\nu}_{jk}(t)$ are real-valued C^∞ functions of t. Thus

$$Z^{\#}_{jk}(t) = i\sum_{\nu} \varsigma^{\nu}_{jk}(t)X^0_{\nu}e^{itX_j}v. \tag{D.66}$$

In view of (D.57), the continuity of X^0_ν on $\mathcal{D}$, and the fact that $\varsigma^{\nu}_{jk}(t)$ are scalars, it follows that $Z^{\#}_{jk}(t)$ is differentiable in t, and the derivatives can be computed using the product rule. Since clearly $(d/dt)\sum_\nu \varsigma^{\nu}_{jk}(t)x_\nu = [x_j, \mathrm{Ad}(\exp tx_j)x_k]$, we obtain

$$\begin{aligned}(d/dt)Z^{\#}_{jk}(t) &= i[X^0_j, \rho(\mathrm{Ad}(\exp tx_j)x_k)]e^{itX_j}v \\ &\quad + i\rho(\mathrm{Ad}(\exp tx_j)x_k)X^0_j e^{itX_j}v \\ &= iX^0_j Z^{\#}_{jk}(t).\end{aligned} \tag{D.67}$$

Now let

$$W(t) = Y^{\#}_{jk}(t) - Z^{\#}_{jk}(t), \tag{D.68}$$

so $W(0) = 0$ and

$$dW/dt = iX^0_j W(t). \tag{D.69}$$

It follows that $\|W(t)\|^2$ is differentiable, and

$$\begin{aligned}(d/dt)\|W(t)\|^2 &= (W'(t), W(t)) + (W(t), W'(t)) \\ &= i(X^0_j W(t), W(t)) - i(W(t), X^0_j W(t)) \\ &= 0.\end{aligned} \tag{D.70}$$

Hence $W(t) = 0$ for all t, so (D.61) and (D.62) coincide for all $v \in \mathcal{D}$. This proves the lemma.

To proceed further, we introduce certain auxiliary norms. For a multi-index $\alpha = (\alpha_1, \dots, \alpha_N)$, $1 \le \alpha_j \le l$, set $|\alpha| = N$, let

$$X^\alpha = X^0_{\alpha_1} \cdots X^0_{\alpha_N} \quad \text{on } \mathcal{D}, \tag{D.71}$$

and, for $v \in \mathcal{D}$, let

$$p_m(v) = \left(\sum_{|\alpha| \le m} \|X^\alpha v\|^2 \right)^{1/2} . \tag{D.72}$$

Let $\overline{\mathcal{D}}_m$ denote the completion of $\mathcal{D}$ with respect to this norm, so $\mathcal{D} \hookrightarrow \overline{\mathcal{D}}_m$ and $\overline{\mathcal{D}}_m \hookrightarrow H$ are continuous injections. It is immediate that $p_{m-1}(X_j^0 v) \le p_m(v)$ for all $v \in \mathcal{D}$, and hence X_j^0 has a unique continuous extension

$$X_j' \colon \overline{\mathcal{D}}_m \to \overline{\mathcal{D}}_{m-1} \tag{D.73}$$

for each positive integer m.

LEMMA D.6. *We have*

$$e^{itX_j} \colon \overline{\mathcal{D}}_m \to \overline{\mathcal{D}}_m, \quad \textit{continuous}, \ 1 \le j \le l_0, \tag{D.74}$$

with e^{itX_j} actually a strongly continuous semigroup on each $\overline{\mathcal{D}}_m$.

PROOF. In view of Lemma D.5, we see that

$$v \in \mathcal{D} \Rightarrow X_{\alpha_1}^0 \cdots X_{\alpha_N}^0 e^{itX_j} v = e^{itX_j} Y_{\alpha_1} \cdots Y_{\alpha_N} v$$

where $Y_\nu = (1/i)\rho(\mathrm{Ad}(\exp -tx_j)x_\nu)$ is a linear combination of $X_1^0, \ldots, X_l^0$. The present lemma is immediate.

The following lemma will justify our operations in the proof of Theorem D.4. Let

$$E_j(t, v) = e^{itX_j} v, \qquad 1 \le j \le l_0. \tag{D.75}$$

LEMMA D.7. *For each $m \ge 1$,*

$$E_j \colon R \times \overline{\mathcal{D}}_m \to \overline{\mathcal{D}}_{m-1} \quad \textit{is a } C^1 \textit{ map.} \tag{D.76}$$

PROOF. The differentiability of E_j follows from the formula (D.57), and the continuity of the derivative of E_j follows from Lemma D.6.

We are now ready to prove Theorem D.4. We begin by defining a function π from G to the set $U(H)$ of unitary operators on H. Suppose

$$g = \exp(t_1 x_{\alpha_1}) \cdots \exp(t_N x_{\alpha_N}) \in G \qquad (1 \le \alpha_\nu \le l_0). \tag{D.77}$$

We want to set

$$\pi(g) = e^{it_1 X_{\alpha_1}} \cdots e^{it_N X_{\alpha_N}}. \tag{D.78}$$

Our main task is to show that (D.78) is independent of the representation (D.77), since each $g \in G$ has one, and indeed, many representations of this form. To do this, it is equivalent to show that, if

$$e = \exp(t_1 x_{\alpha_1}) \cdots \exp(t_N x_{\alpha_N}), \tag{D.79}$$

then

$$e^{it_1 X_{\alpha_1}} \cdots e^{it_N X_{\alpha_N}} = I. \tag{D.80}$$

Most of the work in doing this is accomplished by

LEMMA D.8. *Let $t_\nu(s)$ be C^∞ functions of s and suppose*

$$g = \exp(t_1(s)x_{\alpha_1})\cdots\exp(t_N(s)x_{\alpha_N}) \tag{D.81}$$

be independent of s. Then

$$U(s) = e^{it_1(s)X_{\alpha_1}}\dots e^{it_N(s)X_{\alpha_N}} \tag{D.82}$$

is independent of s.

PROOF. Lemma D.7 implies that, if we set

$$\mathcal{E}(t_1,\dots,t_N,v) = e^{it_1X_{\alpha_1}}\dots e^{it_NX_{\alpha_N}}v, \tag{D.83}$$

then

$$\mathcal{E}: R^N\times\mathcal{D}_{m+N}\to\mathcal{D}_m \quad \text{is a } C^1 \text{ map}, \tag{D.84}$$

and we can use the chain rule to evaluate $(d/ds)\mathcal{E}(t_1(s),\dots,t_N(s),v)$. We obtain, for $v\in\mathcal{D}$,

(D.85)

$$\begin{aligned}(d/ds)U(s)v &= i(dt_1/ds)X^0_{\alpha_1}e^{it_1(s)X_{\alpha_1}}\dots e^{it_N(s)X_{\alpha_N}}v\\ &\quad + e^{it_1(s)X_{\alpha_1}}(i(dt_2/ds)X^0_{\alpha_2})e^{it_2(s)X_{\alpha_2}}\dots e^{it_N(s)X_{\alpha_N}}v\\ &\quad +\cdots\\ &\quad + e^{it_1(s)X_{\alpha_1}}\dots e^{it_{N-1}(s)X_{\alpha_{N-1}}}(i(dt_N/ds)X^0_{\alpha_N})e^{it_N(s)X_{\alpha_N}}v,\end{aligned}$$

and using Lemma (D.5), we push all the terms $i(dt_\nu/ds)X^0_\nu$ to the far left, and obtain

$$(d/ds)U(s)v = A(s)U(s)v, \qquad v\in\mathcal{D}, \tag{D.86}$$

where

$$\begin{aligned}A(s) &= (dt_1/ds)\rho(x_{\alpha_1}) + (dt_2/ds)\rho(\mathrm{Ad}(\exp t_1x_{\alpha_1})x_{\alpha_2})\\ &\quad +\cdots+(dt_N/ds)\rho(\mathrm{Ad}(\exp(t_1x_{\alpha_1})\cdots\exp(t_{N-1}x_{\alpha_{N-1}}))x_{\alpha_N}).\end{aligned} \tag{D.87}$$

Now if we replace ρ by a faithful finite-dimensional representation ρ_0 of $\mathfrak{g}$, which certainly exponentiates to a locally faithful representation π_0 of G, we see that

$$0 = (d/ds)\pi_0(g) = A_0(s)\pi_0(g), \tag{D.88}$$

where $A_0(s)$ is as in (D.87) with ρ replaced by ρ_0. Since $A_0(s)=0$ and ρ_0 is faithful, it follows that the element of $\mathfrak{g}$ to which ρ is applied in (D.87) must vanish; hence $A(s)=0$. This proves Lemma D.8.

We resume the task of showing that (D.79) implies (D.80). Define a path γ from the interval $[0,t_1+\cdots+t_N]$ to G by

$$\gamma(s) = \exp(t_1x_{\alpha_1})\cdots\exp(t_\nu x_{\alpha_\nu})\exp((s-t_1-\cdots-t_\nu)x_{\alpha_{\nu+1}}) \tag{D.89}$$

if $t_1+\cdots+t_\nu\le s\le t_1+\cdots+t_{\nu+1}$. Then (D.79) implies this path is closed. By the device of writing $\exp t_\nu x_{\alpha_\nu} = \exp(t_\nu x_{\alpha_\nu}/K)\cdots\exp(t_\nu x_{\alpha_\nu}/K)$ (K factors), we can suppose the points $\gamma(t_\nu)$, $1\le\nu\le N$, are closely spaced. We can perturb γ

slightly to produce a *smooth* closed curve $\tilde{\gamma}(s)$, with $\tilde{\gamma}(t_\nu) = \gamma(t_\nu)$ for $1 \le \nu \le N$. Assuming G is simply connected, we can produce a smooth contraction $\tilde{\gamma}_\tau(s)$ to the point e; $\tilde{\gamma}_0(s) = \tilde{\gamma}(s)$, $\tilde{\gamma}_1(s) = e$, $\tilde{\gamma}_\tau(0) = \tilde{\gamma}_\tau(t_1 + \cdots + t_N) = e$. Consider the curves

$$\begin{aligned} &\sigma_1(\tau) = \tilde{\gamma}_\tau(t_1) \\ &\sigma_\nu(\tau) = \tilde{\gamma}_\tau(t_1 + \cdots + t_{\nu-1})^{-1}\tilde{\gamma}_\tau(t_1 + \cdots + t_\nu), \qquad 2 \le \nu \le N. \end{aligned} \tag{D.90}$$

We have each $\sigma_\nu(\tau)$, $0 \le \tau \le 1$, taking values in a neighborhood of $e \in G$ which can be arranged to be as small as desired, and

$$e = \sigma_1(\tau)\sigma_2(\tau)\cdots\sigma_N(\tau), \qquad \tau \in [0,1]. \tag{D.91}$$

Recall that

$$\sigma_\nu(0) = \exp(t_\nu x_{\alpha_\nu}), \qquad \sigma_\nu(1) = e. \tag{D.92}$$

Now we claim there exist smooth $s_{\mu\nu}(\tau)$ such that

$$\sigma_\nu(\tau) = \exp(s_{1\nu}(\tau)x_{\beta_1})\cdots\exp(s_{M\nu}(\tau)x_{\beta_M}), \qquad 1 \le \beta_\mu \le l_0, \tag{D.93}$$

for $0 \le \tau \le 1$. This is an immediate consequence of the following simple lemma.

LEMMA D.9. *There exists a smooth map*

$$\Phi\colon R^L \to G \tag{D.94}$$

of the form

$$\begin{aligned} \Phi(s_1,\ldots,s_L) = {} & \exp(b_{11}x_{\gamma_{11}})\cdots\exp(b_{1K_1}x_{\gamma_{1K_1}})\exp(s_1x_{\beta_1})\exp(-b_{1K_1}x_{\gamma_{1K_1}}) \\ & \cdots\exp(-b_{11}x_{\gamma_{11}})\cdots \\ & \cdot\exp(b_{L1}x_{\gamma_{L1}})\cdots\exp(s_Lx_{\beta_L})\cdots\exp(-b_{L1}x_{\gamma_{L1}}) \end{aligned} \tag{D.95}$$

where $b_{jk} \in R$, $1 \le \gamma_{jk} \le l_0$, $1 \le \beta_j \le l_0$, *such that* Φ *has surjective differential at* $(s_1,\ldots,s_L) = 0$.

PROOF. We need merely arrange the quantities $b_{jk}x_{\gamma_{jk}}$ and x_{β_j} such that $\mathrm{Ad}(\exp(b_{j1}x_{\gamma_{j1}})\cdots\exp(b_{jK_j}x_{\gamma_{jK_j}})x_{\beta_j}$, $1 \le j \le L$, span $\mathfrak{g}$ as a linear space.

Having established (D.93), if we plug this into (D.91) and apply Lemma D.8, we deduce that

$$U(\tau) = \prod_{\nu=1}^{N}(e^{is_{\nu1}(\tau)X_{\beta_1}}\ldots e^{is_{\nu M}(\tau)X_{\beta_M}})$$

is independent of τ. Note that some of the $s_{\mu\nu}(\tau)$ are the constants $\pm b_{jk}$ and some are functions $s_j(\tau)$, $1 \le j \le L$, and we can arrange $s_j(1) = 0$, since $\sigma_\nu(1) = e$. The cancellations that accrue at $\tau = 1$ imply $U(1) = I$. Thus $U(0) = I$. We claim we can choose $\Phi = \Phi_\nu$ and $s_{\mu\nu}(\tau)$ so that $U(0)$ is equal to the left side of (D.80). Indeed, if all b_{1k} are taken to be 0 and $x_{\beta_1} = x_{\alpha_\nu}$ in (D.95), with the other $b_{jk}x_{\gamma_{jk}}$, x_{β_k} chosen as indicated in the proof of Lemma D.9, then we can obtain (D.93) with $s_{1\nu}(0)x_{\beta_1} = t_\nu x_{\alpha_\nu}$ and $s_{2\nu}(0) = \cdots = s_{M\nu}(0) = 0$, by constructing a right inverse Ψ_ν of Φ_ν, defined on a neighborhood of e in G, into R^L, such that $\Psi_\nu(e) = 0$ and Ψ_ν maps the one parameter group in G in the direction $x_{\alpha_\nu} = x_{\beta_1} \in \mathfrak{g} = T_eG$ to the line through the first coordinate vector in R^L. This can be arranged by the implicit function theorem. We can finally conclude that (D.79) implies (D.80).

We are now able to say that (D.77)–(D.78) produces a uniquely defined function $\pi\colon G \to U(H)$, and it is automatic from (D.78) that π is a group homomorphism. To check the strong continuity of π and verify (D.59), we will again use Lemma D.9, which guarantees the existence of a C^∞ map

$$\Psi\colon \mathcal{O} \to R^L, \tag{D.96}$$

$\mathcal{O}$ a neighborhood of e in G, such that $\Psi(e) = 0$ and $\Phi \circ \Psi = \text{id}$ on $\mathcal{O}$. By the fact that π is well defined, we can write

$$\begin{aligned}\pi(g) = e^{ib_{11}X_{\gamma_{11}}} \cdots e^{ib_{1K_1}X_{\gamma_{1K_1}}} e^{is_1(g)X_{\beta_1}} e^{-ib_{1K_1}X_{\gamma_{1K_1}}} \cdots e^{-ib_{11}X_{\gamma_{11}}} \\ \cdots e^{ib_{L1}X_{\gamma_{L1}}} \cdots e^{is_L(g)X_{\beta_L}} \cdots e^{-ib_{L1}X_{\gamma_{L1}}}\end{aligned} \tag{D.97}$$

where

$$\Psi(g) = (s_1(g), \ldots, s_L(g)). \tag{D.98}$$

By (D.76) we deduce that

$$v \in \tilde{\mathcal{D}}_{m+B} \quad \Rightarrow \quad g \mapsto \pi(g)v \text{ is a } C^1 \text{ map from } \mathcal{O} \text{ to } \tilde{\mathcal{D}}_m, \tag{D.99}$$

where $B = L + 2(K_1 + \cdots + K_L)$, and we can calculate its derivative by the chain rule. Then the same sort of argument used to establish Lemma D.8 shows the limit in (D.59) exists for all $v \in \mathcal{D} \subset \tilde{\mathcal{D}}_B$, and equals $\rho(x)v$. The proof of Theorem D.4 is complete. Let us remark that $\mathcal{D}$ is a space of smooth vectors for π and $\pi(g)\colon \mathcal{D} \to \mathcal{D}$ for all $g \in G$.

Let us see how Theorem D.4 applies to the representation ω of $hp_2 \approx \text{sp}(n, R)$, given by

$$\omega(Q) = iQ(X, D), \tag{D.100}$$

as in Chapter 1, §4. We take

$$\mathcal{D} = \mathcal{S}(R^n). \tag{D.101}$$

As shown in Chapter 1, §4, $e^{itQ_j(X,D)}$ preserves $\mathcal{S}(R^n)$ for a *basis* Q_j of hp_2. It is even a little easier to see this for polynomials $Q(x, \xi)$ of the form $x_j x_k$ and $\xi_j \xi_k$, which generate $hp_2 = \text{sp}(n, R)$. This provides an alternate method of showing that ω exponentiates to a representation of $\widetilde{\text{Sp}(n, R)}$. Compare Proposition 4.1 of Chapter 1.

From time to time, it has been suggested that it is enough for $\mathcal{D}$ to be a common dense domain of selfadjointness for $(1/i)\rho(x)$, preserved by $\rho(x)$, $x \in \mathfrak{g}$, in order to exponentiate ρ. We note that this is false, and refer to Nelson **[183]** for a counterexample; see also pages 296–296 of Warner **[256]**.

References

1. R. Abraham and J. Marsden, *Foundations of mechanics*, Benjamin, Reading, Mass., 1978.

2. J. Adams, *Lectures on Lie groups*, Benjamin, Reading, Mass., 1969.

3. R. Anderson, *The Weyl functional calculus*, J. Funct. Anal. **4** (1969), 240–267.

4. V. Arnold, *Mathematical methods of classical mechanics*, Springer, New York, 1978.

5. M. Atiyah, *Characters of semi-simple Lie groups*, Lecture Notes, Univ. of Oxford, 1975.

6. M. Atiyah, R. Bott, and A. Shapiro, *Clifford modules*, Topology **3** (1964), 3–38.

7. M. Atiyah and W. Schmid, *A geometric construction of the discrete series for semi-simple Lie groups*, Invent. Math. **42** (1977), 1–62.

8. L. Auslander, *Lecture notes on nil-theta functions*, CBMS Regional Conf. Ser. in Math., No. 34, Amer. Math. Soc., Providence, R.I., 1977.

9. L. Auslander and B. Kostant, *Polarizations and unitary representations of solvable Lie groups*, Invent. Math. **14** (1971), 255–354.

10. L. Auslander and C. Moore, *Unitary representations of solvable Lie groups*, Mem. Amer. Math. Soc., No. 62, Amer. Math. Soc., Providence, R.I., 1966.

11. W. Baldoni Silva, *Unitary dual of* $\mathrm{Sp}(n,1)$, $n \geq 2$, Duke Math. J. **48** (1982), 549–583.

12. V. Bargmann, *Irreducible unitary representations of the Lorentz group*, Ann. of Math. **48** (1947), 568–640.

13. ——, *On a Hilbert space of analytic functions and an associated integral transform*, Comm. Pure Appl. Math. **14** (1961), 187–214.

14. H. Bateman, *Partial differential equations of mathematical physics*, Dover, New York, 1944.

15. R. Beals, P. Greiner, and N. Stanton, *The heat equation and the geometry of CR manifolds*, Bull. Amer. Math. Soc. (N.S.) **10** (1984), 275–276.

16. F. Berezin, *Laplace operators on semi-simple Lie groups*, Amer. Math. Soc. Transl. (2) **21** (1962), 239–339.

17. M. Berger, P. Gauduchon, and E. Mazet, *Le spectre d'une variété Riemanniene*, Lecture Notes in Math., vol. 194, Springer, 1971.

18. P. Bernat, *Sur les representations unitaries des groupes de Lie resolubles*, Ann. École Norm. Sup. **82** (1965), 37–99.

19. L. Bers, F. John, and M. Schechter, *Partial differential equations*, Interscience, New York, 1964.

20. L. Biedenharn and J. Louck, *Angular momentum in quantum physics*, The Racah-Wigner Algebra in Quantum Theory, Vols. 8 and 9, Encyclopedia of Math. and its Applications, Addison-Wesley, Reading, Mass., 1981.

21. R. Blattner, *Quantization and representation theory*, Proc. Sympos. Pure Math., Vol. 26, Amer. Math. Soc., Providence, R.I., 1972, pp. 147–166.

22. ——, *General background*, Harmonic Analysis and Representations of Semi-simple Lie Groups, D. Reidel, Boston, 1980.

23. A. Borel and N. Wallach, *Continuous cohomology, discrete subgroups, and representations of reductive groups*, Princeton Univ. Press, Princeton, N.J., 1980.

24. L. Boutet de Monvel, *Hypoelliptic operators with double characteristics and related pseudodifferential operators*, Comm. Pure Appl. Math. **27** (1974), 585–639.

25. L. Boutet de Monvel, A. Grigis, and B. Helffer, *Parametrixes d'operateurs pseudodifferentiels a caracteristiques multiples*, Astérisque **34–35** (1976), 93–121.

26. L. Boutet de Monvel and F. Treves, *On a class of pseudodifferential operators with double characteristics*, Comm. Pure Appl. Math. **27** (1974), 59–81.

27. F. Bruhat, *Sur les representations induites des groupes de Lie*, Bull. Soc. Math. France **84** (1956), 97–205.

28. J. Brezin, *Unitary representation theory for solvable Lie groups*, Mem. Amer. Math. Soc., No. 79, Amer. Math. Soc., Providence, R.I., 1968.

29. ——, *Harmonic analysis on nilmanifolds*, Trans. Amer. Math. Soc. **150** (1970), 611–618.

30. ——, *Harmonic analysis on compact solvmanifolds*, Lecture Notes in Math., Vol. 602, Springer, 1977.

31. R. Cahn, *The K-types of principal series representations*, manuscript.

32. R. Cahn and M. Taylor, *Asymptotic behavior of multiplicities of representations of compact groups*, Pacific J. Math. **83** (1979), 17–28.

33. E. Cartan, *The theory of spinors*, Dover, New York, 1981.

34. P. Cartier and J. Dixmier, *Vecteurs analytiques dans les representations des groupes de Lie*, Amer. J. Math. **80** (1958), 131–145.

35. J. Cheeger, *Spectral geometry of spaces with cone-like singularities*, Proc. Nat. Acad. Sci. U.S.A. **76** (1979), 2103–2106.

36. J. Cheeger, M. Gromov, and M. Taylor, *Finite propagation speed, kernel estimates for functions of the Laplace operator, and the geometry of complete Riemannian manifolds*, J. Differential Geom. **17** (1982), 15–53.

37. J. Cheeger and M. Taylor, *On the diffraction of waves by conical singularities*, Comm. Pure Appl. Math. **35** (1982), 275–331, 487–529.

38. P. Chernoff, *Essential self-adjointness of powers of generators of hyperbolic equations*, J. Funct. Anal. **12** (1973), 401–414.

39. C. Chevalley, *Theory of Lie groups*, Vol. 1, Princeton Univ. Press, Princeton, N.J., 1946.

40. ——, *Invariants of finite groups generated by reflections*, Amer. J. Math. **77** (1955), 778–782.

41. R. Coifman and G. Weiss, *Analyze harmonique non-commutative sur certaine espaces homogenes*, Lecture Notes in Math., No. 242, Springer, 1971.

42. J. Dieudonné, *Special functions and linear representations of Lie groups*, CBMS Regional Conf. Ser. in Math., No. 42, Amer. Math. Soc., Providence, R.I., 1980.

43. J. Dixmier, *Sur les representations unitaires des groupes de Lie nilpotents.* I, Amer. J. Math. **81** (1959), 160–170; II, Bull. Soc. Math. France **85** (1957), 325–388; III, Canad. J. Math. **10** (1958), 321–348; IV, Canad. J. Math. **11** (1959), 321–344; V, Bull. Soc. Math. France **87** (1959), 65–79; VI, Canad. J. Math. **12** (1960), 324–352.

44. ——, *Representations integrable du groupe de De Sitter*, Bull. Soc. Math. France **89** (1961), 9–41.

45. ——, *Sur les representations unitaires des groupes de Lie resolubles*, Math. J. Okayama Univ. **11** (1962), 1–18.

46. ——, *Les C^* algebres et leurs representations*, Gauthier-Villars, Paris, 1964.

47. ——, *Les algebres d'operateurs dans l'espace Hilbertiens* (*algebres de von Neumann*), Gauthier-Villars, Paris, 1969.

48. V. Dobrev, G. Mack, V. Petkova, S. Petrova, and I. Todorov, *Harmonic analysis on the n-dimensional Lorentz group and its applications to conformal quantum field theory*, Lecture Notes in Physics, No. 63, Springer, New York, 1977.

49. J. J. Duistermaat, *Harmonic analysis on semi-simple Lie groups* (preprint).

50. N. Dunford and J. Schwartz, *Linear operators.* I, II, Interscience, New York, 1958, 1963.

51. H. Dym and H. McKean, *Fourier series and integrals*, Academic Press, New York, 1972.

52. A. Dynin, *An analogue of pseudodifferential operators on the Heisenberg group; symbolic calculus*, Soviet Math. Dokl. **17** (1976), 508–512.

53. ——, *Pseudodifferential operators on Heisenberg groups*, Sem. CIME, Bressanone, Italy, June, 1977.

54. L. Ehrenpreis, *Harmonic analysis on Lie groups*, manuscript.

55. ——, *The use of partial differential equations for the study of group representations*, Proc. Sympos. Pure Math., Vol. 26, Amer. Math. Soc., Providence, R.I., 1972, pp. 317–320.

56. L. Ehrenpreis and J. Mautner, *Some properties of the Fourier transform on semi-simple Lie groups.* I, Ann of Math. **61** (1955), 406–439; II, Trans. Amer. Math. Soc. **84** (1957), 1–55; III, Trans. Amer. Math. Soc. **90** (1959), 431–484.

57. M. Englefield, *Group theory and the coulomb problem*, Wiley-Interscience, New York, 1972.

58. A. Erdelyi, *Asymptotic expansions*, Dover, New York, 1965.

59. A. Erdelyi et al., *Higher transcendental functions*, Vol. I–III (Bateman Manuscript Project), McGraw-Hill, New York, 1953.

60. L. Falicov, *Group theory and its physical applications*, Chicago Lectures in Physics, Chicago Univ. Press, Chicago, Ill., 1966.

61. H. Fegan, *The fundamental solution of the heat equation on a compact Lie group*, J. Differential Geom. **18** (1983), 659–668.

62. G. Folland, *Subelliptic operators on the Heisenberg group*, Proc. Sympos. Pure Math., Vol. 26, Amer. Math. Soc., Providence, R.I., 1973, pp. 359–361.

63. ——, *Introduction to partial differential equations*, Princeton Univ. Press, Princeton, N.J., 1976.

64. G. Folland and E. Stein, *Estimates for the $\overline{\partial}_b$ complex and analysis on the Heisenberg group*, Comm. Pure Appl. Math. **27** (1974), 429–522.

65. H. Freudenthal and H. de Vries, *Linear Lie groups*, Academic Press, New York, 1969.

66. F. G. Friedlander, *Sound pulses*, Cambridge Univ. Press, 1958.

67. R. Gangolli, *Spectra of discrete subgroups*, Proc. Sympos. Pure Math., Vol. 26, Amer. Math. Soc., Providence, R.I., 1973.

68. L. Gårding, *Note on continuous representations of Lie groups*, Proc. Nat. Acad. Sci. U.S.A. **33** (1947), 331–332.

69. ——, *Vecteurs analytiques dans les representations des groupes de Lie*, Bull. Soc. Math. France **88** (1960), 73–93.

70. B. Gaveau, *Principe de moindre action, propagation de chaleur, et estimees sous elliptiques sur certains groups nilpotents*, Acta Math. **139** (1977), 95–153.

71. I. Gelfand, M. Graev, and N. Vilenkin, *Generalized functions*, Vol. 5, Academic Press, New York, 1966.

72. I. Gelfand, M. Graev, and I. Pyatetskii-Shapiro, *Representation theory and automorphic functions*, W. B. Saunders, Philadelphia, Penn., 1969.

73. I. Gelfand, R. Minlos, and Z. Shapiro, *Represenations of the rotation and Lorentz groups and their applications*, Pergamon Press, New York, 1963.

74. I. Gelfand and M. Naimark, *Unitare Darstellungen der Klassischen Gruppen*, Akademie-Verlag, Berlin, 1957.

75. D. Geller, *Fourier analysis on the Heisenberg group. I. Schwartz space*, J. Funct. Anal. **36** (1980), 205–254.

76. ——, *Local solvability and homogeneous distributions on the Heisenberg group*, Comm. Partial Differential Equations **5** (1980), 475–560.

77. H. Georgi, *Lie algebras in particle physics*, Benjamin, Reading, Mass., 1982.

78. S. Gindikin and F. Karpelevic, *Plancherel measure of Riemannian symmetric spaces of non-positive curvature*, Dokl. Akad. Nauk SSSR **145** (1962), 252–255.

79. R. Gillmore, *The classical limit of quantum nonspin systems*, J. Math. Phys. **20** (1979), 891–893.

80. R. Godement, *Sur la theorie des representations unitaires*, Ann. of Math. **53** (1951), 68–124.

81. ——, *A theory of spherical functions*. I, Trans. Amer. Math. Soc. **73** (1952), 496–556.

82. H. Goldstein, *Classical mechanics*, Addison-Wesley, New York, 1950.

83. R. Goodman, *Nilpotent Lie groups, structure and applications to analysis*, Lecture Notes in Math., No. 562, Springer, 1976.

84. ——, *Analytic and entire vectors for representations of Lie groups*, Trans. Amer. Math. Soc. **143** (1969), 55–76.

85. W. Greub, *Multilinear algebra*, Springer, New York, 1978.

86. A. Grossman, G. Loupias, and E. Stein, *An algebra of pseudodifferential operators and quantum mechanics in phase space*, Ann. Inst. Fourier (Grenoble) **18** (1969), 343–368.

87. V. Guillemin and S. Sternberg, *Geometric asymptotics*, Amer. Math. Soc., Providence, R.I., 1977.

88. ——, *The metaplectic representation, Weyl operators, and spectral theory*, J. Funct. Anal. **42** (1981), 128–225.

89. ——, *Symplectic techniques in physics*, Cambridge Univ. Press, 1984.

90. R. Gunning, *Lectures on Riemann surfaces*, Princeton Univ. Press, Princeton, N.J., 1967.

91. N. Xuan Hai, *The Fourier transform on nilpotent Lie groups*, Comm. Partial Differential Equations **8** (1983), 667–691.

92. Harish-Chandra, *Representations of semi-simple Lie groups*. I, Trans. Amer. Math. Soc. **75** (1953), 185–243; II, Trans. Amer. Math. Soc. **76** (1954), 26–65; III, Trans. Amer. Math. Soc. **76** (1954), 234–253; IV, Amer. J. Math. **77** (1955), 743–777; V, Amer. J. Math. **78** (1956), 1–41; VI, Amer. J. Math. **78** (1956), 564–628.

93. ——, *The Plancherel formula for the complex semi-simple Lie groups*, Trans. Amer. Math. Soc. **76** (1954), 485–528.

94. ——, *The Fourier transform on a semi-simple Lie algebra*. I, II, Amer. J. Math. **79** (1957), 87–120, 653–686.

95. ——, *Spherical functions on a semi-simple Lie group*. I, II, Amer. J. Math. **80** (1958), 241–310, 553–613.

96. ——, *Discrete series for semi-simple Lie groups*. I, II, Acta Math. **113** (1965), 241–318; **116** (1966), 1–111.

97. ——, *Harmonic analysis on semi-simple Lie groups*, Bull. Amer. Math. Soc. **76** (1970), 529–551.

98. ——, *Harmonic analysis on real reductive groups*. I, J. Funct. Anal. **19** (1975), 104–204; II, Invent. Math. **36** (1975), 1–55); III, Ann. of Math. **104** (1976), 117–201.

99. D. Hejhal, *The Selberg trace formula for* PSL$(2, R)$, Lecture Notes in Math., Nos. 548, 1001, Springer, 1976, 1983.

100. S. Helgason, *Differential geometry and symmetric spaces*, Academic Press, New York, 1962.

101. ——, *Analysis on Lie groups and symmetric spaces*, CBMS Regional Conf. Ser. in Math., No. 14, Amer. Math. Soc., Providence, R.I., 1972.

102. ——, *Topics in harmonic analysis on homogeneous spaces*, Birkhauser, Boston, 1981.

103. ——, *The Radon transform*, Birkhauser, Boston, 1980.

104. ——, *A duality for symmetric spaces with applications to group representations*, Adv. in Math. **5** (1970), 1–154.

105. ——, *Invariant differential operators and eigenspace representations*, London Math. Soc. Lecture Notes, No. 34, Cambridge Univ. Press, 1979, pp. 236–286.

106. ——, *Wave equations on homogeneous spaces* (preprint).

107. ——, *Fundamental solutions of invariant differential operators on symmetric spaces*, Amer. J. Math. **86** (1964), 565–601.

108. B. Helffer and J. Nourrigat, *Caracterization des operateurs hypoelliptiques homogenes invariant a gauche sur un groupe de Lie nilpotent gradue*, Comm. Partial Differential Equations **4** (1979), 899–958.

109. L. Helms, *Introduction to potential theory*, Wiley-Interscience, New York, 1969.

110. R. Herman, *Lie groups for physicists*, Benjamin, Reading, Mass., 1966.

111. E. Hewitt and K. Ross, *Abstract harmonic analysis*, Springer, New York, I, 1963; II, 1970.

112. E. Hille and R. Phillips, *Functional analysis and semi-groups*, Colloq. Publ., Vol. 31, Amer. Math. Soc., Providence, R.I., 1957.

113. T. Hirai, *On irreducible representations of the Lorentz group of nth order*, Proc. Japan Acad. **38** (1962), 258–262.

114. ——, *The characters of irreducible representations of the Lorentz group of nth order*, Proc. Japan Acad. **41** (1965), 526–531.

115. ——, *The Plancherel formula for the Lorentz group of nth order*, Proc. Japan Acad. **42** (1965), 323–326.

116. ——, *Classification and the characters of irreducible representations of* SU$(p,1)$, Proc. Japan Acad. **42** (1966), 907–912.

117. ——, *The Plancherel formula for* SU(p,q), J. Math. Soc. Japan **22** (1970), 134–179.

118. E. Hobson, *The theory of spherical harmonics*, Chelsea, New York, 1965.

119. L. Hörmander, *The Weyl calculus of pseudodifferential operators*, Comm. Pure Appl. Math. **32** (1979), 359–443.

120. ——, *A class of pseudodifferential operators with double characteristics*, Math. Ann. **217** (1975), 165–188.

121. ——, *The analysis of linear partial differential operators.* I, II, Springer, New York, 1983; III, IV, 1985.

122. ——, *An introduction to complex analysis in several variables*, Van Nostrand, Princeton, N.J., 1966.

123. R. Howe, *On the role of the Heisenberg group in harmonic analysis*, Bull. Amer. Math. Soc. (N.S.) **3** (1980), 821–843.

124. ——, *Quantum mechanics and partial differential equations*, J. Funct. Anal. **38** (1980), 188–254.

125. ——, *Some results of Strichartz and of Rallis and Schiffman*, J. Funct. Anal. **32** (1979), 297–303.

126. ——, *Dual pairs and the classical equations of physics* (preprint).

127. A. Hulanicki, *The distribution of energy in the Brownian motion in the Gaussian field and analytic-hypoellipticity of certain subelliptic operators on the Heisenberg group*, Studia Math. **51** (1976), 165–173.

128. H. Jacobson and M. Vergne, *Wave and Dirac operators and representations of the conformal group*, J. Funct. Anal. **24** (1977), 52–106.

129. N. Jacobson, *Lie algebras*, Interscience, New York, 1965.

130. H. Jacquet and R. Langlands, *Automorphic forms on* GL_2, Lecture Notes in Math., No. 114, Springer, 1970.

131. F. John, *Plane waves and spherical means*, Interscience, New York, 1955.

132. F. Karpelevic, *The geometry of geodesics and the eigenfunctions of the Laplace-Beltrami operators on symmetric spaces*, Trans. Moscow Math. Soc. **14** (1965), 48–185.

133. M. Kashiwara, A. Kowata, K. Minemura, K. Okamoto, T. Oshima, and M. Tanaka, *Eigenfunctions of invariant differential operators on a symmetric space*, Ann. of Math. **107** (1978), 1–39.

134. A. Kirillov, *Elements of the theory of representations*, Springer, New York, 1976.

135. ——, *Unitary representations of nilpotent Lie groups* , Russian Math. Surveys **17** (1962), 53–104.

136. A. Knapp, *Representation theory of semisimple groups*, Princeton Univ. Press, Princeton, N.J., 1986.

137. A. Knapp and B. Speh, *Status of classification of unitary representations*, Lecture Notes in Math., No. 908, Springer, 1982, pp. 1–38.

138. ——, *Irreducible unitary representations of* SU(2, 2), J. Funct. Anal. **45** (1982), 41–73.

139. A. Knapp and E. Stein, *Intertwining operators for semi-simple Lie groups*, Ann. of Math. **93** (1971), 489–578.

140. A. Knapp and G. Zuckerman, *Classification theorems for representations of semi-simple Lie groups*, Lecture Notes in Math., No. 587, Springer, 1977, pp. 138–159.

141. J. Kohn and L. Nirenberg, *An algebra of pseudodifferential operators*, Comm. Pure Appl. Math. **18** (1965), 269–305.

142. B. Kostant, *Quantization and unitary representations*, Lecture Notes in Math., No. 170, Springer, 1970, pp. 87–208.

143. ——, *On the existence and irreducibility of certain series of representations*, Bull. Amer. Math. Soc. **75** (1969), 627–642.

144. T. Kotake and M. Narasimhan, *Regularity theorems for fractional powers of linear elliptic operators*, Bull. Soc. Math. France **90** (1962), 449–472.

145. H. Kraljevic, *Representations of the universal covering group of the group* SU$(n, 1)$, Glas. Mat. **8** (1973), 23–72.

146. R. Kunze and E. Stein, *Uniformly bounded representations and harmonic analysis on the real unimodular group*. I, II, III, Amer. J. Math. **82** (1960), 1–62; **83** (1961), 723–786; **89** (1967), 385–442.

147. S. Lang, $SL_2(R)$, Addison-Wesley, New York, 1975.

148. R. Langlands, *On the classification of irreducible representations of real algebraic groups*, IAM mimeographed lecture notes, 1973.

149. H. B. Lawson and M. L. Michelson, *The geometry of spinors*, Monograph, in preparation.

150. P. Lax and R. Phillips, *Scattering theory*, Academic Press, New York, 1967.

151. ——, *Scattering theory for automorphic functions*, Princeton Univ. Press, Princeton, N.J., 1976.

152. ——, *An example of Huygens' principle*, Comm. Pure Appl. Math. **31** (1978), 415–421.

153. ——, *The asymptotic distribution of lattice points in Euclidean and non-Euclidean spaces*, J. Funct. Anal. **46** (1982), 280–350.

154. N. Lebedev, *Special functions and their applications*, Dover, New York, 1972.

155. J. Leray, *Lagrangian analysis and quantum mechanics*, M.I.T. Press, Cambridge, Mass., 1972.

156. G. Lion and M. Vergne, *The Weil representation, Maslov index, and theta series*, Birkhauser, Boston, 1980.

157. R. Lipsman, *Group representations*, Lecture Notes in Math., No. 388, Springer, 1974.

158. L. Loomis, *Abstract harmonic analysis*, Van Nostrand, Princeton, N.J., 1953.

159. I. MacDonald, *Algebraic structure of Lie groups*, London Math. Soc. Lecture Notes, No. 34, Cambridge Univ. Press, 1979, pp. 91–150.

160. G. Mackey, *Unitary group representations in physics, probability, and number theory*, Benjamin, Reading, Mass., 1978.

161. ——, *The theory of unitary group representations*, Univ. of Chicago Press, Chicago, Ill., 1976.

162. ——, *Imprimitivity for representations of locally compact groups*, Proc. Nat. Acad. Sci. U.S.A. **35** (1949), 537–545.

163. ——, *Induced representations of locally compact groups*. I, II, Ann. of Math. **55** (1952), 101–139; **58** (1953), 193–221.

164. ——, *Unitary representations of group extensions*, Acta Math. **99** (1958), 265–311.

165. ——, *Infinite dimensional group representations*, Bull. Amer. Math. Soc. **69** (1963), 628–686.

166. H. McKean and I. Singer, *Curvature and the eigenvalues of the Laplacian*, J. Differential Geom. **1** (1967), 43–70.

167. J. Mather, *Differentiable invariants*, Topology **16** (1977), 145–155.

168. F. Mautner, *Unitary representations of locally compact groups*, Ann. of Math. **52** (1950), 528–556.

169. A. Melin, *Parametrix construction for right invariant differential operators on nilpotent groups*, Comm. Partial Differential Equations **6** (1981), 1363–1405.

170. R. Melrose and M. Taylor, *Near peak scattering and the corrected Kirchhoff approximation for convex obstacles*, Adv. in Math. **55** (1985), 242–315.

171. A. Menikoff and J. Sjöstrand, *On the eigenvalues of a class of hypoelliptic operators*, Math. Ann. **235** (1978), 55–85.

172. A. Messiah, *Quantum mechanics*, Wiley, New York, 1958.

173. K. Miller, *Hypoellipticity on the Heisenberg group*, J. Funct. Anal. **31** (1979), 306–320.

174. ——, *Parametrices for hypoelliptic operators on step two nilpotent Lie groups*, Comm. Partial Differential Equations **5** (1980), 1153–1184.

175. W. Miller, *Lie theory and special functions*, Academic Press, New York, 1968.

176. S. Minakshisundaram and A. Pleijel, *Eigenfunctions of the Laplace operator on Riemannian manifolds*, Canad. J. Math. **1** (1949), 242–256.

177. C. Moore, *Representations of solvable and nilpotent groups and harmonic analysis on nil and solvmanifolds*, Proc. Sympos. Pure Math., Vol. 26, Amer. Math. Soc., Providence, R.I., 1973.

178. C. Moore and J. Wolf, *Square integrable representations of nilpotent Lie groups*, Trans. Amer. Math. Soc. **185** (1973), 445–462.

179. C. Morrey, *Multiple integrals and the calculus of variations*, Springer, New York, 1966.

180. G. Mostow, *Strong rigidity of locally symmetric spaces*, Princeton Univ. Press, Princeton, N.J., 1973.

181. A. Nachman, *The wave equation on the Heisenberg group*, Comm. Partial Differential Equations **7** (1982), 675–714.

182. E. Nelson, *Operants, a functional calculus for non commuting operators*, Functional Analysis and Related Topics, Springer, New York, 1970, pp. 172–187.

183. ——, *Analytic vectors*, Ann. of Math. **70** (1959), 572–615.

184. ——, *Operator differential equations*, Mimeographed graduate lecture notes, Princeton, 1964.

185. E. Nelson and W. Stinespring, *Representations of elliptic operators in an enveloping algebra*, Amer. J. Math. **91** (1959), 547–560.

186. H. Nussensweig, *High frequency scattering by an impenetrable sphere*, Ann. Physics **34** (1965), 23–95.

187. F. Olver, *Introduction to asymptotics and special functions*, Academic Press, New York, 1974.

188. B. Orsted, *Induced representations and a new proof of the imprimitivity theorem*, J. Funct. Anal. **31** (1979), 355–359.

189. ——, *Conformally invariant differential equations and projective geometry*, J. Funct. Anal. **44** (1981), 1–23.

190. ——, *The conformal invariance of Huygens' principle*, J. Differential Geom. **16** (1981), 1–9.

191. U. Ottoson, *A classification of the unitary irreducible representations of* SU(N, 1), Comm. Math. Phys. **10** (1968), 114–131.

192. T. Parker, *Gauge theories on four dimensional Riemannian manifolds*, Comm. Math. Phys. **85** (1982), 563–602.

193. R. Parthasarathy, *Dirac operators and the discrete series*, Ann. of Math. **96** (1972), 1–30.

194. J. Peetre, *The Weyl transform and Laguerre polynomials*, Le Mathematische, Univ. de Catania Sem. 27, 301–323.

195. L. Pontrjagin, *Topological groups*, Princeton Univ. Press, Princeton, N.J., 1958.

196. W. Poor, *Differential geometric structures*, McGraw-Hill, New York, 1981.

197. L. Pukansky, *Lecons sur les representations des groupes*, Monographie de la Soc. Math. France, no. 2, Dunod, Paris, 1967.

198. ——, *On the characters and the Plancherel formula of nilpotent groups*, J. Funct. Anal. **1** (1967), 255–280.

199. L. Pukansky, *On the unitary representations of exponential groups*, J. Funct. Anal. **2** (1968), 73–113.

200. ——, *Representations of solvable Lie groups*, Ann. Sci. École Norm. Sup. **4** (1971), 464–608.

201. M. Reed and B. Simon, *Methods of mathematical physics*. I–IV, Academic Press, New York, I, 1975; II, 1975; III, 1978; IV, 1978.

202. F. Riesz and B. Sz. Nagy, *Functional analysis*, Ungar, New York, 1955.

203. W. Rossmann, *Analysis on real hyperbolic spaces*, J. Funct. Anal. **30** (1978), 448–477.

204. L. Rothschild and E. Stein, *Hypoelliptic differential operators and nilpotent groups*, Acta Math. **137** (1976), 247–320.

205. W. Rudin, *Function theory on the unit ball of* $\mathbf{C}^n$, Springer, New York, 1980.

206. W. Schmid, *On the characters of the discrete series*, Invent. Math. **30** (1975), 47–144.

207. ——, *Representations of semi-simple Lie groups*, London Math. Soc. Lecture Notes, No. 34, Cambridge Univ. Press, 1979, pp. 185–235.

208. R. Schrader and M. Taylor, *Small* $\hbar$ *asymptotics for quantum partition functions associated to particles in external Yang-Mills potentials*, Comm. Math. Phys. **92** (1984), 555–594.

209. A. Schwartz, *Smooth functions invariant under the action of a compact group*, Topology **14** (1975), 63–68.

210. L. Schwartz, *Theorie des distributions.* I, II, Hermann, Paris, 1957.

211. I. Segal, *Transformations for operators and symplectic automorphisms over a locally compact abelian group*, Math. Scand. **13** (1963), 31–43.

212. ——, *An extension of Plancherel's formula to separable unimodular groups*, Ann. of Math. **52** (1950), 272–292.

213. A. Selberg, *Harmonic analysis and discontinuous groups in weakly symmetric Riemannian spaces with applications to Dirichlet series*, J. Indian Math. Soc. **20** (1956), 47–87.

214. J. Serre, *Lie algebras and Lie groups*, Benjamin, New York, 1965.

215. ——, *Algebres de Lie semi-simples complexes*, Benjamin, New York, 1966.

216. D. Shale, *Linear symmetries of free Boson fields*, Trans. Amer. Math. Soc. **103** (1962), 149–167.

217. B. Simon, *The classical limit of quantum partition functions*, Comm. Math. Phys. **71** (1980), 247–276.

218. J. Sjöstrand, *Parametrices for pseudodifferential operators with multiple characteristics*, Ark. Mat. **12** (1974), 85–130.

219. J. Sniatycki, *Geometric quantization and quantum mechanics*, Springer, New York, 1980.

220. A. Sommerfeld, *Mathematische theorie der diffraktion*, Math. Ann. **47** (1896), 317–374.

221. J. Souriau, *Structure des systemes dynamique*, Dunod, Paris, 1970.

222. B. Speh, *The unitary dual of* $\mathrm{GL}(3,R)$ *and* $\mathrm{GL}(4,R)$, Math. Ann. **258** (1981), 113–133.

223. I. Stakgold, *Boundary value problems of mathematical physics*, Vol. 2, MacMillan, New York, 1968.

224. E. Stein, *Singular integrals and differentiability properties of functions*, Princeton Univ. Press, Princeton, N.J., 1970.

225. ——, *Analysis in matrix spaces and some new representations of* SL(n, C), Ann. of Math. **86** (1967), 461–490.

226. E. Stein and G. Weiss, *Introduction to Fourier analysis on Euclidean spaces*, Princeton Univ. Press, Princeton, N.J., 1971.

227. W. Stinespring, *A semi-simple matrix group is of type* I, Proc. Amer. Math. Soc. **9** (1958), 965–967.

228. R. Strichartz, *Harmonic analysis on hyperboloids*, J. Funct. Anal. **12** (1973), 341–383.

229. ——, *Fourier transforms and non-compact rotation groups*, Indiana Univ. Math. J. **24** (1974), 499–526.

230. ——, *Mean value properties of the Laplacian via spectral theory* (preprint).

231. M. Sugiura, *Unitary representations and harmonic analysis*, Wiley, New York, 1975.

232. C. Sutherland, *Cartan subalgebras, transverse measures, and non-type-*I *Plancherel formulae*, J. Funct. Anal. **60** (1985), 281–308.

233. D. Tartakoff, *Local analytic hypoellipticity of* $\Box_b$ *on nondegenerate Cauchy Riemann manifolds*, Proc. Nat. Acad. Sci. U.S.A. **75** (1978), 3027–3028.

234. M. Taylor, *Pseudodifferential operators*, Princeton Univ. Press, Princeton, N.J., 1981.

235. ——, *Noncommutative microlocal analysis*. I, Mem. Amer. Math. Soc., No. 313, Amer. Math. Soc., Providence, R.I., 1984.

236. ——, *Fourier integral operators and harmonic analysis on compact manifolds*, Proc. Sympos. Pure Math., Vol. 35, Part 2, Amer. Math. Soc., Providence, R.I., 1979, pp. 115–136.

237. ——, *Diffraction of waves by cones and polyhedra*, Proc. Conf. on Analytical and Numer. Approaches to Asymptotic Problems in Analysis, North Holland, Amsterdam, 1981, pp. 235–248.

238. A. Terras, *Fourier analysis on symmetric spaces and applications*. I, Springer-Verlag, New York, 1985.

239. F. Treves, *Introduction to pseudodifferential and Fourier integral operators*, Vols. 1 and 2, Plenum Press, New York, 1980.

240. ——, *Analytic hypoellipticity for a class of pseudodifferential operators*, Comm. Partial Differential Equations **3** (1978), 475–642.

241. M. Tsuchikawa, *On the representations of* SL$(3, C)$. I–III, Proc. Japan Acad. **43** (1967), 852–855; **44** (1968), 127–129; **44** (1968), 130–132.

242. A. Unterberger, *Oscillateur harmonique et operateurs pseudodifferentiels*, Ann. Inst. Fourier (Grenoble) **29** (1979), 201–221.

243. I. Vahutinskii, *Unitary representations of* GL$(3, R)$, Mat. Sb. **75** (1968), 303–320.

244. L. van Hove, *Sur certains representations unitaires d'un groupe infinie de transformations*, Acad. Roy. Belg. Cl. Sci. Mém. 26, no. 6, 1951.

245. B. L. van der Waerden, *Group theory and quantum mechanics*, Springer, New York, 1980.

246. V. Varadarajan, *Lie groups, Lie algebras, and their representations*, Prentice-Hall, New Jersey, 1974.

247. ——, *Infinitesimal theory of representations of semisimple Lie groups*, Harmonic Analysis and Representations of Semisimple Lie Groups, D. Reidel, Boston, 1980.

248. ——, *Harmonic analysis on real reductive groups*, Lecture Notes in Math., No. 576, Springer, 1976.

249. N. Vilenkin, *Special functions and the theory of group representations*, Trans. Math. Monographs, Vol. 22, Amer. Math. Soc., Providence, R.I., 1968.

250. D. Vogan, *Algebraic structure of irreducible representations of semisimple Lie groups*. I, Ann. of Math. **109** (1979), 1–60.

251. ——, *Representations of real reductive Lie groups*, Birkhauser, Boston, 1982.

252. A. Voros, *An algebra of pseudodifferential operators and asymptotics of quantum mechanics*, J. Funct. Anal. **29** (1978), 104–132.

253. N. Wallach, *Harmonic analysis on homogeneous spaces*, Marcel Dekker, New York, 1973.

254. ——, *Symplectic geometry and Fourier analysis*, Math. Sci. Press, Brookline, Mass., 1977.

255. ——, *Cyclic vectors and irreducibility for principal series representations*, Trans. Amer. Math. Soc. **158** (1971), 107–113.

256. G. Warner, *Harmonic analysis on semisimple Lie groups*, Springer, New York, 1972.

257. G. Watson, *A treatise on the theory of Bessel functions*, Cambridge Univ. Press, 1966.

258. A. Weil, *Sur certaines groupes d'operateurs unitaires*, Acta Math. **111** (1964), 143–211.

259. ——, *L'integration dans les groupes topologiques et ses applications*, Hermann, Paris, 1940.

260. A. Weinstein, *Lectures on symplectic manifolds*, CBMS Regional Conf. Ser. in Math., No. 29, Amer. Math. Soc., Providence, R.I., 1977.

261. H. Weyl, *The classical groups*, Princeton Univ. Press, Princeton, N.J., 1956.

262. ——, *Group theory and quantum mechanics*, Dover, New York, 1931.

263. E. Whittaker and G. Watson, *A course of modern analysis*, Cambridge Univ. Press, 1927.

264. E. Wigner, *On unitary representations of the inhomogeneous Lorentz group*, Ann. of Math. **40** (1939), 149–204.

265. J. Wolf, *Foundations of representation theory for semisimple Lie groups*, Harmonic Analysis and Representations of Semisimple Lie Groups, D. Reidel, Boston, 1980.

266. H. Yamabe, *On a deformation of Riemannian structures on compact manifolds*, Osaka J. Math. **12** (1960), 21–37.

267. K. Yosida, *Functional analysis*, Springer, New York, 1965.

268. D. Zelobenko, *Compact Lie groups and their representations*, Trans. Math. Monographs, Vol. 40, Amer. Math. Soc., Providence, R.I., 1973.

269. A. Zygmund, *Trigonometrical series*, Cambridge Univ. Press, 1959.

Index